Soils of
the British Isles

Soils of
the British Isles

B.W. AVERY

C·A·B INTERNATIONAL

C·A·B International
Wallingford
Oxon OX10 8DE
UK

Tel: Wallingford (0491) 32111
Telex: 847964 (COMAGG G)
Telecom Gold/Dialcom: 84: CAU001
Fax: (0491) 33508

© C·A·B International 1990. All rights reserved. No part of this publication may be reproduced in any form or by any means, electronically, mechanically, by photocopying, recording or otherwise, without the prior permission of the copyright owners.

British Library Cataloguing in Publication Data
Avery, B. W. (Brian William)
Soils of the British Isles.
1. Great Britain. Soils
I. Title
631.4'941

ISBN 0-85198-649-8

Typeset by Enset Photosetting, Midsomer Norton, Bath
Printed in Great Britain
at the University Press, Cambridge

Contents

Preface	xi
Chapter 1 Introduction: Soil and its Variability	1
1.1 The Nature of Soil	1
1.2 Processes in the Soil and the Horizons that Result	3
1.2.1 The Organic Cycle	3
1.2.2 Structural Reorganization	4
1.2.3 Weathering and Translocation of Inorganic Constituents	4
1.2.4 Profile Development Sequences	8
1.3 Factors Affecting Soil Variation in the British Isles	9
1.3.1 Parent Material and Age Factors	9
1.3.2 Climate	15
1.3.3 Relief	18
1.3.4 Biotic Factors: Vegetation and Man	21
1.4 Scales and Types of Soil Variation	25
1.5 Soil Classification	26
1.5.1 Purposes and Limitations	26
1.5.2 Kinds of Classification	28
1.5.3 Development of General Classifications	28
1.5.4 Application of Numerical Methods	31
1.6 Soil Surveys	33
1.6.1 Purpose and Nature	33
1.6.2 Soil Map Scales and Survey Intensity Levels	35
Chapter 2 Soil Surveys and Soil Classification in the British Isles	37
2.1 Historical Introduction	37
2.1.1 The First Soil Surveys	37
2.1.2 The Geological Approach	37
2.1.3 The Introduction of Modern Methods	39
2.1.4 Systematic Soil Surveys (1946–1987)	43
2.2 The Soil Survey of England and Wales	43
2.2.1 Soil Mapping	43
2.2.2 Soil Characterization and Classification	45
2.3 The Soil Survey of Scotland	56
2.3.1 Soil and Land Capability Mapping	56
2.3.2 Soil Characterization and Classification	60

2.4	Forestry Commission Surveys	60
2.5	The National Soil Survey of Ireland	64
	2.5.1 Soil and Land Capability Mapping	64
	2.5.2 Soil Characterization and Classification	65
2.6	Application of the US Soil Taxonomy to British soils	66
2.7	The FAO Classification and the Soil Map of the EEC Countries	71
2.8	The Classification Adopted	73
2.9	Schematic Soil Map	80

Chapter 3 Methods and Definitions — 81

3.1	Introduction	81
3.2	Soil Profile Descriptions	81
3.3	Micromorphological Data	90
3.4	Analytical Data	92
3.5	Horizon Notation	99
3.6	Named Diagnostic Surface Horizons	106
3.7	Named Diagnostic Subsurface Horizons	108
3.8	Other Differentiating Criteria	111
3.9	Key to Major Soil Groups	114

Chapter 4 Lithomorphic Soils — 115

4.1	General Characteristics, Classification and Extent	115
4.2	Lithosols	116
	4.2.1 General Characteristics, Classification and Distribution	116
	4.2.2 Non-calcaric Lithosols	118
	4.2.3 Calcaric Lithosols	118
4.3	Rankers	120
	4.3.1 General Characteristics, Classification and Distribution	120
	4.3.2 Typical (Humic) Rankers	122
	4.3.3 Humic-calcaric Rankers	124
	4.3.4 Organic (Peaty) Rankers	125
	4.3.5 Podzolic Rankers	126
	4.3.6 Ochric Rankers	126
	4.3.7 Ochric-calcaric Rankers	128
4.4	Rendzinas	129
	4.4.1 General Characteristics, Classification and Distribution	129
	4.4.2 Typical (Humic) Rendzinas	132
	4.4.3 Grey Rendzinas	134
	4.4.4 Colluvial Rendzinas	137
	4.4.5 Brown Rendzinas	138
	4.4.6 Pararendzinas	142
4.5	Sandy Regosols	143
	4.5.1 General Characteristics, Classification and Distribution	143
	4.5.2 Non-calcaric Sandy Regosols	146
	4.5.3 Calcaric Sandy Regosols	149

4.6	Rego-alluvial Soils		152
	4.6.1	General Characteristics, Classification and Distribution	152
	4.6.2	Non-calcaric Rego-alluvial Soils	152
	4.6.3	Calcaric Rego-alluvial Soils	155

Chapter 5 Brown Soils — 156

5.1	General Characteristics, Classification and Extent		156
5.2	Sandy Brown Soils		158
	5.2.1	General Characteristics, Classification and Distribution	158
	5.2.2	Typical Sandy Brown Soils	161
	5.2.3	Gleyic Sandy Brown Soils	164
	5.2.4	Calcaric Sandy Brown Soils	165
	5.2.5	Gleyic-calcaric Sandy Brown Soils	168
5.3	Alluvial Brown Soils		169
	5.3.1	General Characteristics, Classification and Distribution	169
	5.3.2	Typical (Loamy) Alluvial Brown Soils	172
	5.3.3	Gleyic Alluvial Brown Soils	173
	5.3.4	Pelo-alluvial Brown Soils	175
	5.3.5	Calcaric Alluvial Brown Soils	177
	5.3.6	Gleyic-calcaric Alluvial Brown Soils	178
	5.3.7	Pelocalcaric Alluvial Brown Soils	180
5.4	Calcaric Brown Soils		181
	5.4.1	General Characteristics, Classification and Distribution	181
	5.4.2	Typical Calcaric Brown Soils	182
	5.4.3	Colluvial Calcaric Brown Soils	185
	5.4.4	Gleyic Calcaric Brown Soils	187
	5.4.5	Stagnogleyic Calcaric Brown Soils	190
	5.4.6	Pelocalcaric Brown Soils	191
5.5	Orthic Brown Soils		194
	5.5.1	General Characteristics, Classification and Distribution	194
	5.5.2	Typical Orthic Brown Soils	195
	5.5.3	Colluvial Orthic Brown Soils	200
	5.5.4	Gleyic Orthic Brown Soils	203
	5.5.5	Stagnogleyic Orthic Brown Soils	206
	5.5.6	Pelo-orthic Brown Soils	208
5.6	Luvic Brown Soils		209
	5.6.1	General Characteristics, Classification and Distribution	209
	5.6.2	Typical Luvic Brown Soils	211
	5.6.3	Gleyic Luvic Brown Soils	217
	5.6.4	Stagnogleyic Luvic Brown Soils	220
	5.6.5	Peloluvic Brown Soils	223
	5.6.6	Chromoluvic Brown Soils	225
	5.6.7	Stagnogleyic Chromoluvic Brown Soils	228
5.7	Podzolic Brown Soils		232
	5.7.1	General Characteristics, Classification and Distribution	232
	5.7.2	Typical Podzolic Brown Soils	235
	5.7.3	Gleyic Podzolic Brown Soils	240
	5.7.4	Stagnogleyic Podzolic Brown Soils	241

Chapter 6 Podzols   243
  6.1 General Characteristics, Classification and Extent   243
  6.2 Non-hydromorphic Podzols   246
      6.2.1 General Characteristics, Classification and Distribution   246
      6.2.2 Humoferric Podzols   248
      6.2.3 Humus Podzols   254
      6.2.4 Luvic Podzols   259
  6.3 Gley-podzols   261
      6.3.1 General Characteristics, Classification and Distribution   261
      6.3.2 Typical Gley-podzols   262
      6.3.3 Humoferric Gley-podzols   265
      6.3.4 Luvic Gley-podzols   266
  6.4 Stagnopodzols   270
      6.4.1 General Characteristics, Classification and Distribution   270
      6.4.2 Ironpan Stagnopodzols   273
      6.4.3 Humus-ironpan Stagnopodzols   280
      6.4.4 Ferric Stagnopodzols   284
      6.4.5 Densipan Stagnopodzols   287

Chapter 7 Gley Soils   290
  7.1 General Characteristics, Classification and Extent   290
  7.2 Sandy Gley Soils   293
      7.2.1 General Characteristics, Classification and Distribution   293
      7.2.2 Typical Sandy Gley Soils   294
      7.2.3 Humic Sandy Gley Soils   297
      7.2.4 Calcaric Sandy Gley Soils   299
      7.2.5 Humic-calcaric Sandy Gley Soils   300
  7.3 Alluvial Gley Soils   300
      7.3.1 General Characteristics, Classification and Distribution   300
      7.3.2 Typical (Loamy) Alluvial Gley Soils   303
      7.3.3 Pelo-alluvial Gley Soils   306
      7.3.4 Humic Alluvial Gley Soils   310
      7.3.5 Calcaric Alluvial Gley Soils   311
      7.3.6 Pelocalcaric Alluvial Gley Soils   316
      7.3.7 Humic Calcaric Alluvial Gley Soils   319
      7.3.8 Saline Alluvial Gley Soils   321
      7.3.9 Acid-sulphate Alluvial Gley Soils   325
  7.4 Calcaric Gley Soils   328
      7.4.1 General Characteristics, Classification and Distribution   328
      7.4.2 Ground-water Calcaric Gley Soils   329
      7.4.3 Stagnocalcaric Gley Soils   332
      7.4.4 Pelocalcaric Gley Soils   333
      7.4.5 Humic Calcaric Gley Soils   336
  7.5 Orthic Gley Soils   339
      7.5.1 General Characteristics, Classification and Distribution   339
      7.5.2 Ground-water Orthic Gley Soils   342

	7.5.3	Stagno-orthic Gley Soils	344
	7.5.4	Pelo-orthic Gley Soils	348
	7.5.5	Humic Orthic Gley Soils	352
	7.5.6	Humic Stagno-orthic Gley Soils	354
7.6	Luvic Gley Soils	359
	7.6.1	General Characteristics, Classification and Distribution	359
	7.6.2	Ground-water Luvic Gley Soils	360
	7.6.3	Stagnoluvic Gley Soils	364
	7.6.4	Chromic Stagnoluvic Gley Soils	371
	7.6.5	Peloluvic Gley Soils	375
	7.6.6	Humic Luvic Gley Soils	378
	7.6.7	Humic Stagnoluvic Gley Soils	380

Chapter 8 Man-made Soils — 385

8.1	General Characteristics, Classification and Extent	385
8.2	Cultosols	385
	8.2.1 General Characteristics, Classification and Distribution	385
	8.2.2 Sandy Cultosols	388
	8.2.3 Loamy and Clayey Cultosols	389
8.3	Disturbed Soils	394
	8.3.1 General Characteristics and Distribution	394
	8.3.2 Restored Opencast Coal Workings	395
	8.3.3 Restored Ironstone Workings	397
	8.3.4 Restored 'Coprolite' Workings	397
	8.3.5 Land Restored after Extraction of Sand, Gravel or Brickearth	398

Chapter 9 Peat Soils — 400

9.1	General Characteristics and Extent	400
9.2	Mire Types	400
9.3	Classification of Peat Soils	406
9.4	Fen Soils	408
	9.4.1 General Characteristics, Classification and Distribution	408
	9.4.2 Raw Fen Soils	410
	9.4.3 Earthy Semi-fibrous Fen Soils	411
	9.4.4 Earthy Sapric Fen Soils	413
	9.4.5 Earthy Acid-sulphate Fen Soils	416
9.5	Bog Soils	418
	9.5.1 General Characteristics, Classification and Distribution	418
	9.5.2 Raw Fibrous Bog Soils	420
	9.5.3 Raw Semi-fibrous Bog Soils	421
	9.5.4 Raw Sapric Bog Soils	424
	9.5.5 Earthy Semi-fibrous Bog Soils	426
	9.5.6 Earthy Sapric Bog Soils	428

References — 429

Index — 450

Preface

This book is an attempt to collate the accumulated knowledge and understanding of soil variation in the British Isles which has stemmed from the extension of field surveys and associated laboratory studies over the last half century. Using a definitional classification scheme devised for the purpose, it describes the salient properties, distribution, environmental relationships and agronomic significance of the main kinds of soil that have been distinguished, illustrates them by data on selected examples, and relates them to current internationally used schemes. Although the primary objective has been to provide an authoritative systematic review for practitioners and students of soil science, the coverage and presentation have been framed with the additional aim of eliciting the interest of workers in allied fields and others concerned with the use, management and conservation of land resources.

The work could not have been accomplished without the continued co-operation and encouragement of past and present members of the three soil-survey organizations in Great Britain and Ireland, particularly in permitting me to reproduce published and unpublished soil-profile data, diagrams and other illustrative material. Among those who have helped in various ways, I am especially indebted to Mr D. Mackney, OBE, Professor P. Bullock, Mr B. Clayden, Mr J.M. Hollis, Mr M.G. Jarvis, Dr R.J.A. Jones, Dr P.J. Loveland, Dr R.W. Payton, Mr E.M. Thomson and Mr P.S. Wright of the Soil Survey of England and Wales (now Soil Survey and Land Research Centre); Mr R. Grant, Professor J.S. Bibby, Mr C.J. Bown and Mr R.E.F. Heslop of the Soil Survey Department of the Macaulay Institute for Soil Research (now Macaulay Land Use Research Institute); and Dr M.J. Gardiner, Dr J. Lee, Mr T. F. Finch and Dr R.F. Hammond of the National Soil Survey of Ireland.

I also thank Dr J.A. Catt and Professor G.F. Mitchell for providing photographs, Dr T. Keane of the Irish Meteorological Service, Dublin, for climatic data, the Forestry Commission of Great Britain and Dr D.A. Jenkins for allowing me to reproduce hitherto unpublished material, Mrs B. Moseley for word processing, Mrs S. Talbot for undertaking the art work and Mr B. Butters for checking the proofs.

B.W. Avery   *Harpenden, 1990*

1
Introduction: Soil and its Variability

1.1 The Nature of Soil

Soil, as the term is used in this book, includes the near-surface layers of the earth in which plants root. Wherever the ground is stabilized beneath a cover of vegetation, the underlying material is subject to modification by a number of interacting processes, including addition, decay and incorporation of organic residues, physical reorganization, weathering of mineral particles or rock fragments, and loss or translocation of soluble or finely dispersed constituents brought about by water movements. The continued action of these processes gives rise to more or less distinct *soil horizons*, roughly parallel to the surface, that differ in such features as colour, texture (particle-size distribution), structure, amount and type of organic matter and degree of root development from each other and from the relatively unaltered rock material below. In some places this can be equated with the *parent material* in which the soil horizons have developed; in others they have evidently formed in different material or been superimposed on a succession of originally distinct layers. There are also situations in which the soil consists wholly or partly of material that has been excavated and replaced in its present position. A *soil profile* is a vertical section through the soil at any particular place and it is generally on the basis of their profile characteristics that different kinds of soil are identified and mapped.

Soil in this sense underlies the land surface as a nearly continuous mantle interrupted only by bodies of open water and naturally or artificially barren ground such as rock outcrops, icefields, shifting sands, beaches, active screes, quarries and mine dumps, which do not support a continuous cover of higher plants. Except where there is consolidated bedrock at shallow depth, its lower boundary is usually gradual. In tropical areas with porous substrata and pronounced wet and dry seasons, perennial grasses and trees root to depths of several metres in search of water. In the British Isles, however, few roots penetrate below 2 m, even where the substratum is sufficiently permeable. Horizons showing appreciable evidence of alteration often extend to about the same depth but can terminate much nearer the surface, especially where erosion is active or where the parent material consists of recently deposited aeolian or water-laid sediment. Under these conditions roots grow into little altered material, the upper part of which forms part of the soil as conceived here. Conversely, rock weathered *in situ* extends in places to depths well below the rooting zone, or a more or less cemented soil horizon forms an impenetrable pan. In the former case the lower boundary of the soil is taken as the lower limit of biological activity;

in the latter the soil is considered to include the compacted horizon and any immediately underlying horizons which contain some organic matter and have distinctive properties indicating that they are genetically related to those above.

Soil, once formed, may be truncated or entirely removed by natural or man-induced erosion, or buried by fresh material. This can take place slowly without destruction of the vegetative cover or result from a rapid episodal influx of sediment such as volcanic ash, blown sand or alluvium in which a new soil may eventually develop. Where burial is progressive, as in peat bogs and other accretionary situations, organic or mineral materials containing more or less decomposed plant remains commonly extend to depths of several metres. Buried soil horizons are not considered part of the present-day soil if they lie below the current rooting zone, though they are of considerable interest in aiding reconstruction of past landscapes and environments.

Soil horizons as they exist in the field consist of variously sized solid particles separated by a more or less interconnected system of voids or pores which are occupied at any one time by air or water and in the larger of which the roots of plants grow. A multitude of other living organisms ranging in size from bacteria to burrowing animals such as earthworms and moles also live in the pore space. The water in the pores contains dissolved salts, gases and organic solutes; the air normally has a higher concentration of carbon dioxide than the atmosphere and can contain traces of other gases produced by microbial metabolism. Both plants and soil-inhabiting organisms withdraw the water and oxygen they need from the pore space, together with mineral nutrients which are dissolved in the pore water or held as exchangeable ions on the bounding solid surfaces.

Excluding living roots and soil organisms, the solid particles are broadly divisible into coarse (> c. 0.01 mm) and fine fractions, the spatial arrangement of which is most readily appreciated by microscopic examination of thin sections cut from blocks of undisturbed soil. Depending mainly on the relative proportions of coarse and fine material, the latter can occur as coats around the larger particles, in more or less porous aggregates between them, or as a matrix in which the coarse particles are embedded. Crystalline mineral grains or rock fragments inherited with little change from the parent material constitute by far the largest part of the coarse fraction in most soils. In certain surface horizons and peats, however, it consists mainly of comminuted plant remains retaining cell structure. The fine fraction includes crystalline mineral particles, chiefly of layer silicates (clay minerals) and iron oxides, which are too small to be resolved by optical microscopy; poorly ordered or cryptocrystalline aluminosilicates and hydrous oxides of iron, aluminium, manganese and silicon; and finely particulate organic matter (humus) derived directly from plant tissue or by microbial synthesis. These constituents occur in very variable proportions and in most soil materials the inorganic and organic components of the fine fraction are so closely bonded that they are difficult to separate completely by either physical or chemical means. All of them, including organic matter, can originate by direct inheritance from sedimentary parent material or as alteration products, and are in turn subject to reorganization or translocation within the soil.

Besides varying vertically and laterally, many soil properties which result from or affect plant growth and biological activity change with the seasons or even daily as well as over longer periods. These labile properties include the numbers of organisms of various kinds, the amounts of organic matter in different stages of decomposition or polymerization, the geometry of the pore space and the distribution of liquid and gaseous phases, the pH and the temperature. The occurrence of such changes, all of which are most pronounced near the surface and become imperceptible at varying depths, further differentiates the soil from the inert rock material beneath it.

This concept of soil as part of a dynamic system embracing organisms and their environment was developed in Russia by V.V. Dokuchaiev (1846–1903) and his associates towards the end of the last century. In 1882 the provincial government of Nizhni-Novgorod engaged Dokuchaiev, a geologist, to classify and map soils for land evaluation and tax assessment purposes. According to Yarilov (1927), he distinguished two stages in the assignment, first the establishment of a natural classification of soils and secondly the grading of classes according to their agricultural potentialities. As the basis for his classification he conceived types of soil not as unorganized mixtures of decaying organic matter and distintegrating rock but as 'independent natural bodies', each with a distinctive succession of horizons reflecting the combined effects of a particular combination of genetic factors. Thus variations in the morphology and composition of soil profiles from place to place were attributed to differences in climate, vegetation and associated organisms, relief, parent material, and age of land form; and soils were considered to evolve with time towards a condition of equilibrium corresponding to a particular vegetational climax.

1.2 Processes in the Soil and the Horizons that Result

1.2.1 The Organic Cycle

Wherever plants occupy the ground, their remains are continually being added to the surface and subsurface horizons, either directly or through animals, and are used as sources of food and energy by the soil-inhabiting organisms. Part of the organic material added to the soil normally undergoes oxidative decomposition (mineralization), with the result that the nitrogen and other nutrient elements it contains become available to nourish succeeding generations of plants, while part is transformed into dark-coloured humus which is more or less incorporated with the mineral particles and decomposes at a much slower rate. The amount of accumulated organic matter at any one time represents the balance between addition of residues and losses by mineralization, and this is governed in turn by the nature of the natural or cultivated vegetation and by other factors, including moisture conditions, aeration, temperature and nutrient supply, which regulate the composition and activity of the soil population.

Differences in the nature and distribution of organic matter accumulated in uncultivated land under well-aerated conditions are indicated by the terms *mor* (or *raw humus*), *mull* and *moder* (e.g. Kubiena, 1953; Barratt, 1964; Duchaufour, 1982). Each of these 'humus forms' normally consists of two or more layers or horizons in which the degree of decomposition of the organic residues increases from the surface downwards (Figure 1.1). In mor the organic accumulation is sharply separated from the mineral soil and is divisible into a litter (L) layer of recently fallen leaves or stems; a fermentation or F layer in which the residues are undergoing active decomposition, mainly through the agency of small Arthropods (mites and springtails) and fungi; and a dark humus or H layer of well-decomposed (humified) material. In typical mull, by contrast, most of the organic matter is incorporated and intimately mixed with the mineral soil by earthworms to form a crumbly organo-mineral (A) horizon, and the litter layer, if present, consists of recognizable leaves or stems which disappear within a year. Moder is intermediate in most respects between mor and mull. Although F and H layers are generally recognizable, they are looser and more porous than in mor and the H layer, which consists largely of fine granular aggregates (faecal pellets) produced by

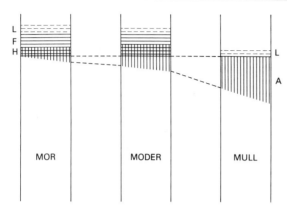

Figure 1.1 Diagrammatic representation of superficial organic-matter profiles.

animals smaller than earthworms, is not clearly separated from the mineral soil below.

The nature of the soil, the climate and the vegetation govern the distribution of these superficial profile forms through their influence on the soil organisms. Thus the soil-ingesting and casting earthworms responsible for mull formation flourish best under equable moisture and temperature conditions in soils that carry broad leaved trees or grasses, are not too acid (pH > 4.5), contain appreciable amounts of clay, and are deep enough to allow them to hibernate and aestivate. Typical mor, on the other hand, occurs under coniferous forest or heath on strongly acid, nutrient-deficient soils, usually coarse textured, and either moder or mor in other situations where internal or external conditions are unfavourable for earthworm activity.

Where persistent waterlogging and consequent poor aeration inhibit mineralization, more or less decomposed plant remains accumulate to form *peat*. In initial stages of peat formation, with plants continuing to root in the underlying mineral layers, the organic and/or organo-mineral surface horizons have clear affinities with those formed under well-aerated conditions and can be designated accordingly as *hydromor*, *hydromoder* or *hydromull* (Duchaufour, 1982). Thicker bodies of peat include layers of differing composition and morphology which reflect changes in the depositional environment rather than progressive modification of plant residues accumulated on or just below the present surface.

The ploughed surface layers (Ap horizons) in arable land resemble mull horizons formed naturally under grass but usually contain fewer earthworms and smaller amounts of organic matter, particularly where annual harvesting of crops is not balanced by regular manurial additions.

1.2.2 Structural Reorganization

The mineral or organic layers in which plants root are also affected by agencies such as seasonal wetting, drying and freezing, the activities of soil animals, and the growth and decay of the roots themselves, that cause changes in the arrangement of solid particles and voids. These include the destruction of sedimentary layering and development of variously shaped aggregates known as *peds*, which are separated by voids or surfaces of weakness (Figure 3.3 p. 87). Granular or crumb-like peds are commonly produced by animal activity and blocky or prismatic peds by swelling and shrinking of clay-rich materials during wetting and drying cycles. In initially water saturated sediments such as intertidal muds, soil structure develops only after periodic drying has caused irreversible loss of part of the water, giving the material a firmer consistence. This process is known as *physical ripening*.

The effects of structural reorganization, like those of the organic cycle, are normally most pronounced near the surface and become progressively less evident with increasing depth.

1.2.3 Weathering and Translocation of Inorganic Constituents

The mineral particles of the soil are subject to

disintegration and transformation or decomposition and it is primarily through these *weathering* processes that nutrient elements such as phosphorus, potassium and calcium are released in forms available to plants. Except in initial stages of soil formation on hard rocks and in arctic or desert regions generally, where physical breakdown predominates, the main agent of weathering is rain water charged with oxygen, carbon dioxide and soluble products of plant decay. As this acidulated water percolates through the soil it causes various reactions to take place at particle surfaces: thus fragments of chalk or limestone are gradually dissolved; ferrous and sulphide ions are oxidized; the layer-lattice structure of micas is transformed by entry of water molecules between the layers and replacement of interlayer potassium by other more hydrated ions; and primary silicate minerals such as feldspars undergo hydrolysis, whereby alkali and alkaline-earth cations and some silica pass into solution.

The rate and type of weathering depend on the minerals present in the parent material, the surface areas exposed as determined by particle size, the quantity, temperature and pH of the water passing through the soil, the oxidation/reduction status within the weathering zone and the effects of organic solutes. Of the common rock-forming minerals, calcite is the least resistant to chemical weathering, followed by ferromagnesian minerals such as olivine, amphiboles and pyroxenes, biotite and chlorite. Plagioclase (calcium-sodium) feldspars are less resistant than orthoclase; muscovite and quartz are both very resistant but muscovite is readily reduced to clay size by physical disintegration and is then more stable than quartz particles of similar size. Although the rates of most weathering reactions increase with temperature, solution of calcium carbonate proceeds most rapidly at temperatures close to freezing point.

Wherever the rainfall exceeds evaporation and hydrologic conditions permit free downward movement of water, soluble products of weathering are *leached* from the soil whilst other products, including clay minerals and iron oxides, tend to accumulate. The alkali and alkaline-earth cations released can be held in exchange sites on the surfaces of clay and humus particles but are relatively easily removed by continued leaching, in which case they are replaced by hydrogen or hydroxy-aluminium ions and the soil reaction (pH) becomes increasingly acid. In subhumid to semi-arid regions, however, calcium freed by weathering within the rooting zone is reprecipitated at greater depths to form nodules or coatings of calcium carbonate. Silica, calcium sulphate (gypsum) or more soluble salts may also be redeposited in this way.

In well drained soils of similar age, changes in composition brought about by weathering and leaching are greatest where the parent material is calcareous or rich in readily weatherable silicate minerals and least where it already consists largely of quartz and/or clay minerals which are relatively stable in the soil environment. On calcareous materials, the amount and nature of the non-calcareous residue which remains after leaching largely determine the composition and thickness of the weathered horizons formed. The effects of silicate weathering are most pronounced in soils associated with old land surfaces in the humid tropics. Prolonged exposure to these conditions causes almost complete removal of alkali and alkaline-earth cations. Much silica is also lost, leaving a weathered mantle (oxic horizon) that consists of more or less hydrated oxides of iron or aluminium, or both, together with variable amounts of 1:1 lattice (kaolinitic) clay and quartz or other very insoluble minerals of sand size, depending on the composition of the parent rock.

In British soils the most obvious effects are partial or complete removal of carbonates, if present, and development of brownish colours reflecting oxidative weathering of iron-bearing minerals and consequent accumulation of amorphous or cryptocrystalline ferric hydroxides (ferrihydrite, goethite). Structurally reorganized subsurface horizons showing one or both of these features are called colour/structure B or cambic horizons. They often also contain more aluminosilicate clay than those below, either as a result of carbonate removal or through physical or chemical weathering of non-calcareous rock fragments or silicate minerals in coarser size grades. In the latter case the extra clay may originate by transformation of the parent mineral without complete destruction of the crystal lattice or by coprecipitation of colloidal decomposition products. The subsurface horizons of some well drained soils have red or reddish colours reflecting the presence of haematite (α Fe$_2$O$_3$), which may be inherited from the parent material or formed by dehydration of ferric hydroxide within the soil. The latter process is called *rubification* (Duchaufour, 1982); its effects are most widely apparent in soils of subtropical and tropical regions with pronounced dry seasons, though its incidence is evidently influenced by parent material and age factors as well as by atmospheric climate (e.g. Schwertmann *et al.*, 1982).

Continued weathering and leaching tend to impoverish the upper soil by removal of basic cations. Concurrently, however, plant roots are constantly extracting nutrient elements and other mobilizable constituents such as silica from lower horizons and incorporating them in their leaves and stems, which are eventually returned to the surface. By this means the vegetation produces a continuous circulation of mineral substances in the soil and so counteracts to some extent their removal by leaching. In many soils earthworms and other burrowing animals also bring material to the surface and so, in effect, also oppose the processes of removal.

Restricted drainage affects the type and intensity of weathering by hindering removal of soluble products and leads in particular to the prevalence of reducing processes. Where water stagnates in the soil for appreciable periods, microorganisms and plant roots utilize the dissolved oxygen faster than it can be replaced by diffusion. Under the anaerobic conditions so produced, accessible ferric ions are readily reduced to the ferrous state by microbial action or by direct reaction with soluble products of plant decomposition (Bloomfield, 1951; Duchaufour, 1982). This process, known as *gleying*, gives rise to subsurface horizons with greyish, greenish or bluish colours contrasting with those of weathered horizons in well aerated soils developed in lithologically similar materials. The reduced iron compounds are relatively soluble and therefore tend to migrate, either in seepage water or in response to capillary suction gradients, and are reoxidized to form ochreous mottles or concretions in horizons which are only partially or intermittently anaerobic. Other polyvalent metals, including manganese, cobalt, nickel and copper, are similarly subject to mobilization and redeposition in seasonally wet soils. As shown in Figure 1.2, the mobility of iron and manganese is influenced by the ambient pH as well as the redox potential (Eh). Thus ferrous iron normally becomes stable under more oxidizing conditions as acidity increases and divalent manganese is more stable than ferrous iron at the same Eh and pH values. Reduction also tends to increase the mobility of phosphorus, particularly in non-calcareous soils where phosphate is often strongly adsorbed on or combined with ferric iron (Glentworth and Dion, 1949; Russell, 1973).

Apart from or in addition to gleying, the principal processes involving mobilization and redeposition of fine soil constituents are clay translocation (*lessivage*) and podzolization, both of which give rise to successive *eluvial* (E) and *illuvial* (B) horizons characterized respectively by losses and gains of the constituents concerned.

Clay translocation is an essentially physical process whereby layer-lattice silicate clays and any iron oxides or humus bonded to them are moved downwards. According to the gener-

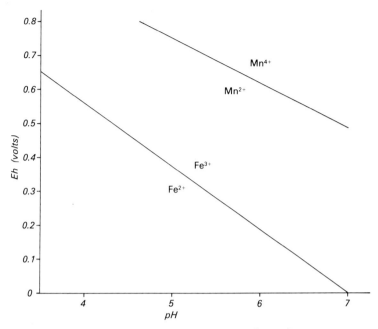

Figure 1.2 Equilibrium curves for Fe^{2+}/Fe^{3+} and Mn^{2+}/Mn^{4+} in relation to Eh and pH (from Duchaufour, 1982).

ally accepted explanation (Brewer and Sleeman, 1970; Soil Survey Staff, 1975; Duchaufour, 1982; McKeague, 1983), rain water penetrating dry or partially dried soil causes disruption of aggregates and dispersion of some clay, which remains in suspension as the water percolates through non-capillary voids. If the lower part of the rooting zone is also dried to some extent, as happens during the growing season whenever evapotranspiration exceeds rainfall and the water table is out of reach, the downward moving water is withdrawn into finer capillary pores and the suspended clay is deposited on the walls of the larger voids, forming distinct coats or pore fillings which are recognizable in thin sections (Figure 3.4, p. 93) and often in the field. Deposition may also result from any marked change in pore-size distribution which restricts water flow. The bodies of illuvial clay so formed may be preserved *in situ* or disrupted and incorporated in the matrix as a result of disturbance by soil animals, frost action, or seasonal swelling and shrinking.

The extent to which clay translocation takes place and eventually gives rise to a distinct illuvial horizon (Bt or argillic horizon) is evidently influenced by several factors (McKeague, 1983). Among those apparently favouring clay movement are a seasonally dry climate; absence of flocculating agents such as finely divided carbonates and amorphous sesquioxides; pH (in water) between about 4.5 and 6.5 or high pH associated with exchangeable sodium; and zero point of net charge different from the soil pH. Mobilization of clay is inhibited in well aggregated organo-mineral surface horizons of mull type, but soluble organic matter may aid dispersion under acid conditions, as does removal of surface oxide films by intermittent gleying. Leaching of carbonates leaves voids which provide favourable sites for redeposition, with the result that a clay-enriched illuvial horizon commonly occurs immediately above the unleached parent material in initially calcareous deposits.

In originally base-deficient (siliceous) or strongly leached soils with mor-forming biotic regimes, sesquioxides (aluminium and/or iron) are translocated independently of silicate clay to form light-coloured eluvial horizons and black to reddish brown or ochreous illuvial horizons. This process, termed *podzolization* (from the Russian word *podzol*, meaning ashy soil), has generally been considered to involve complexation of the polyvalent metal cations by organic acids or soluble polyphenols derived from the slowly decomposing litter, transport of these metal-organic complexes, and their precipitation at some depth as a result of increased metal saturation, changing ionic concentrations, drying, or microbiological decomposition (e.g. De Coninck, 1980; Duchaufour, 1982; Mokma and Buurman, 1982). Support for this genetic model comes from experiments with litter and soil extracts and artificially prepared metal-organic complexes, and from the distribution in soil profiles of aluminium, iron and carbon removed by alkali pyrophosphate solutions or other extractants (Chapter 3, p. 94). Recently, however, evidence that the horizons identified as illuvial usually contain microcrystalline or poorly ordered aluminosilicates (imogolite or proto-imogolite allophane) which can be derived from migrating hydroxyaluminium orthosilicate (proto-imogolite) sols has led Farmer and his co-workers (Farmer *et al.*, 1980, 1983, 1985; Anderson *et al.*, 1982; Farmer, 1984) to conclude that aluminium in particular may be transported independently of organic matter, especially in cool humid environments where the parent material contains weatherable primary silicate minerals and the soil is well drained.

1.2.4 Profile Development Sequences

The variety of soil profile forms resulting from the interaction of these processes over varying periods of time can be placed in morphogenetic sequences which evidently correspond in a general way with their actual evolution under natural or semi-natural vegetation. Two such sequences, widely represented in Lowland Britain on permeable calcareous and non-calcareous parent materials containing appreciable amounts of silicate clay, are shown schematically in Figure 1.3. The designated horizons and the named stages into which the sequences are divided are fully described in the following chapters.

In each case the effects of weathering and leaching are masked in the initial developmental stages by those of organic matter incorporation and structural reorganization; they extend deeper in subsequent stages and are eventually accompanied by differentiation of more or less distinct eluvial and illuvial horizons resulting from clay translocation. In originally calcareous materials this follows decalcification and consequent acidification of the upper horizons, but the organo-mineral surface layer normally remains of mull type so long as roots continue to penetrate the calcareous substratum. Clay translocation is often less clearly marked in originally non-calcareous materials and progressive leaching leads to stronger acidification. Mull is then replaced by moder or mor and horizons attributable to podzolization become increasingly apparent, especially if deciduous forest gives place to heath.

Soils in which downward percolation of water is impeded by a relatively impermeable substratum or by seasonally rising ground water exhibit essentially similar developmental sequences in which the horizons formed are modified to varying degrees by the effects of gleying. Where the soil is continually wet, however, organic matter accumulates to form peat and both oxidative weathering and clay translocation are inhibited. In upland areas with high rainfall and low summer temperatures, progressive impoverishment of the upper horizons and replacement of

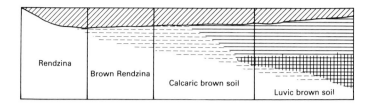

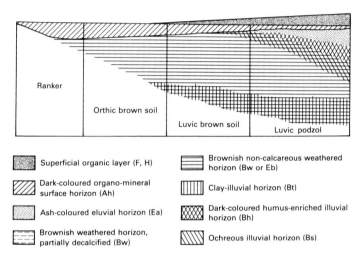

Figure 1.3 Profile development sequences in permeable calcareous and non-calcareous materials.

forest by moorland vegetation cause peat to form even on initially well drained soils. Under these conditions a combination of podzolization and superficial gleying gives rise to a greyish eluvial horizon depleted of iron, followed by a thin cemented illuvial horizon (thin ironpan) enriched in iron and organic matter.

1.3 Factors Affecting Soil Variation in the British Isles

1.3.1 Parent Material and Age Factors

Differences in the particle-size distribution, mineralogical composition and permeability of parent materials influence soil properties both directly and by conditioning the nature and extent of the changes brought about by the various developmental processes. Hard or soft bedrock which has weathered in place forms the parent material of at least the lower part of the soil in many places. Over most of the British Isles, however, the predominant soil-forming materials are Quaternary sediments of glacial, aeolian, alluvial, colluvial or biogenic origin, which were laid down at various stages in the evolution of the present landscape. The mineral superficial deposits, collectively known as drift, vary greatly in composition depending on the source rocks (Figure 1.4) from which they were derived, the extent to which the materials composing them were weathered prior to transport, and the nature of the transporting process. Thus

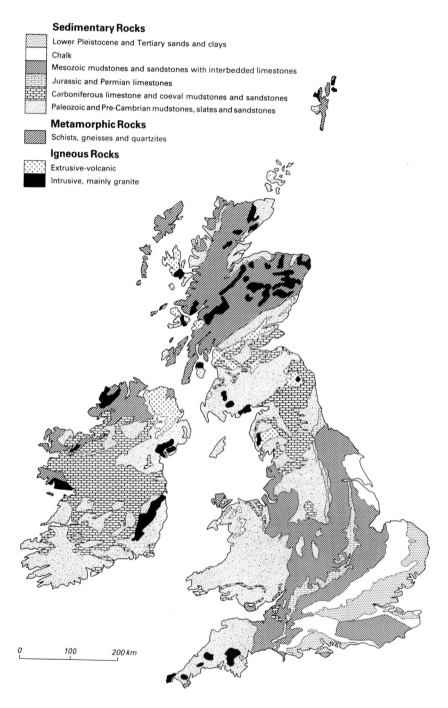

Figure 1.4 Schematic map showing pre-glacial source rocks in the British Isles.

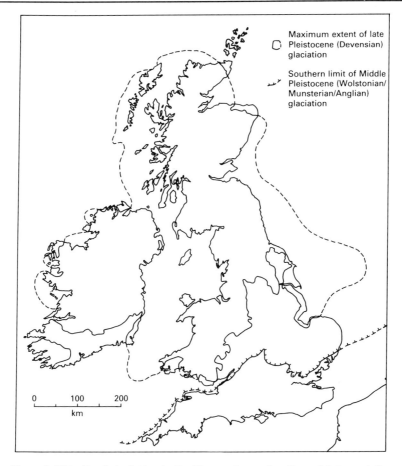

Figure 1.5 Limits of glaciation in the Devensian and earlier cold stages (after Bowen *et al.*, 1986).

aeolian deposits such as loess and blown sand are characteristically well sorted, with narrow ranges in particle size; fluviatile, marine and lacustrine sediments are also sorted to varying degrees and normally show stratification, whereas glacial drift (till) and colluvium resulting from downslope movement of surface materials are typically ill sorted and no more than crudely stratified.

Apart from recent (Holocene) deposits, which are widely distributed but only locally extensive, the main soil-forming materials of Quaternary age originated during Pleistocene cold stages, the last of which terminated about 10,000 years ago (Table 1.1, p. 11). In this and earlier cold periods, ice sheets spread out from the British and Scandinavian mountains (Figure 1.5), displacing and transporting rock and soil as they advanced; melt waters heavily charged with rock waste fed forerunners of present-day rivers which deposited the coarser fractions of their loads to form spreads of gravel and sand, while finer materials settled in temporary lakes resulting from disruption of pre-existing drainage lines. In unglaciated (periglacial) areas, tundra conditions prevailed and stresses caused by repeated freezing and thawing of superficial layers or by thermal contraction of deeply frozen ground led to shattering of exposed bedrock

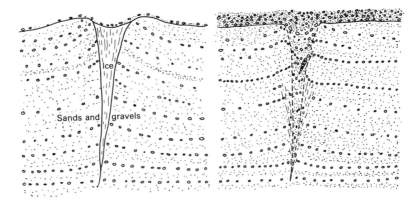

Figure 1.6 Cross-sections of ice-wedge casts showing (a) upturning of adjacent gravel beds and (b) the inward collapse of walls on melting, partly filling the wedge with slumped material before burial beneath further gravel (after Catt, 1986).

and development of cryoturbation structures such as ice-wedge casts (Figure 1.6), involutions (Figure 1.7) and various kinds of patterned ground (Figure 1.8).

Colluviation was also greatly facilitated, resulting in formation of screes and gelifluction deposits (Catt, 1986). Unstratified or crudely stratified accumulations of the latter type, most of which appear to have originated by slow flow of seasonally thawed and hence water saturated material over frozen subsoil, have been grouped with Holocene colluvial deposits as *head* by British geologists (Dines *et al.*, 1940). Often underlying very gentle slopes, they form the upper horizons or all of the soil over large areas, particularly in those parts of England which were never covered by ice (Figure 1.5) or from which glacial drift

Figure 1.7 Involutions in a chalky gelifluxion (head) deposit at Pitstone, Buckinghamshire (Photo by J.A. Catt).

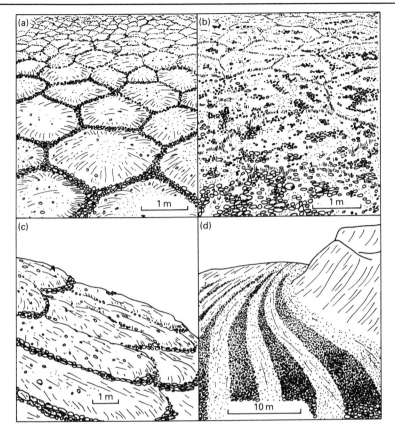

Figure 1.8 Types of patterned ground in periglacial areas (after Catt, 1986): (a) sorted polygons; (b) non-sorted polygons; (c) stone garlands; (d) sorted stripes.

emplaced during one or more of the pre-Devensian cold stages was removed during subsequent phases of valley incision. Parent materials consisting wholly or partly of wind-borne silt (loess) or sand (coversands) that accumulated under cold and relatively dry conditions in the Devensian period or earlier also occur in parts of southern and eastern England, South Wales and the Midlands (Perrin *et al.*, 1974; Catt, 1977, 1978). As Britain lay at the margin of the north European loess and coversand belt, however, and the supply of aeolian sediment was limited, the deposits were mostly thin and have commonly been mixed with subjacent materials by cryoturbation or reworked by gelifluction or fluvial action. Loess and its derivatives have often been described as *brickearth*, but this term has also been applied to fine glaciolacustrine or riverine sediments with little or no loess component.

Judging from palaeontological evidence, the climate became at least as warm as it is today in each of the earlier temperate stages, with the result that scanty tundra vegetation was progressively replaced by forest at all but the highest altitudes. The consequent stabilization of ground surfaces enabled deep reaching soil development to take place but the horizons so created were again subject to frost disturbance, erosion or burial during the following cold stage.

Table 1.1 British and Irish Quaternary stages and main deposits (based on Mitchell *et al.*, 1973, Bowen *et al.*, 1986, and Catt, 1986)

Years b.p.	Temperate stages	Cold stages	Main deposits
10,000	Holocene (Flandrian) (*Littletonian*[a])		Alluvium, colluvium (hillwash), peat, lake marl, tufa
116,000		Devensian (*Midlandian*[a])	Till, glaciolacustrine deposits, gelifluction (head) deposits, river gravels, loess, coversands
128,000	Ipswichian		Lake, river and marine (raised beach) deposits
367,000		Wolstonian[b] (*Munsterian*[a])	Till, glaciolacustrine deposits, gelifluction (head) deposits, river gravels, loess
444,000	Hoxnian (*Gortian*[a])		Lake and marine (raised beach) deposits
		Anglian (*pre-Gortian*[a])	Till, glaciolacustrine deposits, gelifluction (head) deposits, river gravels, loess
	Cromerian		Peat, estuarine sands
		Beestonian	River sands and gravels
	Pastonian		Peat, marine silts, gravels
		Pre-Pastonian	River gravels, marine sands
	Bramertonian		Shelly marine sands (Norwich Crag), river gravels
		Baventian	Marine silts and clays
	Antian		Shelly sands
		Thurnian	Marine silts and clays
	Ludhamian		Shelly marine sands
		Pre-Ludhamian (Pliocene?)	Shelly marine sands (Red Crag)

[a]Irish regional stages
[b]The type Wolstonian succession in the English Midlands comprises glacigenic deposits assigned to the Anglian stage by Bowen *et al.* (1986) and Rose (1987).

These successive phases of landscape stability and instability are reflected in the profiles of present-day soils. Thus the horizons of most British soils in both 'solid' and drift parent materials appear to have formed over the relatively short period that has elapsed since the end of the Devensian cold stage (Table 1.1), but some show additional horizons or features, including rubification (p. 6), which can be attributed to the action of weathering and soil-forming processes during one or more of the preceding temperate stages or even in pre-Quaternary times (Catt, 1979; Avery, 1985). The profiles of those showing no definite evidence of pre-Devensian soil formation can be described as *monocyclic* and

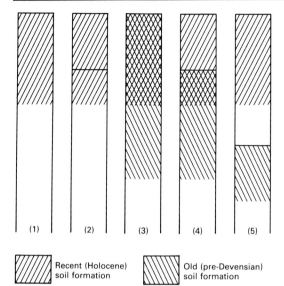

Figure 1.9 Types of profile in uniform and layered parent materials: (1) simple monocyclic; (2) composite monocyclic; (3) simple polycyclic; (4) composite polycyclic (complex); (5) simple monocyclic with underlying buried soil (from Avery, 1985). (Reproduced by permission of John Wiley & Sons, Ltd.)

are further divisible into *simple* profiles in initially uniform materials and *composite* profiles consisting of two or more lithologically distinct layers (Figure 1.9). Genetically more complex profiles of the second type are termed *polycyclic* (cp. Duchaufour, 1982); simple and composite variants can again be distinguished, but the latter, in which the upper horizons have evidently formed in a thin Devensian deposit, are much more common. The material underlying the superficial layer can be a glacial, glaciofluvial or head deposit emplaced before the Ipswichian temperate period (Table 1.1), bedrock deeply weathered *in situ* (e.g. Fitzpatrick, 1963) or a quasi-residual deposit, such as Clay-with-flints (Catt, 1985), incorporating remains of pre-existing sedimentary rocks which have apparently been subjected to repeated phases of temperate weathering and periglacial disturbance.

At the other end of the age scale, soils in alluvial and other deposits of Holocene age, or in sloping sites subject to erosion or colluviation under present-day conditions, normally show shallower and/or less well marked horizonation than those associated with older and more stable ground surfaces on materials of similar composition.

1.3.2 Climate

The main climatic factors affecting the properties and evolution of soils are temperature and effective precipitation, as the excess (if any) of precipitation as rain or snow over evaporation from plants or moist ground is known. Temperature influences plant growth, the activity of soil organisms, and mineral weathering and synthesis through its effect on chemical reaction rates, most of which are roughly doubled with each 10°C rise in temperature between 0° and 40°C. The balance between precipitation and loss of water by evaporation or surface run-off governs the rate of leaching of soluble constituents; it is also the major factor determining the moisture regime of the soil and this in turn affects rates of organic matter decomposition and the incidence of other developmental processes such as gleying and translocation of clay.

Temperature varies with latitude and altitude, and the extent of adsorption and reflection of solar radiation by the atmosphere. Soil temperatures are governed by atmospheric temperatures but also vary with aspect and exposure, the nature of the vegetation, water contents within the soil, the presence or absence of winter snow cover and the colour of the surface where the ground is bare. The mean annual soil temperature at any site remains constant with increasing depth and seldom diverges from the mean annual air temperature by more than 1–2°C, but the seasonal and diurnal fluctuations decrease with depth and the seasonal maxima and minima are attained later in the year than in the atmosphere. Because of the proximity of the Atlantic Ocean and the warming effect of the

Gulf Stream, winters are milder in the British lowlands, especially those close to western seaboards, than in continental areas at the same latitude. On the other hand the summers are cooler and both the length of the growing season during which the mean daily temperature exceeds 5.6°C and the mean annual accumulated temperature over this period decrease more sharply with increasing altitude. An extreme expression of these oceanic climatic characteristics is illustrated in Figure 1.10, which shows the annual temperature cycles at Fort William, Inverness-shire (altitude 10 m) and at the nearby summit of Ben Nevis (1,343 m) over the same 13-year period, with data for Moscow (146 m) at about the same latitude added for comparison. The decrease in summer temperature with altitude, combined with heavier rainfall and increased windiness, gives rise to marked zonation of vegetation and soil on the British and Irish mountains, despite their relatively modest height. In striking contrast to the Alps, the 'tree line' is everywhere below 600 m; signs of active cryoturbation appear in soils above this level in the Scottish Highlands and elsewhere, and occur nearly everywhere at around 900 m (Birse, 1980).

Effective precipitation controls the moisture regime of the soil in conjunction with other factors, including the permeability of horizons or layers within the rooting zone, their capacity to retain water that can be used by plants in rainless periods, and the presence or absence of ground water of extraneous origin. Alternative indices of climatic wetness are the *potential water deficit* (PWD) (Green, 1964) and the *maximum potential soil moisture deficit* (Hodgson, 1976), usually abbreviated to *moisture deficit* (MD), both of which are based on estimates of potential evapotranspiration (PE). This was defined by Penman (1948) as the water transpired by a short green crop such as grass which completely covers the ground and is adequately supplied with water around the roots. It can be computed from

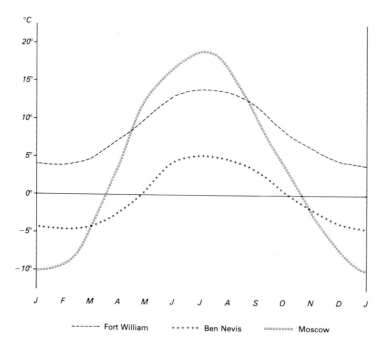

Figure 1.10 Annual temperature cycles at Fort William, the summit of Ben Nevis (Omond, 1910) and Moscow, USSR.

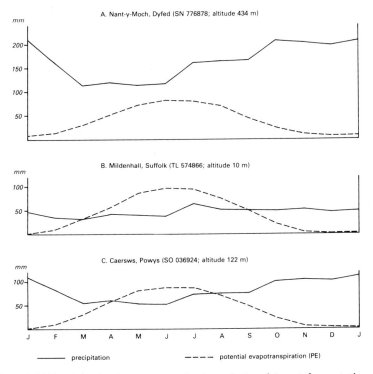

Figure 1.11 Precipitation/evapotranspiration relationships at three stations in England and Wales, based on data from Rudeforth (1970), Corbett (1973) and Lea (1975).

meteorological data (MAFF, 1967; Jones and Thomasson, 1985) and PWD values are then obtained by summing the amounts (if any) by which the mean PE (in mm) during each calendar month exceeds the mean monthly rainfall over the same period of years. MD is computed for a station by calculating the maximum potential water deficit for each year separately and averaging the results. The values obtained are higher than those of PWD and give a better representation of the climatically determined water regime to the extent that yearly variations are taken into account. The actual evaporative loss from the soil is normally larger under forest (e.g. Rutter and Fourt, 1965) than under grass and is considerably smaller where the ground is bare. It also falls below the calculated PE value once the reserve of easily available water (Hall *et al.*, 1977) in the rooting zone is depleted.

As mean annual rainfall varies much more than PE within the British Isles, ranging from less than 550 mm in parts of Eastern England to as much as 5,000 mm on the higher western mountains, the ranges in both effective precipitation and MD are correspondingly large. This is shown graphically in Figure 1.11, which is based on mean monthly rainfall and PE values for three Welsh and English stations. At Nant-y-Moch, Dyfed, the effective precipitation (rainfall—PE) is some 1,450 mm annually and rainfall exceeds evapotranspiration in every month of most years, so that the soil is always moist or wet throughout. At Mildenhall, by contrast, PE exceeds rainfall from April to September in an average year and is only 15 mm less than precipitation on an annual basis. Except where roots have access to ground water, however, the actual evapotranspiration from grass falls well short

Table 1.2 Climatic regimes in the British Isles

Climatic regime	Mean annual accumulated temperature above 5.6°C (day–degrees)	Average maximum potential soil-moisture deficit (MD) or approximately equivalent potential water deficit (PWD)
Subhumid temperate	> 1375	MD > 125 mm (PWD > 100 mm)
Humid temperate	> 1375	MD 50–125 mm (PWD 25–100 mm)
Perhumid temperate	> 1375	MD < 50 mm (PWD < 25 mm)
Humid (oro[a]) boreal	675–1375	MD 50–125 mm (PWD 25–100 mm)
Perhumid (oro[a]) boreal	675–1375	MD < 50 mm (PWD < 25 mm)
(Perhumid) oroarctic[a]	< 675	MD < 50 mm (PWD < 25 mm)

[a] Altitude greater than 150 m.

of the potential value from June onwards under these conditions, especially where the capacity of the soil to retain easily available water is small, but the normally even distribution of rainfall ensures that reserves are seldom completely exhausted for more than a week or so. Caersws is a site of intermediate humidity at which the effective precipitation is about 440 mm; PE normally exceeds rainfall during May, June and July but the resulting soil-moisture deficit is generally made good by October.

Classifications and maps based on climatic components affecting vegetation, land-use capability and soil development have been produced for Scotland by Birse (1971) and for England and Wales by Bendelow and Hartnup (1980). Although the classifications differ in detail, both include divisions based on thermal regime (mean annual accumulated atmospheric temperature above 5.6°C), moisture regime (PWD or MD), exposure to wind and oceanicity, which were mapped using standard altitudinal temperature-lapse rates, rainfall maps and observations of vegetation to extend data from available stations. A simplified version of the Scottish system was proposed by Birse (1976) and a similar scheme consisting of six climatic regime classes based on accumulated temperature and potential soil moisture deficit only, with altitudinal subclasses (Table 1.2), is adopted in this book. Figure 1.12 shows generalized boundaries of the climatic zones so defined in the British Isles as a whole.

1.3.3 Relief

The configuration of the land affects the evolution of the soil by causing local differences in climate and ground-surface stability, and more particularly by giving rise to *hydrologic sequences* of soils with differing moisture regimes. Thus undulating landscapes on permeable materials commonly have freely draining soils in upper slope and hilltop positions and progressively less well drained and hence increasingly gleyed soils in lower positions where the water table delimiting the subterranean zone of permanent saturation approaches the surface (Figure 1.13). Other kinds of hydrologic sequence occur where there are relatively impermeable soil or subsoil layers at varying depths, causing disposal of the effective precipitation to take place by surface run-off or lateral subsurface flow rather than by downward percolation. Poorly drained soils are then associated with level plateaux as well as with footslopes

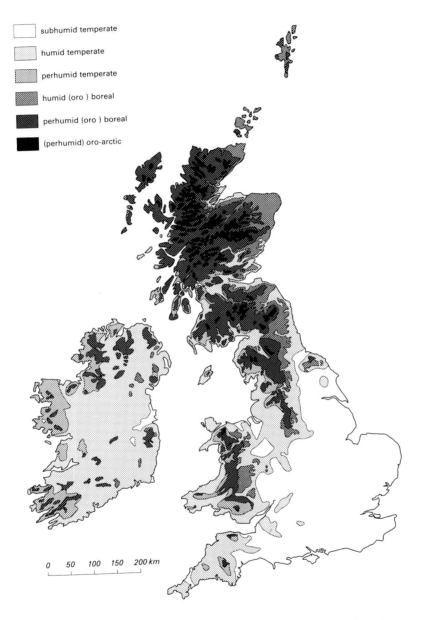

Figure 1.12 Schematic map of climatic regimes in the British Isles, based on data from Birse and Dry (1970), Bendelow and Hartnup (1980), Jones and Thomasson (1985) and Irish Meteorological Service, Dublin.

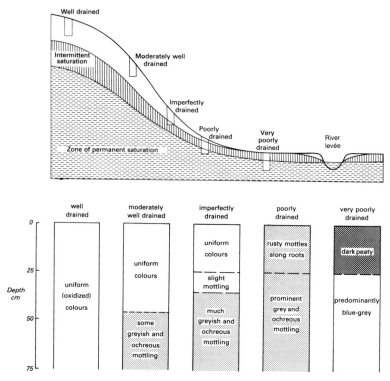

Figure 1.13 A hydrologic soil sequence in permeable parent material; after Batey, 1971, and White, 1987)

and depressions, and seepage water emerges as springs in places where the upper boundary of the impermeable layer is intersected by a slope.

Milne (1935a, b) called these and other topographically related patterns of soil variation *catenas* and distinguished two types as follows.

In one the parent material does not vary, the topography having been modelled out of a single type of rock at both the higher and the lower levels. In this case the soil differences are brought about by differences in drainage conditions, combined with some differential reassortment of eroded material and the accumulation at lower levels of soil constituents chemically leached from higher up the slope . . . In the other the topography has been carved out of two superposed formations, so that the upper one (A) now forms the capping of the hills and ridges and the lower one (B) is exposed further down the slopes. Here a geological factor is added to the conditions that make for soil differences. In such circumstances we may have as the top member of the catena an old soil or its denuded remnants (on formation A); younger soils below (on formation B with some admixture of secondary material); and an accumulation of erosion products in the bottomlands.

Both types of catena are well represented in the British Isles, the first on nearly uniform parent materials in landscapes that originated during or after the last (Devensian) cold stage and the second in older (polycyclic) landscapes which characterize many parts of southern England. Two examples of these more complex catenas are shown schematically in Figure 1.14. In each, the soils of lower

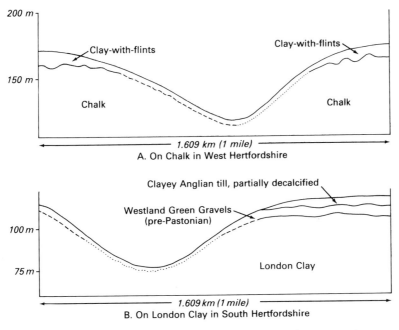

Figure 1.14 Simplified representation of two complex catenas in Hertfordshire (after Thomasson and Avery, 1963).

valley sides are mainly developed in head deposits containing materials derived from the older drift formations which underlie the interfluves. The asymmetry of the valleys has been attributed to the effects of varying insolation on slope development processes under periglacial conditions (Ollier and Thomasson, 1957; Thomasson, 1961).

1.3.4 Biotic Factors: Vegetation and Man

The properties of the soil at any particular place and time are partly determined by the vegetation and other organisms it supports. Replacement of forest by grassland or heath, for example, causes significant changes in its moisture and temperature regimes as well as in the amount and nature of the plant remains deposited on the surface or added to subsurface horizons, all of which affect the composition and activity of the soil population and the processes in which its members are directly or indirectly involved.

In virgin landscapes, variations in vegetation and associated organisms generally accord with variations in climate, parent material and relief, and soil differences reflecting the independent influence of biotic factors are correspondingly limited in scope and extent, occurring chiefly in 'tension zones' between two climatically determined plant communities such as forest and grassland. Wherever man has settled and managed the land, however, the natural ecosystems have been modified or disrupted by his activities and the resulting changes are reflected in soil profiles, the upper parts in particular taking on new characteristics which depend on land use and management practices and more or less mask the effects of the primitive vegetation. When ground is brought into cultivation, the naturally formed soil horizons are mixed to the depth of ploughing and the surface becomes liable to accelerated erosion. Under intensive agricultural systems, the organic-matter content and nutrient status of the soil are largely determined by the opposing effects of regular crop removal and additions of organic manure or mineral fertilizers, and bear little relation to the equilibrium values attained under the original vegetation. Application of lime to acid soils raises their pH, and artificial drainage of wet land, or irrigation of dry land, causes equally profound changes in the soil moisture regimes. Replanting of deciduous woods with exotic conifers also affects the evolution of the soil, mainly through differences in the litter produced.

The climatic, vegetational and cultural history of the British Isles during the 8,000 years which elapsed between the last glaciation and the Roman colonization of southern Britain have been reconstructed by relating analyses of pollen preserved in peat, organo-mineral sediments and soils to other palaeontological and stratigraphic data and archaeological findings (Godwin, 1975; Mitchell, 1976; Bridges, 1978). Figure 1.15 summarizes the generally accepted sequence of events, including the rapid postglacial rise in sea level which finally severed Britain from continental Europe and later minor fluctuations which governed the evolution of vegetation and soils in low-lying coastal areas such as the fenlands of eastern England. Inland from the coasts, increasing warmth during the early part of the Holocene stage led to replacement of tundra vegetation by forests of birch, pine and hazel. These gave place in turn to mixed broadleaf forest with alder, oak, elm and lime as major components, which appears to have spread to all but the coldest parts of northern Britain by about 6,000 years B.P. At the same time the Mesolithic inhabitants were already promoting localized clearances through grazing and burning. Clearances were greatly extended during the succeeding Neolithic, Bronze Age and Iron Age periods, which also saw the introduction and extension of tillage based initially on the use of hand implements. Evidence of prehistoric cultivation of easily worked soils is provided by pollen analyses indicating the prevalence of weed species, widely distributed 'Celtic fields', lynchets (cultivation terraces) and other artefacts produced by ploughing, often in high lying sites remote from more recent settlements, and the occurrence in valleys of contemporary alluvial and colluvial deposits apparently reflecting increased soil erosion on the slopes above. The vegetation that replaced the virgin forest in uncultivated areas included various grassland and heath communities, the composition of which varied in accordance with local climatic and soil conditions and the incidence of grazing. From about the beginning of the Iron Age onwards, the combined effects of climatic deterioration, deforestation and progressive soil impoverishment led to widespread extension of peat-forming (moorland) vegetation in perhumid upland areas (Figure 1.12) and cultivation was restricted accordingly to less humid and more temperate lowland localities.

The effects of these and subsequent changes in vegetation, land use and management on soil evolution have been reviewed by Bridges (1978) and Davidson (1982) and are treated in relation to particular soil groups in the chapters which follow. Of the developments during the centuries before the Industrial Revolution made its impact, the most important was the introduction of the mouldboard plough, which enabled Romano-British and Anglo-Saxon cultivators to bring the predominantly heavy soils of the English lowlands into regular agricultural use, with the result that the rural settlement pattern recorded in the Domesday Book of 1086 is recognizably similar to that of today. Later came widespread expansion of sheep grazing to meet the demands of the wool trade and gradual replacement of the exhaustive open-field system of arable farming by fertility-conserving rotations, involving alternation of cereal and root crops with grass-clover leys and repeated applications of animal manure,

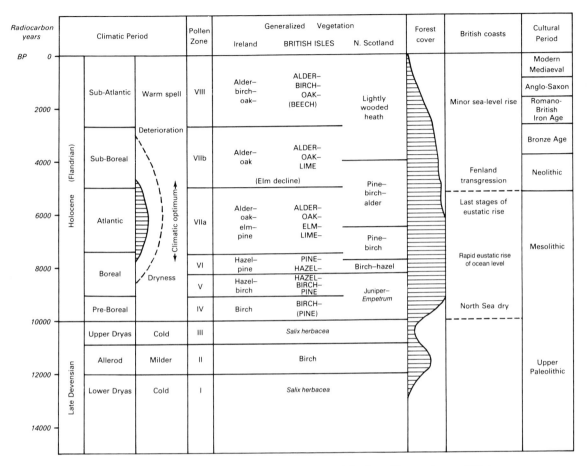

Figure 1.15 Correlation table showing the main events of the Late Devensian and Holocene (Flandrian) periods in the British Isles (after Godwin, 1975, and Bridges, 1978).

which remained in use in some places until after the second world war. Deforestation proceeded at varying but generally increasing rates throughout this period until by the end of the eighteenth century no more than about 5 per cent of the British Isles remained wooded. Subsequent plantings by estate owners reversed the decline, particularly in Great Britain, but extensive felling during the 1914–18 war left timber stocks seriously depleted. In order to remedy this deficiency, the Forestry Commission was instituted in 1919 to establish and manage new plantations. Since that time the forested area of Great Britain has increased to 10 per cent, of which approximately two-thirds is privately owned, but only about 6 per cent of Ireland is covered by woods. However, as nearly all the new plantations are coniferous, with Sitka spruce, Scots pine and Norway spruce as the principal species, and many pre-existing woods have also been replanted with conifers, the proportionate area of semi-natural broadleaf woodland is now no more than 2 per cent.

The large increase in population which began around the middle of the eighteenth century stimulated the reclamation and improvement of land, especially by artificial drainage. Besides providing much of the necessary capital, the accompanying Industrial Revolution led to the invention and gradual introduction of tile drains, new labour saving agricultural machinery and artificial fertilizers. From the 1870s onwards, however, declining grain prices due to foreign competition evoked a steady conversion of arable land to grass in most areas; and in Ireland, where population pressure was greatest and undue reliance had been placed on the potato crop as a means of subsistence, the disastrous famine of 1846–51 was accompanied and followed by large-scale emigration. During and since the second world war, the proportion of the total agricultural area devoted to tillage crops again increased and is now nearly as great as it was around the middle of the nineteenth century. Approximate percentages of this and other broad land-use categories in Great Britain and Ireland are given in Table 1.3. The marked variations it reveals are closely related to climate and topography. Although grass for dairying and livestock production remains by far the most important crop overall, tillage crops are considerably more extensive in the subhumid lowlands of eastern England, and the large proportion of rough grazing in Scotland reflects the predominance of uplands and mountains with humid to perhumid, oroboreal and oro-arctic climatic regions (Figure 1.12).

Despite continuing diversion of farmland to urban and other uses, and a large reduction in the labour force, the introduction of improved crop varieties, herbicides and pesticides, heavy fertilizer applications, more intensive animal husbandry and ever-increasing mechanization have together resulted in a rather more than twofold increase in agricultural production from the British Isles during the post-war period. This 'second agricultural revolution' has affected the more intensively used lowland soils chemically and physically. In terms of nutrient status, they are generally more fertile than they have ever been, but adverse changes in their physical condition resulting from continuous cultivation and the use of heavy machinery have given cause for concern, particularly in wet years (MAFF, 1970). The most serious and far-reaching effect, which is now well documented (Morgan, 1977, 1980; Colborne and Staines, 1985; Burnham and Pitman, 1987) has been a

Table 1.3 Major land-use categories in Great Britain[a] and Ireland[b] (1985–86)

	England	Wales	Scotland	Northern Ireland	Republic of Ireland	British Isles
Total land area (km^2)	129,671	20,636	77,075	13,482	68,890	309,754
Agricultural land %	74.9	80.6	76.5	79.3	81.3	77.3
Tillage land %	34.3	4.4	8.7	5.7	7.1	18.6
Temporary grass[c] %	6.8	8.2	5.7	17.0	8.4	7.1
Permanent grass %	23.6	42.2	7.8	39.7	65.8	50.5
Rough grazing %	9.1	25.2	53.8	14.7		
Other cropland %	1.1	0.6	0.5	2.2	< 0.05	0.8
Woodland %	7.3	11.8	14.2	1.8	5.6	8.7
All other land %	17.8	7.6	9.3	18.9	13.1	14.3

[a]MAFF (1988)
[b]Commission of the European Communities (1985a)
[c]less than five years old.

Figure 1.16 Gully erosion on arable land in south-east Somerset (Photo by S.J. Staines).

marked increase in the incidence of water erosion on susceptible soils (Figure 1.16.).

1.4 Scales and Types of Soil Variation

The impress of the various controlling factors on soil variation is perceptible over a wide range of geographic scales. Beckett (1967) and Beckett and Webster (1971) examined lateral changes in soil variability as determined by measurements of the variance of particular properties within sample areas of progressively increasing size. Studies of this type show that as much as half the total variation in some properties can occur within a horizontal distance of one metre. Thus variation in physical, chemical and biological properties affected by earthworm activity, ped formation, stones, roots, etc. may already be pronounced within areas of 100 cm^2 but may not increase much more as the sample area is increased. However, total variability increases more or less stepwise as causes such as tree throw, ridge-and-furrow cultivation, or lateral segregation of near-surface materials as in patterned ground, become effective over distances of 1–10 m; and other changes in parent material, drainage conditions or land-use history (e.g. between fields or fields and woods) over greater distances. Such changes may be *continuous*, with no well defined inflections in the lateral rate of change, or *discontinuous*, as where two distinct but relatively homogeneous bodies of soil adjoin. They can also be categorized as systematic or non-systematic (random) according to whether an orderly or recurrent pattern of change is discernible (Wilding and Drees, 1983).

Most landscapes exhibit systematic soil changes which are partly continuous and partly discontinuous. Systematic continuous changes are exemplified by progressive long-range variations in such properties as amount and distribution of organic matter and inorganic products of weathering, horizon development and thickness, which can be attributed to the effects of changing climate. Others of more local significance include those related to gradational changes in parent material, drainage or duration of undisturbed horizon development in landscape units such as floodplains, fans, pediments, areas of loess deposition and metamorphic aureoles. Many catenary sequences on hillslopes are also continuous (e.g. Walker *et al.*, 1968). Others, reflecting sharp changes in parent material or ground-surface age in 'mixed' or polycyclic catenas of Milne's second type (p. 20), include well-marked discontinuities. Bodies of soil bounded by systematic discontinuous changes are *natural soil-landscape units* (Knox, 1965) in the sense that they are discrete and independent of the observer. Where their boundaries can be consistently related to surface features they afford an economical and effective basis for soil mapping, providing that the units are large enough to be represented at the scale used.

Short-range systematic variation characterized by discontinuous soil changes at more or less regular intervals has been recorded in many parts of the British Isles, particularly in association with fossil patterned ground of periglacial origin (Figure 1.17a), jointed limestone substrata (Figure 1.17b), interstratified hard and soft sedimentary rocks in erosional situations (Figure 1.17c) and marshland with intricate creek patterns (Figure 1.17d). Especially in patterns of the first two types, the subsurface discontinuities are normally unrelated to surface relief and can therefore escape notice, but their presence is revealed on air photos taken when the ground is bare or during dry periods when crops are subject to moisture stress (e.g. Evans, 1972). As they are unresolvable at any ordinary map scale, they often have to be represented as compound units (p. 34) in soil surveys (e.g. Corbett, 1973; Seale, 1975b).

Superimposed on any systematic variations are unsystematic or random variations, which can account for a large proportion of the total variability (e.g. Walker *et al.*, 1968). Common sources of such variations are parent material heterogeneity and biotically related or cultural features such as tree throw, animal droppings, infilled excavations and other disturbances resulting from human activity. Processes operating within the soil itself can also give rise to apparently random variation over short distances, as exemplified by tongued or irregular horizon boundaries related to irregular dissolution of calcareous substrata. However, when these or similar features are examined in detail, for example along trenches, a systematic pattern of variation is sometimes discernible.

1.5 Soil Classification

1.5.1 Purposes and Limitations

As in other natural sciences, the essential purpose of soil classification is to organize existing knowledge so that the properties and relationships of different kinds of soil can be systematically recalled and communicated (Cline, 1949). This involves creating classes at one or more levels of abstraction, either by subdividing a universe or by grouping individuals. Membership of classes is determined by differentiating criteria, which may be intrinsic properties but can also be formulated in terms of external features, genetic factors or processes.

Figure 1.17 Systematic short-range soil variation patterns revealed by air photography (Cambridge University Collection: copyright reserved). (a) Polygon and stripe patterns of periglacial origin in bare ground on chalk- sand drift near Brettenham, Norfolk. (b) Bedrock jointing patterns on Jurassic limestone near Raunds, Northamptonshire; the dark lines represent deep soils extending down cracks in the underlying rock. (c) Patterns on contrasting rock types at Tickencote, Leicestershire. (d) Relic creek patterns in fenland between Chatteris and Ramsey, Cambridgeshire.

Unlike populations of plants and animals, however, the soil universe presents a continuum of variation and does not consist of discrete individuals with inborn distinguishing characteristics that facilitate empirical recognition of natural groups. There is therefore strictly no such thing as 'a soil' (Knox, 1965) and both the choice of criteria and the imposition of class limits necessarily entail arbitrary judgements. Also, since it is impossible in practice to study all parts of the soil mantle, classification has to be based on information derived from examination of samples, the dimensions of which are governed by operational considerations.

When the soil as it exists in the field is the universe under consideration, the objects classified are effectively soil profiles extended to three dimensions in order to permit adequate description and sampling of the constituent horizons. In England and Wales, for example, the basic profile unit is taken to have a maximum cross-sectional area of about 1 m^2 and a maximum depth of 1.5 m (Hodgson, 1976; Avery, 1980). With the aim of basing classification on larger bodies of soil, the Soil Conservation Service of the US Department of Agriculture defined the *pedon* as a sampling unit and the *polypedon* as a unit of classification (Soil Survey Staff, 1975). The pedon has a horizontal area that can range from 1 to 10 m^2 to take account of intermittent cyclic horizons which recur at linear intervals of 2 to 7 m, and the polypedon consists of contiguous similar pedons bounded on all sides by 'not soil' or by pedons of unlike character. As polypedons can seldom be fully characterized, however, and their limits depend on the criteria used to classify them, this attempt to define an articifical individual has failed to gain general acceptance.

1.5.2 Kinds of Classification

Soils, soil horizons or soil materials can be classified using any attribute or set of attributes appropriate to the purpose for which the classification is made.

A specific or single-purpose classification is made with a specific aim in mind and based on one or very few relevant attributes. Examples are classification of soil profiles based on properties related to natural drainage (Figure 1.13), the textural classification represented in Figure 3.2 (p. 84) and division of 'available' P and K contents of topsoils into a small number of classes for agricultural advisory purposes.

A general (i.e. general-purpose) classification, by contrast, is intended to serve as many purposes as possible, foreseen or unforeseen, and is therefore based in principle on similarities in a wide range of properties, certain of which are chosen as differentiating criteria. Classifications of this kind have also been termed *natural* (e.g. Gilmour, 1951; Kubiena, 1958; Muir, 1969) but the degree of naturalness (or overall usefulness) that can be achieved is limited by the extent to which different properties are interrelated. As noted by Mulcahy and Humphries (1967), the value of a general soil classification depends on the selection as criteria of attributes that are significantly correlated with many others. If a property relevant to a particular objective varies independently, a general system cannot predict its distribution. In British agricultural soils, for example, properties such as pH, exchangeable cation contents and organic matter status often correlate hardly at all with other, less easily altered properties.

1.5.3 Development of General Classifications

Since Dokuchaiev introduced the soil profile as a unit of study, numerous general classifications in which the classes are defined partly or entirely on this basis have been published. Those used extensively in one or more countries were reviewed recently by Clayden (1982) and those that have been applied to

Table 1.4 Classification of soils by V.V. Dokuchaiev, 1900 (after Soil Survey Staff, 1960)

	Zones	Soil types
Class A. Normal, otherwise dry land vegetative or zonal soils		
I	Boreal	Tundra (dark brown) soils
II	Taiga	Light grey podzolized soils
III	Forest-steppe	Grey and dark grey soils
IV	Steppe	Chernozem
V	Desert steppe	Chestnut and brown soils
VI	Aerial or desert zone	Aerial soils, yellow soils, white soils
VII	Subtropical and zone of tropical forest	Laterite or red soils
Class B. Transitional soils		
VIII	Dry land moor-soils or moor-meadow soils	
IX	Carbonate containing soils (rendzinas)	
X	Secondary alkali soils	
Class C. Abnormal soils		
XI	Moor-soils	
XII	Alluvial soils	
XIV	Aeolian soils	

British and Irish soils are fully described in the following chapter.

Dokuchaiev's original classification of Russian soils had two categories, the higher with three classes and the lower with thirteen (Table 1.4). The normal, transitional and abnormal classes were subsequently renamed zonal, intrazonal and azonal soils by Sibertsiev (1901), his closest collaborator. *Zonal soils* were conceived as 'mature' soils with profile characteristics reflecting the predominating influence of regional climate and vegetation; *intrazonal soils* as those with well-marked profile features attributable to the modifying influences of local factors such as poor drainage or parent material resistant to zonal processes; and *azonal soils* as those with weakly developed (immature) profiles formed in freshly deposited sediments or ground surfaces newly exposed by erosion. Classes in the second category were called *soil types*, the concepts of which entailed specific developmental processes and environmental factors as well as distinctive profile morphology.

The early Russian work stimulated the study of soils in the field throughout the world. It provided a gallery of broad soil classes related to climatic and vegetational zones on a continental scale and led to the development of more elaborate multicategorical classifications on similar lines, first in Russia (Tiurin, 1965) and subsequently elsewhere. In both the United States and Britain, however, this was preceded by the establishment of narrower classes designed for use in detailed soil surveys. In the USA, where a national soil survey was started in 1899, physiographic regions were identified as soil provinces and soils from similar parent materials within a province as soil series, each of which was divided into soil types on a textural basis. From the 1920s onwards, under the leadership of C.F. Marbut and C.E. Kellogg, soil series were redefined in terms of observable profile characteristics and allocated to *great soil groups* corresponding approximately to the Russian genetic soil types (Simonson, 1964). In later versions of this system (Baldwin *et al.*, 1938;

Thorp and Smith, 1949), the great soil groups were arranged in suborders identified in 'genetico-geographic' terms (e.g. light-coloured podzolized soils of the timbered regions; hydromorphic soils of marshes, swamps, seep areas and flats) and these in turn were grouped into orders of zonal, intrazonal and azonal soils as first conceived by Dokuchaiev and Sibertsiev.

When attempts were made to apply the zonal concept in regions with more varied parent materials and landscape histories than the continental plains in which it originated, its shortcomings as a worldwide basis for soil classification became increasingly apparent. Thus it seemed that the influence of present-day climate and natural vegetation on soil variation had been overstressed at the expense of other factors, especially when viewed from west European countries where little truly natural vegetation remains, or from tropical or subtropical areas such as Australia where many of the soils appeared to have acquired their most distinctive charcteristics under environmental conditions markedly different from those now obtaining. Hence, while retaining great soil groups or genetic soil types as a central category, the various classifications which were eventually developed for use in such regions, including Britain (Clarke, 1940), Europe as a whole (Kubiena, 1953), France and its former colonies (Aubert and Duchaufour, 1956; CPCS, 1967), Germany (Mückenhausen, 1965; Ehwald et al., 1966) and Australia (Stace et al., 1968), placed more emphasis on internal characteristics such as degree and type of weathering and horizon development, distribution and type of organic matter (humus forms) and features due to gleying than on environmental factors as such.

Recent trends in general soil classification are exemplified by the comprehensive system (Soil Taxonomy) developed by the USDA Soil Conservation Service (Soil Survey Staff, 1960, 1975), the Netherlands system of de Bakker and Schelling (1966), Northcote's (1971) factual key for identification of Australian soils, the revised Canadian (Canada Soil Survey Committee, 1978) and England and Wales (Avery, 1973, 1980) schemes, and the binomial system of soil units designed for use in the FAO–Unesco (1974) soil map of the world. Although these new classifications are intended to convey current understanding of genetic processes and environmental relationships, and resemble the earlier genetic systems in this respect, they are *definitional* rather than *typological* (Gilmour, 1962) insofar as identification (i.e. allocation to a class) is based on the presence or absence of particular soil properties, or on limiting property values, rather than by reference to central concepts. In every case the main criteria include specified 'diagnostic horizons' such as the mollic epipedon and the argillic horizon (Soil Survey Staff, 1975); other morphological features such as colours attributable to gleying; and physical, chemical or mineralogical properties of the soil material within specified depths. Defined soil-moisture and soil-temperature regimes are also used in the US Taxonomy. In most of the earlier systems emphasis was placed on soils under natural or semi-natural vegetation and cultivated soils were either ignored or classified according to characteristics they were presumed to have had in the virgin condition. In the newer systems they are classified according to their existing properties, but in order to ensure that the placement of uncultivated soils is unaffected within short periods by changes in land use, relatively permanent properties are used as differentiating criteria and superficial organic or organo-mineral horizons readily obliterated by ploughing are disregarded.

Although the US and FAO systems in particular were both designed to accommodate all known kinds of soil and are therefore widely used for international reference purposes, more or less revised versions of pre-existing national classifications have been retained for internal use in most countries and even the three soil survey organizations in the British

Isles continue to use different schemes. This failure to achieve international agreement can be attributed in part to the natural desire of the scientists and governing bodies concerned to classify the soils of their own countries as conveniently and meaningfully as possible, and to avoid frequent changes likely to confuse users. Fundamentally, however, the diversity of systems reflects the basic difficulty of applying the concept of general or natural classification to a universe which is not only initially ill-defined and continuously variable in space and time but also displays a low degree of inherent clustering (Webster, 1968). Thus, although differences in parent material or geomorphic factors frequently give rise to contrasting kinds of soil within limited areas, both research (e.g. Arkley, 1976) and experience indicate that if all the soil profiles ever sampled were represented as points in a hyperspace model with as many axes as there are independent variables, the resulting distribution pattern would reveal few well-defined frequency minima. To the extent that this is the case, any attempt to divide the whole universe into a limited number of classes inevitably entails separation of soils which differ only in the attribute used for differentiation and certain closely related properties. A further consequence is that ranking of criteria to define classes in successive categories of a hierarchical system necessarily emphasizes certain relationships at the expense of others which may well be considered more important in particular areas or for particular purposes.

The latter limitation can be countered to some extent by adopting a co-ordinate system (Avery, 1968; Webster, 1968) in which two or more independent classifications using different sets of attributes are combined to form a framework for creating more narrowly defined classes. Schemes of this type have been devised for use in Russia (Rozov and Ivanova, 1967), Belgium (Tavernier and Maréchal, 1962), East Germany (Ehwald et al., 1966) and Scotland (Glentworth, 1962). Despite its shortcomings, however, a hierarchical class structure aids memory and facilitates systematic nomenclature and construction of identification keys, which feature in all the recently developed definitional systems. It also has distinct advantages for purposes such as this book, where the prime requirement is a meaningful basis for collating available information about the soils of a region at varying levels of generalization.

1.5.4 Application of Numerical Methods

During and since the 1960s, access to computers has encouraged the application of numerical methods to soil classification problems. As comprehensively reviewed by Webster (1977), the techniques that have been employed include ordination, construction of hierarchical systems by numerical means, and methods of analysing the dispersion of a population which can be used to improve or optimize pre-existing classifications. All of them involve objective rather than intuitive evaluation of similarities between soil profiles or other sampling units on the basis of as many measured or number coded characters as possible.

In ordination methods such as principal-component and principal-co-ordinate analyses, the relations between individuals represented as points in a multidimensional space are reduced to manageable proportions by rearranging them mathematically along one or a few principal axes which account for as much of the total variance as possible. Their distribution along any two of the three principal axes can then be displayed graphically. Scatter diagrams of this kind serve to reveal the structure of the population sampled, and the contribution made by particular variables to the overall pattern of variation can also be assessed.

In numerical classification as originally developed for use in biological systematics by Sokal and Sneath (1963), degrees of resem-

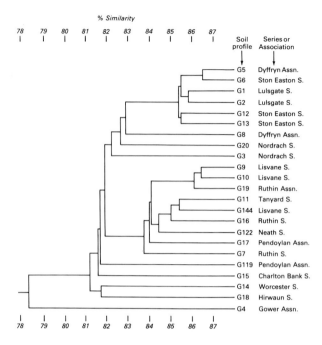

Figure 1.18 Dendrogram of soil-profile similarities based on average horizon similarities (from Rayner, 1966).

blance between individuals or classes are computed and the resulting similarity matrix used to construct a hierarchy or dendrogram (e.g. Rayner, 1966; Grigal and Arneman, 1969; de Gruijter, 1977). A basic problem in applying this and other numerical methods to soil profiles is how to deal with their inherent vertical variation. One approach, used by Rayner in his pioneering study of 23 Glamorganshire profiles representing previously mapped soil classes (Figure 1.18), is to treat horizons recognized in the field as separate entities initially and calculate similarities between all of them as a basis for comparing the profiles to which each belongs. Another is to describe or measure all properties at specific depths and compare the profiles accordingly (e.g. Cuanalo de la C. and Webster, 1970).

However, a 'natural' hierarchy can only be constructed if the population exhibits 'nested clustering' (Sokal and Sneath, 1963) and soil populations rarely do. Webster therefore favours the use of multiple discriminant and canonical variate analyses to aid the development of non-hierarchical classifications which are optimal in the sense that between-class differences are maximized and within-class differences minimized. 'The methods involve creating a classification, evaluating some function of within-class variance as a criterion of goodness, transferring individuals between groups, and evaluating the criterion afresh. The process is repeated until no further improvement seems possible' (Webster, 1975). The classification may be based on computation of resemblances among the sample data studied or on intuitive recognition of 'core classes' to which other individuals are allocated.

Although the multivariate methods that have been applied are clearly valuable as exploratory tools, especially in designing classifications suitable for use in soil surveys of limited areas, numerical classification of soil profiles presents problems essentially similar to those attending the construction of general-purpose systems by more subjective means. That of dealing with vertical anisotropy has already been mentioned. Others which affect the results obtained include the selection and weighting of properties on which measurements of similarity are based, the use of coded characters, the choice of clustering or sorting procedures, and the need to include large numbers of both soil individuals and attributes in the analysis if the bias inherent in the initial selection is to be reduced to an acceptable level. An additional problem of practical import arises from the fact that the classes produced are normally *polythetic* (Sokal and Sneath, 1963), implying that no single attribute is either sufficient or necessary to confer class membership. Thus, although they are 'natural' in respect of individuals and properties included in the original study, allocation of new members is not a simple matter and it can be very difficult to construct an identification key.

1.6 Soil Surveys

1.6.1 Purpose and Nature

The general aim of soil surveys is to collect and organize information about soil variation on an area basis. As stated by Beckett and Burrough (1971), a soil survey of an area can be expected to answer any or all of the following questions.

1(a) What kinds (i.e. classes) of soil are present?
 (b) In what proportions do they occur?
 (c) What proportions are occupied by soil with particular properties or particular ranges in one or more properties?
2(a) What is the soil class at any site of interest?
 (b) What are the properties of the soil at any site of interest?
3(a) Where can soil of a particular class be found?
 (b) Where can soil with particular properties be found?

Questions of the first type can be answered most efficiently by identifying the soil or determining the properties of interest at a sufficient number of points within the area, using a statistically sound sampling procedure. However, most soil surveys also aim to answer questions of the second and third types by the production of a soil map showing how and where the soil varies.

Although soil surveys play an essential scientific role in increasing knowledge and understanding of soil variability, they have generally been undertaken for more practical purposes and can be categorized accordingly as special-purpose or general-purpose in the same way as soil classifications.

Special-purpose surveys are made to meet specific objectives, such as designing an irrigation or land reclamation scheme, and include 'single property' surveys in which the results are displayed as point-symbol or isoline maps (Webster, 1977).

A *general-purpose survey* is intended to serve as a systematic inventory of the soil resources of an area, providing information that can be used or interpreted for a wide range of purposes. The information is customarily displayed as a soil map showing the generalized distribution of soil series or other classes of soil, the differentiating and accessory characteristics of which are described in the map legend or in an accompanying explanatory text. In making such maps, the soil is identified in auger borings or small pits at randomly or purposively sited sampling points and the boundaries are located by extrapolation or by some combination of direct field inspection and indirect evidence of soil changes afforded by changes in land form, vegetation or appearance on air photos. Insofar as the primary purpose of the map is to predict soil properties at sites of interest, its utility depends on the precision and accuracy of the statements made about the map units (Beckett, 1968). Precision refers in this context to the inclusiveness or breadth of the statements and accuracy to their correctness. As these desirable attributes are in practice conflicting, the task of the surveyor is to judge the best compromise between creating many map units in order to increase precision and increasing the sampling effort needed to delineate and describe them accurately.

Nearly all the surveys that have been made by the Soil Survey of England and Wales, the Soil Survey of Scotland and the National Soil Survey of Ireland are of this type. In common with those undertaken in the USA and several other European countries, they were commissioned by governments with the primary aim of making predictions concerning land-use potential and soil management, especially for growing crops but also for non-agricultural purposes. In order to make such predictions, the information has to be interpreted for particular applications, usually by deriving special classifications or maps from the basic survey data (e.g. Jarvis and Mackney, 1979).

In order to facilitate orderly transfer of information from one area to another, the map units in general-purpose national or supranational surveys are normally based on a multi-categorical soil classification system and all the currently used definitional systems (p. 30) were designed with this end in view. Because of the limitations imposed by map scale and sampling density, the bodies of soil represented by map delineations normally include profiles conforming to more than one class, even at scales as large as 1:2,500. Conventionally, however, a soil-map unit is identified by the name of a single class when it is predicted that more of the soil in every delineation of the unit conforms to that class than to any other, and that unconforming inclusions belong to one or more conceptually adjacent, and hence similar, classes or occupy a limited (e.g. < 15 per cent) proportionate area. Relatively homogeneous (simple) units of this type are called *consociations* in the United States (Soil Survey Staff, 1983). Units that include significant proportions of dissimilar kinds of soil, or in which no single profile class can be regarded as dominant, are termed compound and identified by the names of two or more constituent classes, by the additional terms *complex* or *association*, or both.

Butler (1980) criticized the use of predefined soil classes to identify soil-map units on the basis that unless the units are defined 'to suit the country' they may be 'difficult to map and require the subdivision of areas that common sense requires to be undivided'. On the other hand a local classification designed to obviate this difficulty will necessarily be linked to particular patterns of soil variation that are seldom exactly repeated elsewhere, and experience has shown that harmonizing such classifications presents equally difficult problems. Moreover, since lateral soil variation in most landscapes is at least partly continuous (p. 25), arbitrary limits have to be imposed in order to delineate areas that can be grouped and identified as single units in the map legend.

As indicated above, the information collected in general-purpose surveys has traditionally been presented in the form of a soil-series or soil-association map, legend and explanatory text. In recent years, however, the introduction of computerized data-handling methods (Rudeforth, 1982) has enabled the original field and laboratory data to be stored and retrieved for presentation in forms tailored to meet particular requirements. Soil or land information systems incorporating wide ranges of soil and soil-related data have been set up in several countries with this end in view, and computer compatible data collection procedures (e.g. Hodgson, 1976) have been developed for the purpose. These developments have led in turn to increased use of quantitative methods for evaluating soil spatial variability, on the one hand to aid construction of the mapping legend before routine survey is begun and on the other to enable statements about the composition of map units in terms of soil classes or properties to be made with known confidence (Webster, 1977).

1.6.2 Soil Map Scales and Survey Intensity Levels

In planning a soil survey, the purpose it is intended to serve dictates the minimum map scale, as the smallest area about which separate information can be provided is limited by cartographic considerations to one represented by about 0.25 cm^2 on the map. This in turn determines the average inspection density needed to maintain an acceptable standard of reliability. It also sets limits to the precision of the map units and the appropriateness of the different mapping procedures referred to below as grid survey, free survey and physiographic survey. When boundary location is based entirely on data recorded at sampling points, at least one and preferably more profile observations are needed in every area corresponding to 0.25 cm^2 on the published map. Fewer may be used where the map units can be correlated with relief or other surface features observed on the ground or inferred by air-photo interpretation, but the extent to which the overall inspection density can be reduced without serious loss of accuracy clearly depends on the complexity of the soil pattern, its external expression and the precision of the legend units.

Table 1.5 lists the areas represented by 0.25 cm^2 on maps at various scales. These are grouped as large, medium and small, with divisions at approximately 1:35,000 and 1:150,000. In this way large-scale maps comprise those in which the boundaries of individual fields or other land management units can be shown, and small-scale maps those in which the minimum sized delineation is commonly larger than the average sized farm. In nearly all systematic soil surveys,

Table 1.5 Soil-map scales and areas represented by 0.25 cm^2 at each scale

Map scale		Area represented by 0.25 cm^2
large	1:2,500	0.016 ha
	1:10,000	0.25 ha
	1:25,000	1.56 ha
medium	1:50,000	6.25 ha
	1:100,000	25 ha
small	1:250,000	1.56 km^2
	1:500,000	6.25 km^2
	1:1,000,000	25 km^2

however, the published map has been produced at a smaller scale than that of the maps or air photos on which field observations were recorded and boundaries plotted.

Large-scale soil maps are normally based on detailed or very detailed (intensive) surveys in which the soil is examined at frequent intervals with the aim of delineating units that are as homogeneous as possible at the chosen scale.

Very detailed surveys are usually made only to investigate the soil patterns of small sample areas in detail or for special purposes which justify the high costs involved. Inspection densities are commensurate with publication at scales of 1:10,000 or larger, mapped boundaries are based on observations made at equally spaced points (*grid survey*) or at regular intervals along closely spaced traverses, and the information may be displayed either as a soil-class map or as one or more single property maps. Grid sampling is favoured for such surveys because it is normally unbiased and gives the best possible representation for a given number of sampling points. It is also particularly suitable for the application of mathematical interpolation techniques such as kriging, which provides optimal estimates of property values at unvisited points (Webster, 1977; Burgess *et al.*, 1981). However, its efficacy as a basis for boundary location falls off more or less steadily as the grid spacing is increased, depending on the lateral rate of change of soil properties (e.g. Webster and Cuanalo de la C., 1975).

Detailed surveys include those made for publication at scales ranging from 1:15,000 to 1:25,000. They are intended for application at field level and are particularly useful in areas where soil-related management problems are known to occur. The mapped units are typically identified in terms of soil series or divisions (phases) of series and boundaries are delineated using some form of *free survey* (Steur, 1961; Beckett, 1968). In this procedure an essential prerequisite is the establishment of a provisional mapping legend following preliminary investigations in which the main kinds of soil in the survey area are identified and relationships between their distribution and surface features, chiefly land form, are assessed. Profile observations are then made at more or less regular intervals along purposively aligned traverses, partly to verify the expected relationships and partly to position boundaries, which are plotted in the field as mapping proceeds. As long as the legend units have some external expression, albeit variable, they can normally be mapped at least as accurately by this method as by grid survey with fewer observations. Insofar as the field inspections are purposively rather than randomly located, however, they cannot be relied upon to give statistically valid estimates of intra-unit variation. This disadvantage can be met by a combination of methods in which purposively located observations are used to aid boundary location and regularly spaced observations (probability sampling) to estimate unit composition. As noted by Steur (1961), a procedure of this kind offers the best means of producing a soil-class map in which maximal reduction of intra-unit variance is combined with an ascertainable degree of accuracy.

Soil maps at medium and small scales may be derived directly from semi-detailed or reconnaissance surveys of hitherto unsurveyed ground or compiled, entirely or in part, from pre-existing maps based on more detailed surveys. In either case they are intended to serve as resource inventories for use at farm, district or regional level depending on the scale, and as bases for locating areas meriting more precise investigation. The map units are inevitably less homogeneous than those which can be shown at larger scales and at 1:150,000 or smaller are normally identified as *soil associations* characterized by particular combinations of soil series or broader classes such as great soil groups.

In *semi-detailed surveys,* defined as those in which inspection densities are commensurate with publication at medium scales, the mapping procedures generally used are essentially the same as in detailed surveys but more reliance is placed on remote sensing, pre-existing geological maps or other external clues to soil changes in locating boundaries.

In less intensive *reconnaissance surveys,* a procedure distinguished as physiographic mapping by Beckett (1968) has commonly been employed, particularly for exploratory purposes in undeveloped areas where variations in native vegetation visible on air photos can be related to soil changes. This method differs from free survey in that delimitation of map units (soil associations or land systems) is based entirely on remote sensing and the limited field effort is directed to elucidating the nature and variability of the soil within each unit by studying small sample areas or transects.

2

Soil Surveys and Soil Classification in the British Isles

2.1 Historical Introduction

2.1.1 The First Soil Surveys

According to Muir (1961a), a systematic collection of information on British soils was proposed as long ago as 1665 by a committee of the Royal Society, as follows:

> The several kinds of soyls of *England*, being supposed to be, either Sandy, Gravelly, Stony, Clayie, Chalky, Light-mould, Heathy, Marish, Boggy, Fenny or Cold weeping ground, information is desired, what kind of soyls your Country doth most abound with, and how each of them is prepared when employed for *Arable*?

However, the first attempts to produce soil maps and to provide reasonably detailed accounts of agronomically significant soil variations were made by authors of the 'General Views on Agriculture' commissioned on a county basis by the newly constituted Board of Agriculture at the end of the eighteenth century (e.g. Vancouver, 1794, 1808; Young, 1804). Some of the maps produced at this time represent the broad distribution of soil types identified as sandy, clayey, loamy, chalky, peaty or brashy, etc. (Figure 2.1); others appear to reflect land use rather than soil differences or are based on what would now be called geomorphological distinctions.

Some forty years later, a system of land evaluation in which soils were mapped and rated according to their financial value for crop production was introduced in Ireland as a basis for farm taxation (Lee, 1980).

2.1.2 The Geological Approach

The nineteenth century saw great advances in geological knowledge and contemporary attempts to classify soils, then regarded as unorganized mixtures of disintegrated rock and decaying organic matter, relied largely on the nature of the underlying deposits. Thus each geological formation was assumed to give rise to a particular kind of soil, and these were (and often still are) described accordingly as Chalk soils, Old Red Sandstone soils, Boulder Clay soils, Alluvial soils, and so on. Many of the explanatory memoirs produced by the UK Geological Survey between the mid-century and the 1930s include notes on soils; a few also contain soil texture maps based on the geological mapping (e.g. Berry *et al.*, 1930). The completion of the primary 1 inch to 1 mile survey of Ireland was followed by publication of a detailed account of the 'soil geology' (Kilroe,

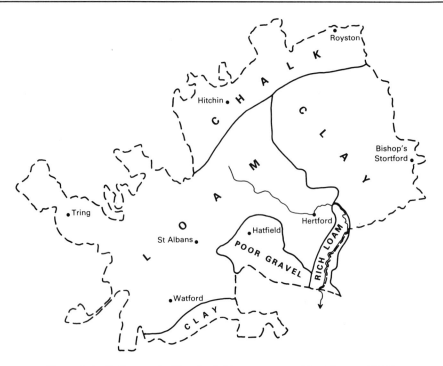

Figure 2.1: Soil types in Hertfordshire as mapped by Young (1804).

1907) accompanied by a very intricate 10 miles to the inch (1:633,360) map of the superficial deposits, so further strengthening the assumption that correlations between geology and soil were good enough to allow the drift maps in particular to serve as soil maps.

This approach was used in a number of surveys of English counties conducted by agricultural chemists from the turn of the century onwards (Luxmoore, 1907; Hall and Russell, 1911; Newman, 1912; Robinson, 1912; Temple, 1929; Pizer, 1931). The most thorough and influential of these surveys, which focused attention on relations between soils and agriculture, was that by Hall and Russell of Kent, Surrey and Sussex. Although soils were not mapped as such, numerous topsoil and subsoil samples were collected on each geological formation and subjected to mechanical and chemical analyses with the aim of determining their physical character and manurial requirements. Crop and stock distribution maps were also constructed and broad correlations drawn between soil types and land use.

The availability of detailed geological maps covering nearly all of the British Isles, the prestige enjoyed by the Geological Survey, and the evident association of marked changes in soil with many of the mapped boundaries, together tended to delay recognition, even among soil scientists, that classifying soils according to their geological origin has serious limitations. The most fundamental defect arises from the fact, first fully appreciated by Dokuchaiev and his followers in Russia, that the same parent material can give rise to various kinds of soil depending on the conditions of development. Also, as Robinson (1912) noted in his survey of Shropshire, a geological map on which sedimentary formations are differentiated primarily by stratigraphic age or mode of origin is not *ipso facto* a parent material map, as substrata distinguished in this way often vary considerably in composition from place to place and many soils are formed entirely or partly in thin but distinct superficial deposits customarily ignored by geologists. A further and equally important limitation, shared by any system of soil classification based on a single genetic factor, is the absence of any formal provision for grouping intrinsically similar soils on different formations.

2.1.3 The Introduction of Modern Methods

The methods of soil survey and classification now in general use were introduced during the inter-war years, by which time the earlier Russian literature had become accessible and the idea that soils could be classified most effectively on the basis of their own characteristics was gaining ground.

These developments were inspired in large measure by G.W. Robinson, who combined the offices of adviser, teacher and research worker at the University College of North Wales, Bangor, from his first appointment as advisory agricultural chemist in 1912 until his untimely death in 1950. It was he (Robinson, 1924) who introduced the word *pedology* to describe the scientific study of soils as independent natural bodies and produced the first English textbook on the subject, the first edition appearing in 1932 and a much revised third edition seventeen years later (Robinson, 1949). Both he and W.G. Ogg, who became director of the Macaulay Institute for Soil Research on its foundation at Aberdeen in 1930, were strongly influenced by visits to the United States (e.g. Ogg, 1920), where systematic soil mapping had been in progress since 1899. Mainly as a result of these contacts and visits to Britain by two American soil surveyors, Dr J.O. Leach in 1924 and Dr Linwood Lee in 1929, British soil survey methods evolved along much the same lines as in the USA and *soil series* identified by locality names were adopted as basic units of classification and mapping. The first British reference to soil series is in an account of the soils and crops of the market-garden district of Bedfordshire by Rigg (1916), who mapped the soils of the area on a geological basis and identified the resulting divisions as Oxford Clay series, Boulder Clay series, etc., in approximate accordance with the original American usage of the term (Simonson, 1964). By the 1920s, however, soil series were being defined more restrictively in terms of profile characteristics, and it was in this guise that the concept gained acceptance, initially in England and Wales and subsequently throughout the British Isles.

Immediately after the 1914–18 war, plans were made to develop soil survey in connection with the agricultural advisory services, then provided through certain universities and agricultural colleges. Aided by grants from the Ministry of Agriculture, assistants were appointed for the purpose at a number of advisory centres; regular Soil Survey Conferences were also organized, the main object of which was to develop and improve methods of soil mapping. In 1925 G.W. Robinson and his colleagues at Bangor began the systematic mapping of Welsh soils, at first on a geological and textural basis, but a Progress Report for 1927–29 (Robinson *et al.*, 1930) explains the subsequent adoption of a classification similar to that used in the USA as follows.

> Briefly, the underlying consideration is that the character of the soils is determined both by the nature of the parent material and also by the methods of formation, and in order to secure the best grouping of the observed facts we arrange our soils into series, each consisting of a set of soils derived from the same or similar parent material under the same conditions of formation, showing a general similarity in the soil profile, i.e. in the vertical succession of horizons, and only differing among themselves in texture. Each series, following the American practice, is named after the locality where it was first studied or where it

attains its most extensive development... for convenience of statement we have grouped them together into suites, each suite comprising soils derived from the same or similar geological material and yielding its different series by variations in the mode of occurrence and formation of the constituent soils.

The Monmouth suite, for example, included soils derived from coarse to medium grained Old Red Sandstone rocks and embraced freely draining 'sedentary' soils (Monmouth series); freely draining soils in drift (Castleton series) and soils in drift with impeded drainage (Frog Moor series). Soils from fine-grained Lower Palaeozoic rocks formed the Powys suite and were subdivided on a similar basis with the addition of the Conway series consisting of poorly drained alluvial soils.

Free-survey procedures (Chapter 1, p. 36) developed in this and other mapping projects started at around the same time in Shropshire (Morley Davies and Owen, 1934), the Vale of Evesham (Osmond et al., 1949) and elsewhere have been followed with minor variations in nearly all subsequent detailed soil surveys in Great Britain. Series were mapped on a field-to-field basis, using auger borings to examine and identify profiles at roughly regular intervals; boundaries were located partly by interpolation between borings and partly according to predicted relationships between soil characteristics and relief or other external features, depending on the nature of the landscape, and plotted in the field on 6 inch to 1 mile (1:10,560) OS maps. Pits were dug at representative sites on each series in order to describe the profile in detail and samples collected from each horizon recognized for particle-size and chemical analyses.

Laboratory measurements of particle-size distribution were improved and standardized during this period by general adoption of the so-called International Method with size limits at 0.002, 0.02, 0.2 and 2mm, which Robinson (1934) had done much to develop, but no attempt was made to relate the results systematically to the farmer's terms such as sand, light loam, medium loam, heavy loam and clay then used to describe soil texture in the field. Chemical determinations made to aid characterization of soil series included loss on ignition and later organic carbon to evaluate organic matter content; calcium carbonate; pH; exchangeable cations and cation-exchange capacity; and analyses of the clay fraction, results of which were used to assess the nature of the weathering complex and to confirm or reveal differential eluviation of sesquioxides (e.g. Robinson, 1928, 1930; Ogg, 1935).

By 1939, soils had been mapped on a profile basis in various widely scattered areas south and north of the Scottish border, though much of the accumulated data remained unpublished. In lowland England, the potential value of a soil survey in aiding advisory work on specialized crops, especially fruit, was appreciated at an early stage and surveys of parts of the Fen country (Wright and Ward, 1929), Herefordshire (Wallace et al., 1931), Kent (Furneaux, 1932; Bane and Jones, 1934), Berkshire (now Oxfordshire) and Hampshire (Kay, 1934, 1939), and the Vale of Evesham (Osmond et al., 1949), were located with this end in view. No soil map was made in the earlier investigations in Herefordshire and Kent, but careful identification and characterization of soil series was shown to provide a good basis for assessing the influence of soil factors on the growth, performance and disease susceptibility of different varieties of fruit. Other surveys, particularly in Wales and the West Midlands, were in areas where their application was less immediate. In Scotland, soil mapping was begun in the Lothians and Aberdeenshire before 1930. Thereafter it was mainly conducted from the Macaulay Institute and much of the surveying during the following decade was done on forest land with the aim of establishing correlations between the performance of planted conifers and soil type (e.g. Muir, 1934; Muir and Fraser, 1939), but a large block

Table 2.1: Differentiating characteristics of genetic soil groups and subtypes (after Kay, 1939 and Clarke, 1940)

Genetic soil group	Sub-groups	Subtypes[a] (variants or intergrades)
Brown Earths 1. Free drainage 2. No vertical differentiation of silica and sesquioxides 3. No natural free $CaCO_3$ in soil horizons	*Brown earths of high base status* Only slightly acid and become neutral with depth *Brown earths of low base status* with tendency to acidity throughout the profile	*Creep or colluvial soils* Dependent on topography for their development *Brown earths with gleyed B and C horizons* With slight gleying in lower horizons.
Podzolized soils (Podzols) 1. Bleached (grey) layer under surface raw humus or peat (can be absent in cultivated soils) 2. Underlying yellow to rusty-coloured accumulation layer 3. Vertical differentiation of silica and sesquioxides in clay fraction	*Normal podzolized soils* With raw humus surface *Peaty podzolized soils* With peaty surface	*Concealed podzols* With no bleached layer *Podzolized soils with gleying* With signs of gleying in B or C horizon
Gley soils Gleying as shown by the presence of greenish, bluish-grey, rusty or yellowish spots or mottling	*Non-calcareous surface-water gley soils* Non-peaty surface; gleying decreases progressively with depth and may be absent in lower horizons *Non-calcareous ground-water gley soils* Non-peaty surface; gleying essentially present in lower layers *Gley-calcareous soils* Calcareous throughout profile *Gley-podzolized soils* Bleached (grey) layer under surface raw humus; B horizon thin or absent and gleying below *Peaty gley-podzolized soils* Similar to gley-podzolized soils but with peaty surface *Peaty gley soils* Completely gleyed, with peat surface	

Table 2.1: Contd.

Genetic soil group	Sub-groups	Subtypes[a] (variants or intergrades)
Calcareous soils Contain primary $CaCO_3$ in soil horizons and are base-saturated	*Grey calcareous soils (rendzina type* Very dark greyish to almost white (when cultivated) surface horizon over calcareous parent material *Red and brown calcareous soils* shallow red or brown soils containing fragmentary calcareous rock	*Calcareous soils with gleyed B and C horizons* With slight gleying in lower horizons
Organic soils At least 20 cm of waterlogged organic matter (peat) at surface	*Basin peat* Soligenous, i.e. formed under the influence of excessive or stagnant ground water	*Fen (including carr)* With calcareous or base-rich ground water *Acid low moor* With acid or base-poor ground water
	Moss peat Ombrogenous, i.e. formed under the influence of rainfall	*Raised moss* Formed over basin peat *Hill peat (blanket moss)* Formed on hill tops and slopes
Undifferentiated alluvium Mineral soils in recent alluvial deposits (individual series, if identified, may be allocated to other groups)		

[a]Subtypes entitled 'leached soils from calcareous parent materials' and 'slightly to strongly podzolized soils' were also distinguished.

of agricultural land in central Aberdeenshire was also mapped.

In 1930, a Soils Correlation Committee was appointed to visit the areas in which mapping was in progress with the object of promoting uniformity in methods, classification and nomenclature. This was replaced in 1936 by a Soil Survey Executive Committee with more authority but similar functions. These bodies, together with annual Field Meetings, helped to establish a standard system for describing soil profiles and recording information on maps, which was eventually issued as part of the first Soil Survey Field Handbook (Clarke, 1940). This also contained guidelines for differentiating soil parent materials according to their mode of origin (e.g. weathering *in situ*, till, colluvium) and lithology, and definitions of genetic soil groups, subgroups and subtypes (Table 2.1). The latter classification, in which three of the groups are named by terms of Russian (podzol, gley) or German (brown earth from *Braunerde*) origin, was compiled by a committee of surveyors as a basis for grouping soil series and colouring soil maps. It was first used in a soil survey report by Kay (1939), and essentially the same class concepts were employed by Tansley (1939) in his comprehensive account of 'The British Islands and their Vegetation' published at about the same time.

In 1939, the Soil Survey of England and Wales was formally recognized; G.W. Robinson was appointed director and the six surveyors then employed came under his direct supervision, but continued to work from their own centres. Following the outbreak of war, however, most of them were deployed as advisors in connection with the food production campaign so that little soil mapping was done for a while. It was resumed from 1942 onwards, when additional staff were recruited. Several 'reconnaissance' surveys were made in England and Scotland during these years, mainly to meet increasing demands for soil information which arose from the development and reorganization of town and country planning. The procedure that came to be adopted in such surveys (e.g. Robinson, 1948) was to select sample areas on a geological basis, map them in detail in order to identify soil series and establish their mode of occurrence, and then use the information so gained to construct a smaller scale map showing the distribution of compound soil-landscape units (soil associations), each characterized by a particular combination of geographically associated series.

2.1.4 Systematic Soil Surveys (1946–1987)

Since the end of the Second World War, systematic soil surveys have been conducted in Great Britain by the Soil Survey of England and Wales and the Soil Survey of Scotland, and from 1961 onwards by the Site Survey Section of the Forestry Commission. When the agricultural advisory services were reorganized in 1946, the Soil Survey of England and Wales was placed under the administration of the Lawes Agricultural Trust with headquarters at Rothamsted Experimental Station, Harpenden, Herts, whilst the Scottish Survey remained as a section of the Macaulay Institute for Soil Research. During the following 25 years, which saw a tenfold increase in staff and activity, the work of both Surveys was co-ordinated by a Soil Survey Research Board appointed by the Agricultural Research Council (ARC). Under new arrangements introduced in 1974, however, this was disbanded and their programmes controlled directly by the ARC (later AFRC) in consultation with the Ministry of Agriculture, Fisheries and Food (MAFF) and the Department of Agriculture and Fisheries for Scotland (DAFFS) respectively, which became the sole funding bodies. Since 1985, financial support from these quarters has been reduced and the Soil Survey of England and Wales has now (1987) joined the Cranfield Institute of Technology as the renamed Soil Survey and Land Research Centre.

In Ireland, the work of Gallagher and Walsh (1942) set a basis for the classification of the soils of the country along modern lines but it was not until the establishment of the National Soil Survey as a research department within the Agricultural Institute (An Foras Taluntais) in 1959 that systematic soil mapping began in the Republic. Little detailed mapping has been done in Northern Ireland.

2.2 The Soil Survey of England and Wales

2.2.1 Soil Mapping

For twenty years after the reorganization of the Soil Survey of England and Wales in 1946, its resources were mainly directed to the production of 1 inch to 1 mile (1:63,360) maps showing the distribution of soil series or complexes of series. Following long-standing

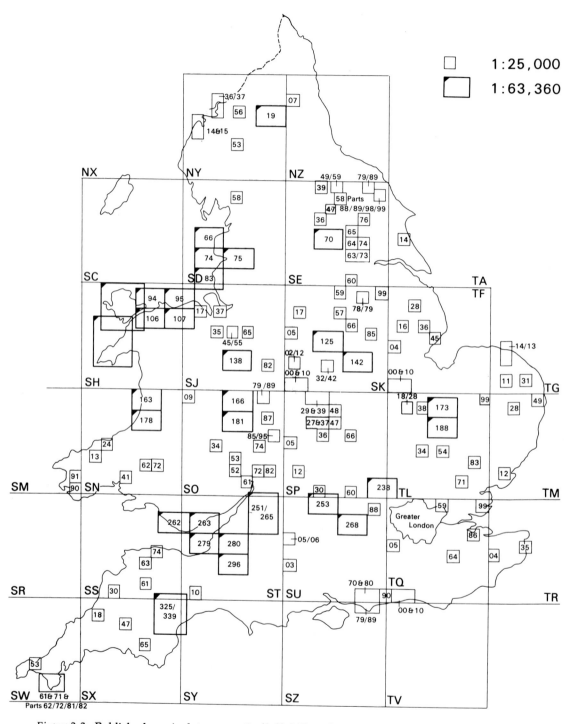

Figure 2.2: Published one inch to one mile (1:63,360) and 1:25,000 soil maps based on detailed field surveys in England and Wales.

practice by the Geological Survey, the maps were based on detailed field surveys at 6 inch to 1 mile (1:10,560) or, latterly, 1:25,000, published on Ordnance Survey third edition sheets (Figure 2.2) and accompanied by explanatory Memoirs, 23 of which were eventually produced, some covering more than one map sheet. Each contains accounts of the geography, geology, climate, vegetation and land use of the area surveyed, and of the survey methods employed, followed by descriptions of the map units and their agronomic significance, together with analytical data on representative profiles. Other publications based on surveys completed during this period include 1:25,000 soil-series maps of specially selected areas such as the West Sussex Coastal Plain (Hodgson, 1967) and Romney Marsh (Green, 1968), and soil association maps of the West Midlands (Mackney and Burnham, 1964) and of the counties of Hertfordshire (Thomasson and Avery, 1963) and Lancashire (Hall and Folland, 1970) at scales of 10 miles to the inch (1:625,000) and 4 miles to the inch (1:250,000) respectively, all with explanatory texts.

In 1967 a new mapping programme was started whereby 10 × 10 km sample areas in each county not already surveyed were mapped in detail for publication at 1:25,000 and as a basis for constructing county, regional and national soil-association maps. By 1987 105 1:25,000 sheets (Figure 2.2) had been published with explanatory Records. These and the earlier 1:63,360 maps based on detailed surveys together cover about 20 per cent of England and Wales. A number of land-use capability maps (see p. 59) and other interpretative maps were also produced.

A 1:1,000,000 soil-association map of the two countries with an explanatory legend was compiled by Avery et al. (1975), and in 1979 the Survey began a five-year project to make an improved national map at the scale of 1:250,000 and to characterize the map units in appropriate detail, with emphasis on agricultural interpretations. The resulting map sheets, which show the boundaries of 296 soil associations, identified in all but three cases by the most frequently occurring soil series (Mackney et al., 1983), were published four years later and six regional bulletins describing the associations in each region in 1984. In making the map, previously unsurveyed ground was covered by recorded observations in purposively located small pits or auger borings at an average frequency of 250 per 100 km^2. Soil profiles were also described and sampled for analysis at all 5 × 5 km national grid intersects to form a systematic National Soil Inventory. The knowledge gained from these observations and from previous detailed surveys of similar terrain was used in conjunction with air photos and geological maps to position map boundaries and characterize the associations.

After this project was completed, a special survey of lowland peat soils based on systematic sampling at 0.5 km intervals was undertaken (Burton and Hodgson, 1987) and detailed mapping for publication at 1:50,000 or 1:25,000 was resumed, but the latter programme has now been suspended following withdrawal of support by MAFF.

2.2.2 Soil Characterization and Classification

Apart from the increased use of air-photo interpretation (Carroll et al., 1977) and of statistical methods for evaluating soil variability (Webster, 1977; Rudeforth, 1982), basic mapping procedures have changed little since the inter-war years, but soil characterization and identification have been placed on a sounder basis, partly through progressive improvements in field description (Soil Survey of Great Britain 1960; Hodgson, 1976) and the introduction of additional or improved laboratory methods (Avery and Bascomb, 1982), and partly through the adoption of an improved multi-categorical classification system (Avery, 1973, 1980; Clayden and Hollis, 1984). Probably the most important single improvement was the adoption of USDA textural classes (Soil Survey Staff, 1951) defined by limiting proportions of sand (0.05–2 mm), silt (0.002–0.05 mm) and clay (< 0.002 mm) sized particles (Figure 3.2a, p. 84). The last 20 years have also seen major advances in the interpretation of soil survey information for agricultural and other purposes (Bibby and Mackney, 1969; Mackney, 1974; Thomasson, 1975; Jarvis and Mackney, 1979).

In the surveys initiated between 1946 and 1972, soil series were differentiated on a parent material and profile basis as hitherto and allocated to the broader classes listed in Table 2.1 with various modifications and additions. For example, the division of the brown earths according to base status was abandoned because it was found difficult to apply in long cultivated land and subgroups of *sols bruns* and *sols lessivés* based on the presence or absence of a subsoil horizon of clay accumulation as in the French classification of Aubert and Duchaufour (1956) were substituted (e.g. Avery, 1964; Mackney and Burnham, 1964). Similarly, lowland soils of the podzol group were divided into humus podzols, humus-iron podzols and gley-podzols in accordance with Kubiena's (1953) classification of the soils of Europe, and an additional subgroup of peaty gleyed podzols (Crompton, 1956) was introduced to accommodate commonly occurring upland soils with a peaty surface and a bleached subsurface horizon overlying a thin iron-cemented pan. A number of other kinds of soil not distinguished in the original scheme were also recognized and named by terms already in use elsewhere (e.g. rankers, warp soils, humic gley soils) or coined for the purpose (e.g. immature soils; grey siliceous soils).

By the late 1960s, the number of soil series identified had greatly increased. Each was characterized in terms of genetic soil group and subgroup, texture, and the lithology and geological origin of the parent material, but there was no clearly stated systematic basis for differentiating them. As described by Clayden and Hollis (1984), 'it was possible for a surveyor, after consultation with his colleagues, to expand the concept of a series, or to create a new one if it was felt that the profiles under consideration were sufficiently different'. Thus although parent-material lithology was recognized as being the most important differentiating characteristic, in practice different soil series were identified on lithologically similar materials if they occurred in different areas and the materials were derived from rocks of differing stratigraphic age. This pragmatic approach, coupled with the piecemeal introduction of new concepts and differing interpretations of class boundaries, particularly between gley soils and other genetic groups, gave rise to considerable inter-regional correlation problems. Effort was therefore directed to reconstructing the classification system, especially following the development in the United States (Soil Survey Staff, 1960) and the Netherlands (de Bakker and Schelling, 1966) of new definitional schemes (Chapter 1, p. 30) in which class limits at all categorical levels were defined precisely in terms of intrinsic soil properties.

A comprehensive system for England and Wales embodying this principle was eventually devised by the present author (Avery, 1973, 1980) and used with minor modifications in the legends of soil maps published from mid-1975 onwards. Soil is considered for the purpose to include any unconsolidated material directly below a ground surface. As indicated in Chapter 1 (p. 28) the things classified are soil profiles, considered as three-dimensional samples of the soil mantle about 1 m^2 in cross-sectional area, that extend from the ground surface (or a buried surface within 30 cm) to a maximum depth of 1.50 m. Classes are differentiated by characteristics that can be evaluated in the field or inferred with reasonable assurance from field examination, either by comparison with analysed samples of 'bench-mark' soils or by reference to geological data, and that are relatively permanent.

Table 2.2: Classification of soils in England and Wales (Avery, 1980)

Major soil group	Soil group	Soil subgroup
1. *Terrestrial raw soils* Mineral soils with no diagnostic horizons or disturbed fragments of such horizons, unless buried beneath a recent deposit more than 30 cm thick	1.1 *Raw sands* Non-alluvial, sandy 1.2 *Raw alluvial soils* In recent alluvium 1.3 *Raw skeletal soils* With coherent or fragmented bedrock at 30 cm or less 1.4 *Raw earths* In naturally occurring, unconsolidated, non-alluvial loamy or clayey materials 1.5 *Man-made raw soils* In artificially emplaced materials	(none recognized)
2. *Raw gley soils* Gleyed mineral soils with no diagnostic surface horizon and/or unripened (soft and muddy) at 20 cm or less	2.1 *Raw sandy gley soils* In sandy material 2.2 *Unripened gley soils* In loamy or clayey alluvium	(none recognized)
3. *Lithomorphic soils* Soils with a distinct, humose or peaty topsoil and no diagnostic subsurface horizon, i.e. normally with bedrock or little altered unconsolidated material within 30 cm depth	3.1 *Rankers* Non-calcareous, over non-calcareous, non-alluvial substratum (excluding sands) or hard limestone 3.2 *Sand-rankers* In non-calcareous, non-alluvial sandy material 3.3 *Ranker-like alluvial soils* In non-calcareous recent alluvium 3.4 *Rendzinas* Calcareous, with an extremely calcareous non-alluvial substratum (excluding sands) 3.5 *Pararendzinas* With a moderately calcareous non-alluvial substratum (excluding sands) 3.6 *Sand-pararendzinas* In calcareous, non-alluvial sandy material	3.11 Humic rankers 3.12 Gleyic rankers 3.13 Brown rankers 3.14 Podzolic rankers 3.15 Stagnogleyic rankers 3.21 Typical sand-rankers 3.22 Podzolic sand-rankers 3.23 Gleyic sand-rankers 3.31 Typical ranker-like alluvial soils 3.32 Gleyic ranker-like alluvial soils 3.41 Humic rendzinas 3.42 Grey rendzinas 3.43 Brown rendzinas 3.44 Colluvial rendzinas 3.45 Gleyic rendzinas 3.46 Humic gleyic rendzinas 3.47 Stagnogleyic rendzinas 3.51 Typical pararendzinas 3.52 Humic pararendzinas 3.53 Colluvial pararendzinas 3.54 Stagnogleyic pararendzinas 3.55 Gleyic pararendzinas 3.61 Typical sand-pararendzinas 3.62 Gleyic sand-pararendzinas

Table 2.2: Contd.

Major soil group	Soil group	Soil subgroup
	3.7 *Rendzina-like alluvial soils* In calcareous recent alluvium, lake marl or tufa	3.71 Typical rendzina-like alluvial soils 3.72 Gleyic rendzina-like alluvial soils 3.73 Humic gleyic rendzina-like alluvial soils
4. *Pelosols* Slowly permeable non-alluvial clayey soils with a distinct topsoil, weathered or argillic B horizon and no non-calcareous gleyed subsurface horizon within 40 cm depth	4.1 *Calcareous pelosols* With a calcareous substratum and no argillic B horizon	4.11 Typical calcareous pelosols
	4.2 *Non calcareous pelosols* With a non-calcareous substratum and no argillic B horizon	4.12 Typical non-calcareous pelosols
	4.3 *Argillic pelosols* With an argillic B horizon	4.13 Typical argillic pelosols
5. *Brown soils* Other mineral soils with a weathered or argillic B horizon and no gleyed sub-surface horizon within 40 cm depth	5.1 *Brown calcareous earths* Non-alluvial, loamy or clayey, with a weathered B horizon in or over calcareous material	5.11 Typical brown calcareous earths 5.12 Gleyic brown calcareous earths 5.13 Stagnogleyic brown calcareous earths 5.14 Colluvial brown calcareous earths
	5.2 *Brown calcareous sands* Non-alluvial, sandy, with a weathered B horizon in or over calcareous material	5.21 Typical brown calcareous sands 5.22 Gleyic brown calcareous sands
	5.3 *Brown calcareous alluvial soils* With a weathered B horizon in calcareous loamy or clayey recent alluvium	5.31 Typical brown calcareous alluvial soils 5.32 Gleyic brown calcareous alluvial soils 5.33 Pelogleyic brown calcareous alluvial soils
	5.4 *Brown earths (sensu stricto)* Non-alluvial, loamy or clayey, with a weathered B horizon in non-calcareous material	5.41 Typical brown earths 5.42 Stagnogleyic brown earths 5.43 Gleyic brown earths 5.44 Ferritic brown earths 5.45 Stagnogleyic ferritic brown earths 5.46 Gleyic ferritic brown earths 5.47 Colluvial brown earths
	5.5 *Brown sands* Non-alluvial, sandy, with a weathered or argillic B horizon in non-calcareous material	5.51 Typical brown sands 5.52 Gleyic brown sands 5.53 Stagnogleyic brown sands 5.54 Argillic brown sands 5.55 Gleyic argillic brown sands

Table 2.2: Contd.

Major soil group	Soil group	Soil subgroup
	5.6 *Brown alluvial soils* With a weathered B horizon in non-calcareous loamy or clayey recent alluvium	5.61 Typical brown alluvial soils 5.62 Gleyic brown alluvial soils 5.63 Pelogleyic brown alluvial soils
	5.7 *Argillic brown earths* Loamy or clayey, with an ordinary argillic B horizon	5.71 Typical argillic brown earths 5.72 Stagnogleyic argillic brown earths 5.73 Gleyic argillic brown earths
	5.8 *Paleo-argillic brown earths* Loamy or clayey, with a paleo-argillic B horizon	5.81 Typical paleo-argillic brown earths 5.82 Stagnogleyic paleo-argillic brown earths 5.83 Gleyic paleo-argillic brown earths
6. *Podzolic soils* Mineral soils with a podzolic B horizon (Bs, Bh and/or thin ironpan)	6.1 *Brown podzolic soils* With a Bs horizon only and no albic E or gleyed subsurface horizon	6.11 Typical brown podzolic soils 6.12 Humic brown podzolic soils 6.13 Paleo-argillic brown podzolic soils 6.14 Stagnogleyic brown podzolic soils 6.15 Gleyic brown podzolic soils
	6.2 *Humic cryptopodzols* With a humose or peaty topsoil, humose Bh and no albic E horizon, thin ironpan or gleyed subsurface horizon	6.21 Typical humic cryptopodzols 6.22 Ferri-humic cryptopodzols
	6.3 *Podzols (sensu stricto)* With an albic E and/or Bh horizon and no peaty topsoil, thin ironpan or gleyed subsurface horizon	6.31 Humo-ferric podzols 6.32 Humus podzols 6.34 Paleo-argillic podzols
	6.4 *Gley-podzols* With a gleyed subsurface horizon and no thin ironpan	6.41 Typical gley-podzols 6.42 Humo-ferric gley-podzols 6.43 Stagnogley-podzols
	6.5 *Stagnopodzols* With a peaty topsoil and/or gleyed albic E over a thin ironpan or Bs horizon	6.51 Ironpan stagnopodzols 6.52 Humus-ironpan stagnopodzols 6.53 Hardpan stagnopodzols 6.54 Ferric stagnopodzols
7. *Surface-water gley soils* Non-alluvial, non-podzolic soils with a non-calcareous gleyed	7.1 *Stagnogley soils* With a distinct topsoil	7.11 Typical stagnogley soils 7.12 Pelo-stagnogley soils 7.13 Cambic stagnogley soils 7.14 Paleo-argillic stagnogley soils 7.15 Sandy stagnogley soils

Table 2.2: Contd.

Major soil group	Soil group	Soil subgroup
subsurface horizon, that is slowly permeable; gleyed E and/or B horizons below a humose or peaty topsoil and little or no gleying in underlying horizons; or both.	7.2 *Stagnohumic gley soils* With a humose or peaty topsoil	7.21 Cambic stagnohumic gley soils 7.22 Argillic stagnohumic gley soils 7.23 Paleo-argillic stagnohumic gley soils 7.24 Sandy stagnohumic gley soils
8. *Ground-water gley soils* Other non-podzolic soils with a distinct, humose or peaty topsoil and a gleyed subsurface horizon within 40 cm depth	8.1 *Alluvial gley soils* With a distinct topsoil, in loamy or clayey recent alluvium	8.11 Typical alluvial gley soils 8.12 Calcareous alluvial gley soils 8.13 Pelo-alluvial gley soils 8.14 Pelo-calcareous alluvial gley soils 8.15 Sulphuric alluvial gley soils
	8.2 *Sandy gley soils* Sandy, with a distinct topsoil and no argillic B horizon	8.21 Typical sandy gley soils 8.22 Calcareous sandy gley soils
	8.3 *Cambic gley soils* Non-alluvial, loamy or clayey, with a distinct topsoil and no argillic B horizon	8.31 Typical cambic gley soils 8.32 Calcaro-cambic gley soils 8.33 Pelo-cambic gley soils
	8.4 *Argillic gley soils* With a distinct topsoil and an argillic B horizon	8.41 Typical argillic gley soils 8.42 Sandy argillic gley soils
	8.5 *Humic-alluvial gley soils* With a humose or peaty topsoil, in loamy or clayey recent alluvium	8.51 Typical humic-alluvial gley soils 8.52 Calcareous humic-alluvial gley soils 8.53 Sulphuric humic-alluvial gley soils
	8.6 *Humic-sandy gley soils* Sandy, with a humose or peaty topsoil and no argillic B horizon	8.61 Typical humic-sandy gley soils 8.62 Calcareous humic-sandy gley soils
	8.7 *Humic gley soils (sensu stricto)* Non-alluvial, loamy or clayey, with a humose or peaty topsoil	8.71 Typical humic gley soils 8.72 Calcareous humic gley soils 8.73 Argillic humic gley soils
9. *Man-made soils* Other mineral soils with a thick man-made	9.1 *Man-made humus soils* With a thick man-made A horizon	9.11 Sandy man-made humus soils 9.12 Earthy man-made humus soils

Table 2.2: Contd.

Major soil group	Soil group	Soil subgroup
A horizon and/or a disturbed subsurface layer	9.2 *Disturbed soils* Without a thick man-made A horizon	(non recognized)
10. *Peat (organic) soils* Soils having more than 40 cm of organic material within the upper 80 cm, or more than 30 cm of organic material resting on bedrock or extremely stony material	10.1 *Raw peat soils* Without an earthy topsoil or ripened mineral surface layer 10.2 *Earthy peat soils* With an earthy topsoil or a ripened mineral surface layer	10.11 Raw oligo-fibrous peat soils 10.12 Raw eu-fibrous peat soils 10.13 Raw oligo-amorphous peat soils 10.14 Raw eutro-amorphous peat soils 10.21 Earthy oligo-fibrous peat soils 10.22 Earthy eu-fibrous peat soils 10.23 Earthy oligo-amorphous peat soils 10.24 Earthy eutro-amorphous peat soils 10.25 Earthy sulphuric peat soils

The system is hierarchical, with classes termed major soil groups, soil groups, soil subgroups and soil series defined at four successive categorical levels. Those in the three higher categories are named connotatively using terms derived for the most part from other European classifications or from the Soil Taxonomy developed by the USDA Soil Conservation Service. They are listed in Table 2.2, together with summarized definitions of the major groups and groups. These are distinguished by broad differences in the composition or origin of the soil material within specified depths and the presence or absence of diagnostic surface and subsurface horizons defined in terms of thickness and/or depth and internal properties. The same or similar criteria are employed in the classification of British soils adopted for the present purpose (Table 2.14, p. 73). They are fully defined in Chapter 3, along with those used in differentiating subgroups.

Major groups, 5, 6, 7, 8 and 10 (Table 2.2) correspond more or less closely to previously recognized genetic groups or subgroups (Table 2.1). Major groups 1 and 2 accommodate so called raw or 'immature' soils occurring in unstable sites such as coastal sand dunes and sparsely vegetated intertidal flats. They lack distinct pedogenic horizons and differ in this respect from the lithomorphic soils (major group 3), which include those already classed as rendzinas and rankers on calcareous and non-calcareous rocks respectively. Soils profoundly modified by human activity (man-made soils) and slightly gleyed clayey soils (pelosols) derived from soft argillaceous rocks and Pleistocene deposits of similar composition are also set apart in the highest category. The latter innovation, along with the separation of sandy and alluvial groups in the second category, was introduced with the overall aim of dividing the soils of the country into a manageable number of classes which would be more homogeneous for practical objectives than groups based on genetic horizons alone, particularly when used to characterize units on a small-scale map (Avery *et al.*, 1975).

The 118 subgroups, which were intended to convey useful information about soil variability at a more detailed level of generalization, are of two main kinds. Each of those named as *typical* represents the central concept of the group to which it belongs but is not necessarily the most extensive; others comprise transitional profile forms (e.g. stagnogleyic brown earths; pelo-stagnogley soils) or variants characterized by particular diagnostic horizons (e.g. humic rendzinas; humus podzols). Soils in recent colluvial deposits are also segregated at this level.

As outlined by Avery (1973), soil series are conceived as divisions of subgroups distinguished by lithologic characteristics of soil and substratum not used for differentiating classes in higher categories. From the time that the new system was adopted, established series were redefined accordingly and lithologic differentiating criteria were progressively standardized, but stratigraphic distinctions, exemplified by lithologically similar brown earths of the Denbigh and Highweek series in materials derived from fine grained Lower Paleozoic and Devonian rocks respectively, were retained in order to maintain continuity with previous work in the main areas of occurrence. Following the initiation of the national 1:250,000 mapping project in 1979, however, all the series identified in earlier surveys were reviewed and the systematic differentiating criteria that had been developed were tested and finalized. As a result of this nationwide correlation exercise, many series distinguished only by the differing stratigraphic ages of pre-Quaternary source rocks were united and others created to fit the prescribed criteria. These are fully defined in a monograph by Clayden and Hollis (1984), which also contains an index of all the series named in previous survey publications and a key to those currently recognized.

The series definitions are based on selected characteristics of the soil or rock material within specified depths, including stoniness and mode of origin (substrate type), texture (particle-size grouping), and the presence or absence of distinctive mineralogical or mineralogically related properties. Although most of the properties used are directly measurable, the traditional distinction between 'drift soils' in Quaternary deposits and 'sedentary' soils consisting wholly or partly of material weathered in place from soft pre-Quaternary formations is retained.

Five broad substrate (parent material) types and ten subtypes are recognized and subdivided using additional differentiae based mainly on the lithology of rock substrata or included stones, or, in peat soils, their botanical composition (Table 2.3).

Further divisions are based on the predominant texture of the soil, or of two texturally contrasting layers if present within the reference section. The particle-size groups and subgroups defined in Chapter 3 (p. 84) are used for this purpose (Table 2.4). Thus mineral soils without contrasting layers are designated sandy (coarse, medium or fine), loamy (coarse loamy, coarse silty, fine loamy or fine silty) or clayey, and those with contrasting layers coarse loamy over clayey, clayey over coarse loamy or sandy, etc. As already noted, predominantly sandy and clayey soils are mostly separated in higher categories, and it is mainly the medium textured soils that are further subdivided at series level.

In addition, soils in which the fine material (<2 mm) within limited depths has specified mineralogical or mineralogically related properties are separated as distinct series from otherwise similar soils. The differentiating characteristics recognized, most of which originate from particular kinds of parent material, are listed in Table 2.5.

For ease of communication, soil series are still named from the place where they were first described but each now has a standardized definition consisting of the soil subgroup symbol (Table 2.2) and a combination of terms indicating its substrate type, texture, and any other diagnostic properties. Examples of series definitions applicable to soils of differing substrate type are given below.

Lithoskeletal soils

Bangor series: 3.11. Loamy or peaty; lithoskeletal acid crystalline rock.
Andover series: 3.43. Silty; lithoskeletal chalk.

Table 2.3: Substrate types and subtypes and associated soil series differentiae (after Clayden and Hollis, 1984)

Substrate type	Substrate subtype	Associated differentiae
Soils with a lithoskeletal substrate Bedrock or angular skeletal material (containing more than 35 per cent stones by volume) within 80 cm depth	*Lithoskeletal soils* With bedrock or angular skeletal material occupying at least half the upper 80 cm and no surface layer more than 30 cm thick with less than 16 per cent stones *Soils over lithoskeletal material* With a non-skeletal surface layer at least 40 cm thick, or more than 30 cm thick if it contains less than 16 per cent stones	acid crystalline rock basic crystalline rock ultra-basic crystalline rock acid schist basic schist chert, quartzite or quartzitic sandstone ironstone limestone chalk mudstone, shale (or slate) siltstone or silty shale sandstone
Soils with a gravelly substrate Gravel (more than 35 per cent water-rounded stones by volume) starting above and extending below 80 cm depth	*Gravelly soils* With gravel occupying at least half the upper 80 cm and no loamy or clayey^a surface layer more than 30 cm thick with less than 16 per cent stones	acid crystalline stones basic crystalline stones ultrabasic crystalline stones chalk stones limestone stones very hard siliceous stones sandstones, siltstones, mudstones or slates
	Soils over gravel With a loamy^a or clayey^a surface layer at least 40 cm thick or more than 30 cm thick if it contains less than 16 per cent stones	non-calcareous gravel calcareous gravel
Soils with a soft pre-Quaternary substrate Weathered or little altered, soft (non-skeletal) pre-Quaternary rock within 80 cm depth	*Soils in thin drift over a soft pre-Quaternary substrate* With an upper drift layer of contrasting texture or mineralogy *Soils passing to a soft pre-Quaternary substrate* With no upper drift layer of contrasting texture or mineralogy	weathered acid crystalline rock weathered basic crystalline rock weathered ultrabasic crystalline rock weathered acid schist weathered basic schist soft chalk clay, clay shale or soft mudstone sand or soft sandstone loam (or soft sandstone, shale or siltstone) interbedded strata, specified accordingly

Table 2.3: Contd.

Substrate type	Substrate subtype	Associated differentiae
Soils in thick drift Mineral Quaternary deposit at least 80 cm thick, or overlying an organic layer, with no skeletal or texturally contrasting gravelly layer starting above and extending below 80 cm depth	*Soils in thick drift* (including recent alluvium[a] and colluvium[a], glacial and glacio-fluvial or river drift and head deposits) drift with limestones	river alluvium marine alluvium lake marl or tufa[a] recent colluvium[a] stoneless drift chalky drift (with chalk stones) drift with siliceous stones
Soils in peat (peat soils) Peat at least 40 cm thick, starting at the surface or within 40 cm depth, or at least 30 cm of peat resting directly on bedrock or skeletal material	*Soils in peat (undifferentiated)* Peat without a lithoskeletal substratum or interstratified mineral layers *Soils in peat over lithoskeletal material* With bedrock or angular skeletal material within 80 cm depth *Soils in peat with mineral layers layers* With interstratified mineral layers between 30 and 90 cm depth	*Sphagnum* peat *Eriophorum–Sphagnum* peat *Molinia* peat grass-sedge peat sedimentary peat[a] humified peat[a]

[a]Textural and other terms defined in Chapter 3.

Soils over lithoskeletal material
　Denbigh series: 5.41. Fine loamy over lithoskeletal mudstone and sandstone or slate.
　Banbury series: 5.44. Ferruginous fine loamy over lithoskeletal ironstone.
Gravelly soils
　Southampton series: 6.34. Sandy-gravelly; very hard siliceous stones.
　Baschurch series: 6.11. Loamy-gravelly; sandstones, siltstones, mudstones or slates.
Soils over gravel
　Sutton series: 5.71. Fine loamy over calcareous gravelly.
　Hurst series: 8.41. Coarse loamy over non-calcareous gravelly.

Soils in drift over a soft pre-Quaternary substrate
　Holidays Hill series: 6.43. Sandy drift over fine loamy or fine silty material passing to loam (clay and sand).
　Lawford series: 7.12. Clayey drift passing to clay or soft mudstone.
Soils with a soft pre-Quaternary substrate
　Windsor series: 7.12. Swelling-clayey passing to brownish clay.
　Harwell series: 5.71. Grey siliceous fine loamy or fine silty over clayey, passing to interbedded clay and sandstone.
Soils in thick drift
　Wisbech series: 8.12. Coarse silty; marine alluvium.

Table 2.4: Textural grouping for soil series differentiation (after Clayden and Hollis, 1984)

Substrate type or subtype	Reference section	Classes used for series differentiation
Lithoskeletal and gravelly soils	Upper 30 cm	Particle-size groups (sandy, loamy or clayey) in all soil subgroups except brown rendzinas
Soils over gravel	From the surface to the top of the gravelly layer	Particle-size subgroups (sandy, coarse loamy, coarse silty, fine loamy, fine silty or clayey)
Soils over lithoskeletal material	From the surface to bedrock or the top of the very stony layer	1. Particle-size groups in lithomorphic soils except brown rendzinas, stagnopodzols and stagnohumic gley soils
Soils with soft pre-Quaternary substrates or in thick drift	Upper 80 cm	2. Particle-size subgroups in all other soil groups
Soils in peat	30 to 90 cm or the 60 cm directly above a mineral substrate, or, if less than 60 cm thick, from the surface to the top of the mineral substrate	Peat, sandy peat and loamy peat (Hodgson, 1976, p. 21)

Crewe series: 7.12. Reddish clayey; stoneless drift.
Newport series: 5.51. Sandy; drift with siliceous stones.
Willingham series: 3.72. Loamy-carbonatic over peaty; lake marl or tufa.
Beccles series: 7.11. Fine loamy over clayey; chalky drift.

Thus, from the time that soil series were first introduced as a convenient means of identifying delineations on soil maps, the concept has been progressively refined and is now restricted in England and Wales to narrowly defined kinds of soil profile representing generally small but specific variations within subgroups. A clear distinction therefore has to be drawn between the profile classes (*taxa*) and soil-map units identified by the same locality names, which normally include profiles conforming to other series because of the limitations imposed by map scale and sampling density (Chapter 1, p. 34). More than 700 soil series as now defined have been recognized in England and Wales to date, of which more than 400 are listed as major or minor components of soil associations in the legend of the national 1:250,000 map.

In addition to the prescribed subgroup and series divisions, supplementary divisions termed *phases* are introduced to meet the practical objectives of soil mapping. These are defined in terms of internal (e.g. depth to hard rock; topsoil texture; current soil water regime) or external (e.g. slope) attributes in conjunction with, or independently of, the systematic classification.

Table 2.5: Distinctive mineralogical or mineralogically related properties used in differentiating soil series (after Clayden and Hollis, 1984)

Distinctions applied to soils of all particle-size groups

Carbonatic	Extremely calcareous (Chapter 3, p. 85).
Ferruginous	Containing more than 2 per cent free (dithionite-extractable) iron, and the ratio of percentage free iron to percentage clay is more than 0.25.
Glauconitic	At least 5 per cent glauconite and greenish hue.
Serpentinitic	Sand fraction composed largely of magnesium-rich serpentine minerals and the ratio of exchangeable Mg to Ca is more than 2.
Saline	(Chapter 3, p. 96) Containing significant amounts of salts more soluble than gypsum
Sulphidic	Waterlogged material that contains oxidizable sulphides and becomes extremely acid when drained.
Sulphuric	Oxidized sulphidic material, usually containing yellow deposits of jarosite.

Distinctions applied only to loamy and clayey soils

Grey siliceous	Non-carbonatic, non-glauconitic, greyish or olive-coloured material in which the ratio of free iron to clay is less than 0.05.
Reddish	Red or reddish colour with Munsell hue 5YR or redder and chroma 3 or more, inherited from haematite-bearing parent rock.

Distinctions applied only to clayey soils

Swelling-clayey	Large shrink-swell capacity reflecting the presence of expansible clay minerals: the coefficient of linear extensibility (COLE: Soil Survey Staff, 1975) is 0.09 or more or linear shrinkage (British Standards Institution, 1975) is more than 13 per cent.
Brownish	Dominantly brownish substratum resulting from oxidation of pyrite.
Kaolinitic	Clay fraction consisting predominantly of kaolinite (Avery and Bullock, 1977).

2.3 The Soil Survey of Scotland

2.3.1 Soil and Land Capability Mapping

Systematic mapping of the soils of Scotland at a scale of 1:25,000 for publication at 1:63,360 was started in 1947 by staff of the Macaulay Institute for Soil Research under the direction of Dr R. Glentworth, who remained in charge until 1976. After initial work in Aberdeenshire and Banffshire, the Survey was soon active in each of the advisory regions served by the three Scottish agricultural colleges and became a fully fledged department of the Institute in 1959. Attention was concentrated on the arable areas, though hill land was also surveyed when it formed part of a particular map sheet.

In the early work in north-east Scotland, Glentworth introduced the *association* as originally conceived by Ellis (1932) in Canada as the primary basis for grouping and identifying the soil-map units (Glentworth and Dion, 1949). According to this restricted usage of the term, which can be equated with Robinson's *suite* (p. 40) and the first kind of catena recognized by Milne (Chapter 1, p. 20), an association comprises a range of soil series that are developed in similar parent material within the same climatic zone and differ mainly in hydrologic (natural drainage) properties related to their position in the landscape (Figure 1.13, p. 20). In the legends of the first 1:63,360 maps produced (Glentworth, 1954), each association is identified by a locality name and its constituent series by connotative terms referring to drainage class and soil depth. For example

Table 2.6: Sample map legend showing relationships between soil associations, genetic soil groups and soil series as mapped by the Soil Survey of Scotland (After Glentworth, 1962)

ASSOCIATION	PARENT MATERIAL	SERIES								SKELETAL SOILS
		Brown forest soils		(Humus-) iron podzols			Peaty podzols	Non-calc. gleys	Peaty gleys	
		Freely drained	Imperfectly drained	Freely drained	Imperfectly drained		Freely drained below B_1	Poorly drained	Very poorly drained	
TARVES	Drifts from intermediate or mixed acid and basic igneous rocks	TR Tarves	TL Thistlyhill	TN Tillypronie			PS Pressendye	PD Pitmedden	PK Pettymuck	
TIPPERTY	Red lacustrine silts and clays from Old Red Sandstone sediments		TP Tipperty					BX Birness	BQ Dorbs	
HATTON	Drifts from Old Red Sandstone conglomerates	CA Chapelden	MV Middlehill	HN Hatton			WY Windyheads	BZ Blackrie	GM Garthfield	
COUNTESSWELLS	Drifts from granite and granitic gneiss	RM Raemoir		CW Countesswells	DS Dess		CR Charr	TV Terryvale	DM Drumlasie	CWZ
FOUDLAND	Drifts from slates and phyllites			FD Foudland	MA Mairlenden		SI Suie	FH Fisherford	SQ Shanquhar	FDZ
STRICHEN	Drifts from acid schists	FV Fungarth	BI Baikies	ST Strichen			GR Gaerlie	AE Anniegathel	HY Hythie	STZ

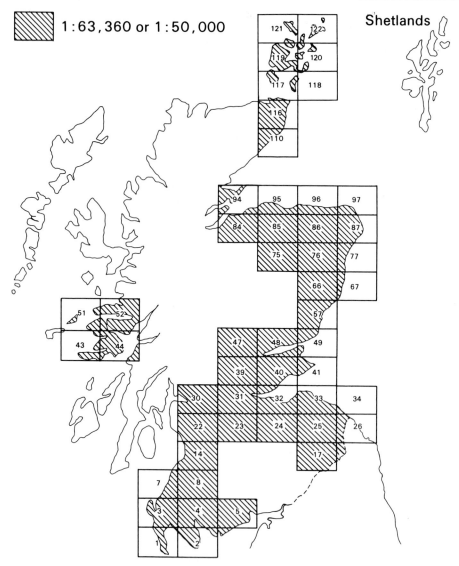

Figure 2.3: Medium-scale (1:63,360 or 1:50,000) soil maps published by the Soil Survey of Scotland.

the Foudland association in drift derived mainly from Dalradian slates and phyllites is divided into shallow freely drained, intermediate freely drained, deep freely drained, poorly drained, and very poorly drained series. In subsequent surveys (e.g. Muir, 1956; Mitchell and Jarvis, 1956) the association concept was broadened to embrace all the soils in materials of similar geological origin, irrespective of climatic zone. The component series, each defined in terms of drainage class and genetic soil group or subgroup (Table 2.6), were also given locality names as in England and Wales and elsewhere.

By 1987 about half the country, including most of the arable land, had been covered and

36 soil maps published (Figure 2.3) together with eleven explanatory memoirs, each covering two or more map sheets. More detailed *ad hoc* surveys had also been made of smaller areas, usually for advisory or planning purposes (e.g. Heslop and Bown, 1969). Within associations, shallow stony (skeletal) soils over bedrock are not identified as named series and are commonly mapped with one or more associated series as components of soil complexes, particularly in areas of rocky land with many short slopes. Organic soils are indicated in the map legends as peat (basin or blanket) and weakly developed (immature) soils in recent alluvium and blown sands are similarly set apart as 'miscellaneous soils'. Despite differences in classification (see below) and presentation, however, the maps convey essentially the same kind of information as those produced by the Soil Survey of England and Wales at the same scale.

In 1966, in response to pressure on land created by the siting of new towns in the central lowlands, the Survey started to make land-use capability maps and these were subsequently compiled and published along with the soil maps as a regular part of the programme. The land-use capability classification used during this period was originally devised in the United States (Klingebiel and Montgomery, 1961) and modified for application under British conditions by Bibby and Mackney (1969). It grades land into seven classes according to physical attributes, including soil, climatic and topographic factors, that limit its adaptability for agricultural use to varying degrees. Thus land with few or no physical limitations is placed in class 1 and land with more or less severe limitations that restrict its use to pasture, forestry or recreation in classes 4–6. Subclasses indicate the kind or kinds of limitation affecting land use, including wetness, other adverse soil features, gradient and soil pattern limitations, climatic limitations and liability to erosion.

In 1978 the survey was authorized to acquire the extra information needed to publish soil and land capability maps of the whole country at a scale of 1:250,000 within five years. This was achieved by making extensive use of air-photo interpretation techniques to speed reconnaissance mapping of the Highlands and other areas not already covered by systematic medium-scale surveys. The resulting maps were published as seven overlapping sheets in 1982 and explanatory handbooks for each area by 1984. The soil maps show the boundaries of 580 numbered map units, the first four of which consist predominantly of recent alluvial soils and organic (peat) soils. All the others are represented as divisions of named associations as in the previous 1:63,360 surveys, each division being characterized by a particular combination of component major soil subgroup(s) (Table 2.7) and land form (slope and rockiness). The types of vegetation found on each unit are also listed in the map legend, using a classification designed for the purpose (Soil Survey of Scotland, 1984). As on the corresponding maps of England and Wales, the colours assigned to the units, for example browns for brown earths, pink, orange or red for podzols and blues or greens for gley soils, indicate the general nature of the dominant soils. The use of subgroups rather than series names to identify the soil types present can be justified as being more appropriate in relation to the map scale and sampling density, especially in the hill lands, and conveys information about them more directly. On the other hand it can be argued that the large number of geologically based associations recognized has led to undue proliferation of units without a commensurate reduction of intra-unit variation in properties of interest.

The accompanying interpretative maps showing land capability for agriculture are based on an improved version of the earlier classification (Bibby and Mackney, 1969), amended to take the requirements of users into account (Bibby *et al.*, 1982). Following the acceptance by DAFFS of land capability assessment using this system as essential information to be considered in planning land use, the chief task assigned to the Survey after the 1:250,000 project had been completed was to compile updated land capability maps covering the principal agricultural areas at the 1:50,000 scale. Despite progressive reductions in staffing levels, the additional medium-scale soil mapping needed to implement this programme is now (1987) in its final stages.

2.3.2 Soil Characterization and Classification

The field and laboratory methods (Chapter 3, pp 81–99) that have been used to characterize the soils are described in the explanatory texts accompanying the published maps. Terminology for describing soil profiles in the field was standardized during the 1950s (Muir, 1956) and is largely based on the USDA Soil Survey Manual (Soil Survey Staff, 1951). Systematic analyses of samples from selected profiles, carried out by other departments of the Macaulay Institute, include measurements of loss on ignition, particle-size distribution, exchangeable cations, pH, organic carbon, total nitrogen, and total and acetic acid-soluble P_2O_5. Silica/sesquioxide ratios in clay fractions, amounts of iron and aluminium removed by various extracting solutions, total and acetic acid-soluble trace element contents, and the mineralogical composition of clay and fine sand separates, were also determined on smaller numbers of samples in some or all of the surveys.

The classification used as a basis for differentiating soil series within associations has also remained essentially unchanged since the 1950s, though additional divisions have been introduced at intervals to accommodate kinds of soil not recognized in earlier surveys. Table 2.7 shows the latest version, which was compiled to meet the requirements of the national 1:250,000 mapping programme. As in the current England and Wales classification (Table 2.2), profile classes are defined at three categorical levels above that of the soil series, but the Scottish system is typological rather than definitional in character (Chapter 1, p. 30). Also, although the traditional concepts of calcareous soils, brown earths, podzols, gley soils and organic soils (Table 2.1) are retained in both schemes, there are considerable differences in detail. For example, no provision is made in the Scottish system for distinguishing soils with clay-illuvial B horizons from those without, or so-called brown podzolic soils from normal brown earths; alluvial soils are segregated from otherwise similar soils at divison level, and no separations based on texture are made in any of the higher categories.

As in other systems, soil series comprise soils with similar horizons developed in similar parent materials, but they are differentiated in Scotland as members of named associations rather than as divisions of subgroups defined by specified profile characteristics. Most of the associations recognized are distinguished by the lithology, or lithology and stratigraphic age, of the pre-Quaternary rocks from which the soil parent materials are apparently derived. Others, like the Tipperty association in reddish lacustrine silts and clays derived from Old Red Sandstone rocks and the Eckford association in glaciofluvial sands of similar derivation, are confined to water sorted sediments with particular lithologies. In associations which include soils in superficial deposits of widely varying texture, separate series have been distinguished on this basis. For example, the poorly drained noncalcareous gley soils occurring in the extensive Ettrick association in materials derived mainly from Lower Palaeozoic greywackes and shales are divided into two series, the Ettrick series in fine textured (clay loam to clay) till and the Littleshalloch series in coarser textured and more stony drift.

2.4 Forestry Commission Surveys

In 1961 the Forestry Commission formed a small site survey section to investigate the

Table 2.7: Classification of Scottish soils (Soil Survey of Scotland, 1984)

Division	Major soil group	Major soil subgroup
1. Immature soils	1.1 Lithosols	
	1.2 Regosols	1.21 Calcareous regosols
		1.22 Non-calcareous regosols
	1.3 Alluvial soils	1.31 Saline alluvial soils
		1.32 Mineral alluvial soils
		1.33 Peaty alluvial soils
	1.4 Rankers	1.41 Brown rankers
		1.42 Podzolic rankers
		1.43 Gley rankers
		1.44 Peaty rankers
2. Non-leached soils	2.1 Rendzinas	2.11 Brown rendzinas
	2.2 Calcareous soils	2.21 Brown calcareous soils
3. Leached soils	3.1 Magnesian soils	3.11 Brown magnesian soils
	3.2 Brown earths	3.21 Brown forest soils
		3.22 Brown forest soils with gleying
	3.3 Podzols	3.31 Humus podzols
		3.32 Humus-iron podzols
		3.33 Iron podzols
		3.34 Peaty podzols
		3.35 Subalpine podzols
		3.36 Alpine podzols
4. Gleys	4.1 Surface-water gleys	4.11 Saline gleys
		4.12 Calcareous gleys
		4.13 Magnesian gleys
		4.14 Non-calcareous gleys
		4.15 Humic gleys
		4.16 Peaty gleys
	4.1 Ground-water gleys	4.21 Calcareous gleys
		4.22 Non-calcareous gleys
		4.23 Humic gleys
		4.24 Peaty gleys
		4.25 Subalpine gleys
		4.26 Alpine gleys
5. Organic soils	5.1 Peats	5.11 Eutrophic flushed peat
		5.12 Mesotrophic flushed peat
		5.13 Dystrophic flushed peat
		5.14 Dystrophic peat

usefulness of soil and site surveys in forestry (Toleman and Pyatt, 1974). After four years during which methods were developed and tested under a wide range of silvicultural and environmental conditions, the section was expanded to extend the work on a routine basis throughout the national forests and on land destined for afforestation. Mapping has been done at a scale of 1:10,560 or 1:10,000 and explanatory reports issued for each forest surveyed. These include silvicultural recommendations or predictions concerning choice of species, cultivation and drainage, fertilization at planting or during the rotation, weed

Table 2.8: Forestry Commission Soil Classification (Pyatt, 1982)

Soil group		Soil type
A. *The main mineral and shallow peaty soils (peat <45 cm thick)*		
Soils with well aerated subsoil	Brown earths	Typical brown earth
		Basic brown earth
		Upland brown earth
		Podzolic brown earth
	Podzols	Typical podzol
	Ironpan soils	Intergrade ironpan soil
		Ironpan soil
		Podzolic ironpan soil
Soils with poorly aerated subsoil	Ground-water gley soils	Ground-water gley
	Peaty gley soils	Peaty gley
		Peaty podzolic gley
	Surface-water gley soils	Surface-water gley
		Brown gley
		Podzolic gley
B. *Peatland types (peat 45 cm or more thick)*		
Flushed peatlands	*Juncus* bogs (Basin bogs, including fens)	*Phragmites* bog
		Juncus articulatus or *acutiflorus* bog
		Juncus effusus bog
		Carex bog
	Molinia bogs (Flushed blanket bogs)	*Molinia-Myrica-Salix* bog
		Tussocky *Molinia* bog; *Molinia-Calluna* bog
		Tussocky *Molinia-Eriophorum vaginatum* bog
		Non-tussocky *Molinia-Eriophorum vaginatum-Trichophorum* bog
		Trichophorum-Calluna-Eriophorum-Molinia bog (weakly flushed blanket bog)
Unflushed peatlands	Sphagnum bogs (Flat or raised bogs)	Lowland Sphagnum bog
		Upland Sphagnum bog
	Calluna-Eriophorum-Trichophorum bogs (Unflushed blanket bogs)	*Calluna* blanket bog
		Calluna-Eriophorum vaginatum blanket bog
		Trichophorum-Calluna blanket bog
		Eriophorum blanket bog
	Eroded bogs	Eroded (shallow hagging) bog
		Deeply hagged bog
		Pooled bog
C. *Other soils*		
	Man-made soils	Mining spoil, stony or coarse textured
		Mining spoil, shaly or fine textured
	Calcareous soils (soils on limestone rock)	Rendzina (shallow soil)
		Calcareous brown earth
		Argillic brown earth (clay enriched subsoil)

Table 2.8: Cont.

Soil group	Soil type
Rankers and skeletal soils (Rankers = shallow soils over bedrock Skeletal = excessively stony)	Brown ranker Gley ranker Peaty ranker Rock Scree Podzolic ranker
Littoral soils	Shingle Dunes Excessively drained sand Sand with moderately deep water table Sand with shallow water table Sand with very shallow water table

competition and windthrow hazard for each of the site types recognized. Although the maps and reports have not been formally published, they provide the best available information on the soils of considerable areas of Upland Britain which have not yet been mapped in detail by the national soil survey organizations.

Details of the special-purpose soil classification developed for use in these surveys are given by Pyatt (1970). It was subsequently revised and extended, and the latest version (Table 2.8) includes further minor changes in structure and nomenclature. This two-category scheme was designed specifically for application in forestry and more particularly to aid the establishment and management of exotic conifer plantations in upland areas where most of the soils overlie consolidated non-calcareous rocks and many consist wholly or partly of peat. The mineral and shallow peaty soil groups and types recognized have equivalents or close counterparts in higher categories of the systems used by the Soil Survey of England and Wales (Table 2.2) and the Soil Survey of Scotland (Table 2.7), though some are differently named. For example, the ironpan soils can be equated with stagnopodzols (Avery, 1980) and peaty podzols (Soil Survey of Scotland, 1984) respectively. The peatlands, however, are divided into flushed (minerotrophic) and unflushed (ombrotrophic) bog types that are identified by the natural or semi-natural plant communities they support rather than by internal characteristics as in the England and Wales system. According to Toleman (1973), this traditional approach was adopted because it can easily be operated in the field by relatively inexperienced staff and provides a useful guide to differences in nutrient status and afforestation potential (Chapter 9, p. 419).

In applying the system, the country is divided on a geological and climatic basis into a number of site regions and subregions, each characterized by a relatively narrow range of land forms, soil types, climate and tree-growth rates. Within each region, the site types identified and mapped are based on the national soil classification and the soil types listed in Table 2.8A are further subdivided into phases based on properties such as depth of rootable soil over bedrock or indurated material, the thickness of surface peat, if present, and the presence of potentially competing ericaceous vegetation. The site types identified in one such region, embracing the forests on soils derived from slaty and shaly

Palaeozoic rocks in north and mid-Wales, are described in detail by Pyatt (1977).

2.5 The National Soil Survey of Ireland

2.5.1 Soil and Land Capability Mapping

From its inception in 1959, the National Soil Survey of Ireland has engaged in systematic mapping of soil series and complexes of series on a county basis. Field observations have been recorded and soil boundaries delineated on six inch to one mile (1:10,560) maps, which are reduced to a scale of two miles to the inch (1:126,720) for publication. By 1984 maps of eight counties and two special regions (West Cork and West Mayo) had been published with explanatory bulletins (Figure 2.4) and surveys were well advanced elsewhere, but reductions in staff have since caused the work to be curtailed. Numerous more detailed sur-

Figure 2.4: Two miles to one inch (1:126,720) soil maps published by the National Soil Survey of Ireland.

veys of smaller areas, mostly experimental farms, have also been made.

The soil information is used in conjunction with climatic and topographic data to classify the land according to its suitability for tillage crops, pasture or forestry. A six-class system similar in principle to the land-use capability classification of Bibby and Mackney (1969) was employed for this purpose in early surveys of the Counties of Wexford (Gardiner and Ryan, 1964), Limerick (Finch and Ryan, 1966) and Carlow (Conry and Ryan, 1967) but was replaced in later work by one entailing separate five-class ratings for tillage and grassland (e.g. Finch and Gardiner, 1977). Those for grassland are supported by measurements of pasture productivity and response to fertilizers on the most extensive soil series, and potential productivity for forestry was also investigated.

Following production of a generalized soil map of the whole country in 1969, a ten-year programme involving reconnaissance studies of areas not covered by the medium-scale surveys was started with the aim of producing an improved version. This was achieved in 1980 with the publication of a 1:575,000 soil association map together with a bulletin describing the associations and their land-use potential (Gardiner and Radford, 1980). A peatland map at the same scale with an accompanying bulletin was also published (Hammond, 1981). Although both maps include Northern Ireland, the information on this province was derived from various sources other than systematic soil surveys and is therefore less precise and probably less reliable than that on the Republic.

The 44 numbered soil associations recognized are grouped in the map legend into five physiographic divisions distinguished by altitude and relief, namely: (1) Mountain and Hill, mostly above 500 m; (2) Hill, mainly between 150 and 365 m; (3) Rolling Lowland; (4) Drumlins; and (5) Flat to Undulating Lowland, mostly below 100 m. Within these divisions, each association comprises soils developed in a limited range of parent materials and is further characterized by the principal and associated great soil groups (Table 2.9) it contains and their proportionate extent. The accompanying bulletin includes a chapter on potential land use of Irish soils with sections on the range of uses to which the different soils are suited (six-class system), their major limitations, and the extent, kind and quality of marginal land and land suitable for tillage. Tables show the area and proportionate extent of each class of land on provincial and county bases.

2.5.2 Soil characterization and Classification

The methods and terms used to describe soil profiles in the field (Chapter 3, pp. 81–90) are largely derived from the USDA Soil Survey Manual (Soil Survey Staff, 1951). Like the corresponding British publications, the bulletins accompanying the 1:126,720 and 1:575,000

Table 2.9: Great Soil Groups in Ireland (after Gardiner and Radford, 1980)

Zonal soils	1. Podzols (including peaty podzols) 2. Brown podzolic soils 3. Brown earths (including acid brown earths and brown earths of medium to high base status) 4. Grey-brown podzolic soils 5. Blanket peats
Hydromorphic soils	6. Gleys (including ground-water, surface-water, and peaty podzolic gleys) 7. Basin peats
Calcimorphic soils	8. Rendzinas
Azonal soils	9. Regosols (including non-hydromorphic alluvial soils) 10. Lithosols

soil maps contain descriptions of representative profiles supplemented by analytical data. Particle-size distribution, exchangeable cations, cation-exchange capacity, pH, total neutralizing value, carbon, nitrogen and free iron oxide have been determined on a routine basis, and trace-element contents (total and/or acid extractable) and clay mineralogy on selected samples. As in Scotland, the trace-element data have been used to aid delineation of 'problem areas', particularly those subject to cobalt and copper deficiencies and areas in which soils and pasture herbage contain abnormally large amounts of selenium or molybdenum potentially harmful to animals.

The system of classification adopted as the basis for identifying soil-map units is a simplified modification of that formerly used in the United States (Thorp and Smith, 1949). Only two categories, great soil group (Table 2.9) and soil series, are formally recognized but variants such as peaty podzols, acid brown earths and peaty gleys have in fact been distinguished within particular great groups and are set apart in the legend of the national soil association map. Similarly, the legend of the 1:575,000 peatland map features divisions of the blanket and basin peats into mire types and subtypes (Table 9.1, p. 401). Both these subdivisions and the undivided great groups are closely paralleled in the higher categories of one or both of the systems used in Great Britain, though the class limits are not always congruent and some, particularly the grey-brown podzolic soils (argillic brown earths) are differently named. As in the corresponding categories of the Scottish system (Table 2.7), the great groups of mineral soils are distinguished primarily by the kind and degree of expression of genetic horizons in the profile and there are no textural separations like those in the England and Wales classification (Table 2.2). A further point of resemblance to Scottish practice is that the divisions made in the soil-series category, which are based mainly on differences in the composition and geological origin of the parent materials, are defined by reference to type profiles rather than by systematic differentiating criteria.

2.6 Application of the US Soil Taxonomy to British Soils

Although intended primarily as a basis for making and interpreting soil surveys in the United States, the Soil Taxonomy developed by the Soil Conservation Service of the US Department of Agriculture (Soil Survey Staff, 1975) was designed to achieve unambiguous and meaningful identification of all known kinds of soil. It has therefore been widely used as an international reference system and is regarded by many, including the present author, as the best currently available for the purpose. Its application to British soils was investigated in detail by Ragg and Clayden (1973) and is further reviewed in the following chapters. The structure of the system and the nature of the differentiating properties are outlined below, leaving the reader to consult the definitive text for details of the very large number of classes distinguished. In order to identify a soil correctly, familiarity with the diagnositic criteria and careful reference to the keys provided are essential. A number of minor amendments (Soil Survey Staff, 1987) have been introduced since 1975 but they hardly affect the placement of British soils.

The Soil Taxonomy is a hierarchical definitional classification consisting of six categories: order, suborder, great soil group, subgroup, family and series. Classes in the subgroup and higher categories are named by an entirely new set of terms in which formative elements derived mainly from Greek and Latin sources indicate relationships between classes at successive categorical levels, as illustrated by the following example:

Order: Inceptisols
Subgroup: Och*repts*
Great group: Eutroch*repts*
Subgroup: Rendollic *Eutrochrepts*

Table 2.10: Orders and suborders of Soil Taxonomy (Soil Survey Staff, 1975, 1987)

1. *Entisols.* Weakly developed mineral soils
   1.1 *Aquents* – wet or artificially drained Entisols with gley (hydromorphic) features
   1.2 *Arents* – others with fragments of soil horizons (artificially disturbed soils)
   1.3 *Psamments* – others in sandy materials
   1.4 *Fluvents* – others in recent alluvium or colluvium
   1.5 *Orthents* – other Entisols, mainly in erosional sites

2. *Vertisols.* Clay soils that crack widely and deeply during some part of the year
   2.1 *Xererts*[a] – dry for long periods in summer and moist or wet in winter
   2.2 *Torrerts*[a] – usually dry
   2.3 *Uderts*[a] – usually moist
   2.4 *Usterts*[a] – others with one or more dry periods

3. *Inceptisols.* Moderately developed soils of humid or subhumid regions, mostly with a cambic horizon, umbric epipedon, fragipan or duripan.
   3.1 *Aquepts* – periodically wet or artificially drained Inceptisols with gley (hydromorphic) features
   3.2 *Andepts* – others, mainly in volcanic ash, with low bulk density and exchange complex dominated by amorphous material
   3.3 *Plaggepts* – others with a plaggen epipedon
   3.4 *Tropepts*[a] – others in tropical climates
   3.5 *Ochrepts* – others in mid to high latitudes, mainly with an ochric epipedon and a cambic horizon and/or fragipan
   3.6 *Umbrepts* – others in mid to high latitudes, mainly with an umbric epipedon

4. *Aridisols*[a]. Soils of deserts and semi-deserts (with an aridic moisture regime)
   4.1 *Argids*[a] – with an argillic or natric horizon
   4.2 *Orthids*[a] – others, without an argillic or natric horizon

5. *Mollisols.* Base-rich soils with a mollic epipedon (mainly of the steppes)
   5.1 *Albolls*[a] – periodically wet or artificially drained Mollisols with albic and argillic horizons
   5.2 *Aquolls* – other periodically wet or artificially drained Mollisols with gley (hydromorphic) features
   5.3 *Rendolls* – others over extremely calcareous materials
   5.4 *Xerolls*[a] – others with a long dry period in summer
   5.5 *Borolls*[a] – others in cold climates (with frigid, cryic or pergelic temperature regime)
   5.6 *Ustolls*[a] – others with one or more dry periods
   5.7 *Udolls* – others in humid or subhumid regions

6. *Spodosols.* Soils with a spodic horizon, or a placic horizon overlying a fragipan
   6.1 *Aquods* – periodically wet or artificially drained Spodosols, normally with gley (hydromorphic) features
   6.2 *Ferrods*[a] – others with a spodic horizon containing little organic carbon
   6.3 *Humods* – others with a spodic horizon containing little iron in some part
   6.4 *Orthods* – other Spodosols

7. *Alfisols.* Soils with an argillic or kandic horizon of moderate to high base status and normally an ochric epipedon
   7.1 *Aqualfs* – periodically wet or artificially drained Alfisols with gley (hydromorphic) features
   7.2 *Boralfs*[a] – others in cold climates (frigid or cryic temperature regime)
   7.3 *Ustalfs*[a] – others in subhumid or semi-arid regions with one or more dry periods
   7.4 *Xeralfs*[a] – others with a long dry period in summer
   7.5 *Udalfs* – others in humid regions

8. *Ultisols.* Soils with an argillic or kandic horizon of low base status in mid to low latitudes
   8.1 *Aquults* – periodically wet or artificially drained Ultisols with gley (hydromorphic) features
   8.2 *Humults*[a] – others that are rich in organic matter
   8.3 *Udults* – others in humid regions
   8.4 *Ustults*[a] – others in warm regions with high rainfall and one or more dry periods
   8.5 *Xerults*[a] – others with a long dry period in summer

Table 2.10: Cont.

9. *Oxisols.*[a] Soils with an oxic horizon or with plinthite within 30 cm depth
  9.1 *Aquox*[a] – periodically wet or artificially drained Oxisols with plinthite or gley (hydromorphic) features
  9.2 *Torrox*[a] – others in arid regions
  9.3 *Humox*[a] – others in humid regions with much organic matter and low base status
  9.4 *Ustox*[a] – others with one or more dry periods
  9.5 *Orthox*[a] – others in humid regions
	10. *Histosols.* Soils consisting mainly of organic materials (peat soils)
	  10.1 *Folists*[a] – never saturated with water for more than a few days
	  10.2 *Fibrists* – others consisting mainly of little decomposed plant remains (fibrous peat)
	  10.3 *Hemists* – others consisting mainly of partially decomposed plant remains (semi-fibrous peat)
	  10.4 *Saprists* – others consisting mainly of strongly decomposed organic material

[a] Absent or rare in the British Isles.

Table 2.11: Diagnostic surface horizons – epipedons (Soil Survey Staff, 1975)

1. *Mollic epipedon.* Thick dark-coloured surface horizon with more than 0.6 per cent organic carbon (1 per cent organic matter) after mixing to a depth of 18 cm and base saturation of 50 per cent or more

2. *Anthropic epipedon.* Surface horizon with properties of a mollic epipedon except for large amounts of acid-soluble phosphorus resulting from long-continued use by man

3. *Umbric epipedon.* Surface horizon with properties of a mollic epipedon except that base saturation is less than 50 per cent

4. *Histic epipedon.* Peaty surface horizon that contains at least 8 to 16 per cent organic carbon depending on clay content and is saturated with water for at least 30 consecutive days in most years unless artificially drained

5. *Plaggen epipedon.* Man-made surface horizon 50 cm or more thick produced by long-continued manuring

6. *Ochric epipedon.* Surface horizon too pale in colour, too low in organic matter, or too thin to be a mollic, anthropic, umbric or plaggen epipedon, or is both massive and hard when dry

Orders. The ten orders (Table 2.10) are distinguished by the presence or absence of specified diagnostic horizons (Tables 2.11 and 2.12) and other definitive properties reflecting the operation of particular sets of genetic processes or factors regarded as significant on a global scale. As might be expected (Chapter 1, p. 31), their definitions are generally complicated. Thus, whilst all podzols and related soils having a spodic horizon are Spodosols, this order also includes soils with a placic horizon (thin ironpan) but no spodic horizon. The argillic horizon, characterized by illuvial accumulation of silicate clay, is a diagnostic feature of both Alfisols and Ultisols but the latter differ in having subsoils with low base status indicative of advanced weathering and leaching. Argillic horizons can also occur in the Mollisol and Aridisol orders characteristic of steppes and deserts or semi-deserts respectively, and mollic epipedons are similarly not restricted to Mollisols.

Suborders. The orders are divided into 47 suborders (Table 2.10), most of which are based on differences in soil-moisture regime that influence current soil processes and are important for plant growth. Moisture regimes designated aquic, aridic, torric, udic, ustic and xeric are defined in terms of ground-water

Table 2.12: Diagnostic subsurface horizons (Soil Survey Staff, 1975, 1987)

1. *Argillic horizon.* Illuvial horizon in which silicate clay has accumulated to a significant extent.
2. *Kandic horizon*[a]. Similar to argillic horizon but contains low-activity clays (CEC <16 me/100 g clay)
3. *Agric horizon.* Illuvial horizon, below a ploughed layer, in which silt, clay and humus have accumulated to a significant extent.
4. *Natric horizon*[a]. Argillic horizon with prismatic or columnar structure and 15 per cent or more saturation with exchangeable sodium.
5. *Sombric horizon*[a]. Illuvial horizon in which humus has accumulated neither in association with aluminium, as in spodic horizons, nor with exchangeable sodium, as in natric horizons.
6. *Spodic horizon.* One in which active amorphous materials, composed of organic matter and aluminium with or without iron, have precipitated.
7. *Placic horizon.* Thin, black to dark reddish pan cemented by iron, iron and manganese, or an iron-organic complex; generally 2–10 mm thick.
8. *Cambic horizon.* Loamy or clayey horizon altered by pedogenesis, as indicated by structure, colour, gley features, or evidence of clay formation or removal of carbonates.
9. *Oxic horizon*[a]. Strongly altered horizon, at least 30 cm thick, that consists of a mixture of hydrated oxides of iron or aluminium, or both, with variable amounts of low-activity (mainly kaolinitic) clay and resistant minerals such as quartz.
10. *Duripan*[a]. Horizon cemented by silica to the extent that dry fragments do not slake in water.
11. *Fragipan.* Horizon of high bulk density, brittle when moist and hard or very hard when dry; dry fragments slake in water.
12. *Albic horizon.* One from which clay and free iron oxides have been removed, to the extent that its colour is determined by the colours of uncoated sand and silt particles.
13. *Calcic horizon.* One in which calcium, or calcium-magnesium, carbonate has accumulated; it is at least 15 cm thick and contains at least 5 per cent by volume of identifiable secondary carbonates.
14. *Gypsic horizon.* One in which calcium sulphate has accumulated to an equivalent extent.
15. *Petrocalcic horizon*[a]. An indurated calcic horizon that is cemented to the extent that dry fragments do not slake in water.
16. *Petrogypsic horizon*[a]. An indurated gypsic horizon that is cemented to the extent that dry fragments do not slake in water.
17. *Salic horizon*[a]. One containing a secondary enrichment of salts more soluble than gypsum; it is at least 15 cm thick.
18. *Sulfuric horizon.* Mineral or organic horizon that has both a pH less than 3.5 (1:1 in water) and yellowish mottles of jarosite.

[a] Absent or rare in the British Isles.

levels and the average annual duration of periods when a specified part of the rooting zone (the moisture control section) is wholly or partly dry (Chapter 3, p. 85). Thus gleyed soils that are saturated with water for substantial periods when the soil temperature is above 5°C are considered to have an aquic moisture regime and soils with a xeric moisture regime occur in areas with Mediterranean climates characterized by a long dry period in summer. Most well-drained British soils have a udic moisture regime in which dry periods, if any, are of short duration.

Great groups. The chief properties used for differentiation in this central category include:

1. Presence or absence of diagnostic horizons, including natric, placic and calcic horizons, fragipans and duripans (Table 2.12), which are not differentiating at order or suborder level.
2. Soil-temperature (pergelic, cryic, frigid, mesic, thermic or hyperthermic) and soil-moisture regimes.
3. Base status.

The use of base status is illustrated by the

segregation of Eutrochrepts and Dystrochrepts within the Ochrept suborder, which includes most of the soils that have been called brown earths and brown calcareous soils in Britain. Eutrochrepts are calcareous or at least 60 per cent base saturated within limited depths whereas the Dystrochrepts have a lower base status. This distinction reflects the traditional division of the brown earths into base-rich and base-deficient subgroups (Table 2.1).

Subgroups. This category, absent in earlier US schemes, was introduced to separate central concepts of great groups (e.g. Typic Eutrochrepts) from intergrades to other great groups, suborders or orders (e.g. Rendollic Eutrochrepts). There are also 'extragrade' subgroups, such as Lithic Eutrochrepts having hard bedrock within 50 cm depth, that are not transitional to other named kinds of soil.

Families. At this level subgroups are divided on the basis of properties, including particle-size distribution within the rooting zone but below plough depth (c. 25–100 cm), mineralogy, temperature and depth of soil penetrable by roots, considered to affect their agronomic behaviour and responses to management practices. The families so differentiated are named accordingly. For example the coarse-loamy, mixed, mesic family of Typic Eutrochrepts comprises variants in which the *particle-size class* is coarse-loamy, the *mineralogy class* is mixed and the *soil-temperature class* is mesic. The particle-size and mineralogical criteria are similar but not identical to those used for differentiating series within subgroups in the England and Wales classification (p. 52).

Series. The series, of which about 10,500 had been recognized in the United States by 1975, is the lowest category of the system. Most of them were originally defined as map units in accordance with earlier concepts but they are now all differentiated within families on the basis of minor variations in profile morphology and composition. As hitherto, each established series is identified by a locality name.

Reactions to the system by pedologists in a number of countries have been reviewed by Ragg and Clayden (1973), Cline (1980) and Clayden (1982). The main features which have been criticized include:

1. The rejection of all previously used soil names and the unpalatability of the new nomenclature.
2. The choice and ordering of the differentiating criteria, particularly the absence of a hydromorphic class at the highest categorical level and the weight attached to the argillic horizon.
3. The use of differentiae, including soil-moisture and soil-temperature regimes, which are based on quantitative data unavailable in many parts of the world or applicable only to type profiles or pedons which have been studied in detail, implying that the apparent precision of the system can be illusory.

Ragg and Clayden (1973) concluded that the unambiguous nomenclature, the carefully defined diagnostic horizons and other criteria, and the emphasis placed on properties not readily altered by normal cultivation, are valuable attributes which make Soil Taxonomy particularly suitable for international use. On the other hand they considered that its application as a practicable basis for soil surveys in Britain presented serious problems which can be summarized as follows:

1. Too many classes above the family level.
2. The diagnostic criteria are generally too complex and some, particularly those for the spodic horizon, require amendment for meaningful and consistent application to British soils.
3. Difficulties resulting from the use at high categorical levels of diagnostic criteria involving base status or base status combined with topsoil colour and thickness, as in the definitions of the mollic and umbric epipedons. (Table 2.11).

Criteria of the latter kind, which are used in distinguishing Mollisols from other orders, Alfisols from Ultisols, and Eutrochrepts from Dystrochrepts, are especially difficult to apply consistently in agricultural land, partly because percentage base saturation cannot be measured in the field but mainly because both base status and the relevant topsoil characters are strongly influenced in originally non-calcareous soils by local (field to field) differences in past and present land use, liming and other ameliorative measures. The distinction between mollic and anthropic epipedons (Table 2.11), which requires that the latter must contain more than 250 ppm P_2O_5 soluble in 1 per cent citric acid, is also extremely difficult to apply, particularly since readily soluble phosphorus contents vary greatly over short distances (Beckett and Webster, 1971).

2.7 The FAO Classification and the Soil Map of the EEC Countries

In addition to or instead of the US Soil Taxonomy, the classification on which the legend of the 1:5,000,000 Soil Map of the World (FAO 1974–78) was based has also been widely used for international communication in recent years (e.g. Duchaufour, 1978; de Bakker, 1979; Birse, 1980; FitzPatrick, 1986).

This system was compiled by the Food and Agricultural Organization of the United Nations (FAO) in consultation with an international advisory panel of soil scientists as part of a project intended to provide a uniform basis for worldwide transfer of information on soil resources. As originally published with a diagnostic key (FAO-Unesco, 1974), it consists of 106 classes termed soil units which are defined at two categorical levels. The 26 higher classes (or orders) are listed in Table 2.13, together with abbreviated definitions and correlations with American suborders. Some are identified by 'traditional' names like Chernozem, Podzol and Rendzina; others by newly coined names such as Luvisol and Acrisol, and Histosols and Vertisols by names imported from the US Soil Taxonomy. All but three are subdivided in the second category into adjectivally named units. The Podzols, for example, are segregated into Placic, Gleyic, Humic, Ferric, Leptic and Orthic Podzols. Although the differentiating criteria are largely derived from the US system and many of the separations are consequently comparable, there are also numerous divergences, particularly regarding the use of soil-moisture and soil-temperature criteria. The only FAO classes defined in this way are the Xerosols and Yermosols, both being required to have an aridic moisture regime, and *gelic* units (e.g. Gelic Gleysols) characterized by the presence of permafrost. Definitions of the other units are based mainly on the presence or absence of diagnostic horizons or other 'static' properties and also diverge in some cases from those adopted in the American system. Thus the podzol group is restricted to soils having a spodic B horizon and the distinction between Luvisols and Acrisols, which mirrors that between Alfisols and Ultisols (Table 2.10) and is similarly defined in terms of percentage base saturation at specified depths, entails placement of many Acrisols in the Alfisol order.

Young (1979) suggested that the FAO classification was the most valuable part of the Soil Map of the World Project, particularly in view of the evident disparity between the precision of the soil units and the quality of the information on which the maps of most countries were based. As a medium for international communication, it has the advantage of being considerably simpler than the US system. On the other hand, many of the separations are equally difficult to apply consistently and the lack of divisions based on climatic factors severely limits their practical significance on a global scale.

Basically the same system, modified by the introduction of a third category, is used in the legend of the recently produced 1:1,000,000 soil-association map of the EEC countries. This was compiled by a committee of European soil scientists led by Professor R. Taver-

Table 2.13: Soil units in the highest category of the FAO classification (FAO–Unesco, 1974)

1. *Fluvisols.* Weakly developed soils in recent alluvium or colluvium; lacking a diagnostic B horizon and a mollic A horizon (Fluvents and some Aquents, Orthents and Psamments).
2. *Gleysols.* Soils with gley (hydromorphic) features in materials other than recent alluvium or colluvium; lacking an argillic or spodic B horizon, high salinity and Vertisol characteristics (including Aquents, Aquepts, Aquolls and some Aquods).
3. *Regosols.* Weakly developed non-hydromorphic soils in unconsolidated materials other than recent alluvium or colluvium; having no diagnostic horizons other than an ochric A horizon (Orthents and Psamments).
4. *Lithosols.* Very shallow soils with hard rock at less than 10 cm depth (Orthents).
5. *Arenosols.* Non-hydromorphic sandy soils with weakly expressed subsurface horizons but no diagnostic argillic or spodic B horizon (mainly Psamments).
6. *Rendzinas.* Soils with a mollic A horizon over extremely calcareous material (Rendolls).
7. *Rankers.* Non-hydromorphic soils with an umbric A horizon no more than 25 cm thick and no diagnostic subsurface horizon (Umbrepts).
8. *Andosols.* Non-hydromorphic soils, mainly in volcanic ash, with low bulk density and exchange complex dominated by amorphous material (Andepts).
9. *Vertisols.* Cracking clay soils (Vertisols).
10. *Solonchaks*[a]. Saline soils in materials other than recent alluvium (Saline Aquepts, Aridisols and Aquolls).
11. *Solonetz*[a]. Soils with a natric B horizon and no abrupt textural change at its upper boundary (Alfisols and Mollisols with natric B horizons).
12. *Yermosols*[a]. Non-saline desert soils with a very weak ochric A horizon (Aridisols).
13. *Xerosols*[a]. Non-saline desert soils with an ochric A horizon containing more organic matter (Aridisols).
14. *Kastanozems*[a]. Non-hydromorphic, non-saline soils with a dark brown mollic A horizon and a calcic or gypsic horizon or soft concentrations of secondary carbonates (some Ustolls and Borolls).
15. *Chernozems*[a]. Non-hydromorphic, non-saline soils with a very dark mollic A horizon and a calcic or gypsic horizon or soft concentrations of secondary carbonates (mostly Borolls).
16. *Phaeozems.* Soils with a mollic A horizon and no calcic or gypsic horizon or soft concentrations of secondary carbonates; lacking gley (hydromorphic) features unless an argillic B horizon is present (mainly Udolls).
17. *Greyzems*[a]. Non-saline soils with a very dark mollic A horizon and bleached coatings on aggregate surfaces, usually with an argillic B horizon (some Borolls and Aquolls).
18. *Cambisols.* Non-hydromorphic soils with an ochric or umbric A horizon and a cambic B horizon, or an umbric A horizon more than 25 cm thick (Ochrepts, Tropepts and some Umbrepts).
19. *Luvisols.* Soils with an argillic B horizon of medium to high base status; lacking a mollic A horizon and a bleached eluvial (E) horizon that overlies a slowly permeable B horizon or tongues into the B horizon (mainly Alfisols).
20. *Podzoluvisols.* Soils having an argillic B horizon with an irregular or tongued upper boundary (mainly Alfisols).
21. *Podzols.* Soils with a spodic B horizon (Spodosols).
22. *Planosols.* Hydromorphic soils with an abrupt textural change between a bleached (albic) E horizon and a slowly permeable horizon, usually an argillic or natric B horizon (some Aqualfs and Aquults).
23. *Acrisols.* Soils with an argillic B horizon of low base status; lacking a mollic A horizon and a bleached eluvial (E) horizon that overlies a slowly permeable B horizon or tongues into the B horizon (mainly Ultisols).
24. *Nitosols*[a]. Other soils with an argillic B horizon extending to a depth of 150 cm or more (some Alfisols and Ultisols).
25. *Ferralsols*[a]. Soils with an oxic B horizon (Oxisols).
26. *Histosols.* Soils composed dominantly of organic materials (Histosols).

[a] Absent or rare in the British Isles.

nier of the University of Ghent and published by the Commission of the European Communities (1985b). Like the Soil Map of the World, it was designed for interpretation in conjunction with climatic data as a basis for land evaluation. Each of the 312 map units is represented in the legend as an association of named soil units occurring within the limits of a mappable physiographic entity. In addition to the dominant and associated kinds of soil, contrasting soil components covering less than 10 per cent of the mapped area are listed as inclusions, the textural class (coarse, medium, medium fine, fine or very fine) of the dominant soil and the dominant slope class (level, sloping, moderately steep or steep) are given, and phases are used to indicate the common occurrence of agronomically significant accessory characteristics such as hard rock or indurated layers at shallow depths, surface stoniness, and salinity or alkalinity. Unfortunately, however, the value of the map for the purpose intended is vitiated by inadequate international correlation as well as by the uneven reliability of the information on which it is based and the use of criteria which, though precisely defined, could not be applied confidently in many situations.

2.8 The Classification Adopted

The classification of British soils used in the following chapters is based on the England and Wales system (Table 2.2) but incorporates various modifications which take account of experience gained in applying it. In the first place, the basic concept of soil (Chapter 1, pp. 1–3) is restricted to ground-surface layers that support or are capable of supporting a more or less closed cover of higher plants and contain appreciable amounts of accumulated organic matter, implying that 'raw soils' not meeting this requirement are excluded from consideration. This and other changes in the number, arrangement and nomenclature of classes in the higher categories are summarized below. Some have been made with the aim of simplifying the system, particularly for the purpose of this book. Others entail changes in the 'weighting' assigned to certain differentiating criteria, including predominantly clayey texture, slow permeability, and the presence or absence of a humose or peaty topsoil.

The major soil groups, groups and subgroups recognized are listed in Table 2.14 and the differentiating criteria employed are defined in Chapter 3, which concludes with a key to major soil groups. Each succeeding chapter deals with soils of a single major group and includes keys for identification of the constituent soil groups and subgroups. Although frequent references are made to individual soil series identified by one or other of the three national soil survey organizations, the much smaller number of separations in the subgroup category are applicable to the whole of the British Isles and represent a more appropriate level of generalization for the present purpose.

Major soil groups. In accordance with the more restrictive concept of soil, the major groups of terrestrial raw soils and raw gley soils (Table 2.2) are eliminated and all the soils that lack gleyed subsurface horizons and have little altered mineral material or bedrock at shallow depth are classed as lithomorphic or manmade. Similarly, the clayey soils set apart as pelosols by Avery (1980) are grouped as brown soils or gley soils and the ground-water and surface-water gley soils are united in a single class, as in the former British system (Table 2.1) and the highest categories of those currently used in Scotland and Ireland (Tables 2.7 and 2.9). Finally, most of the soils hitherto classed as brown podzolic are grouped with the brown soils and the remaining podzolic soils as podzols.

Soil groups. The number of soil groups is reduced from 43 to 23, partly through elimination of the raw soil groups and relegation of the surface-water/ground-water, pelosol and humic/non-humic separations to the subgroup category, and partly by uniting certain groups with all differentiating characteristics

Table 2.14: The classification adopted

Major soil group	Soil group	Subgroup
Lithomorphic soils	Lithosols	Non-calcaric lithosols
		Calcaric lithosols
	Rankers	Typical (humic) rankers
		Humic-calcaric rankers
		Organic (peaty) rankers
		Podzolic rankers
		Ochric rankers
		Ochric-calcaric rankers
	Rendzinas	Typical (humic) rendzinas
		Grey rendzinas
		Colluvial rendzinas
		Brown rendzinas
		Pararendzinas
	Sandy regosols	Non-calcaric sandy regosols
		Calcaric sandy regosols
	Rego-alluvial soils	Non-calcaric rego-alluvial soils
		Calcaric rego-alluvial soils
Brown soils	Sandy brown soils	Typical sandy brown soils
		Gleyic sandy brown soils
		Calcaric sandy brown soils
		Gleyic-calcaric sandy brown soils
	Alluvial brown soils	Typical (loamy) alluvial brown soils
		Gleyic alluvial brown soils
		Pelo-alluvial brown soils
		Calcaric alluvial brown soils
		Gleyic-calcaric alluvial brown soils
		Pelocalcaric alluvial brown soils
	Calcaric brown soils	Typical calcaric brown soils
		Colluvial calcaric brown soils
		Gleyic calcaric brown soils
		Stagnogleyic calcaric brown soils
		Pelocalcaric brown soils
	Orthic brown soils	Typical orthic brown soils
		Colluvial orthic brown soils
		Gleyic orthic brown soils
		Stagnogleyic orthic brown soils
		Pelo-orthic brown soils
	Luvic brown soils	Typical luvic brown soils
		Gleyic luvic brown soils
		Stagnogleyic luvic brown soils
		Peloluvic brown soils
		Chromoluvic brown soils
		Stagnogleyic chromoluvic brown soils
	Podzolic brown soils	Typical podzolic brown soils
		Gleyic podzolic brown soils
		Stagnogleyic podzolic brown soils

Table 2.14: Cont.

Major soil group	Soil group	Subgroup
Podzols	Non-hydromorphic podzols	Humoferric podzols
		Humus podzols
		Luvic podzols
	Gley-podzols	Typical gley-podzols
		Humoferric gley-podzols
		Luvic gley-podzols
	Stagnopodzols	Ironpan stagnopodzols
		Humus-ironpan stagnopodzols
		Ferric stagnopodzols
		Densipan stagnopodzols
Gley soils	Sandy gley soils	Typical sandy gley soils
		Humic sandy gley soils
		Calcaric sandy gley soils
		Humic-calcaric sandy gley soils
	Alluvial gley soils	Typical (loamy) alluvial gley soils
		Pelo-alluvial gley soils
		Humic alluvial gley soils
		Calcaric alluvial gley soils
		Pelo-calcaric alluvial gley soils
		Humic-calcaric alluvial gley soils
		Saline alluvial gley soils
		Acid-sulphate alluvial gley soils
	Calcaric gley soils	Ground-water calcaric gley soils
		Stagnocalcaric gley soils
		Pelocalcaric gley soils
		Humic calcaric gley soils
	Orthic gley soils	Ground-water orthic gley soils
		Stagno-orthic gley soils
		Pelo-orthic gley soils
		Humic orthic gley soils
		Humic stagno-orthic gley soils
	Luvic gley soils	Ground-water luvic gley soils
		Stagnoluvic gley soils
		Chromic stagnoluvic gley soils
		Peloluvic gley soils
		Humic luvic gley soils
		Humic stagnoluvic gley soils
Man-made soils	Cultosols	Sandy cultosols
		Loamy and clayey cultosols
	Disturbed soils	
Peat soils	Fen soils	Raw fen soils
		Earthy semi-fibrous fen soils
		Earthy sapric fen soils
		Earthy acid-sulphate fen soils
	Bog soils	Raw fibrous bog soils
		Raw semi-fibrous bog soils
		Raw sapric bog soils
		Earthy semi-fibrous bog soils
		Earthy sapric bog soils

Table 2.15: Approximate relations between the classification adopted and the US and FAO systems

Classification adopted	US Soil Taxonomy (Soil Survey Staff, 1975, 1987)	FAO-Unesco (1974)
Lithomorphic soils		
Lithosols	Very shallow (Lithic) Orthents	Lithosols
Rankers	Mainly shallow (Lithic) Umbrepts and Orthents; some Lithic Histosols and Hapludolls	Mainly Rankers and Regosols (Dystric or Eutric); some Lithic Histosols and Phaeozems
Rendzinas	Rendolls and calcareous Orthents; some Udifluvents	Rendzinas and Calcaric Regosols; some Fluvisols
Sandy regosols	Mainly Psamments; some Orthents	Sandy Regosols
Rego-alluvial soils	Fluvents, Orthents or Psamments in recent alluvial deposits	Fluvisols or Regosols in recent alluvial deposits
Brown soils		
Sandy brown soils	Mainly Psamments; some sandy-skeletal Orthents and sandy Umbrepts and Hapludolls	Mainly Arenosols; some sandy Phaeozems and Humic Cambisols
Alluvial brown soils	Mainly Fluventic Ochrepts; some Udifluvents and Hapludolls	Mainly Cambisols; some Fluvisols and Phaeozems
Calcaric brown soils	Mainly Eutrochrepts; some Hapludolls and Udifluvents	Mainly Calcic Cambisols; some Phaeozems and Fluvisols
Orthic brown soils	Mainly non-calcareous Ochrepts; some Umbrepts, Udifluvents and Hapludolls	Mainly Cambisols; some Phaeozems and Fluvisols
Luvic brown soils	Mainly Udalfs; some Udults and Argiudolls	Luvisols, Acrisols and Podzoluvisols; some Luvic Phaeozems
Podzolic brown soils	Mainly Dystrochrepts or Fragiochrepts; some Orthods and Andepts	Mainly Dystric Cambisols; some Leptic Podzols and Andosols
Podzols		
Non-hydromorphic podzols	Humods and Orthods, excluding Placic subgroups	Orthic and Humic Podzols

Table 2.15: Cont.

Classification adopted	US Soil Taxonomy (Soil Survey Staff, 1975, 1987)	FAO-Unesco (1974)
Gley-podzols	Mainly Haplaquods and Sideraquods	Gleyic Podzols
Stagnopodzols	Placaquods and Placaquepts; some Placic Humods, Sideraquods and Humaquepts	Placic Podzols and some Gleyic Podzols and Humic Gleysols
Gley soils		
Sandy gley soils	Psammaquents, sandy Haplaquolls and Humaquepts; some Haplaquents	Sandy Gleysols
Alluvial gley soils	Fluvaquents, Hydraquents, Fluventic Haplaquolls and Humaquepts; some Haplaquepts	Fluvisols and Mollic or Humic Gleysols in recent alluvial deposits
Calcaric gley soils	Calcareous Haplaquepts and Haplaquolls	Calcaric and Mollic Gleysols
Orthic gley soils	Mainly non-calcareous Aquepts; some Haplaquolls	Non-calcareous Gleysols
Luvic gley soils	Mainly Aqualfs and Aquepts; some Aquults and Argiaquolls	Mainly Gleyic Luvisols or Acrisols and Planosols
Man-made soils		
Cultosols	Plaggepts; some Mollisols or related soils with anthropic epipedons	(no related unit)
Disturbed soils	Mainly Arents	(no related unit)
Peat Soils		
Fen soils	Histosols, mainly euic or sulfic (including Sulfihemists and Sulfohemists)	Eutric and some Dystric Histosols
Bog soils	Histosols, mainly dysic	Dystric Histosols

78 *Chapter 2*

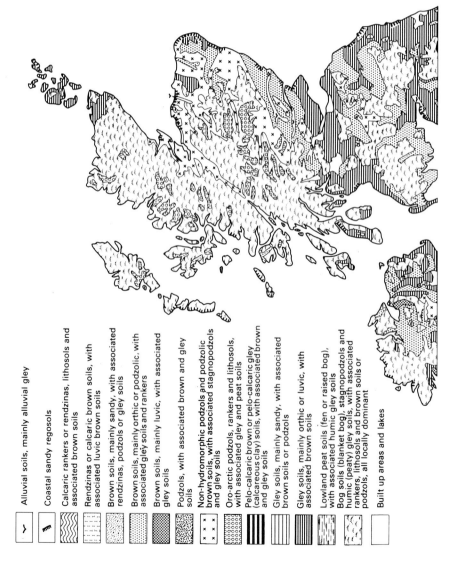

Figure 2.5: Schematic soil map of the British Isles.

except one in common (e.g. sand-rankers and sand-pararendzinas; argillic and paleo-argillic brown earths; podzols and gley-podzols; alluvial and humic-alluvial gley soils). In some cases the name of one of the previously recognized soil groups (e.g. rendzinas; podzols; sandy gley soils; alluvial gley soils) is applied to the enlarged group. Other groups are named or renamed using terms derived from the FAO classification. Thus very shallow lithomorphic soils over hard bedrock are grouped as lithosols and those in sandy deposits other than recent alluvium as sandy regosols. Brown soils and gley soils with textural (argillic) B horizons are distinguished as *luvic*, and those with colour/structure (cambic) B horizons in calcareous and non-calcareous materials other than recent alluvium as *calcaric* and *orthic* respectively. *Luvic* is used in preference to *argillic* because it signifies eluviation as well as or rather than clayiness as such; *calcaric* in preference to *calcareous* because the soils so named are not necessarily calcareous throughout; and *orthic* (from Greek *Orthos* meaning right or true) in preference to *cambic* because cambic horizons (Table 2.12) also occur in soils of other groups. The 'man-made humus soils' are renamed *Cultosols* (Mückenhausen, 1985) and the traditional terms *Fen* and *Bog* are re-introduced as soil group names in conformity with a revised classification of the peat soils (Chapter 9, p. 407).

Subgroups. Of the 94 subgroups recognized, 67 are exact or nearly exact equivalents of corresponding classes in the original scheme, though many are differently named to accord with the changes in the number and nomenclature of soil groups. Others are combinations or divisions of previously defined subgroups and one, the saline alluvial gley soils, is distinguished by a property not used in England and Wales for differentiation at this categorical level.

Climatic phases. The classification differs from the US Soil Taxonomy and certain other systems intended for global application in having no divisions based on soil moisture or soil temperature regimes. Even within the British Isles, however, both these characteristics vary greatly and the variations are not consistently associated with the more readily determinable 'static' properties adopted as differentiating criteria. The moisture regimes of certain gley soils, for example, may be fundamentally modified by artificial drainage without causing any marked change in profile morphology, and podzols with similar horizonation but very different temperature and moisture regimes occur in sites as climatically diverse as the drier English lowlands and the summits of Scottish mountains. In order to take account of climatic variations affecting current soil processes, plant growth, and land use capability, supplementary *climatic phases* are distinguished and applied to the soil-profile classes at any categorical level. The six climatic regimes distinguished for this purpose, each comprising a combination of specified moisture and thermal regimes, are defined in Table 1.2 (p. 18).

Correlation with the US Soil Taxonomy and the FAO System. Table 2.15 relates the 23 soil groups recognized to suborders or other taxa in the US Soil Taxonomy and to soil units in the FAO system. Because of divergencies resulting from the use of different differentiating criteria, or of the same or similar criteria at different categorical levels, only the lithosol group has an exact counterpart in either of the globally applicable schemes. Generally, therefore, it is only possible to indicate the class or classes to which most of the soils in each group belong. Specific relationships and divergencies are treated in Chapters 4–9.

2.9 Schematic Soil Map

The accompanying schematic soil map (Figure 2.5) is derived from the 1:1,000,000 map of the EEC countries with minor modifications. It shows the generalised distribution of 15 broad soil associations defined in terms of dominant soil groups or subgroups and in one case by thermal regime.

3
Methods and Definitions

3.1 Introduction

The salient properties, distribution and agronomic characteristics of the main kinds of soil identified in the British Isles are described in the following six chapters. These generalized accounts are supplemented by data on 187 representative soil profiles, 151 from England and Wales, 22 from Scotland and 14 from Ireland, which were recorded in the course of soil surveys or other pedological investigations from the 1950s onwards. In all but six cases, the field descriptions are accompanied by laboratory data on samples taken from visually recognizable horizons; some of the profiles were also sampled for micromorphological studies, results of which are referred to in the text and presented systematically in the original publications. This chapter gives details of the methods and terms used, and the analytical procedures employed, followed by definitions of the diagnostic horizons and other differentiating criteria which are used in defining major soil groups, groups and subgroups in the classification adopted.

3.2 Soil Profile Descriptions

Nearly all the profiles cited were described and sampled in freshly dug pits and the remainder in pre-existing exposures or in vertical cores of undisturbed soil 15 cm in diameter extracted by a Proline soil coring machine (Avery and Bascomb, 1982). Characteristics of the profile site are listed first and followed by descriptions of the depth and clarity, colour, composition, structure, onsistence and other properties of successive horizons, using the standard terms defined below. As the original descriptions were made at widely differing times by staff of different organizations, and different terms were used for some properties, they have been edited to conform with the prescribed terminology as closely as possible. Those issuing from the Soil Survey of England and Wales since 1974 were based on the Soil Survey Field Handbook (Hodgson, 1976) and are generally more detailed and comprehensive than the others. The methods and terms used in Scotland and Ireland are summarized in the survey publications from which the profiles are derived; they are based largely on the USDA Soil Survey Manual (1951), as were those used in England and Wales before 1974 (Soil Survey of Great Britain, 1960).

Site Description

The site is described by its locality, national grid reference (UK only), climatic regime (p. 18), altitude, slope in degrees, land use and vegetation.

Depth and Clarity of Horizons

The depths of horizon boundaries are measured from the surface of the soil, excluding fresh litter (L layer). Boundaries are described as *smooth* if uniform in depth, *wavy* if undulating, and *irregular* if punctuated by pockets deeper than their width.

The clarity of a boundary is *sharp* if the transition zone is less than 0.5 cm thick, *abrupt* if between 0.5 and 2.5 cm, *clear* if between 2.5 and 6 cm, *gradual* if between 6 and 13 cm, or *diffuse* if it is more than 13 cm thick.

Colour

Horizon colours are determined by comparison with Munsell Soil Color Charts (Munsell Color Company Inc., Baltimore, Maryland 21218, USA) and designated accordingly. The colour recorded is normally that of a freshly broken fragment in the moist state; in some instances the colour of air-dry soil or of a moist sample rubbed between finger and thumb is also noted. In mottled horizons the chief colours are identified and the abundance and size of mottles described as follows:

Mottle abundance (percentage of the surface or matrix described)

Few < 2 per cent
Common 2–20 per cent
Many 20–40 per cent
Very many > 40 per cent

Mottle size

Very fine < 1 mm
Fine 1–2.5 mm
Medium 5–15 mm
Coarse > 15 mm

The same terms are used to describe the abundance and/or size of other features such as concretions or soft concentrations of pedogenic origin, for example of calcite, gypsum, or oxides of iron and/or manganese.

Organic Matter Content

Organic soil materials (Figure 3.1) are defined as having:

1. More than 12 per cent organic carbon (c. 20 per cent organic matter) if the mineral fraction contains no clay (< 2 μm); or
2. More than 18 per cent organic carbon (c. 30 per cent organic matter) if the mineral fraction contains 60 per cent or more clay; or
3. More than the proportionate organic carbon content if the clay content of the mineral fraction is intermediate.

As organic matter is much less dense than mineral particles, it constitutes most of the volume of even the least organic of these materials. The variable limits accord with the observation that a given proportion of organic matter modifies the physical properties of sandy material more than it does those of a clay.

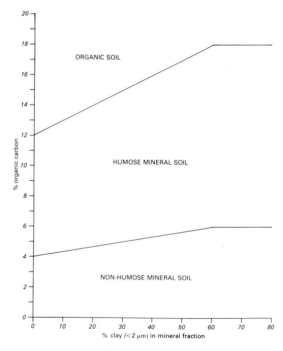

Figure 3.1: Limits of organic and humose soil materials.

Organic layers accumulated under wet conditions are described as *peat* and others as more or less decomposed litter or simply as organic material.

Mineral horizons, consisting of material with less organic matter, are described as *humose* if the organic carbon content exceeds 4–6 per cent (c. 8–12 per cent organic matter), again depending on clay content as shown in Figure 3.1.

Texture (Particle-size Class) of the Mineral Fraction

Mineral horizons are assigned to textural classes based on the particle-size distribution of the inorganic fraction smaller than 2 mm (the fine earth). Texture is assessed in the field by working a moistened sample between finger and thumb and checked in representative profiles by particle-size analysis (p. 93) in the laboratory. It is perhaps the most important single property of a mineral soil horizon. Together with organic-matter content and stoniness, it strongly influences structure and consistence, and the capacity to retain water and plant-available nutrients.

Two textural classifications based on different particle-size grades (Table 3.1) have been used in soil surveys in the British Isles. The first, which originated in the United States (Soil Survey Staff, 1951), has twelve basic classes (Figure 3.2a) defined by limiting percentages of sand (50 μm–2 mm), silt (2–50 μm) and clay (< 2 μm) sized particles. This has been used in Scotland and Ireland and was also employed in England and Wales prior to 1974, when it was replaced by an eleven-class system (Figure 3.2b) based on the particle-size grades of the British Standards Institution (1975), which places the limit between sand and silt fractions at 60 μm. In each of these schemes the basic sand, loamy sand and sandy loam classes are divided into subclasses according to the particle-size distribution within the sand fraction, though those based on the USDA size grades have been little used in the British Isles. Subclasses in the England and Wales system are defined as follows:

fine: more than two-thirds of the sand fraction (60 μm–2 mm) between 60 and 200 μm.
medium: less than two-thirds between 60 and 200 μm and less than one-third coarser than 600 μm.
coarse: more than one-third coarser than 600 μm.

Horizon textures according to both systems are included in the profile descriptions which follow. The England and Wales description is given first, followed by the USDA textural class in parentheses.

Table 3.1: Size limits of particle-size fractions

United States Department of Agriculture (Soil Survey Staff, 1951)		Soil Survey of England and Wales (Hodgson, 1976)	
Name of fraction	Diameter range (μm)	Name of fraction	Diameter range (μm)
sand: very coarse	1,000–2,000	sand: coarse	600–2,000
coarse	500–1,000	medium	200–600
medium	250–500	fine	60–200
fine	100–250	silt: coarse	20–60
very fine	50–100	medium	6–20
silt	2–50	fine	2–6
clay	< 2	clay	< 2

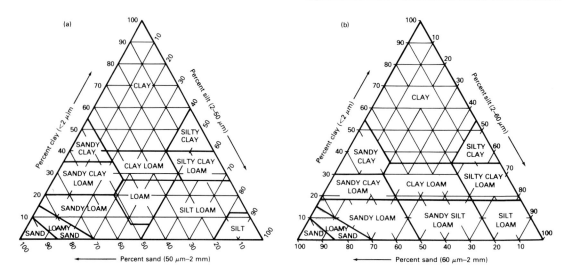

Figure 3.2: Textural (particle-size) classes according to (a) United States Department of Agriculture; (b) Soil Survey of England and Wales.

As a basis for differentiating soil series and classes in higher categories, soil materials containing less than 35 per cent stones by volume are assigned to a smaller number of particle-size groups and subgroups in the US (Soil Survey Staff, 1975) and England and Wales (Avery, 1980; Clayden and Hollis, 1984) classifications. For this purpose very fine sandy materials with few particles coarser than 100 μm are grouped with materials rich in silt, which they resemble in agronomically important properties such as water retention (Hall *et al.*, 1977) and erodibility (Wischmeier and Mannering, 1969). The particle-size groups and subgroups referred to in this and the following chapters are defined as follows:

Sandy Sands and loamy sands with more than 50 per cent of the fine earth coarser than 100 μm.
Loamy Other soil materials with less than 35 per cent clay in the fine earth.
  Coarse Loamy Less than 18 per cent clay and 15 per cent or more coarser than 100 μm.
  Fine loamy 18–35 per cent clay and 15 per cent or more coarser than 100 μm.
  Coarse silty Less than 18 per cent clay and less than 15 per cent coarser than 100 μm.
  Fine silty 18–35 per cent clay and less than 15 per cent coarser than 100 μm.
  (Coarse and fine silty subgroups are bracketed as *silty*)
Clayey More than 35 per cent clay in the fine earth.

Soil materials containing more than 35 per cent stones by volume and appreciable amounts of interstitial fine earth are described as *sandy-skeletal, loamy-skeletal* and *clayey-skeletal*.

Carbonates

The following terms are prefixed to the textural class names in descriptions of horizons that contain 1 per cent or more calcium carbonate or calcium-magnesium carbonate (dolomite) in the fine earth as determined by laboratory analysis.

	CaCO₃ equivalent (%)
slightly calcareous	1–5
moderately calcareous	5–10
very calcareous	10–40
extremely calcareous (carbonatic)	> 40

Where no analyses were made, field estimates based on degree of effervescence with acid (Hodgson, 1976, p. 56) are given, or the horizon is described simply as calcareous.

Stoniness

In the profiles from England and Wales, and in others on which appropriate information was recorded, proportions by volume of mineral fragments coarser than 2 mm (stones) are described as follows:

stoneless (rare stones)	< 1 per cent
very slightly stony (few stones)	1–5 per cent
slightly stony (common stones)	6–15 per cent
moderately stony (many stones)	16–35 per cent
very stony (abundant stones)	36–70 per cent
extremely stony	> 70 per cent

In other cases less precise terms such as *stony* or *gravelly* are used. Extremely stony horizons can consist of water-laid gravel or angular rock fragments with little interstitial fine earth.

Size, shape and kind of stone are also described in most instances. Size classes by diameter (cm) are:

very small (grit)	0.2–0.6	large	6–20
small (gravel)	0.6–2	very large	20–60
medium (gravel)	2–6	boulders	> 60

Soil-water State

This term refers to the moisture condition of a horizon when the profile is described. The three water states recognized are defined as follows and are identifiable in the field with varying degrees of precision:

dry – containing no water held at suctions less than 15 bars (1,500 kPa).

moist – containing water held at suctions between 0.01 bar (1 kPa) and 15 bars (1,500 kPa).

wet – containing water held at a suction of 0.01 bar (1 kPa) or less; wet soil is normally below or not far above a water table and usually contains visible water films, but these may be confined to widely spaced fissures or channels if the horizon is otherwise very slowly permeable.

Where no indication of soil-water state is given, it can generally be assumed that the horizon was moist at the time of sampling.

Soil Structure

In its broadest sense, the term *soil structure* refers to the physical organization of soil materials as expressed by the arrangement of solid particles and voids. Field descriptions of soil structure place emphasis on the degree of development, size and type (shape and arrangement) of naturally formed aggregates (peds) that are separated from each other by voids or surfaces of weakness and persist through cycles of wetting and drying. The occurrence and nature of *macropores* such as holes, tubes and burrows within peds or in apedal horizons are also noted.

The following terms are used to describe degree of ped development:

Apedal (structureless): no observable aggregation; *massive* if coherent; *single grain* if incoherent.

Weak: peds barely observable in place and incompletely separated by voids or natural surfaces; in some cases the soil breaks easily when gently disturbed into a mixture of entire or broken peds and apedal fragments; in others the ped surfaces present are strongly *adherent* and separate with difficulty.

Moderate: peds evident but not distinct in undisturbed soil; when disturbed the soil breaks into a mixture of entire and broken peds with comparatively little unaggregated material.

Strong: peds distinct in undisturbed soil, adhere weakly to one another and mostly remain entire when the soil is gently disturbed.

Terms describing the type and size of peds are defined in Table 3.2 and Figure 3.3 depicts the types recognized. Impersistent aggregates (clods or fragments) formed by cultivation in ploughed surface horizons are not always readily distinguishable from peds and are described using similar terms.

Consistence

The consistence of soil material is described by terms that refer to its resistance to breaking or deformation when force is applied (strength) and the way in which it fails. As both characteristics are closely related to soil-water state, a description of consistence has little meaning unless the soil-water state is specified or is implied by the term used.

Soil strength is assessed in either the moist (at or rather below field capacity) or the dry state by the resistance to crushing of a specimen cube about 3 cm across and described by terms derived from the USDA Soil Survey Manual (Soil Survey Staff, 1951). These are defined in Table 3.3, together with related terms currently used by the Soil Survey of England and Wales (Hodgson, 1976). In most of the profile descriptions which follow, only assessments of strength in the moist state are given and determinations of stickiness and plasticity applicable to wet cohesive soil are omitted.

Whether a soil sample breaks or deforms in response to an applied force is generally determined by its water state, structure and co-

Table 3.2: Types and sizes of peds and fragments

Size	Type (shape and arrangement)				
	Platelike with one dimension (the vertical) limited and much less than the other two; faces mostly horizontal	Prismlike with two dimensions (the horizontal) limited and considerably less than the vertical; vertical faces well defined and vertices angular	Blocklike (polyhedral or spheroidal) with three dimensions of the same order of magnitude		
			Blocks or polyhedrons with plane or curved surfaces that are casts of the moulds formed by the faces of surrounding peds		Spheroids or polyhedrons with plane or curved surfaces having slight or no accommodation with the faces of surrounding peds
			Faces flattened; most vertices angular	Mixed rounded and flattened faces with many rounded vertices	
	Platy	*Prismatic*	*Angular blocky*	*Subangular blocky*	*Granular*[a]
Fine	<2 mm	< 20 mm	< 10 mm	< 10 mm	<2 mm
Medium	2–5 mm	20–50 mm	10–20 mm	10–20 mm	2–5 mm
Coarse	5–10 mm	50–100 mm	20–50 mm	20–50 mm	5–10 mm
Very coarse	>10 mm	> 100 mm	> 50 mm	> 50 mm	>10 mm

[a] Including subrounded or irregular porous aggregates originally described as crumbs (Soil Survey Staff, 1951).

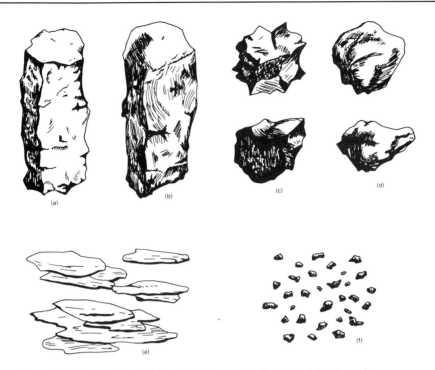

Figure 3.3: Types of ped (after Soil Survey Staff, 1951). (a) prismatic; (b) columnar; (c) angular blocky; (d) subangular blocky; (e) platy; (f) granular (crumb).

hesiveness. The following additional terms are used to denote particular types of failure.

Brittle – a very moist or wet specimen fractures abruptly rather than deforming gradually when pressure is applied; lower horizons that are compact and brittle have been described as *indurated* (weakly, moderately or strongly indurated) by the Soil Survey of Scotland (e.g. Glentworth and Muir, 1963); they are usually cemented to some extent (see below).

Smeary – a moist sample exhibits the properties of a thixotropic gel in becoming fluid when pressure is applied, causing it to feel smeary or slippery.

Fluid – the soil in place is wet and tends to flow between the fingers when a handful is squeezed; it is described as *slightly fluid* if most of it remains in the hand after exerting full pressure, or as *very fluid* if it flows easily and little or no residue is left.

Cementation or Induration

A soil horizon is described as very weakly, weakly or strongly cemented (Table 3.3) if it is firm or stronger and brittle when moist and an air-dry 3 cm specimen does not slake after immersion in water for one hour. Cementation is normally caused by bonding agents other than layer-lattice silicate clays, for example calcium carbonate, silica, iron or manganese oxides, organic matter in combination with aluminium and/or iron (ortstein), or poorly ordered aluminous materials.

Coats or Cutans

These are modifications in the composition

and/or arrangement of the soil constituents associated with natural interfaces in soils. They can result from absolute concentration of particular constituents, relative concentration of coarser constituents by removal of fine material, or re-arrangement of primary particles *in situ*. Those observable in the field and recorded in profile descriptions include clay coats (argillans), sand or silt coats (skeletans), sesquioxidic coats (sesquans, mangans or ferrans), organic coats (organans), carbonate coats (calcans) and stress-oriented coats (stress cutans). The latter reflect *in situ* modification and differ from the adjacent matrix in density and in the degree of orientation of clay-size particles. They normally result from pressures caused by swelling and have smooth polished faces which may be grooved or striated where lateral movement (slickensiding) has occurred. More precise information on the nature and distribution of coats is obtained by microscopic examination of thin sections (p. 90).

Concentrations Within the Soil

These are concentrations of particular constituents, including calcium carbonate, gypsum, and oxides of iron and manganese, which occur as discrete bodies embedded within the soil. They generally result from pedogenic processes but may also be inherited from the parent rock. Where the occurrence of such bodies has been noted in profile descrip-

Table 3.3: Soil strength classes

Condition of failure of 3 cm cube or ped	Consistence (USDA)		Strength class (Hodgson, 1976)	Cemented material
	Air-dry	Moist (field capacity)		
No specimen can be obtained (soil material incoherent)	Loose	Loose	Loose	
Specimen crushes or breaks when very slight force applied by thumb and forefinger	Soft	Very friable	Very weak	
Specimen crushes or breaks when slight force applied by thumb and forefinger	Slightly hard	Friable	Moderately weak	
Specimen crushes or breaks when moderate force applied by thumb and forefinger	Slightly hard	Firm	Moderately firm	Very weakly cemented
Specimen crushes or breaks when strong force applied by thumb and forefinger	Hard	Very firm	Very firm	Very weakly cemented
Specimen cannot be crushed or broken by thumb and forefinger but can be crushed by squeezing between the hands	Very hard	Extremely firm	Moderately strong	Weakly cemented
Specimen cannot be crushed or broken by hand but can be crushed or broken underfoot on a hard flat surface	Extremely hard	Extremely firm	Very strong	Weakly cemented
Specimen cannot be crushed or broken underfoot			Rigid	Strongly cemented

tions, their nature, composition, size and shape are recorded. Three main kinds are distinguished, as follows:

Crystals – macroscopic crystals, particularly of gypsum.

Nodules and concretions – more or less rounded bodies, normally cemented, that are sufficiently coherent to be separated from the matrix; concretions differ from nodules in having a concentric internal structure.

Soft concentrations – contrast with the surrounding soil in colour and structure but cannot be readily separated as discrete bodies, though some have clearly defined boundaries.

Roots

In the profiles from England and Wales the frequency and size of roots in each horizon are generally described in terms of the classes defined in Table 3.4. In those from Scotland and Ireland, where different terms have been used, the original information on root distribution is reproduced. Medium and coarse roots are further categorized as woody, fleshy or rhizomatous.

Table 3.4: Frequency of roots per 100 cm² of profile face (after Hodgson, 1976)

Frequency class	Fine roots (<2 mm)	Medium (2–5 mm) and coarse (>5 mm) roots
Rare or none	<1	<1
Few	1–10	1 or 2
Common	10–25	2–5
Many	25–200	>5
Abundant	>200	

Plant Remains and Degree of Decomposition of Organic Materials

In peaty horizons the nature of any visually identifiable plant remains is noted and the organic material is described as fibrous, semi-fibrous or amorphous (humified or sedimentary) peat according to its degree of decomposition. This is determined in water-saturated (undrained) peats by the von Post method (Table 3.5) or otherwise from estimates of rubbed fibre content and degree of humification as assessed by solubility in sodium pyrophosphate solution. Fibres are defined as fragments of plant tissue large enough to be retained on a 212 μm sieve, excluding wood fragments larger than 2 cm that cannot be crushed in the hand. Rubbed fibre is the fraction that remains after rubbing a sample about ten times between thumb and forefinger. Degree of humification is assessed by mixing a sample with sodium pyrophosphate solution, inserting a strip of white filter paper, and noting the Munsell value and chroma of the colour resulting from absorbance of the extract (Hodgson, 1976, p. 94). The difference between value and chroma is termed the pyrophosphate index (Canada Soil Survey Committee, 1978).

Fibrous peat consists largely of well preserved plant remains which are relatively durable and readily identifiable in the field (H 1–3 on the von Post scale); its rubbed fibre content is at least 75 per cent, or at least 40 per cent if the pyrophosphate index is 5 or more. It is usually reddish or yellowish and the colour changes markedly when firmly pressed or on drying.

Semi-fibrous peat consists mainly of partially decomposed plant remains (H 4–6 on the von Post scale) or is a mixture of little decomposed and strongly decomposed materials, that does not meet the requirements of either fibrous or amorphous peat. It is commonly dark brown or reddish brown when wet and darkens on exposure.

Amorphous peats contain few or no visually identifiable plant remains other than resistant woody fragments (H 7–10 on the von Post scale if undrained) and less than 15 per cent

Table 3.5: Modified version of the von Post scale for assessing the degree of decomposition of peat (after von Post, 1924)

In this field test a sample of wet peat is squeezed in the closed hand and the colour of the liquid expressed, the proportion extruded between the fingers, and the nature of the plant residues are observed.

H1 — undecomposed; plant structure unaltered; yields only clear colourless water.

H2 — almost undecomposed; plant structure distinct; yields only yellowish brown water.

H3 — very weakly decomposed; plant structure distinct; yields distinctly turbid water but no peat escapes between fingers; residue not pasty.

H4 — weakly decomposed; plant structure distinct; yields strongly turbid water but no peat escapes between fingers; residue somewhat pasty.

H5 — moderately decomposed; plant structure recognizable though rather indistinct; yields muddy water and some peat escapes between fingers; residue very pasty.

H6 — strongly decomposed; plant structure rather indistinct but clearer in the squeezed residue than in the undisturbed peat; about one-third escapes between fingers; residue strongly pasty.

H7 — strongly decomposed; plant structure indistinct and few remains identifiable; about half escapes between fingers.

H8 — very strongly decomposed; plant structure very indistinct; about two-thirds escapes between fingers; residue almost entirely resistant remains such as root fibres and wood.

H9 — almost completely decomposed; plant structure almost unrecognizable; nearly all escapes between fingers as a uniform paste.

H10 — completely decomposed; plant structure unrecognizable; all escapes between fingers.

rubbed fibre. They include strongly decomposed (humified) peat and sedimentary peat (e.g. organic lake mud, coprogenous earth or gyttja), both of which may be rich in mineral matter.

Humified peat has a pyrophosphate index of 3 or less; it is usually black and the colour changes little on drying.

Sedimentary peat is a fresh-water sediment, often laminated, that contains residues of underwater or floating aquatic plants modified by aquatic animals; where it remains water saturated and is little humified, it is typically greyish or olive brown in colour and gives a pyrophosphate index greater than 3.

3.3 Micromorphological Data

Where reference is made to micromorphological features of soil horizons, the data were obtained by microscopic examination of representative thin sections prepared from specially collected, undisturbed samples after drying and impregnation with a polyester resin (Bascomb and Bullock, 1982). The nature, size and spatial distribution of soil components identifiable in thin sections have been described and/or quantified using the concepts and terms introduced by Brewer (1964) with minor modifications (Table 3.6; Bullock, 1982; Bullock and Murphy, 1974, 1979). This information is used in conjunction with analytical data to supplement field descriptions and aid identification of pedogenic horizons. Its value lies in the light it casts on microstructure and the arrangement of particular constituents within the soil fabric, which is often irregular. Although the individual particles composing the fine material (plasma) cannot be resolved, its composition can be inferred within limits and interpretation of its birefringence and other optical properties in plane polarized light and between crossed nicols enables genetically significant differences in the degree of preferred

Table 3.6: Micromorphological components of soil

BASIC COMPONENTS OF THE SOIL FABRIC

SKELETON GRAINS
Mineral grains and resistant siliceous and organic bodies larger than about 10 μm.

PLASMA
Fine material consisting mostly of clay and silt-size particles that cannot be resolved under the microscope; it includes clay minerals, sesquioxides and organic matter capable of or having been moved, reorganized or concentrated by pedologic processes.

Plasma concentrations: units characterized by concentration of some fraction of the plasma.

Plasma separations: units characterized by a significant change in the arrangement of plasma constituents, particularly the orientation of clay-size particles.

VOIDS
Spaces between soil particles or aggregates.

Simple packing voids: resulting from random packing of single grains.

Compound packing voids: resulting from packing of compound units (e.g. peds) which do not accommodate each other.

Vughs: relatively large irregular voids, not normally interconnected.

Vesicles: unconnected voids whose walls consist of smooth, simple curves.

Channels: voids of cylindrical form.

Planes: voids that are planar according to the ratios of their principal axes.

Table 3.6: Contd.

PEDOLOGICAL FEATURES
Recognizable units, including features inherited from the parent material, that are distinguished from the enclosing soil material for any reason such as origin (deposition as an entity), differences in concentration of some fraction of the plasma, or differences in arrangement of the constituents.

CUTANS
Modifications in composition or arrangement of constituents at natural surfaces (e.g. of peds or skeleton grains) in soil materials.

Argillans, Ferrans, Organans, etc.: characterized by absolute concentrations of particular plasma constituents.

Skeletans: characterized by relative or absolute concentration of skeleton grains

Matrans: characterized by ill-sorted cutanic material resembling the matrix of overlying horizons.

Stress cutans: characterized by *in situ* modification of the plasma attributable to internal stress.

PEDOTUBULES
Tubular features, normally originating by infilling of faunal or root channels, that consist of soil material (skeleton grains or plasma plus skeleton grains) and have relatively sharp external boundaries.

Table 3.6: Contd.

GLAEBULES	
Three-dimensional features, usually prolate to equant in shape, that are embedded in the soil matrix.	*Nodules:* concentrations of particular constituents (e.g. iron oxides, calcium carbonate) with regular shapes, abrupt boundaries and undifferentiated internal fabric. *Concretions:* similar to nodules but with a generally concentric fabric. *Segregations:* irregularly shaped concentrations of particular constituents, often unevenly distributed, with abrupt to diffuse boundaries. *Papules:* sharply defined concentrations of clay minerals showing strongly preferred orientation (continuous or lamellar fabric).
MISCELLANEOUS PEDOLOGICAL FEATURES	
	Crystallaria: single crystals or arrangements of crystals of plasma constituents (e.g. gypsum) that do not enclose other soil material. *Faecal pellets:* regularly shaped bodies consisting of organic matter, or organic matter intimately mixed with mineral particles, that evidently originated as animal droppings.

orientation of the silicate clay particles to be determined. This in turn aids the recognition of clay coats (argillans) and intrapedal concentrations resulting from movement and redeposition (illuviation) of clay-size material during the course of soil formation, as illustrated in Figure 3.4.

Microscopically, illuviation argillans are characterized by strong optical continuity, strongly preferred orientation, a sharp boundary with adjacent material and often a laminated appearance. The intrapedal concentrations, including papules (Table 3.6), have similar properties but are unrelated to voids or grain surfaces. Many if not most of them appear to have originated as argillans which have been disrupted and incorporated into the matrix by soil fauna, shrinking and swelling or other pedoturbation processes, or which totally infilled voids.

3.4 Analytical Data

Samples collected from soil horizons include (1) disturbed (fragmental or loose) samples for particle-size, chemical and mineralogical analyses; and (2) undisturbed samples for measurements of density and water-release characteristics. Disturbed samples are normally air-dried, screened through a 2 mm sieve, and subsampled for individual laboratory determinations. Data on particle-size distribution (mineral horizons only), organic matter content (organic carbon and/or loss on ignition) and pH are given for most of the horizons sampled. Other determinations quoted include percentages of total nitrogen, calcium carbonate (where present), iron, aluminium and carbon removed by specified extractants, and sulphur, in the oven-dry soil < 2 mm; exchangeable cations and cation-exchange capacity; electrical conductivity of saturation extracts; analyses of clay (< 2 μm) separates, and physical measurements on undisturbed samples.

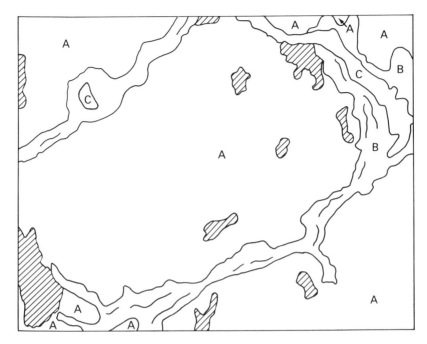

Figure 3.4: Schematic representation of a soil thin section (Findlay, 1965, Plate I) showing deposits of translocated clay in the B horizon of a luvic brown soil (Nordrach series) in north Somerset. A. Fine sand and silt-sized quartz in a brown clay matrix (skeleton grains and plasma with few voids); the shaded areas are manganiferous segregations. B. Reddish brown, strongly oriented, iron-stained clay infilling pores and fissures (ferri-argillans); in places there are indications of layering parallel to fissure walls. C. Voids where clay has incompletely filled the fissures.

The analytical procedures employed and the uses made of the results are briefly reviewed below. Further details of laboratory methods are given in the original publications and those used by the Soil Survey of England and Wales since 1974 are described by Avery and Bascomb (1982).

Particle-size Analysis

Unless otherwise stated, particle-size distribution was determined by the pipette method following removal of organic matter with hydrogen peroxide, dispersion by overnight shaking with alkaline sodium hexametaphosphate solution and separation of the sand fractions by wet sieving. Calcareous samples from profiles 32, 35, 65, 153, 158 and 159 were pretreated with hydrochloric acid to remove carbonates before determination of the size fractions. Where amounts of fine clay (< 0.2 μm) are also reported, they were measured by centrifugation of the dispersate (Avery and Bascomb, 1982). All percentages are based on oven-dry peroxidized soil < 2 mm.

On other profiles, including all those from Scotland, clay (< 2 μm) and silt (2–50 μm) contents were determined by a modification of the hydrometer method (Bouyoucos, 1927, 1951), using sodium hydroxide or sodium hexametaphosphate as the dispersant; sand contents were estimated by difference and no peroxide pretreatment was given. Hence, although an allowance is made for organic matter based on ignition loss, results for samples containing significant amounts are less reliable than those obtained by the standard pipette method because dispersion is unlikely to be complete.

Particle-size analyses are used to verify field estimates of texture, identify lithological discontinuities in profiles, and help elucidate soil processes such as parent-material weathering and clay translocation. They have to be interpreted with caution if the soil contains partially weathered fragments of soft rock such as chalk or sandstone, or aggregates of fine constituents which resist dispersion by conventional procedures.

Organic Matter and Nitrogen

Organic carbon, loss on ignition, or both, were determined in one or more horizons of all the profiles sampled for analysis. Except where otherwise stated, the organic carbon contents were obtained by wet oxidation with potassium dichromate (Walkley and Black, 1934; Tinsley, 1950). The ratio of total organic matter to organic carbon varies with degree of humification but is generally about 1.9 in well-humified mineral surface horizons (Allison, 1965; Broadbent, 1965). An estimate of organic matter content can also be obtained by subtracting the amount of combined water eliminated from the clay fraction at temperatures above 110°C (about 10 per cent of the clay percentage) from the loss on ignition value, which is normally corrected for loss of carbon dioxide from carbonates, if present.

Where the total nitrogen content is also reported, it was determined by Kjeldahl digestion followed by estimation of the ammonia liberated. The carbon/nitrogen ratio (C/N) gives a useful indication of the degree of decomposition of the organic matter. Well-decomposed humus in soils of temperate regions has a C/N ratio around 10 whereas less decomposed organic layers give values ranging up to 50 or more.

Calcium Carbonate

This was determined on < 2 mm samples of calcareous horizons, and in some profiles also on < 2 μm clay separates (active calcium carbonate), using a calcimeter (Bascomb, 1961) which measures the carbon dioxide evolved on treatment with acid. As magnesium carbonate may also be present, the results are reported as 'calcium carbonate equivalent'.

Extractable Iron, Aluminium and Carbon

Several different chemical extractants, including potassium or sodium pyrophosphate at pH 10, acid ammonium oxalate at pH 3, and reagents based on sodium dithionite, have been used in recent years to assess the nature and amounts of iron and aluminium compounds other than layer-lattice clays in soil horizons. Although the substances removed are in no case sharply defined, it seems that most of the 'sesquioxides' extracted by alkaline pyrophosphate exist in the soil as metal-organic complexes or poorly crystallized oxyhydroxides closely associated with peptizable organic matter (Bascomb, 1968; McKeague et al., 1971; Jeanroy and Guillet, 1981; Adams et al., 1987). In addition to these constituents, acid oxalate dissolves allophanic material (allophane and/or imogolite) in which aluminium is linked to silica (Parfitt and Henmi, 1982; Farmer et al., 1983) and removes some of the iron in maghemite (γ Fe_2O_3) and lepidocrocite (γ FeOOH) but little from well-crystallized goethite (α FeOOH) or haematite (α Fe_2O_3) (Schwertmann, 1964, 1973; McKeague and Day, 1966). Nearly complete dissolution of iron oxides and oxyhydroxides, which can be pedogenic or lithogenic in origin, is achieved by extraction with dithionite, either under acid conditions (Deb, 1950) or in admixture with sodium citrate at a near neutral pH (Mehra and Jackson, 1960; Holmgren,

1967), which reduces loss of iron from layer silicates such as chlorite.

Of the three extractants named, pyrophosphate and/or dithionite have been widely used in the British Isles and acid oxalate to a lesser extent, mainly in special investigations (e.g. Loveland and Bullock, 1976; Farmer et al., 1983; Adams et al., 1987). Amounts of iron, aluminium and carbon extracted by overnight shaking with cold 0.1 M potassium pyrophosphate solution (Bascomb, 1968) were measured in horizon samples from 40 of the profiles which follow. Estimates of 'total free iron' in these and other profiles were obtained by one of the following methods, which generally give similar results.

1. Extraction by sodium dithionite buffered at pH 3.8 with ammonium acetate (Avery and Bascomb, 1982); reported as dithionite-extractable Fe.
2. Extraction by citrate-dithionite buffered at pH 7 with sodium bicarbonate (Mehra and Jackson, 1960) or with sodium citrate only (Holmgren, 1967); reported as citrate-dithionite extractable Fe.
3. Extraction by potassium pyrophosphate followed by extraction of the residue with dithionite as in method 1 or (on Scottish samples) dithionite buffered with sodium bicarbonate at pH 7 as described by Mitchell et al. (1971); reported as total extractable Fe.

Sulphur

Total sulphur (sulphate-S+oxidizable S) contents of samples from four of the profiles from England and Wales were obtained by oxidizing with hydrogen peroxide followed by extraction with hydrochloric acid as described by Bloomfield (1972). The primary purpose of this determination is to aid identification of actual or potential acid-sulphate soils (p. 111).

Cation Exchange Data

Data on total exchangeable bases (TEB) and cation-exchange capacity (CEC) expressed as milli-equivalents (me) per 100 g oven-dry soil are given for profiles of non-calcareous soils from Scotland and Ireland and some of those from England and Wales, together with the corresponding percentage base saturation values. The TEB figures were obtained by summing amounts of exchangeable calcium, magnesium, potassium and sodium determined in a neutral normal ammonium acetate leachate, or (in some Irish samples) by direct determination of the extracted bases using the titration procedure of Metson (1956). Unless otherwise stated, cation-exchange capacity was measured as follows:

1. In profiles from Scotland by summation of the exchangeable bases (Ca, Mg, K, Na) and exchange acidity ('exchangeable hydrogen') determined by potentiometric titration of a neutral normal barium acetate leachate to pH 7 (Parker, 1929).
2. In profiles from Ireland by leaching with barium chloride solution buffered at pH 8, displacing the adsorbed barium by a second leaching with calcium chloride and determining barium in the leachate (Mehlich, 1948).
3. In profiles from England and Wales by summation of the exchangeable bases and exchange acidity (at pH 11 approx.) determined as described by Bascomb (1982). In this method, originated by Mados (1943), 10 g of soil are shaken with 100 ml 0.2 N ammonium hydroxide and formaldehyde is added, followed by barium chloride solution. After centrifugation the supernatant liquid is titrated with 0.1 N ammonium hydroxide, adding 2 ml in excess, and then back titrated with 0.1 N hydrochloric acid.

Although cation-exchange data are significant in relation to both soil fertility and soil genesis, the CEC and percentage base saturation values obtained are more or less strongly influenced by the pH and composition of the extracting or displacing solutions used. The different methods employed therefore yield different results, particularly for 'soils with variable charge' in which the net negative charge responsible for retention of exchangeable cations is pH-dependent.

Soil Reaction (pH)

This was determined on samples from England and Wales in 1:2.5 suspensions of soil in water and in 0.01 M calcium chloride solution. In Scotland and Ireland routine measurements have been made only in water. The lower values obtained in calcium chloride suspensions are less dependent on the soil/solution ratio and less affected by seasonal fluctuations of electrolyte concentration in the soil as it exists in the field. They are also presumably closer to the pH of the solution immediately adjacent to plant roots and particle surfaces (Russell, 1973). The following terms are used to describe soil reaction:

	pH (in water)
strongly acid	< 4.5
moderately acid	4.5–5.5
slightly acid	5.6–6.5
neutral	6.6–7.5
alkaline	> 7.5

Electrical Conductivity

The electrical conductivity (in siemens per metre, S/m) of the extract obtained from a paste of soil and water, the so-called saturation extract, is related to the amount of salts more soluble than gypsum and is used as a standard measure of salinity, saline soil being defined as having a conductivity of the saturation extract greater than 0.4 S/m at 25°C. British soils meeting this requirement are mainly confined to coastal or estuarine marshes affected by saline groundwater and it is only on samples from such localities that conductivity measurements (Avery and Bascomb, 1982) have been made.

Analyses of Clay Separates

The data on 12 clayey profiles from England and Wales include determinations of cation-exchange capacity and non-exchangeable potassium content (% K_2O) on clay (< 2 μm) separates pretreated with hydrogen peroxide to remove organic matter (Bullock and Loveland, 1982). These are used in conjunction with semi-quantitative estimates of clay minerals by X-ray diffractometry to assess the kinds and proportionate amounts of layer-lattice minerals present. Following Avery and Bullock (1977), six broad clay mineralogical classes have been recognized, as follows:

Kaolinitic : clays with < 3.5% K_2O, CEC < 25 me/100 g and more kaolin than any other identifiable mineral group.
Smectitic : other clays with CEC ≥ 45 me/100 g and more smectite than vermiculite.
Vermiculitic : other clays with CEC ≥ 45 me/100 g and more vermiculite than smectite.
Micaceous : other clays with ≥ 3.5 per cent K_2O
Chloritic : other clays with more chlorite than any other identifiable mineral group.
Mixed : other clays.

Density, Water Retention and Porosity (Mineral Soils)

Data on the bulk density, packing density, retained water capacity, available water capacity and air capacity of selected horizons are given for 37 profiles of mineral soils from England and Wales and on density only for two. The values are derived from measurements made on triplicate core samples of undisturbed soil as described by Hall *et al.* (1977). The water retained by each sample is measured at suctions of 0.05, 0.1 and 0.4 bars on sand and kaolin

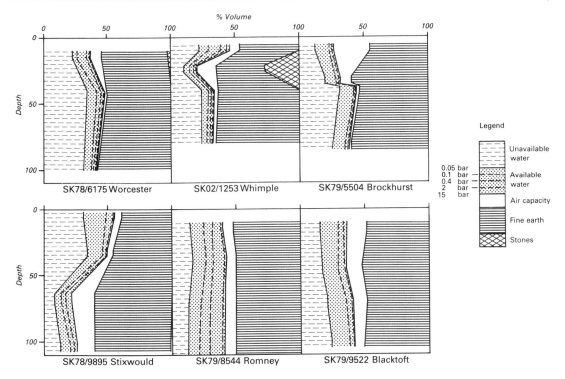

Figure 3.5: Diagrammatic representation of water-release characteristics in representative profiles of six soil series belonging to five different subgroups in north Nottinghamshire (from Reeve and Thomasson, 1981). Worcester *peloluvic brown soil*; Whimple *stagnogleyic luvic brown soil*; Brockhurst *stagnoluvic gley soil*; Stixwould *pelo-alluvial gley soil*; Romney and Blacktoft *gleyic-calcaric alluvial brown soils*.

tension tables and at 2 and 15 bar in pressure-membrane cells. The total oven-dry mass of the original core and the proportion of stones (> 2 mm), if present, are then determined. Particle density is measured for all topsoils and for subsurface horizons with appreciable organic matter or unusual mineralogy; for most subsoils a value of 2.65 g cm^{-3} is assumed.

The results of water retention and air capacity measurements are normally expressed on a volumetric basis and are conveniently represented in graphic form, as illustrated in Figure 3.5. Classes used to describe soil physical characteristics are given in Table 3.7.

Bulk density (g cm^{-3}), reported on a stone-free basis, is the apparent density of the soil as it exists in the field at the time of sampling. This is normally done in winter or spring when the water state is near field capacity.

Packing density (g cm^{-3}) is a derivative of fine-earth bulk density calculated by adding 0.009 × gravimetric clay percentage to the initial figure. Originally called *Lagerungsdichte* (Renger, 1971), it takes account of the proportion of pores finer than 0.2 μm (i.e. those holding water at suctions greater than 15 bars) associated with clay particles; it is both easier to estimate in the field (Hodgson, 1976) and more closely related to macroporosity and hydraulic conductivity than the unmodified bulk density values.

Retained water capacity (% vol.) is the volumetric water content at 0.05 bar suction and is approximately equivalent to the volume of pores < 60 μm. In most soils in lowland Britain it also approximates to the minimum water content during the winter period when rainfall exceeds evapotranspiration and, as such, to field capacity (Webster and Beckett, 1972). In surface soils it is governed mainly by clay and organic matter contents (Hollis et al., 1977) and in subsurface horizons by clay content and bulk density. The relation between retained water capacity and the plastic limit is a major factor influencing the workability of soils and the risk of poaching by stock under given climatic conditions.

Available water capacity (% vol.) is the volume of water retained at suctions between 0.05 and 15 bars (wilting point for most plants). Amounts of 'easily available water' are calculated similarly, using the water retained at 2 bar suction (Figure 3.6). In surface horizons available water is strongly correlated with organic carbon content; in subsurface horizons it generally increases with decreasing density and with increasing proportions of coarse silt and very fine sand (60–100 μm) particles. Available and easily available water contents can be summed to various depths according to crop and used in conjunction with meteorologically based estimates of potential soil-water deficit (Chapter 1, p. 16) to assess the relative droughtiness of soils (Hall et al., 1977; Thomasson, 1979).

Air capacity (% vol.) is the volumetric percentage of air-filled macropores (approximately > 60 μm) at 0.05 bar suction, which corresponds approximately to the field capacity condition when excess moisture has drained away under the influence of gravity. It decreases with increases in clay content and packing density and ranges accordingly from near zero in dense clayey subsoils to 30 per cent or more in loose sand. Measured or estimated air capacity values are used to define porosity classes (Table 3.7). In general, the more porous the soil the greater is its permeability. Slightly porous horizons with high packing density are relatively impervious and impede downward or horizontal redistribution of water. Very slightly porous horizons (air capacity less than 5 per cent) can be expected to have horizontal hydraulic conductivities (K sat.) less than 10 cm/day (Thomasson and Youngs, 1975).

Table 3.7: Class limits for soil physical data

Class	Retained water (% vol.)	Available water (% vol.)	Air capacity (porosity) (% vol.)
Very small	0–10	0–5	0–5 (very slightly porous)
Small	10–20	5–10	5–10 (slightly porous)
Moderately small	20–30	10–15	10–15 (moderately porous)
Moderately large	30–40	15–20	15–20 (very porous)
Large	40–50	20–25	20–25 (extremely porous)
Very large	>50	>25	>25

Class	Bulk density (g cm^{-3})	Packing density (g cm^{-3})
Very low	<20	
Low	0.20–0.80	<1.40
Medium	0.80–1.30	1.40–1.75
High	1.30–1.80	<1.75
Very high	>1.80	

Physical Analyses of Organic (Peat) Soils

Data on bulk density (g cm^{-3}), saturated water content (per cent oven-dry soil) and rubbed fibre content (per cent ash-free dry matter or by volume) of organic horizons are given for six profiles of peat soils, four from England and Wales using the methods described by Bascomb *et al.* (1977) and two from Ireland (Hammond, 1981). All three properties are related to degree of decomposition as assessed in the field (p. 89), saturated water and rubbed fibre contents being markedly greater and bulk densities significantly lower in fibrous peats than in those that are strongly decomposed.

3.5 Horizon Notation

In order to compare and classify soil profiles, the various kinds of horizon comprising them are defined and designated using the letter and number notation explained below. This is based on properties that differentiate horizons from those above or below, or from a presumed parent material, the primary aim being to group those that regularly occupy similar positions in profiles and reflect the same kind or degree of alteration by pedogenic processes. As in the closely comparable system currently employed by the Soil Survey of Scotland (1984), however, the term *horizon* is applied to all morphologically distinct layers, with or without characteristics attributable to soil formation.

Major horizons are denoted by capital letters. Those consisting of organic material (p.x) are designated L, F, H or O; unconsolidated mineral horizons A, E, B or C; and hard bedrock R. Intensely gleyed (reduced) horizons are indicated by the additional symbol G, other specific differentiating characteristics by one or more lower case suffixes (e.g. Ah, Btg) and lithological discontinuities (see below) by numerical prefixes. Transitional or mixed kinds of horizon, where recognized, are denoted by two juxtaposed capitals (e.g. AB, BCg) or by a bracketed suffix, for example Bt(g), if the horizon as a whole is intermediate in character; or by two capitals separated by an oblique stroke (e.g. Eb/Bt; Bt/Cu) if it contains discrete parts with properties of different major horizons. Subhorizons qualifying for the same letter notation are denoted by numerals placed after the letter designation in vertical sequence (e.g. Ah1, Ah2, etc.). Although soil horizons can usually be identified in the field, laboratory determinations are necessary before some can be designated positively.

A *lithologic discontinuity* is a significant vertical change in parent material, normally evidenced by a difference in particle-size distribution (including stone content) or mineralogy that cannot be attributed to soil formation. Horizons comprising the uppermost parent-material layer are not specifically designated in profile descriptions; those comprising succeeding layers are numbered 2, 3, etc., consecutively with depth (e.g. Ap, Bg, 2Cg, 3Cg).

Pedogenic horizons that have apparently been buried by deposition of additional mineral material and little modified subsequently are denoted by the prefix b. The upper surface of a buried profile may or may not coincide with a lithologic discontinuity. Thus the notation for a buried Ah horizon in material similar to the overlying deposit is bAh and that for a buried Ah horizon in dissimilar material is 2bAh.

In any particular profile, the horizons that show evidence of pedogenic modification are referred to collectively as the *solum*, and a B horizon together with any overlying E horizon is called a *sequum*. Some profiles show two sequa, one below the other, which evidently reflect successive phases of horizon differentiation rather than burial of an entire or truncated former profile by a younger deposit in which the upper horizons developed. These are termed *bisequal*.

Litter Layers and Organic Horizons

L Fresh litter deposited during the previous annual cycle; it is normally loose and the original plant structures are little altered.

F,H Organic horizons that originated as litter deposited or accumulated at the surface of the soil and are seldom saturated with water for more than a month at a time.

    F Partly decomposed or comminuted litter, remaining from earlier years, in which some of the original plant structures are visible to the naked eye.

    H Well decomposed litter, often mixed with mineral particles, in which the original plant structures cannot be seen; it normally contains numerous animal droppings (faecal pellets).

O Organic horizons consisting of more or less humified peat; they are saturated with water for long periods or artificially drained.

    Of O horizon consisting predominantly of fibrous peat.

    Om O horizon consisting predominantly of semi-fibrous peat.

    Oh O horizon consisting predominantly of humified peat.

    Oe O horizon consisting predominantly of unhumified or partially humified sedimentary peat (coprogenous earth).

    Op Peaty surface horizon mixed by cultivation; it is normally humified.

Mineral Horizons

A Mineral horizon formed at or near the surface and characterized by intimate incorporation of humified organic matter, mixing by cultivation or both. It normally contains at least 0.6 per cent organic carbon (1 per cent organic matter) and is darker (lower colour value and/or chroma) than the underlying horizon.

    Ah Uncultivated A horizon in which organic matter has been intimately mixed with the mineral fraction as a result of biological activity.

    Ap Surface mineral horizon that has evidently been mixed by cultivation; it normally has an abrupt or sharp lower boundary and may incorporate all or part of a pre-existing E, B or C horizon.

    Ahg
    Apg A horizon having a dominant moist chroma of 1 or less, or 2 if accompanied by ferruginous mottles, that can be attributed to reduction and segregation or removal of iron conditioned by periodic saturation with water (gleying); mottles are normally associated with root channels.

Although surface mineral horizons usually have properties of E or B as well as of A, the designation A is given precedence. Dark subsurface horizons of transitional character, in which properties of A are subordinate to, or partly mask, those of B or C are designated AB or AC respectively.

E Mineral subsurface horizon that contains less organic matter and/or extractable iron and/or silicate clay than the horizon below, normally as a result of removal of one or more of these constituents; it has a moist colour value of 4 or more, dry value 5 or more, or both. An E horizon is normally differentiated from an overlying A or organic horizon by higher colour value and smaller organic matter content and from an **underlying B** by higher colour value (especially when dry), lower chroma, smaller clay content, or by some combination of these properties.

    Ea E horizon that is seldom saturated with water, lacks ferruginous mottles or nodules, and has a colour mainly determined by the colours of uncoated sand or silt particles; if the moist value is 6 or more or the hue 5YR or redder, the chroma (moist or dry) is 3 or less; otherwise it is less than 3.

Eb E horizon that contains enough evenly distributed free iron oxide to give it a dominantly brownish colour with chroma of 3 or more when moist; it normally overlies a Bt horizon containing significantly more clay and may otherwise resemble a Bw horizon of similar composition.

An Eb horizon with ferruginous mottles or other evidence of gleying too slight to qualify for the suffix g is designated Eb (g).

Eg E horizon having a dominant moist
Eag chroma less than 3, or a chroma of 3 or 4 and distinctly higher value and yellower hue (Figure 3.6) than the main colour of the underlying horizon, that can be attributed to reduction and segregation or removal of iron conditioned by saturation with water in the presence of organic matter (gleying).

The horizon is designated Eg if it contains common or many ferruginous mottles, or Eag if there are few or no mottles.

EB The designation EB denotes a relatively
E/B homogeneous horizon of transitional character in which properties of E are superimposed on those of B throughout the soil mass. If discrete parts can be identified separately as E and others as B, the designation E/B is used, with appropriate suffixes. For example, heterogeneous horizons in which material with Eb or Eg characteristics surrounds columnar upward extensions of a Bt or Btg horizon, or wholly encloses small bodies with Bt characteristics, are designated Eb/Bt or Eg/Btg respectively.

B Mineral subsurface horizon that formed below an A or E horizon, lacks rock structure (including fine stratification in soft sediments) and shows evidence of one or both of the following:

1. Illuvial concentration of humus, iron, aluminium or silicate clay, normally in some combination.
2. Alteration of the original material involving solution and removal of carbonates; formation, liberation or residual accumulation of silicate clays or oxides; formation of granular, blocky or prismatic peds; or (usually) some combination of these.

Some soils have more than one kind of B horizon, either in vertical succession, or separated by an E horizon (bisequal profile).

Bh B horizon that contains translocated humus in combination with aluminium, or aluminium and iron, as coats on mineral grains or in discrete pellet-like aggregates of sand or silt size; the organic carbon content is at least 0.6 per cent (1 per cent organic matter), of which a major proportion is extracted by pyrophosphate. If the total- (or dithionite-) extractable iron content is 0.5 per cent or more, the colour value and chroma are both 3 or less.

A Bh horizon usually underlies a lighter coloured Ea and can be cemented (ortstein); where it directly underlies an A horizon (Ah or Ap), it contains significantly more pyrophosphate-extractable Fe + Al, the ratio of pyrophosphate-extractable carbon to total carbon is larger, or both.

Bf Sharply defined, black to dark reddish brown, cemented B horizon, less than 10 mm thick (thin ironpan or placic horizon) that is enriched in extractable iron and organic carbon; it is commonly wavy, or even involute, and normally lies between an Ah, Eag or Bh and an ungleyed B or BC horizon.

Bs B horizon enriched in amorphous (poorly ordered) sesquioxidic material by illuviation and/or biochemical weathering of silicates *in situ*. It has a moist colour value and/or chroma of 4 or more; contains at least 0.5 per cent dithionite- (or total) extractable Fe and at least 0.4 per cent pyrophosphate-extractable Fe + Al, amounting to more than 5 per cent of the measured clay content; and normally reacts positively to the Fieldes and Perrott (1966) sodium fluoride test for hydroxy-aluminium. (This is most conveniently performed in the field by placing a strip of filter paper impregnated with phenolphthalein in a watch glass and a 10 mg sample of soil on the paper, followed by one drop of saturated sodium fluoride solution. A positive reaction is indicated by the development within two minutes of a pronounced red spot on the test paper as seen through the underside of the watch glass.)

Some Bs horizons have sesquioxidic coats on mineral grains and these can be cemented, but those of loamy texture are normally friable and have very porous microfabrics consisting wholly or partly of fine granular (pellety) micro-aggregates (Bullock and Clayden, 1980). A Bs horizon with mottles indicative of gleying is designated Bs (g).

Bt   B horizon that contains translocated silicate clay; it normally underlies an E (usually Eb) or Ap horizon but can also occur below a Bh or Bs or as part of a buried profile. It meets the following requirements:

1. If any part of an E horizon remains and there is no lithologic discontinuity at its base, the Bt contains more total clay than the E as follows:
   (a) If the E horizon has less than 15 per cent clay, the B contains at least 3 per cent more; e.g. E – 10 per cent, Bt – 13 per cent or more.
   (b) If the E has more than 15 per cent clay, the ratio of the clay percentage in the B horizon to that in the E is 1.2 or more; e.g. E – 25 per cent, Bt – 30 per cent or more.
2. If peds are absent, as in most sandy loam or coarser materials, a Bt horizon has strongly oriented clay coating and bridging the sand grains and in some pores.
3. If peds are present, a Bt horizon has one or both of the following:
   (a) Clay coats (argillans) on ped faces or in pores, either throughout or in the lower part where peds and fissures are most stable, and/or intrapedal clay concentrations (pore fillings, papules or disrupted argillans) amounting to more than 1 per cent of a representative cross section.
   (b) More than 35 per cent clay and evidence of pressures caused by swelling (i.e. stress-oriented coats or slickensides; plasma separations in the matrix), accompanied by uncoated sand or silt grains on ped faces in the overlying horizon, a ratio of fine to total clay greater than in overlying and underlying horizons, or both.
4. If there is a lithologic discontinuity between E and B horizons, or if only an Ap overlies the B, a Bt horizon need only meet requirement 2 or 3.

In clayey Bt horizons (provision 3b), illuvial clay is difficult to distinguish, even in thin sections, from clay reorganized as a result of seasonal swelling and shrinking, and it may constitute only a very small proportion of the total clay.

Bw B horizon that shows evidence of alteration *in situ* (by weathering, leaching and/or structural reorganization) under well aerated conditions but does not qualify as Bh, Bs or Bt. It normally lies between an A horizon and a less altered BC or C, or rests directly on hard bedrock (R), but can also overlie a Bg, a buried profile or an E horizon in a lower sequum. In either case it is differentiated by colour, structure, or both, and is usually but not necessarily brown or reddish, with chroma of 3 or more and value 4–5. It lacks fragipan characteristics (see below) and has granular, blocky or prismatic peds if the clay content is large enough to cause volume changes on wetting and drying.

To qualify as Bw rather than BC or C, the horizon must meet one or more of the following requirements:

1. Stronger chroma, redder hue or larger clay content than the underlying horizon.
2. Less than 40 per cent $CaCO_3$ and less calcareous than the underlying horizon, which may contain redeposited (secondary) calcium carbonate.
3. Hue 10YR or redder, moist value 3–4 and chroma 3 or more, or moist value 5 and chroma 4 or more, and either
   (a) granular, blocky or prismatic structure differing from that of the underlying horizon; or
   (b) sand grains with ferruginous coats or fine intergranular aggregates giving a colour differing from that of the underlying horizon.

A Bw horizon with mottles or other evidence of gleying too slight to qualify for the suffix g is designated Bw (g).

Bg B horizons that have greyish colours
Btg and/or ferruginous mottles attributable to reduction and segregation of iron caused by periodic saturation with water in the presence of organic matter (gleying). Ferruginous segregations normally occur as mottles within peds rather than as coats on skeletal grains or bordering voids. A Bg horizon lacks fragipan characteristics (see below); has blocky or prismatic peds, more clay and/or less $CaCO_3$ than the underlying horizon, or both; and meets one of the following colour requirements (Figure 3.6).

1. Moist chroma of 1 or less dominant on ped faces, or in the matrix if peds are absent.
2. Moist chroma of 2 or less dominant on ped faces, or in the matrix if peds are absent, accompanied by mottles of higher chroma and redder hue. (If ped faces have organic coats with value of 4 or less, there must be greyish mottles or matrix colours within peds.)
3. Moist value 5 and chroma 3, or moist value 6 or more and chroma 4 or less, dominant on ped faces or in the matrix as above, if either:
   (a) there are common or many ferruginous mottles; or
   (b) the horizon has a dominant hue of 5YR or redder inherited from reddish (haematitic) parent material and there are common or many greyish and/or brownish mottles within peds.

If the horizon also qualifies as Bt, the designation Btg is used.

Bgf *See* Cgf (p. 105).

BC A BC horizon is transitional in character
B/C (structure, degree of weathering, etc.) between a B and a C horizon and usually underlies a B (Bw, Bg, Bs, Bt or Btg), but the same designation is also applied to morphologically similar subsurface horizons in profiles without a B horizon or in which a B is only intermittently developed. A BC horizon can show features, including greyish colours accompanied by ferruginous mottles (BCg), accumulation of redeposited calcium carbonate (BCk), or fragipan characteristics (BCx), absent in overlying and underlying horizons. A lower horizon containing translocated clay is also designated BC (BCt or BCtg) if the immediately overlying horizon has more clay and otherwise qualifies as Bw or Bg.

Heterogeneous horizons in which thin layers or discrete bodies of varying shape with B characteristics are interbedded with, or surrounded by, C horizon material are designated B/C with appropriate suffixes (e.g. Bt/Cu).

C Unconsolidated or weakly consolidated mineral horizon that is comparatively unaffected by pedogenic processes other than:

1. Reduction, or reduction and segregation or removal, of iron conditioned by saturation with water in the presence of organic matter (gleying).
2. Secondary accumulation of carbonates or more soluble salts.
3. Development of fragipan characteristics (see below).

Cu Unconsolidated C horizon without evidence of gleying, accumulation of carbonates or more soluble salts, or fragipan characteristics; it is normally apedal but may show fine stratification.

Cr Weakly consolidated C horizon that is dense enough to prevent penetration by roots, except along cracks with an average horizontal spacing of 10 cm or more. The coherent material is continuous, or coarsely fractured without significant displacement, and can be dug with difficulty with a spade when moist. Soft rocks such as mudstone, siltstone and chalk, and some exceptionally hard and dense glacial tills, are included.

CG Intensely gleyed ('reduced') C horizon that contains at least 0.3 per cent dithionite-extractable iron but has few or no ferruginous mottles; it is more or less permanently saturated with water and has a dominant chroma of 1 or less, a hue bluer or greener than 10Y, or both.

Some CG horizons contain readily oxidizable ferrous compounds,

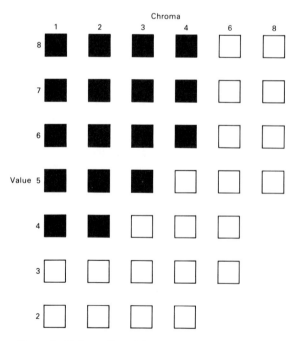

Figure 3.6: Munsell ped-face or matrix colours diagnostic of gleyed subsurface horizons when accompanied by ferruginous mottles (after Avery, 1980).

including sulphide (pyrite) and phosphate (vivianite), which change colour on exposure to air; those in recent alluvial deposits normally have a fluid consistence (p. 87).

Cg Gleyed C horizon that does not qualify as CG; it is saturated with water for long periods or artificially drained and has either

1. A greyish matrix colour and ferruginous mottles as specified for Bg (Figure 3.6); or
2. A colour due to uncoated sand grains that can be attributed to reduction and removal of iron.

Peds are absent or very weakly developed and ferruginous segregations, if present, occur around root or animal channels or bordering other voids where air can enter from above, rather than as mottles within the matrix.
(Grey Pleistocene or pre-Pleistocene sediments showing little or no evidence of post-depositional modification are designated Cu or Cr, and slowly permeable mottled horizons transitional in colour and structure between an overlying Bg or Btg and an underlying less mottled C are designated BCg rather than Cg.)

Cgf C horizon that is periodically saturated with water or artificially drained and has:

1. A dominantly brownish or ochreous colour with hue 10YR or redder and chroma of 3 or more; and/or
2. Abundant ferruginous or ferrimanganiferous coats, nodules, or soft concentrations.

Horizons of this kind occur in permeable materials that are affected by fluctuating groundwater or in which the water table has been lowered. The apparent accumulation of iron oxides can be attributed to an alternation of reducing and oxidizing conditions, or to oxidation and precipitation of iron transported from elsewhere in a reduced state. Only a small proportion of the total (or dithionite-extractable) free iron is extracted by pyrophosphate and there is normally no associated accumulation of extractable aluminium or carbon.

The same suffixes are used with B (Bgf) or BC (BCgf), depending on the position of the iron-enriched horizon in the profile and other features distinguishing it from those above and below.

R Hard bedrock that is continuous, except for cracks with an average horizontal spacing of 10 cm or more. It has a hardness of 3 or more (Moh's scale) and is too hard to dig with a spade, though the softer types can be chipped or scraped.

Additional Lower Case Suffixes

k Used with BC or C to denote horizons that consist of lake marl or tufa (p. 110) or contain redeposited (secondary) calcium carbonate as concretions or nodules, coats, tubular formations or soft concentrations, together occupying at least 1 per cent of the volume.

m Used with B or BC to denote continuously or nearly continuously cemented horizons, other than Bf, that are strong enough to resist root penetration except along cracks. The cementing agent may be some combination of organic matter with aluminium or aluminium and iron (Bhm), oxides of iron or iron and manganese (BCgfm, Cgfm), amorphous inorganic constituents containing silica and/or aluminium (BCm), or calcium carbonate (BCkm, Ckm).

x Used with E, B or BC to denote compact but uncemented horizons (fragipan or densipan) having the following properties:

1. High or very high bulk density (Table 3.7, p. 98).
2. Hard or very hard when dry and firm or very firm and distinctly brittle when moist (a 3 cm specimen cube fractures suddenly when pressure is applied rather than deforming gradually).
3. An air-dry specimen slakes when immersed in water.
4. Massive, coarse or very coarse prismatic, or platy structure.
5. Low organic matter content and few or no roots, except in widely spaced cracks.

A *fragipan* (Soil Survey Staff, 1975, pp. 42–44) is normally loamy and underlies a less dense E or B horizon at depths ranging from 25 to 100 cm but can occur directly below an A in eroded soils. Most fragipans have a roughly polygonal pattern of vertical cracks surrounding coarse or very coarse prisms, within which the structure is massive or platy. They are usually mottled, with grey or greyish colours on the prism faces, and clay coats (argillans) lining cracks and in pores.

Horizons that have fragipan features, meet the definition of Bt, and contain significantly more clay than an overlying E or A horizon are designated Btx (or Btgx if the horizon also qualifies as Bg); others with common or many clay coats are designated BCtx or BCtgx, and those without as BCx, BCgx or Cx, depending on the degree of gleying and/or other evidence of alteration, in comparison with horizons above and below.

Most fragipan-like horizons in British soils, including those described as indurated layers by the Soil Survey of Scotland (1984), contain vesicular pores and have coats of fine material on the upper surfaces of stones. These features, along with associated platy or lenticular and coarse prismatic structures, are believed to have originated under periglacial conditions when the ground was deeply frozen (Fitz-Patrick, 1956; Lozet and Herbillon, 1971; Payton, 1980, 1988; van Vliet-Lanoë, 1985). Other features, including bleaching of prism faces, mottles and clay coats, may have been produced later.

Some of the layers originally described as indurated, particularly in podzolic soils of north-east Scotland (Romans, 1974), fail to slake in water and hence qualify as cemented (m) rather than fragic (x). Studies of similar indurated subsoils (duric horizons) in British Columbia soils by McKeague and Sprout (1975) and McKeague and Kodama (1981) suggest that they are cemented by amorphous (poorly ordered) aluminosilicates, including imogolite (cp. Farmer *et al.*, 1980). There is also evidence (Bridges and Bull, 1983) that weak bonding by cryptocrystalline silica is at least partly responsible for the brittleness of some subsoil pans.

Dense, light-coloured eluvial horizons (Eax) with physical properties similar to those of fragipans were described as *densipans* by Smith *et al.* (1975).

y Used with BC or C to denote the presence of crystalline calcium sulphate (gypsum), normally as a pedogenic accumulation resulting from oxidation of pyrite in the unaltered parent material.

3.6 Named Diagnostic Surface Horizons

Prominent A Horizon

This is a dark coloured A horizon (or a sequence of A and AB horizons) that has the following properties after the surface 18 cm, or the whole thickness if there is bedrock (R or Cr) at 10–18 cm, are mixed as by ploughing.

1. Both broken and crushed samples have a Munsell chroma less than 3.5 when moist

and a colour value darker than 3.5 when moist and 5.5 when dry.
2. Its thickness is:
   (a) 10–18 cm if it rests directly on bedrock; or
   (b) at least 18 cm and more than one-third the thickness of the solum where this is less than 75 cm; or
   (c) more than 25 cm if the solum is more than 75 cm thick, the textural class is sand or loamy sand, or the organic matter content decreases irregularly with depth.
3. Soil structure is such that the horizon is not both massive and hard or very hard when dry.

Prominent A horizons include *mollic, umbric* and *anthropic epipedons* (Table 2.11, p. 68) as defined in the US Taxonomy (Soil Survey Staff, 1975, pp. 14–17).

Thick Man-made A Horizon

This is an A horizon, usually but not necessarily prominent, that has been artificially thickened by regular use of manure containing mineral matter, unusually deep cultivation accompanied by addition of organic manure only, or incorporation of human occupation residues. It is at least 40 cm thick or overlies bedrock at a lesser depth, has a moist colour value of 4 or less and chroma of 3 or less throughout its depth, and generally contains artefacts such as pieces of brick or pottery. Humified organic matter is intimately mixed with the mineral fraction in all subhorizons.

This diagnostic horizon corresponds to the thick A1 horizon of de Bakker and Schelling (1966) and includes *plaggen epipedons* and some *anthropic epipedons* (Soil Survey Staff, 1975, pp. 16–18). As naturally formed A horizons more than 40 cm thick also occur, particularly in recent alluvial or colluvial deposits where biological incorporation of organic matter has kept pace with slow sedimentation (cumulic A horizons), the mode of origin may have to be inferred from external evidence, including the location and lateral extent of the horizon in relation to surface relief and existing or former field boundaries, and available information concerning former land-use practices and settlement sites.

Humose Topsoil

This is a dark-coloured A horizon, or a sequence of H or Oh and Ah horizons, that meets the following requirements over a thickness of more than 15 cm, or 10–15 cm if directly over bedrock (R or Cr):

1. Moist rubbed colour with both value and chroma less than 3.5.
2. Humose (p. 83), or partly humose and partly organic.

Most humose topsoils, particularly those that are cultivated (humose Ap), also qualify as prominent A horizons.

Peaty Topsoil

This is a layer of peat 10–40 cm thick overlying mineral soil or bedrock. It consists of a single O horizon (Of, Om, Oh or Op) or a succession of such horizons and normally starts at the surface or beneath a thin layer of little decomposed litter (L, F), but may be buried by a mineral layer up to 30 cm thick. In uncultivated soils it generally includes an Oh horizon which is denser and more plastic than well aerated H horizons of similar composition and breaks into hard angular blocks when dried.

The *histic epipedon* (Soil Survey Staff, 1975, pp. 17–18) is essentially similar but is defined in rather more exclusive terms.

When a peaty topsoil is cultivated, it is commonly transformed into a humose topsoil as a result of admixture with underlying mineral soil, increased mineralization of organic matter, or both.

Earthy Topsoil

This is a ripened (non-fluid) peaty surface layer (Oh or Op) that is at least 20 cm thick and contains less than 15 per cent of visible plant remains (fibres) other than resistant woody fragments. It overlies an organic subsurface horizon and normally has a distinct granular or subangular blocky structure.

An earthy topsoil is nearly always present in organic (peat) soils which have been drained, or drained and cultivated. Where the soil has not been cultivated, it can underlie a less decomposed F horizon and consistent identification demands particular care in distinguishing living roots from undecomposed or partially decomposed plant residues.

3.7 Named Diagnostic Subsurface Horizons

Albic E Horizon

This is an E horizon (Ea, Eag or Eg) at least 5 cm thick, in which coats on mineral particles are absent, very thin or discontinuous. It has a moist colour value of 4 or more, a dry value of 5 or more, or both. If the value is 6 or more when moist or 7 or more when dry, or if the hue is 5YR or redder, the chroma is 3 or less; otherwise it is less than 3.

An albic E horizon (*albic horizon* in the US Taxonomy) normally overlies a B containing more iron+aluminium but can contain more silicate clay than the horizon below as a result of increased weathering of rock fragments or clay-forming minerals in coarser size grades.

An *albic densipan* (Eax) is an albic E horizon with very high bulk density (Table 3.7, p. 98); its consistence is very firm or firmer when moist but fragments slake when immersed in water.

Podzolic B Horizon

This is a B horizon that meets the following requirements:

1. Qualifies as Bh or Bs, or comprises a Bh and a Bs in vertical succession.
2. Starts within 1.20 m depth; underlies an albic E, A, H or O horizon and extends below 15 cm depth, excluding fresh or partially decomposed litter (L, F).
3. At least 10 cm thick if there is no overlying albic E and it consists only of a Bs; at least 3 cm thick if it consists only of a Bh horizon below an albic E.

In addition to the properties specified as diagnostic, podzolic B horizons have a number of associated properties as follows:

1. High pH-dependent CEC. The difference between CEC measured at pH 8 and at the pH of the soil is usually at least 8 me/100 g.
2. Unless cemented, they have a large retained water capacity in relation to their texture.
3. The ratios of pyrophosphate-extractable carbon to total organic carbon and of pyrophosphate-extractable iron to dithionite-extractable iron are normally at least 0.3 and more than 0.5 in well expressed podzolic B horizons.
4. Large phosphorus retention capacity.
5. High fluoride reactivity (Loveland and Bullock, 1976).

The podzolic B horizon is conceptually analogous to the *spodic horizon* (Soil Survey Staff, 1975, pp. 29–32) but is defined in less restrictive terms (Avery *et al.*, 1977).

Thin Ironpan

This is a Bf horizon (p. 101) that underlies an albic E, A, O or Bh horizon and acts as a barrier to water and roots. Where the pan is discontinuous or partly disrupted by cultivation, it extends over at least half of a horizontal distance of one metre.

Placic horizons (Soil Survey Staff, 1975, p. 33) include thin ironpans as defined here, together with similar but less common pans which are more than 10 mm thick or in which the cementing agent is some combination of iron and manganese oxides rather than of iron and organic matter.

Argillic B Horizon

This is a Bt or Btg horizon (p. 102), or a succession of such horizons, that meets the following requirements:

1. At least 7.5 cm thick if it is sandy loam or finer in texture; at least 15 cm thick if it consists of one or more loamy sand bands (as in certain sandy soils); and at least one-tenth as thick as the total thickness of all overlying horizons.
2. Extends below 30 cm depth or starts within 1.20 m.
3. Immediately underlies an A, E or podzolic B horizon and contains more clay unless it directly underlies an Ap.
4. No overlying layer of recent colluvium (p. 114) or alluvium (p. 113) 40 cm or more thick.

The *argillic horizon* (Soil Survey Staff, 1975, pp. 19–27) is essentially the same but is defined in slightly less restrictive terms.

An argillic B horizon is described as *chromic* if it also meets one or both of the following colour requirements:

1. If the textural class is sandy clay loam or finer, dominant matrix colour with chroma more than 5 (moist or dry) and hue 7.5YR or redder; if coarser the hue is redder than 7.5YR.
2. Many coarse mottles with chroma more than 5 and hue 5YR or redder, or common to many mottles with hues redder than 5YR.

Horizons of this kind in which the reddish colours are evidently not inherited directly from a red or red-mottled pre-Quaternary rock have been called *paleo-argillic B horizons* in England and Wales (Avery, 1980, 1985), implying that the profiles in which they occur are complex or polycyclic in origin (Figure 1.9, p. 15). As indicated in Chapter 1 (p. 15), they are typically developed in Wolstonian or older deposits and examples that have been studied in detail (e.g. Bullock, 1974; Bullock and Murphy, 1979; Chartres, 1980) show distinctive micromorphological features (papules or disrupted argillans) which, together with the strong colours, can be attributed to pedological modification of the original materials during one or more pre-Devensian interglacial periods when the climate was at least as warm as it is today, followed by disturbance under periglacial conditions. As subsurface horizons retaining the impress of pre-Devensian soil formation cannot always be distinguished by particular colour and structural characteristics, however, the term *chromic* is used rather than *paleo-argillic* in this account.

Colour/Structure B Horizon (Weathered B Horizon of Avery, 1980)

This is a Bw horizon (p. 103) that differs from the underlying horizon in colour, structure, or both, is at least 5 cm thick, and extends to at least 30 cm below the surface.

Cambic horizons (Soil Survey Staff, 1975, pp. 33–36) unaffected or little affected by gleying are included, together with other weakly expressed B horizons that fail to qualify as cambic either because they are sandy (sand or loamy sand) or because they do not meet other requirements. The depth limit was fixed at 30 cm, rather than 25 cm as specified for the cambic horizon, because many if not most British soils with B horizons of this kind have been ploughed to a depth of 25 cm and a minimum thickness of 5 cm was considered necessary if the horizon is to be consistently identified as a diagnostic feature.

Gleyed Subsurface Horizon

This is a subsurface horizon with grey or mottled colours attributable to reduction or reduction and segregation or removal of iron in the presence of organic matter. It is wet for substantial periods in most years or formed under wet conditions and meets the following requirements:

1. Qualifies as Eag, Eg, Bg, Btg, BCg, Cg or CG and lacks fragipan characteristics (p. 106).
2. At least 5 cm thick and extends below 30 cm depth.
3. Starts within 40 cm depth and/or directly below a prominent A horizon (p. 106), a humose or peaty topsoil (p. 107), or a Bh horizon (p. 101).

These criteria are basically similar to those used in defining aquic (gleyed or hydromorphic) suborders in the US Taxonomy but differ in two significant respects. The first is the standard depth limit, which is placed at 40 cm rather than 50 cm with the object of excluding cultivated soils that show little or no evidence of current or former wetness in the horizon immediately beneath the ploughed layer. On the other hand the colour criteria (Figure 3.6) are intentionally less restrictive on the basis that mottled E or B horizons with value/chroma ratings of 5/3 or 6/4 on ped faces or in the matrix are commonly saturated with water for considerable periods, particularly in originally reddish parent materials (Robson and Thomasson, 1977; Franzmeier et al., 1983; Vepraskas and Wilding, 1983).

A gleyed subsurface horizon can also qualify as an argillic B (Btg) horizon or a cambic horizon. According to the definition used in the US system (Soil Survey Staff, 1975, pp. 33–36), a gleyed cambic horizon directly underlies an A or O horizon; extends to at least 25 cm depth; has soil structure or lacks rock structure in at least half the volume; lacks fragipan characteristics; and has colours that meet criteria 1 or 2 for a Bg horizon (p. 103) and do not change on exposure to air. It also has one or more of the following:

1. A regular decrease in organic carbon content with depth and less than 0.2 per cent at 1.25 m depth or immediately above a sandy-skeletal substratum at less than 1.25 m.
2. Cracks 1 cm or more wide at 50 cm depth in most years.
3. An overlying prominent (mollic or umbric) A horizon or peaty topsoil (histic epipedon).

Thus a loamy or clayey Eag, Eg, Bg or Cg horizon, or some combination of these, may qualify as cambic but most gleyed subsurface horizons in recent alluvium (p. 113) are excluded because they fail to meet requirement 1.

Hydrocalcic Subsurface Horizon

This is an extremely calcareous Ck or Ckg horizon that starts immediately below a prominent (mollic) A horizon or peaty topsoil, or within 40 cm depth, and consists of Holocene lake marl or tufa accumulated under wet conditions. Lake marl is a subaqueous sediment in which much of the carbonate is biogenic, including animal shells and accumulations by aquatic plants such as algae and stoneworts. Tufa is a similar deposit containing calcium carbonate precipitated from lime-rich water, particularly in areas where springs emerge.

Slowly Permeable Subsurface Horizon

This is a subsurface horizon that is at least 15 cm thick, starts within 80 cm depth, and acts as a significant barrier to water movement when the soil is saturated or nearly so. It is defined in precise terms as having a saturated hydraulic conductivity (K_s) less than 10 cm per day. In the absence of measured K_s values, it is identified by the following properties:

1. Massive, platy or prismatic structure, or angular blocky structure that is coarse, weakly developed, or both.
2. Firm or stronger consistence when moist.
3. Few or widely spaced (< 0.5 per cent) visible macropores (Hodgson, 1976, p. 45);

any fissures present when the horizon is dried close more or less completely when it is wet.
4. Greyish or mottled colours attributable to gleying in and/or immediately above the horizon; if peds are present, they usually have greyish faces and ochreous internal mottling (Bg, Btg or BCg horizon), though these features may be absent in reddish parent materials containing haematite which resists reduction, or on slopes where excess water is removed laterally.
5. High packing density, very small air capacity (Table 3.7, p. 98), or both.

Slowly permeable subsurface horizons which are loamy in texture usually have fragipan characteristics (p. 106), or meet all requirements of a fragipan except brittleness.

The above physical and morphological properties can be induced in surface and subplough layers by injudicious cultivations or other practices. Profiles in which upper horizons have been compacted in this way are not considered to have a slowly permeable subsurface horizon if underlying horizons are relatively permeable.

3.8 Other Differentiating Criteria

Abrupt Textural Change (cp. Soil Survey Staff, 1975, p. 47)

This is a considerable increase in clay content between an A or E horizon and an underlying slowly permeable horizon, normally Btg or 2Bg. Where the A or E has less than 20 per cent clay, the clay content doubles within a vertical distance of 8 cm or less. If the clay content of the upper horizon exceeds 20 per cent, it increases by at least 20 per cent (e.g. from 22 to 42 per cent) within 8 cm and is at least twice as great in some part of the underlying horizon.

Acid-Sulphate Characteristics

Acid-sulphate soils (Bloomfield and Coulter, 1973; Bloomfield and Zahari, 1982) are those that have developed extreme acidity as a result of oxidation of sulphides, chiefly pyrite (FeS_2). Sulphidic materials (Soil Survey Staff, 1975, p. 63) occur in coastal or estuarine marshes and other waterlogged environments where sediments rich in organic matter have accumulated in water containing sulphates in solution. Under the prevailing anaerobic conditions, iron in the sediment and sulphate ions are reduced by bacterial action and pyrite is the main end product of these processes. If the soil is subsequently drained, oxidation of the pyrite generates sulphuric acid and the pH consequently falls rapidly unless enough carbonates are also present to neutralize the acid formed (giving gypsum). Otherwise the acid reacts with the soil material to give iron and aluminium sulphates, particularly jarosite (a potassium or sodium iron sulphate), which is segregated to form conspicuous yellow mottles or coatings (Plate IV.7).

In this account a soil is considered to have acid-sulphate characteristics if there is a mineral or organic horizon within 80 cm depth with the following properties:

1. pH ($CaCl_2$ 1:2.5) 3.5 or less or pH (H_2O 1:2.5) 4.0 or less.
2. Segregations of jarosite (hue 2.5Y or yellower and chroma 6 or more) or gypsum, at least 0.5 per cent sulphur (sulphate-S + oxidizable S), or both.

A *sulphuric horizon* (Soil Survey Staff, 1975, p. 47) is required to have both jarosite and a pH (1:1 in water) less than 3.5.

Sulphidic materials (potential acid-sulphate soils) are identified by measuring the pH of a sample after it has been allowed to oxidize without becoming dry for two months or more.

Artificially Reworked Layer

This is a layer of artificially rearranged or

transported soil material. It evidently consists wholly or partly of materials derived from pre-existing O, A, E, B or BC horizons which have been more or less completely mixed or occur as discrete fragments without discernible natural order. It may overlie an undisturbed soil horizon or substratum, or a layer of artificially disturbed waste material such as mining spoil or domestic refuse that shows no sign of modification by normal pedogenic processes.

Layers of this kind occur in places where soil has been excavated and replaced after mineral extraction or in connection with civil engineering projects. They can also result from deep earth moving operations causing irregular admixture of materials from different horizons (soil mixing).

Gleyic Features

A soil that lacks a gleyed subsurface horizon (p. 109) is described as having gleyic features if there is a Bg, Btg, BCg or Cg horizon starting between 40 and 80 cm depth, or if a subsurface horizon within 60 cm shows one or more of the following:

1. Common or many ferruginous mottles or segregations unaccompanied by ped-face or matrix colours with the value/chroma ratings specified for a Bg horizon (p. 103, Figure 3.6).
2. Matrix hue 5YR or redder and chroma of 4 or more inherited from a reddish (haematitic) parent material, accompanied by common or many mottles, or ped faces with colours at least one unit lower in chroma and at least 2.5 units yellower in hue.
3. In other horizons with dominant chroma of 4 or more, common or many mottles, or ped-face colours, distinctly lower in chroma and yellower in hue than the matrix.

There should be evidence that these features are associated with current wetness if the soil has not been artificially drained, as variegated colours may also be inherited unchanged from a pre-Quaternary rock, result from irregular oxidation of iron-bearing minerals under well aerated conditions, or reflect the juxtaposition of brightly coloured strongly orientated clay bodies and zones depleted of clay (Eb/Bt horizon).

Pelo-characteristics

A soil with the following properties is considered to have pelo-characteristics (from the Greek word *Pelos*, meaning clay):

1. At least half the upper 80 cm, or at least 30 cm if bedrock supervenes, is clayey (more than 35 per cent clay).
2. No overlying horizon that is more than 15 cm thick and contains less than 30 per cent clay.
3. A B horizon that has smooth-faced blocky, prismatic or wedge-shaped peds and extends to more than 30 cm depth.
4. A gleyed subsurface horizon or gleyic features, a slowly permeable subsurface horizon, or both.

Following long-standing usage in West Germany (Mückenhausen, 1965, 1985), clayey soils that meet these requirements and lack a paleo-argillic B horizon (p. 109) are set apart as pelosols or pelosol-intergrades in the England and Wales classification and are separated at subgroup level in that adopted here. Many of those in the English lowlands resemble Vertisols and Vertic subgroups of other orders in the US Taxonomy in showing marked seasonal variations in macroporosity and distinctive structural features, including slickensided pressure faces inclined at oblique angles, which result from shrinkage and swelling caused by changes in moisture content.

The criteria used to distinguish both Vertisols and Vertic subgroups stipulate the occurrence of cracks (Figure 3.7) at some period in most years that are 1 cm or more wide at 50 cm depth and extend upward to the surface or to

Figure 3.7: Cracks in a peloluvic gley soil in glaciolacustrine clay near Dunstall, Staffordshire, June 1975 (Photo by R.J.A. Jones).

the base of an Ap. Soils in Vertic subgroups (e.g. of Aquepts, Ochrepts, Aqualfs and Udalfs) are also required to meet specified coefficient of linear extensibility (COLE) and potential linear extensibility (PLE) limits (Soil Survey Staff, 1975, p. 242). Reeve *et al.* (1980) showed that nine out of nineteen profiles representing major clayey soil series in England met the COLE and PLE criteria and that their shrinkage potentials were well correlated with the cation-exchange capacity and mica-smectite content of the clay fractions expressed on a whole soil basis. They also cited evidence that cracks more than 1 cm wide often extend to depths between 50 and 60 cm depth in such soils during dry summers like those of 1975 and 1976. Without further studies, however, the extent to which British soils with *pelo-*characteristics meet the prescribed cracking requirement is uncertain, particularly since the size and spacing of cracks in soils with the same shrinkage potential is clearly influenced by local differences in land use and soil-water regime as well as by climatic conditions.

Recent Alluvial Origin

Marine, fluviatile and lacustrine sediments of Holocene age, together with calcareous tufa, are grouped for the present purpose as *Recent Alluvium*. Soils in these deposits are generally characterized by one or more of the following properties:

1. Organic carbon content that decreases irregularly with depth or remains greater than 0.2 per cent to a depth of 1.25 m (Soil Survey Staff, 1975). The radiocarbon age (mean residence time) of the organic matter is less than 10,000 years BP.
2. Periodic inundation, unless protected from flooding by stop banks or other preventive measures.
3. Fine stratification in some part.
4. An unripened horizon with fluid consistence less than 1.25 m below the surface.

The first criterion is primarily applicable to recent deposits that are loamy or clayey and at least 1.25 m thick. Sandy or sandy-gravelly (sandy-skeletal) layers, and older mineral deposits beneath thinner alluvium, often contain smaller amounts of organic matter.

A sedimentary layer with none of the properties listed is considered as recent alluvium only if it can be dated as Holocene by stratigraphic, palaeontological or archaeological evidence.

Recent Colluvial Origin

As defined here, *recent colluvium* is an unstratified or crudely stratified deposit that has accumulated during Holocene times by rainwash or downslope creep, particularly as a result of accelerated erosion of pre-existing soils following clearance of vegetation by man. Such deposits occur in dry valley bottoms, on lower slopes, and against artificial obstructions such as hedge banks. Soils in recent colluvium are identified by some combination of the following characteristics:

1. Gradual or irregular decrease in organic carbon content with increasing depth, as in most recent alluvial deposits.
2. Inclusion of charcoal, or of pottery or other artefacts (cp. thick man-made A horizon, p. 107).
3. Weakly differentiated horizons: a Bw horizon, if present, is differentiated primarily by structure, shows little or no evidence of weathering *in situ* and normally fails to qualify as cambic (see colour/structure B horizon, p. 109); if calcareous, the $CaCO_3$ content decreases or remains nearly constant with increasing depth. The colluvium may have been emplaced by rainwash and subsequently modified by structure-forming agencies, or by soil creep in which the original structure was to some extent preserved (Kwaad and Mücher, 1979).
4. Stratigraphic relationships; the colluvium normally thickens downslope and rests unconformably on an older deposit which commonly retains pre-existing soil horizons. A buried A horizon may be present, or the base of the colluvium may be marked by a lag gravel or stone pavement truncating a buried profile.

3.9 Key to Major Soil Groups

1. Soils that have:
   (a) more than 40 cm of organic material (p. 82) within the upper 80 cm, excluding fresh litter (L) and living moss; or
   (b) more than 30 cm of organic material resting directly on bedrock (R or Cr) or skeletal (very or extremely stony) material; and
   (c) no overlying mineral horizon that has a colour value of 4 or more and extends below 30 cm depth.
   Peat soils (Chapter 9)

2. Other soils that have:
   (a) a thick man-made A horizon (p. 107); and/or
   (b) an artificially reworked layer (p. 111) that is more than 40 cm thick or overlies bedrock or other artificially displaced material at a lesser depth.
   Man-made soils (Chapter 8)

3. Other soils that have one or more of the following:
   (a) an albic E horizon (p. 108) over a podzolic B horizon (p. 108).
   (b) a podzolic B horizon that includes a Bh horizon (p. 101) at least 3 cm thick, a cemented Bs (p. 101), or both.
   (c) a thin ironpan (p. 108) that immediately underlies an E, A or Bh horizon at less than 80 cm below the mineral soil surface.
   Podzols (Chapter 6)

4. Other soils that have a gleyed (p. 109) or hydrocalcic (p. 110) subsurface horizon.
   Gley soils (Chapter 7)

5. Other soils that have a podzolic (p. 108), argillic (p. 109) or colour/structure (p. 109) B horizon, or a prominent A horizon (p. 106) more than 30 cm thick in recent alluvium or colluvium.
   Brown soils (Chapter 5)

6. Other soils
   Lithomorphic soils (Chapter 4)

4
Lithomorphic Soils

4.1 General Characteristics, Classification and Extent

This major group comprises most or all of the soils that have been grouped as lithomorphic in England and Wales (Avery 1973, 1980), as immature soils and rendzinas by the Soil Survey of Scotland (1984), and as lithosols, regosols and rendzinas in Ireland (Gardiner and Radford, 1980). Their essential common feature is the presence of bedrock or little altered regolith at shallow depth. In terms of the differentiating characteristics listed in Chapter 3, they are distinguished from brown soils and podzols by the absence of a diagnostic B horizon or a prominent (cumulic) A horizon more than 30 cm thick, from gley soils by the absence of a gleyed or hydrocalcic subsurface horizon, and from peat soils by the absence of an organic surface layer of the requisite thickness. The substratum may have been emplaced by man, but soils that have a thick man-made A horizon or consist wholly or partly of artificially displaced earthy material derived from pre-existing soil horizons are excluded.

This broad class of shallow or weakly developed soils is conceptually comparable with the Entisol order in the US Taxonomy, but excludes the strongly gleyed Aquents and deep sandy soils (Psamments) with brownish B horizons and embraces some Mollisols (Rendolls), Inceptisols (Umbrepts) and shallow peaty soils (Lithic Histosols) with prominent mineral or organic surface horizons thick enough to qualify as mollic, umbric or histic epipedons respectively. According to the FAO (1974) system, it comprises Lithosols, Regosols, Rankers, Rendzinas and some Fluvisols and Histosols.

Although these soils are distributed throughout almost the entire range of environmental conditions represented in the British Isles, they are restricted to young ground surfaces affected by recent erosion or deposition and to other places where Holocene pedogenesis has been constrained by the nature of the substratum, as on chalk (rendzinas) and hard siliceous rocks (lithosols and rankers). Whichever has been the main factor limiting profile development, they commonly occupy small areas bounded in most places by bodies of soil conforming to other major groups and in others by rock outcrops or bodies of raw unconsolidated material without perceptible soil horizons. According to the 1:250,000 soil survey of England and Wales (Mackney *et al.*, 1983), associations consisting dominantly of lithomorphic soils, together with bare rock, unvegetated screes and raw coastal sands, cover about 7 per cent of the total surface area. No strictly comparable assessment is possible for Scotland, but available information indicates that the corresponding proportionate area is around 15 per cent. In Ireland, judging from the 1:1,000,000 soil-association map, it is approximately 6 per cent.

Some lithomorphic soils are uncultivable and the use range and productive capacity of others are restricted to varying degrees by shallowness, stoniness or rockiness, small available water capacity, or liability to erosion or flooding, in addition to any climatic and topographic limitations. Those resulting from recent colonization of bare ground by natural or introduced vegetation are of particular interest to ecologists concerned with plant succession or with rehabilitation of man-made excavations and waste tips. Rates of natural colonization and consequent soil horizon development vary greatly, depending on the supply of plant propagules, the inclination of the surface, and physical and chemical properties of the substrate, including its erodibility and the occurrence in certain waste deposits of toxic constituents such as heavy metals.

The five groups into which these soils have been divided for the present purpose are defined in accordance with the following diagnostic key. Corresponding classes in the England and Wales classification are listed for comparison, but the correlations are not exact, first because the conceptual limit between soil and 'not soil' has been defined differently (Chapter 2, p. 46), and second because certain extremely calcareous soils classed as lithomorphic by Avery (1980) are grouped as gley soils in the system adopted here. The number of subgroups recognized is also smaller because 'semi-hydromorphic' variants transitional to gley soils, which are inextensive and difficult to distinguish consistently, have not been set apart.

Key to Soil Groups

1. Lithomorphic soils with hard rock (R) less than 10 cm below the surface (excluding fresh litter and living moss)
   Lithosols (raw skeletal soils and very shallow rankers and rendzinas)
2. Other lithomorphic soils with a C horizon in recent alluvium (p. 114) more than 30 cm thick
   Rego-alluvial soils (ranker-like alluvial soils; typical rendzina-like alluvial soils)
3. Other lithomorphic soils with a sandy (p. 84) C or E horizon extending below 30 cm depth
   Sandy regosols (sand-rankers and sand-pararendzinas)
4. Other lithomorphic soils with an A horizon in and/or immediately overlying calcareous material (excluding added lime)
   Rendzinas (rendzinas and pararendzinas, excluding those with gleyed or hydrocalcic subsurface horizons)
5. Other lithomorphic soils
   Rankers (rankers)

4.2 Lithosols

4.2.1 General Characteristics, Classification and Distribution

The term *lithosol* was first used in the United States (Thorp and Smith, 1949) to denote 'an azonal group of soils having an incomplete solum or no clearly expressed soil morphology and consisting of a freshly or incompletely weathered mass of hard rock or hard rock fragments'. As in the FAO (1974) system and that used by the Soil Survey of Scotland (1984), it is restricted in this account to very shallow soils with hard bedrock (R) less than 10 cm below the surface. In England and Wales, where no division at that depth has been made, those with a distinct organic or organo-mineral surface horizon 5 cm or more thick have been considered as ultra-shallow rankers or rendzinas and those without as raw skeletal soils.

Small unmappable bodies of soil conforming to this group occur in intricate association with rock outcrops and deeper soils, chiefly rankers, on rugged mountain summits in the Scottish Highlands and Islands, in enclaves of similarly mountainous terrain in the English Lake District (Figure 4.1), North Wales and Ireland; and in numerous smaller rocky areas at lower altitudes. The thin soil layer may

Figure 4.1: Bare rock, lithosols and rankers near the summit of Coniston Old Man, Cumbria (photo by T. Parker).

result from colonization of exposed rock by plants and consequent accumulation of organic residues, or from truncation by erosion of originally deeper mineral or organic soil. As shown in Figure 4.2, bare rock surfaces are usually first colonized by lichens, and subsequently by mosses, growth of which hastens physical and chemical weathering and so helps higher plants capable of withstanding the harsh edaphic conditions to gain a foothold. Remains of successive generations of these pioneer plants are broken down by

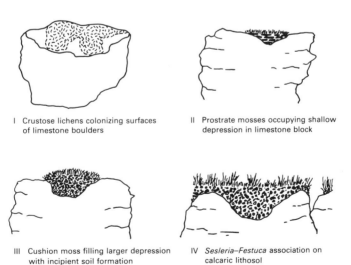

Figure 4.2: Stages in plant colonization and soil formation on hard limestone (from Bullock, 1971).

invading animals and microorganisms which gradually increase in abundance and variety. The organic matter is at first irregularly distributed and only partially decomposed and any finely divided humus that accumulates consists mainly of droppings of small Arthropods, chiefly mites (*Acarina*), Collembola and insect larvae, which play a major role in comminuting and ingesting the residues. Further development depends on the extent to which fine mineral particles are released by weathering of the rock or introduced from other sources, and on other factors, including climate and base status, that affect the evolution and activity of the soil population. Thus in cool wet upland sites on acidic rocks resistant to weathering, little mineral matter is added, organic matter accumulates in increasing amounts, and lithosols give place to peaty rankers (p. 125) in places where the arbitrary depth limit of 10 cm is exceeded. In other situations, especially on base-rich rocks which weather relatively easily, formation of a granular mull-like Ah horizon in which organic and mineral fractions are intimately mixed is promoted by accumulation of fine particles and invasion by earthworms.

In the classification used for the 1:1,000,000 soil map of the EEC countries (Commission of the European Communities, 1985b), lithosols are divided into calcaric, eutric (basic) and dystric (acidic) subgroups based on the composition of the underlying rocks, but only calcaric and non-calcaric subgroups are distinguished here. Few of these thin soils have been described and sampled for analysis and both examples recorded below come close to exceeding the prescribed depth limit.

Key to subgroups

1. Lithosols on calcareous rock

    Calcaric lithosols

2. Other lithosols

    Non-calcaric lithosols

4.2.2 Non-calcaric Lithosols

These are the lithosols on hard non-calcareous rocks. As indicated above, some have thin Ah horizons but many consist of F and H (or O) horizons only and hence qualify as Histosols according to the criteria adopted in the US and FAO systems. The following profile on Silurian greywacke in Cumbria is of the latter type.

Profile 1 (Furness and King, 1972)
Location : Killington, Cumbria (SD 596887)
Climatic regime : humid temperate
Altitude : 186 m   Slope: 1°
Land use and : rough grazing with heather
vegetation   (*Calluna vulgaris*) and sheeps
    fescue (*Festuca ovina*)

Horizons (cm)
L/F   Mat of moss, grass, heather litter and lichens, 1 cm thick.
0–9   Black humified organic material;
H(or Oh)   weak medium granular; very friable; abundant fine roots; sharp boundary.
9+   Greyish brown (10YR5/2) slightly
R   weathered greywacke.

Analytical data

Horizon	H
Depth (cm)	0–9
loss on ignition %	74
organic carbon %	38
nitrogen %	1.52
C/N ratio	25
pH (H$_2$O)	4.3
(0.01 M CaCl$_2$)	3.5

Classification
England and Wales (1984: humic ranker; loamy or peaty, lithoskeletal mudstone, sandstone or slate (Skiddaw series).
USDA (1975): Lithic Medisaprist (or Folist); dysic, mesic, micro-.
FAO (1974): Lithosol.

4.2.3 Calcaric Lithosols

These occur sporadically on Carboniferous and other hard limestones in rocky land that is

Figure 4.3: Lithosols, rankers or rendzinas, deeper brown soils (in foreground), bare rock and scree on Carboniferous limestone in Littondale, North Yorkshire (copyright Geoffrey N. Wright).

either steeply sloping, with frequent crags and screes (Figure 4.3), or consists for the most part of more gently sloping surfaces with much bare limestone pavement (Figure 4.4), as in the Burren district of County Clare (Finch *et al.*, 1971) and the Craven district of North Yorkshire (Bullock, 1971; R.A. Jarvis *et al.*, 1984). Particularly in karst landscapes of the latter type, the mineral fractions of the soils are largely derived from a thin and discontin-

Figure 4.4: Limestone pavements with associated calcaric lithosols and humic-calcaric rankers near Malham, North Yorkshire (British Geological Survey: NERC copyright).

uous cover of glacial or aeolian drift, and since the massive limestone weathers mainly by solution and climatic conditions promote strong leaching, they generally contain little or no finely divided calcium carbonate. Even those shallow enough to qualify as lithosols can therefore be acid, but a mull-type Ah horizon is formed wherever there is an appreciable thickness of loamy mineral material. The following representative profile, described and sampled under herb-rich pasture in the Burren, has a granular humose Ah horizon 9 cm thick.

Profile 2 (Finch *et al.*, 1971)

Location	: Killinaboy, County Clare
Climatic regime	: humid temperate
Altitude	: 45 m Slope: 3°
Vegetation	: semi-natural herb-rich grassland

Horizons (cm)

0–9 Ah Very dark brown (10YR 2/2) humose calcareous clay loam (USDA loam to clay loam); strong fine granular; friable; abundant roots; abrupt irregular boundary.

9+ R Carboniferous limestone bedrock.

Analytical data

Horizon	Ah
Depth (cm)	0–9
sand 200 μm–2 mm %	9
50–200 μm %	16
silt 2–50 μm %	48
clay < 2 μm %	27
organic carbon %	11.2
nitrogen %	0.76
C/N ratio	15
pH (H$_2$O)	7.1

Classification
  Ireland (1980): Lithosol/rendzina intergrade (Burren series).
  England and Wales (1984): grey rendzina?
  USDA (1975): Lithic Udorthent; fine–loamy, mixed (calcareous), mesic.
  FAO (1974): Lithosol.

4.3 Rankers

4.3.1 General Characteristics, Classification and Distribution

Kubiena (1953) introduced the name *Ranker* (from the Austrian word *Rank*, meaning a steep mountain slope) to describe more or less acid soils on non-calcareous rocks, including those then classed as lithosols or regosols in the United States (Thorp and Smith, 1949), that have a distinct organo-mineral (A) or organic surface horizon but lack a B horizon of significant thickness. The term has since been assigned various narrower meanings in European classifications (e.g. CPCS, 1967; Mückenhausen *et al.*, 1977) and is restricted in the FAO (1974) scheme to variants having a very dark coloured, base-deficient (umbric) A horizon 10–25 cm thick. As used here it embraces mineral and organic lithomorphic soils 10 cm or more thick on hard or soft non-calcareous rocks, together with shallow non-calcareous soils over hard limestone. Rankers are defined in nearly the same way by the Soil Survey of Scotland (1984) but similar soils in Ireland have been grouped as lithosols following earlier American usage of that term. According to the US Taxonomy, the mineral variants include Umbrepts and Udorthents and those with organic horizons only are Histosols.

Most soils of this group overlie more or less fragmented bedrock within 30 cm depth. Others are in screes and very stony morainic or water-laid deposits of Pleistocene age, in eroded sites on soft non-calcareous rocks, and in waste materials resulting from quarrying or mining operations. The main areas of occurrence (Figure 2.5, p. 78), nearly all with humid or perhumid climatic regimes, are in mountainous terrain and glacially eroded rocky land of lower relief on hard non-calcareous rocks (Figure 4.5).

Rankers are also locally important components of intricate soil patterns on hard Carboniferous limestone in areas such as Mendip,

Figure 4.5: Rocky land on Lewisian gneiss near Kinlockbervie, western Sutherland (Cambridge University Collection: copyright reserved).

the Craven district of North Yorkshire (Figure 4.4) and the Burren in County Clare. Individual occurrences are generally of small extent and those over hard bedrock are usually interspersed with rock outcrops and boulders or unvegetated screes as well as with lithosols and patches of deeper mineral or organic soil. On the higher Scottish mountains and others that lie within the oro-arctic zone, they appear as components of patterned ground (Chapter 1, p. 12) in which profile development has been modified or curtailed by active cryoturbation (Figures 4.6 and 6.3). There is often no continuous organic or humose surface horizon in such sites, the ground consisting largely of stones and boulders, though organic matter can occur at greater depths as irregular accumulations between the stones or as relics of former topsoils which have been disrupted or buried by coarse detritus as a result of solifluxion or colluviation. Even at the highest altitudes, however, rankers give place in relatively stable areas of mountain-top detritus to oro-arctic (cryic) humus podzols (Chapter 6, p. 254) or other soils in which a diagnostic subsurface horizon can be discerned.

Because of limitations imposed by some combination of slope, adverse climate and shallowness or rockiness, most of the land in which rankers predominate is uncultivable and bears semi-natural grassland, heath or moss-dominated vegetation used for extensive grazing or recreation, alternating at lower altitudes with fragments of semi-natural woodland and coniferous plantations. Where cultivation is possible, as on slaty rocks in parts of Devon and Cornwall, land capability is limited by stoniness, droughtiness and liability to accelerated erosion.

Key to Subgroups

1. Rankers with a peaty topsoil (p. 107) or an organic surface layer of equivalent thickness that is seldom saturated with water
   Organic (peaty) rankers

2. Other rankers with an albic E horizon (p. 108)
   Podzolic rankers

Figure 4.6: Aonach Beag and Carn Dearg Mheadonach, Ben Nevis range. Frost-patterned ground is evident on the summits and low-angle slopes (copyright Aerofilms Ltd).

3. Other rankers with a prominent A horizon (p. 106)
   3.1 with a non-calcareous substratum
   Typical (humic) rankers
   3.2 with a substratum of hard limestone
   Humic-calcaric rankers
4. Other rankers
   4.1 with a non-calcareous substratum
   Ochric rankers
   4.2 with a substratum of hard limestone
   Ochric-calcaric rankers

4.3.2 Typical (Humic) Rankers

This subgroup comprises the shallow, acid, dark coloured soils on non-calcareous rocks to which the name *Ranker* is restricted in the FAO (1974) system, together with less common variants having a prominent (mollic) A horizon that is more than 50 per cent base-saturated. Typical rankers so defined occur in association with orthic (p. 194) or podzolic (p. 232) brown soils on rocky slopes, rounded summits and stabilized screes and occasionally on stony morainic or fluvial deposits. They are mostly under semi-natural grassland, woodland or dry heath in humid upland and coastal localities, and those that are cultivated appear in some cases to have formed from shallow podzols or podzolic brown soils as a result of accelerated erosion, ploughing and manuring. Arable soils of this kind, in which a humose Ap rests directly on an indurated BC or C horizon, have been recorded by Glentworth and Muir (1963) and Futty and Dry (1977) in north-east Scotland.

Development of a prominent humic A horizon depends under cool, humid or perhumid conditions on a combination of ground-surface stability with other factors, including free drainage and weatherable parent rock, which favour deep reaching incorporation of humified organic matter rather than peat formation. In predominantly perhumid areas such as western Scotland (Bibby et al., 1982), where most soils have peaty surface horizons, typical rankers and associated brown soils are notably commoner and extend to higher altitudes on basic or intermediate crystalline rocks than on resistant acidic rocks and quartzose sandstones.

Data on two representative profiles are reproduced below, the first under grazed woodland on hard reddish pre-Cambrian siltstone in the Welsh Borderland and the second under old grass on slightly weathered dolerite in South Devon (Plate I.1). The percentage base saturation values show that the thick dark A horizon of profile 4 is a mollic epipedon. This soil therefore qualifies as a Mollisol (Udoll) in the US Taxonomy and as a Phaeozem in the FAO (1974) system.

Profile 3 (Mackney and Burnham, 1966)

Location	: Ratlinghope, Shropshire (SO 396965)
Climatic regime	: humid oroboreal
Altitude	: 275 m Slope: 14°
Land use and vegetation	: very open broadleaf woodland of beech and sycamore with few oaks; complete grass cover (grazed) with a little gorse (*Ulex europaeus*)

Horizons (cm)

L/F		Thin mat of grass stems, beech and sycamore leaves.
0–3		Black (10YR 2/1) loamy organic material containing very small mudstone fragments; fine granular; very friable; abundant fine roots; abrupt boundary.
3–18	Ah	Very dark brown (10YR 2/2) to very dark greyish brown, moderately stony clay loam (USDA loam); many hard mudstone fragments; fine and very fine granular; friable; common coarse to fine roots, mainly woody; earthworms present; gradual boundary.
18–35	AC	Fragmented hard mudstone with dark reddish brown (5YR 2/2) granular coarse sandy loam in interstices; few roots, mainly fine; gradual boundary.
35–40+	R	Nearly vertically bedded hard purplish mudstone (Longmyndian).

Analytical data

Horizon	H	Ah	AC
Depth (cm)	0–3	3–18	18–35
sand 200 μm–2 mm %		19	60
50–200 μm %		11	10
silt 2–50 μm %		46	21
clay < 2 μm %		24	9
loss on ignition %	47	18	17
organic carbon %	22.5	8.0	
nitrogen %	1.58	0.58	
C/N ratio	14	14	
pH (H$_2$O)	4.8	4.5	4.7

Classification

England and Wales (1984): humic ranker; loamy or peaty, lithoskeletal mudstone and sandstone or slate (Skiddaw series).

USDA (1975): Lithic Haplumbrept; loamy-skeletal, mixed, mesic.

FAO (1974): Ranker; medium textured (lithic phase).

Profile 4 (Clayden, 1964)

Location	: Christow, Devon (SX 836842)
Climatic regime	: humid temperate
Altitude	: 131 m Slope: 6° convex (ridge crest)
Land use	: long-term grass

Horizons (cm)

0–28	Ah	Dark brown (7.5YR 3/2), slightly stony, humose clay loam (USDA loam); common subangular fragments of dolerite and few of shale; strong fine granular; friable; abundant fine roots; earthworms common; clear irregular boundary.
28–50+	BC/Cu	Weathered and shattered dolerite (diabase) dominant; overall colour brown (7.5YR 4/4) but variegated

with weak red (10R 4/3), olive-brown and strong brown; common corestones; common fine roots and earthworm channels with casts.

Analytical data

Horizon	Ah	Ah
Depth (cm)	0–13	13–28
silt[a] 2–50 μm %	43	31
clay[a] < 2 μm %	13	22
loss on ignition %	16.2	11.9
TEB (me/100 g)	25.6	17.9
CEC[b] (me/100 g)	28.4	22.7
% base saturation	90	79
pH (H_2O)	6.3	6.0
(0.01M $CaCl_2$)	5.8	5.3

[a] by hydrometer method
[b] by leaching with N ammonium acetate at pH 7 and determining the ammonia displaced by a second leaching with sodium chloride.

Classification
England and Wales (1984): humic ranker; loamy or peaty, lithoskeletal basic crystalline rock (Preseli series).
USDA (1975): Typic Hapludoll; loamy-skeletal, mixed, mesic.
FAO (1974): Haplic Phaeozem; medium textured.

4.3.3 Humic-calcaric Rankers

These are similar soils having a prominent dark-coloured A horizon, normally humose, that rests directly on hard limestone. They differ from humic rendzinas (p. 131) in containing no more than traces of finely divided calcium carbonate and are mainly if not entirely restricted to humid or perhumid upland areas on Carboniferous or older limestones, where they occur sporadically in association with lithosols, bare limestone pavement and deeper brown soils. Although they generally have a higher base status than their counterparts on non-calcareous rocks, they can be strongly acid, especially in places where the vegetation includes ericaceous species (limestone heath). The following representative profile was described and sampled under semi-natural bent-fescue pasture in Dyfed. Data on similar soils in the Pennines have been recorded by Bullock (1971), Johnson (1971) and Hollis (1975).

Profile 5 (Wright, 1981)

Location	: Llangadog, Dyfed (SN 734190)
limatic regime	: perhumid oroboreal
Altitude	: 490 m Slope: 1°
Land use and vegetation	: rough grazing; sheeps fescue (*Festuca ovina*) dominant, with frequent common bent-grass (*Agrostis capillaris*) and wild thyme (*Thymus drucei*)

Horizons (cm)

0–14 Ah1 Very dark greyish brown (10YR 3/2) stoneless humose clay loam (USDA loam); strong fine granular; friable; many fine roots; abrupt smooth boundary.

14–23 Ah2 Abundant, medium and large, angular and subangular limestone fragments with very dark greyish brown humose clay loam (USDA loam) matrix; many fine roots; abrupt boundary.

23+ R Carboniferous limestone bed-rock.

Analytical data

Horizon	Ah1
Depth (cm)	0–14
sand 600 μm–2 mm %	2
200–600 μm %	11
60–200 μm %	22
silt 2–60 μm %	45
clay < 2 μm %	20
organic carbon %	6.3
pH (H_2O)	5.4
(0.01M $CaCl_2$)	4.7

Classification
England and Wales (1984): humic ranker; loamy or peaty, lithoskeletal limestone (Wetton series).
USDA (1975): Lithic Haplumbrept; fine loamy, mixed, mesic.
FAO (1974): Ranker; medium textured (lithic phase).

4.3.4 Organic (Peaty) Rankers

These consist essentially of an acid organic layer 10–30 cm thick over massive or fragmented bedrock, with no intervening subsurface horizon of significant thickness. Though generally regarded as rankers by European pedologists (e.g. Duchaufour, 1978, 1982), they are classed as Histosols (peat soils) in the US and FAO systems. The depth limit of 30 cm accords with current survey usage in England, Wales and Ireland rather than with that recently adopted by the Soil Survey of Scotland (1984), which permits up to 50 cm of organic material over rock.

Complex soil landscape units consisting of rocky ground with organic rankers as major or minor components occur on hard non-calcareous rocks in all the perhumid mountainous areas of the British Isles and extend practically to sea level in parts of northern Scotland (Figure 4.5) and western Ireland. All examples recorded have peaty surface horizons (peaty rankers) and variants in which the organic layer is seldom saturated with water (Folists in the US Taxonomy) are evidently of very limited extent in the British Isles. In relatively dry (unflushed) sites at lower altitudes, where heather moor and *Nardus* grassland are common vegetation types (Birse and Robertson, 1976), the peaty layer is usually well humified (Oh), especially on base-rich rocks. Where it is waterlogged for longer periods it is commonly less decomposed (Om) and the semi-natural vegetation includes hydrophilous species such as cross-leaved heath (*Erica tetralix*) and heath rush (*Juncus squarrosus*), with bog-moss (*Sphagnum* spp.) and bog asphodel (*Narthecium ossifragum*) in the wettest places.

Data on two representative profiles from northern England are reproduced below. Profile 6 was described and sampled by D. M. Carroll and J. W. Allison on hard gritstone in the eastern Pennines. The climatic regime is humid oroboreal, implying that the very acid soil at this site is subject to regular seasonal drying. Profile 7 in the Lake District is distinctly wetter; it lies just within the oro-arctic zone and the peaty surface layer contains much mineral matter, presumably as a result of cryoturbation.

Profile 6

Location	: Borden, North Yorkshire (SE 038570)
Climatic regime	: humid oroboreal
Altitude	: 236 m Slope: 6° convex
Land use and vegetation	: heather moor managed for grouse (regularly burnt); heather (*Calluna vulgaris*) dominant, with invading bracken (*Pteridium aquilinum*)

Horizons (cm)

L/F	Litter layer, mainly little decomposed heather stems and roots, c. 3 cm thick.
0–21 Om/Oh	Black (5YR 2/1; 5YR 3/2 dry; 5YR 2/2 rubbed) semi-fibrous to amorphous (humified) peat; many fine and few medium woody roots; few bracken rhizomes; fungal mycelia; abrupt wavy boundary.
21–26 Cu(Ea)	Dark greyish brown (10YR 4/2 moist and rubbed) and light brownish grey, slightly stony medium sand (USDA sand); common angular and subangular grit fragments, mainly small; single grain; loose; few fine woody roots and bracken rhizomes; sharp wavy boundary.
26+ R	Yellowish brown (10YR 5/4) hard fractured grit (Millstone Grit Series).

Analytical data

Horizon	Om/Oh	Cu/Ea
Depth (cm)	0–21	21–26
sand 600 μm–2 mm %		9
200–600 μm %		52
60–200 μm %		26
silt 2–60 μm %		12
clay < 2 μm %		1
loss on ignition %	87	
organic carbon %	46	0.4
pH (H$_2$O)	3.0	3.5
(0.01M CaCl$_2$)	2.5	2.9

Classification
  England and Wales (1984): humic ranker; loamy or peaty, lithoskeletal sandstone (Revidge series).
  USDA (1975): Lithic Medihemist; dysic, mesic, shallow.
  FAO (1974): Dystric Histosol (lithic phase).

Classification
  England and Wales (1984): humic ranker; loamy or peaty, lithoskeletal acid crystalline rock (Bangor series).
  USDA (1975): Lithic Cryohemist; dysic, shallow.
  FAO (1974): Dystric Histosol (lithic phase).

4.3.5 Podzolic Rankers

First identified as podzol-rankers by Kubiena (1953), these are non-peaty ranker-like soils with an albic E (Ea) horizon 5 cm or more thick but no diagnostic B horizon. Although recognized at subgroup level in the classifications used by the Soil Surveys of both England and Wales and Scotland, no corresponding map unit has yet been distinguished and no representative profile published, indicating that these soils have a very restricted distribution. Those noted, for example by Bown (1973) and Bibby et al. (1982) in western Scotland, are mainly under upland heath or coniferous woodland on quartz-rich rocks yielding sandy-skeletal detritus on weathering. The characteristic E horizon is overlain in such sites by a mor (L,F,H) layer and passes downwards into more or less fragmented bedrock or extremely stony transported material which usually shows signs of incipient B horizon development in the form of dark coloured or ochreous coatings. In some places the absence of a fully developed illuvial B horizon can be attributed to lateral (downslope) movement of mobilized organic matter and sesquioxides. Soils of this type remain recognizable under cultivation only where the previously formed E horizon was exceptionally thick.

4.3.6 Ochric Rankers

These are weakly developed mineral soils more than 10 cm deep that differ from typical rankers in lacking a prominent A horizon, implying that they are Orthents in the US Taxonomy and Regosols according to the FAO

Profile 7 (R.A. Jarvis et al., 1984)

Location	: Grey Knotts, Borrowdale, Cumbria (NY 223129)
Climatic regime	: perhumid oro-arctic
Altitude	: 677 m Slope: 3°, uneven and rocky
Land use and vegetation	: rough grazing with fine-leaved fescue (*Festuca* spp.) (45%), heather (*Calluna vulgaris*) (25%), bilberry (*Vaccinium myrtillus*) (20%) and bog moss (*Sphagnum* spp.) (10%)

Horizons (cm)
  L/F Fresh and partially decomposed litter, c. 5 cm thick.
  0–25 Black (5YR 2/1 broken and rubbed;
  Om 5YR 3/1 dry) slightly stony semi-fibrous peat (H5); common large sub-rounded rhyolite fragments; wet; many fine roots; gradual wavy boundary.
  25–33 Sub-rounded rhyolite fragments (> 70%)
  Ah with black humose coarse sandy loam (USDA coarse sandy loam) in interstices.
  33+
  R Grey rhyolite, very hard and massive.

Analytical data

Horizon	Om	Ah
Depth (cm)	0–25	25–33
sand 600 μm–2 mm %		39
200–600 μm %		16
60–200 μm %		5
silt 2–60 μm %		31
clay < 2 μm %		9
loss on ignition %	53	25
organic carbon %	31	13
pH (H_2O)	4.3	4.9
(0.01M $CaCl_2$)	3.4	4.1

(1974) system. Soils of this subgroup are widely distributed in the British Isles but have only been well characterized and mapped in humid and subhumid areas at altitudes below about 300 m, where they occur on non-calcareous rocks in close association with deeper brown soils. They are also represented as components of complex soil landscape units in rocky ground at higher altitudes, and on man-made excavations and waste tips. Those with a 'distinct topsoil' (Avery, 1980) are called brown rankers in the England and Wales classification.

The freely draining ochric rankers which overlie more or less fragmented bedrock in lowland situations mostly have brownish or greyish loamy or clayey A horizons less than 25 cm thick. Less well drained variants, distinguished as stagnogleyic rankers by Avery (1980), occupy localized erosional sites, chiefly in agricultural land, on slowly permeable rocks such as mudstone or shale and on compact, slightly weathered till with fragipan features. In these soils seasonal wetness is accompanied by shallow rooting and consequent susceptibility to drought, especially in subhumid areas. The montane variants commonly consist of thin ($<$ 10 cm) or discontinuous organic or humose horizons over shattered rock or scree, though both humic and ochric rankers with mull-like A horizons have been recorded at high altitudes on readily weathered base-rich rocks, for example under dwarf-herb communities on the Ben Lawers massif in the southern Grampians.

Ochric rankers resulting from colonization of non-calcareous waste materials such as colliery spoil by natural or artificially introduced vegetation display comparable variations in degree and kind of horizon development. Vegetational succession and soil development on disused pit heaps in England and Wales were studied by Hall (1957) to aid afforestation. The commonest spoil type he examined consisted mainly of unburnt shale and mudstone, which weather to give a loamy or clayey soil; others were burnt shale or mudstone and washeries waste. The latter contains much finely divided coal and is only very slowly colonized. Hall recognized nine natural vegetation types, ranging from pioneer herb-grass and heathy communities to birch and oak woodland, on spoils of the first type and described the following profile on 100-year-old spoil at Mells colliery in Somerset, with a closed cover of pedunculate oak (*Quercus petraea*).

0–8 cm	Dark brown loam; large crumb structure, mully, spongy, very porous and friable; rare small (2.5 cm) shale fragments; occasional worms; moist; well rooted by grasses and herbs; pH 4.8; fairly distinct from:
8–20 cm	Brown loam; crumb structure deteriorating with depth; occasional small shale fragments and rare larger fragments (5 cm); porous; friable (but less so than above); well rooted; worms; moist; merging into:
20–31 cm	Grey silt loam; frequent large and abundant small shale fragments; structureless; porous, slightly compacted; some rooting, chiefly by oak; moist; merging into blue-grey scarcely weathered shale (probably burnt below).

Other spoils subject to erosion lacked any semblance of soil structure, even after 100 years. The burnt spoils were initially finer in texture with fewer coarse fragments, and appeared to weather more rapidly, though structural development was slower and less marked. Spoils of both types are subject to progressive acidification with time, and this becomes severe where significant amounts of pyrite are present, as in parts of Lancashire and Yorkshire. In recent years, as described by Bradshaw and Chadwick (1980), the soil building process has been accelerated in some places by fertilization, liming and other ameliorative measures, including addition of imported topsoil, peat or sewage sludge.

Data on two representative profiles under grass on naturally occurring parent materials are given below. Profile 8, in cultivated land on Devonian slate with deeper but otherwise similar brown soils nearby, is located on a convex ridge top which has evidently undergone accelerated erosion. Profile 9 is on Carboniferous sandstone in Northumberland.

Profile 8 (Harrod et al., 1976)

Location	: Ugborough, Devon (SX 695583)
Climatic regime	: humid temperate
Altitude	: 173 m Slope: 1° convex
Land use	: grassland

Horizons (cm)

0–18 Ap — Dark brown (7.5YR 4/3) very slightly stony clay loam (USDA clay loam); few very small to medium angular platy slate fragments; moderate fine and medium subangular blocky; friable; many fine roots; gradual boundary.

18–36+ Cu/Cr — Greyish platy slate *in situ*, soft and somewhat weathered, with brown to yellowish brown (7.5–10YR 5/4) earthy material and few roots in interstices.

Analytical data

Horizon	Ap
Depth (cm)	0–18
sand 500 μm–2 mm %	23
200–500 μm %	7
50–200 μm %	5
silt 2–50 μm %	37
clay < 2 μm %	28
organic carbon %	3.1
pH (H$_2$O)	6.3
(0.01M CaCl$_2$)	5.6

Classification
England and Wales (1984): brown ranker; loamy, lithoskeletal mudstone and sandstone or slate (Powys series).
USDA (1975): Typic Udorthent; loamy-skeletal, mixed (non-acid), mesic.
FAO (1974): Eutric Regosol.

Profile 9 (George, 1978)

Location	: Ingoe, Northumberland (NZ 041749)
Climatic regime	: humid oroboreal
Altitude	: 216 m Slope: < 1°
Land use	: long-term grass

Horizons (cm)

0–22 Ap — Dark greyish brown (10YR 4/2) slightly to moderately stony fine sandy loam (USDA fine sandy loam); common sandstone fragments; moderate medium and fine subangular blocky; friable; abundant roots; gradual boundary.

22+ C/R — Fissured weathering sandstone with some fine roots penetrating fissures.

Analytical data

Horizon	Ah
Depth (cm)	0–22
sand 600 μm–2 mm %	3
200–600 μm %	14
60–200 μm %	56
silt 2–60 μm %	16
clay <2 μm %	11
organic carbon %	3.2
pH (H$_2$O)	4.6
(0.01M CaCl$_2$)	3.9

Classification
England and Wales (1984): brown ranker; loamy, lithoskeletal sandstone (Newtondale series).
USDA (1975): Lithic Udorthent; coarse-loamy, mixed, acid, mesic.
FAO (1974): Dystric Regosol; coarse textured (lithic phase).

4.3.7 Ochric-calcaric Rankers

These are shallow, brownish or reddish, non-calcareous soils over hard limestone. They occur in close association with deeper brown soils on all the main Carboniferous limestone outcrops in England and Wales and have also been recorded on Devonian limestone in south Devon (Findlay *et al.*, 1984) and Lower Palaeozoic (Durness) limestone in north-west Scotland (Bibby *et al.*, 1982). The following representative profile, located in rough grazing land on the northern slope of the Mendip Hills, is developed in a thin cover of loess mixed with limestone residuum (Findlay, 1965).

Profile 10 (Findlay, 1965; Ragg and Clayden, 1973)

Location	: Blagdon, Avon (ST 517577)
Climatic regime	: humid temperate
Altitude	: 230 m Slope: 5°
Land use and vegetation	: rough grazing, mainly fescue grasses with some wild white clover (*Trifolium repens*); abundant bracken (*Pteridium aquilinum*) and some gorse (*Ulex europaeus*) and hawthorn (*Crataegus monogyna*) scrub nearby.

Horizons (cm)

0–20 Ah — Dark brown to brown (7.5YR 4/2–5YR 4/3) slightly stony silty clay loam (USDA silt loam); stones mainly of black chert near surface, predominantly dark grey limestone below; granular structure and abundant roots in upper 5 cm; moderate fine subangular blocky and many roots below; friable; sharp irregular boundary.

20+ 2R — Dark grey, massive, micro-crystalline limestone in large cobbles.

Analytical data

Horizon	Ah
Depth (cm)	0–20
sand 200 μm–2 mm %	7
50–200 μm %	10
silt 2–50 μm %	58
clay < 2 μm %	25
organic carbon %	3.9
TEB (me/100 g)	12.2
CEC[a] (me/100 g)	18.7
% base saturation	71
pH (H$_2$O)	6.0
(0.01M CaCl$_2$)	5.5

[a] by leaching with N ammonium acetate at pH 7 and determining the ammonium displaced by a second leaching with sodium chloride.

Classification

England and Wales (1984): brown ranker; loamy, lithoskeletal limestone (Crubin series, formerly Lulsgate).

USDA (1975): Lithic Udorthent; fine silty, mixed (non-acid), mesic.

FAO (1974): Eutric Regosol; medium textured (lithic phase).

4.4 Rendzinas

4.4.1 General Characteristics, Classification and Distribution

The Polish name *Rendzina*, originally applied to 'intrazonal' soils on calcareous rocks by Sibertsiev (1901), is restricted in the FAO (1974) system to those having more than 40 per cent calcium carbonate in or immediately below a mollic A horizon (Rendolls in the US Taxonomy). In Britain and other west European countries, however, its meaning has customarily been extended to embrace closely related soils, including variants classed as Regosols by FAO (1974) and as Orthents in the United States, with less dark coloured greyish or brownish surface horizons. According to the classification adopted here, the lithomorphic soils grouped as rendzinas have formed in materials other than sands and recent alluvial deposits, are at least 10 cm deep, and have finely divided carbonates in and/or directly beneath the topsoil. Those on substrata containing less than 40 per cent CaCO$_3$ are distinguished at subgroup level as pararendzinas but are comparatively rare in the British Isles.

Rendzinas in this broader sense are widespread on chalk, chalky drift, and Jurassic and Permian limestones in southern and eastern England and the Midlands (Figure 2.5, p.78). They also occur locally over Carboniferous limestone in northern England, Wales and Ireland, and in very small areas on limestones and calc-silicate rocks in northern Scotland. Those that remain uncultivated and

Figure 4.7: The escarpment of the South Downs near Ditchling, East Sussex. Grey and brown rendzinas predominate in the cultivated land and typical (humic) rendzinas under old grassland on the scarp face (Cambridge Collection: copyright reserved).

unfertilized support distinctive woodland, scrub and grassland plant communities (Tansley, 1939; Smith, 1980) including a variety of 'calcicolous' species which are rare or absent on non-calcareous soils nearby. As indicated in Figure 4.7, shallow dark coloured soils conforming to the central concept of the group (Plate I.2) are mainly but not exclusively confined to such sites and generally give place in farmland to the much more extensive grey and brown rendzinas, which normally contain less organic matter. The grey rendzinas are mainly on chalk and are usually extremely calcareous throughout, whereas brown rendzinas (Plate I.3) have less calcareous surface horizons containing weathered material of residual or extraneous origin.

Characteristic features of rendzinas, most evident where organic matter levels remain high, are pronounced faunal and microbial activity leading to rapid incorporation and decomposition of plant residues, stable aggregation and consequent free aeration, all of which are favourable to plant growth. Other less favourable physical and chemical properties limit the productive capacity and use range of these soils to varying degrees. Those over hard non-porous limestones have small available water capacities and are more or less stony, making them unsuitable for root crops and others requiring a greater depth of soil. Those on chalk are generally much less drought-susceptible because large amounts of water are retained by all but the hardest chalk strata at suctions well below wilting point (Smith, 1980, Table 3.2). There is also evidence (Wellings, 1984) that it can rise by capillarity from permanently or temporarily saturated layers several metres below the surface, though the ability of plants to extract potentially available water depends on their rooting habit and the degree and depth to which the rock is fragmented.

On rendzinas containing 'active' (clay-size) calcium carbonate, plant nutrition is affected adversely in several ways. In the first place the mineralization of organic nitrogen is retarded (Duchaufour, 1982, p. 43) and phosphorus and minor elements, particularly iron, are liable to be immobilized. In addition, ammonia nitrogen may be lost in gaseous form, and the dominance of calcium in the exchange complex and in the soil solution exacerbates inherent deficiencies in potassium and magnesium. For relatively undemanding crops such as cereals, these conditions can be largely alleviated by fertilization. Crops less tolerant of high pH, including some soft and tree fruits and most conifers, are less responsive, however, and may suffer severe chlorosis, or even early death, which usually results in such cases from some combination of nutrient deficiency, drought, and increased susceptibility to disease, for example the attack of pines and other conifers by the root rotting fungus *Fomes annosus*.

Research and experience in afforesting chalk downland were reviewed by Wood and Nimmo (1962) and later investigations involving the use of mechanical pretreatment (ripping), N, P and K fertilizers and herbicides are described by Fourt (1973, 1975). Beech dominates semi-natural woods on chalk rendzinas and has also been widely planted, usually with a nurse crop of conifers, though its performance on these soils is notably poorer, especially on exposed slopes, than on deeper and less calcareous soils nearby (Brown, 1953). Among the conifers, Corsican and Austrian pines (*Pinus nigra* subsp. *laricio* and *nigra*) have proved most successful.

Key to Subgroups

1. Rendzinas with a prominent A horizon (p. 106) directly over carbonatic (>50 per cent $CaCO_3$) bedrock or extremely calcareous (p. 85) unconsolidated material
   Typical (humic) rendzinas
2. Other rendzinas in recent colluvium (p. 114) more than 40 cm thick
   Colluvial rendzinas
3. Other rendzinas with an A horizon, a thin B, or both, that contains less than 40 per cent $CaCO_3$ (<2 mm fraction); has a moist colour value of 3 or 4, a chroma of 3 or more and a hue of 10YR or redder; and directly overlies carbonatic bedrock or extremely calcareous unconsolidated material
   Brown rendzinas
4. Other rendzinas
   4.1. with carbonatic bedrock or extremely calcareous unconsolidated material directly below the A horizon
   Grey rendzinas
   4.2 with non-carbonatic (<50 per cent $CaCO_3$) bedrock or slightly to very calcareous (p. 85) unconsolidated material directly below the A horizon.
   Pararendzinas

4.4.2 Typical (Humic) Rendzinas

These are the rendzinas *sensu stricto* (Rendolls) with a prominent mollic A horizon. As this nearly always contains enough organic matter to qualify as a humose topsoil (p. 107), the subgroup recognized here as typical corresponds closely to humic rendzinas in the England and Wales system. As already noted, most of these soils are under semi-natural grassland, scrub or woodland, and it is only in a few places, typified by profile 12 below, that their salient characteristics are retained under cultivation. The main areas of occurrence are on the English Chalk in counties south of the Thames, particularly Wiltshire, where much downland under military occupation has remained unploughed. Typical rendzinas also occur locally on Mesozoic and Palaeozoic limestones, limestone screes and other calcareous drift, but generally give place to humic-calcaric rankers (e.g. profile 5, p. 124) on hard, little fragmented bedrock, especially in humid upland localities where any finely divided carbonates produced by physical weathering are quickly lost by leaching.

The relatively large organic matter contents characteristic of these soils are attributable to the action of finely divided calcium carbonate in stabilizing slightly transformed humic compounds and retarding their biodegradation. According to Duchaufour (1982), the calcium-saturated humus of temperate rendzinas contains considerably smaller amounts of highly polymerized humic acids and humins than that of the superficially similar mollic epipedons formed under continental grasslands (chernozemic humus). Where the soil contains significant amounts of silicate clay and other conditions favour earthworm activity, the A horizon consists of very dark coloured calcareous mull with strongly developed fine subangular blocky or granular structure. It is normally less than 25 cm thick, but can be somewhat thicker in receiving sites as a result of slow colluviation (e.g. Canti, 1983). In shallower, more stony and more drought susceptible variants, particularly on exposed slopes over hard chalk or limestone, earthworms are largely replaced by other invertebrates, such as insect larvae, ants, mites, woodlice and millipedes. These produce A horizons with 'humus forms' termed mull-like rendzina moder and rendzina moder by Kubiena (1953), in which the fine fraction is richer in organic matter and has a finer, predominantly granular (coprogenic) structure of low bulk density. Organic matter also accumulates in large amounts under cooler and more humid climatic regimes, especially over limestone at higher altitudes. The typical rendzinas that retain their identity under these conditions are often non-calcareous at the surface and grade into the limestone rankers described above. Some uncultivated rendzinas on chalk are also non-calcareous in the upper 10–15 cm, particularly in places where the surface horizon is largely loess-derived and is separated from the chalky C horizon by a nearly continuous layer of flints. In extreme cases, typified by the 'chalk-heath soils' described by Perrin (1956), Grubb *et al.* (1969) and Burnham (1983a), the vegetation is characterized by intimate admixture of chalk grassland species with acidophile plants including heathers (*Calluna vulgaris* and *Erica cinerea*), and the upper few centimetres of soil are moderately or even strongly acid.

Data on three representative profiles are given below. The first was described and sampled by the author under ungrazed chalk grassland on the Chiltern escarpment. It overlies flint-free Middle Chalk and the A horizon is extremely calcareous. Profile 12, on Upper Chalk in Wiltshire, is located in gently sloping downland ploughed from old grass during the 1940s and represents a flinty cultivated phase of the same soil series (Icknield). Adequate consolidation of seedbeds is often difficult on 'black puffy' chalk soils of this type, and cereal crops are subject to a copper deficiency disorder known as 'blackening' which can reduce grain yields seriously and is alleviated by foliar or soil applications of copper sulphate (Davies *et al.*, 1971). The third example (profile 13) is developed in rubbly limestone scree mantling a steep valley-side slope in the southern Cotswolds. Similar soils occur on scree slopes over Carboniferous limestone in the Pennines (Johnson, 1971; Bullock, 1971) and elsewhere but generally contain smaller amounts of finely divided calcium carbonate.

Profile 11 (Avery, 1964)

Location	: Ivinghoe, Buckinghamshire (SP 959165)
Climatic regime:	subhumid temperate
Altitude	: 206 m Slope: 15°
Vegetation	: ungrazed chalk grassland dominated by upright brome grass (*Bromus erectus*), with frequent sheeps fescue (*Festuca ovina*) and salad burnet (*Sanguisorba minor*)

Horizons (cm)

0–18 Ah — Dark brown (7.5YR 3/2; 10YR 3/2 dry), slightly stony, extremely calcareous, humose clay loam (USDA loam to clay loam); common, very small to medium, subangular chalk fragments and rare small flints; strong fine and medium granular in upper 5 cm, passing downwards into fine subangular blocky; friable; abundant fine and few medium fleshy roots; clear smooth boundary.

18–33 AC — Brown-stained chalk fragments (c. 70 per cent) with brown to very dark greyish brown extremely calcareous loamy material in interstices; few roots; gradual boundary.

33+ Cu/Cr — White, fissured and fragmented chalk *in situ*, becoming progressively less fissured with increasing depth.

Analytical data

Horizon	Ah
Depth (cm)	0–18
sand 200 µm–2 mm %	12
50–200 µm %	11
silt 2–50 µm %	50
clay <2 µm %	27
organic carbon %	9.4
nitrogen %	0.92
C/N ratio	10
CaCO₃ equiv. %	58
pH (H₂O)	7.7
(0.01M CaCl₂)	7.6

Classification
England and Wales (1984): humic rendzina; loamy, lithoskeletal chalk (Icknield series).
USDA (1975): Typic Rendoll; loamy-skeletal, mesic.
FAO (1974): Rendzina; medium textured.

Profile 12 (Cope, 1976)

Location	: Steeple Langford, Wiltshire (SU 053383)
Climatic regime:	subhumid temperate
Altitude	: 145 m Slope: 1°
Land use	: arable; freshly ploughed barley stubble with furrowed surface.

Horizons (cm)

0–21 Ap — Very dark greyish brown (10YR 3/2), moderately stony (22%), humose calcareous silty clay loam (USDA silt loam); many, mainly medium and small, angular and subangular flints and few small chalk fragments; moderate medium subangular blocky with a few large clods breaking to very fine and fine subangular blocky; friable; few fine and common fine to medium rhizomatous (*Agropyron repens*) roots; abrupt smooth boundary.

21–40 2AC — Very stony (50–70 per cent); medium tabular chalk fragments and few large flints with dark yellowish brown surface staining and fine earth similar to above in interstices; few roots; gradual smooth boundary.

40–60 2Cu — Weakly bedded, medium tabular chalk fragments with few large flints and light yellowish brown (10YR 6/4) earthy material lining fissures; rare fine roots.

Analytical data

Horizon	Ap
Depth (cm)	0–21
sand 200 µm–2 mm %	10
60–200 µm %	3
silt 2–60 µm %	65
clay <2 µm %	22
organic carbon %	8.6

CaCO₃ equiv.		
<2 mm %	17	
<2 µm %	2	
pH (H₂O)	8.1	
(0.01M CaCl₂)	7.4	

Classification

England and Wales (1984): humic rendzina; loamy, lithoskeletal chalk (Icknield series).

USDA (1975): Typic Rendoll; loamy-skeletal, mesic.

FAO (1974): Rendzina, medium textured.

Profile 13 (Findlay, 1976)

Location	: West Yatton, Wiltshire (ST 851761)
Climatic regime:	subhumid temperate
Altitude	: 106 m Slope: 20°
Vegetation	: semi-natural grassland with abundant calcicole herbs

Horizons (cm)

0–5 Ah1	Dark brown (7.5YR 3/2) stoneless, slightly calcareous humose clay (USDA clay); strong very fine subangular blocky; friable; abundant fine roots.	
5–24 Ah2	Dark brown (7.5YR 3/2), moderately to very stony, very calcareous humose clay (USDA clay); many rounded 1–2 cm and common subangular 2–8 cm fragments of oolitic and shelly limestone; strong very fine to fine angular blocky; friable; many fine and common fleshy herb-roots; earthworms present.	
24–30 AC	Angular limestone fragments up to 20 cm across with sparse brown (7.5YR 4/2–4) clay loam matrix; structure where discernible as above; fewer fine roots but medium roots penetrating below this depth.	
30–50 Cu	Angular limestone rubble with 'mealy' calcareous matrix or voids between the stones.	

Analytical data

Horizon	Ah1	Ah2
Depth (cm)	0–5	5–24
sand 200 µm–2 mm %	2	14
50–200 µm %	8	8
silt 2–50 µm %	30	29
clay <2 µm %	60	49
loss on ignition %	29	21
CaCO₃ equiv. %	3.2	18
pH (H₂O)	7.7	7.2
(0.01M CaCl₂)	7.9	7.4

Classification

England and Wales (1984): humic rendzina; clayey, lithoskeletal limestone (Yatton series).

USDA (1975): Typic Rendoll; clayey-skeletal, mesic.

FAO (1974): Rendzina, fine textured.

4.4.3 Grey Rendzinas

These are shallow or weakly developed soils, mostly loamy, with greyish or pale brown A horizons that are extremely calcareous or immediately overlie extremely calcareous material. They are typified by the 'white chalk soils' (chiefly Upton and Wantage series) which cover a considerable total area on chalk and chalky drift in southern and eastern England and evidently result mainly from 'degradation' of former typical or brown rendzinas under continued cultivation, accompanied in places by acclerated erosion. Others are in localized erosional sites on older calcareous rocks, particularly the Permian dolomitic limestones in Yorkshire and the north Midlands (Wetherby series), and in waste materials arising from exploitation of chalk or limestone for building, industrial and agricultural purposes. Variants with ochreous mottles attributable to gleying between 40 and 80 cm depth appear locally in low lying or spring-line sites with impervious substrate at moderate depths. These are grouped with associated calcaric gley soils (Chapter 7, p. 328) as gleyic rendzinas in the England and Wales classification and have not been separated from them in soil surveys.

Although most of these soils are cultivated, they also occur under semi-natural grassland or woodland which was formerly cultivated or occupies steeply sloping or otherwise unstable sites where a very dark coloured A horizon thick enough to qualify as prominent (mollic) has either not formed or been truncated or destroyed by recent erosion, excavation, or disturbance by burrowing animals. Man-made exposures of calcareous rock or rock waste have been produced over the centuries in numerous sites ranging from prehistoric earthworks to the large quarries now in use and the vegetation which colonizes such sites has been much studied, especially on the English Chalk. The nature and rate of soil development have received comparatively little attention, though Locket (1946) recorded signs of organic matter accumulation in the top 5 cm of physically weathered chalk in an abandoned pit only 13 years after it had ceased to be worked. Judging by the ecological investigations reviewed by Smith (1980), it seems that formation of a closed community on chalk detritus may take as much as 50 years, but observations on older exposures show that development of a typical rendzina requires freedom from disturbance over a much longer period. Screes or other skeletal detritus composed of harder limestone are generally colonized more slowly, often by woody species which root in the crevices.

Three representative profiles on fragmented chalk are recorded below. The first, in land which has probably never been cultivated, was described by the author in the steeply sloping side of a scarp-foot coombe in the Chilterns. Along with Box Hill in Surrey, this is one of the few localities where box (*Buxus sempervirens*) grows wild in England. Parts of the slope support dense box wood with typical rendzinas (Icknield series) in places, but the site of the profile was nearly bare and had a thin superficial layer of small chalk stones (lag gravel) resulting from sorting by water erosion which followed death of old trees.

Profile 15 is a typical cultivated 'white chalk soil' on periglacially disturbed Upper Chalk with flints in Wiltshire. The fine earth (<2 mm) fraction of the Ap horizon contains 63 per cent $CaCO_3$, of which 14 per cent is clay-sized. Soils of this type become very sticky when wet and some compaction can occur if they are worked with heavy machinery in spring. The nearly white appearance of the soil surface when dry is due partly to the presence of thin calcitans (films of calcite reprecipitated from solution) which line voids and help to stabilize the structural aggregates (Cope, 1976). Some of these soils retain more than 2.5 per cent organic carbon in the Ap horizons, which then qualify as mollic according to the criteria used in the US and FAO systems, but most of those sampled in southern and eastern England have smaller amounts. As the carbon content criterion cannot be applied in the field, all the chalk soils with A horizons having moist value/chroma ratings of 4/2, 5/3 or paler and no underlying Bw horizon are considered here as grey rendzinas in accordance with the England and Wales classification.

Profile 16 was described and sampled by the author in a spoil bank alongside the Tring cutting, Hertfordshire, which was excavated in Lower Chalk during the construction of the London–Birmingham railway in 1837. The bank has been farmed as part of the adjacent field for many years since, and the 150-year old profile, with 1.7 per cent organic carbon in the Ap horizon, differs in no essential respect from a somewhat eroded profile (No. 27 in Avery, 1964, p. 95) sampled on *in situ* marly chalk nearby.

Profile 14

Location	: Great Kimble, Buckinghamshire (SP 832056)
Climatic regime:	subhumid temperate
Altitude	: 170 m Slope: c. 30°

| Land use and vegetation | : | nature reserve; clearing in box (*Buxus sempervirens*) woodland; sparse (<5 per cent) ground cover including shoots of box and dogwood (*Cornus sanguinea*), stunted fine-leaved fescues (*Festuca* spp.), wild thyme (*Thymus* spp.), woodsage (*Teucrium scorodonia*) and daisy (*Bellis perennis*) |

Horizons (cm)
0–1	Loose superficial layer of subrounded chalk fragments with few of angular flint, mainly small (<2 cm).
1–9 Ah	Brown (10YR 5/3), moderately stony, extremely calcareous silty clay loam (USDA silt loam); many small and medium chalk and few flint fragments; moderate fine subangular blocky, friable; common fine and few medium to coarse woody roots; abrupt irregular boundary.
9–40 AC	Medium and small fragments of chalk (70–80 per cent) and few of flint, with brown soil similar to above in interstices; few woody roots; gradual boundary.
40–45+ Cu	Fragmented chalk with little fine earth.

Classification
England and Wales (1984): grey rendzina; loamy, lithoskeletal chalk (Upton series).
USDA (1975): Typic Udorthent; loamy-skeletal, carbonatic, mesic.
FAO (1974): Calcaric Regosol; medium textured.

Profile 15 (Cope, 1976)

Location	:	Stapleford, Wiltshire (SU 079383)
Climatic regime:	subhumid temperate
Altitude	:	122 m Slope: 4° convex
Land use	:	arable (barley stubble)

Horizons (cm)
0–25 Ap	Brown (10YR 5/3; 10YR 6/2 dry), moderately stony (21 per cent), extremely calcareous silty clay loam (USDA silt loam); common small to medium angular flint and small and very small chalk fragments; moderate fine and very fine subangular blocky; friable when moist, slightly hard when dry; few fine roots and straw fragments; abrupt smooth boundary.
25–35 AC	Fragmentary chalk and few flint fragments (70 per cent) with brown (10YR 5/3) extremely calcareous silty material in interstices; few fine roots; clear smooth boundary.
35–60 Cu	Tabular chalk fragments with a white chalky matrix and very few large flints; few yellowish patches; few fine roots.

Analytical data

Horizon	Ap
Depth (cm)	0–25
sand 200 μm–2 mm %	8
50–200 μm %	3
silt 2–60 μm %	65
clay <2 μm %	24
organic carbon %	2.3
CaCO₃ equiv.	
<2 mm %	63
2–60 μm %	33
<2 μm	14
pH (H₂O)	8.0
(0.01M CaCl₂)	7.7

Classification
England and Wales (1984): grey rendzina; loamy, lithoskeletal chalk (Upton series).
USDA (1975): Typic Udorthent; loamy-skeletal, carbonatic, mesic.
FAO (1974): Calcaric Regosol; medium textured.

Profile 16

Location	:	Tring, Hertfordshire (SP 937138)
Climatic regime:	subhumid temperate
Altitude	:	135 m Slope: 1° convex
Land use	:	arable (cereal stubble)

Horizons (cm)
| 0–18 Ap | Greyish brown (2.5Y 5/2), moderately stony, extremely calcareous silty clay loam (USDA silt loam); many subangular and tabular argillaceous chalk |

	fragments, mainly small; moderate fine subangular blocky with some granular; friable; common fine roots; worm holes and casts; abrupt irregular boundary.
18–50 Cu	Light grey (2.5Y 7/2), very to extremely stony, extremely calcareous sandy silt loam (USDA loam) with inclusions of soil similar to above in channels to 40 cm; abundant, large, irregularly arranged argillaceous chalk fragments, some with black coats and few yellowish stains; weak fine subangular blocky; fine earth friable; few fine roots concentrated in earthy pockets.

Analytical data

Horizon	Ap
Depth (cm)	0–18
sand 200 µm–2 mm %	5
60–200 µm %	6
silt 2–60 µm %	63
clay <2 µm %	26
organic carbon %	1.4
CaCO$_3$ equiv.	
<2 mm %	72
<2 µm %	11
pH (H$_2$O)	8.2
(0.01M CaCl$_2$)	7.7

Classification
England and Wales (1984): grey rendzina; loamy, lithoskeletal chalk (Upton series).
USDA (1975): Typic Udorthent; loamy-skeletal, carbonatic, mesic.
FAO (1974): Calcaric Regosol; medium textured.

4.4.4 Colluvial Rendzinas

These are deep rendzina-like soils, including the colluvial rendzinas and pararendzinas of Avery (1973, 1980), with greyish subsurface horizons in which organic matter contents decrease gradually or irregularly with increasing depth. They are widely distributed in small, usually elongate, areas in dry valley bottoms and at the base of slopes on the English Chalk but have not so far been identified elsewhere. The parent deposits, often called slopewash or ploughwash, have been dated as Holocene (Flandrian) and differentiated from the Devensian head or melt-water sediments which they commonly overlie by archaeological, palaeontological (molluscan) and radiocarbon evidence (Kerney, 1963; Kerney et al., 1964; Evans, 1966; Valentine and Dalrymple, 1976; Bell, 1983). Unlike the older deposits, the upper surface of which is marked in places by a buried A horizon, the recent colluvium shows little or no evidence of post-depositional decalcification and is typically most calcareous at the surface, reflecting progressive truncation by erosion of the slope soils from which it was derived. Soils of this and other colluvial subgroups recognized here are grouped with alluvial soils as Fluvisols (Fluvents) in the FAO and US systems.

The following representative profile was described and sampled by R.D. Green in the floor of a dry valley indenting the escarpment of the North Downs near Wye, Kent. It has a weakly expressed A horizon passing into a greyish BC horizon which contains about 1 per cent organic carbon throughout and has a moderately well developed subangular blocky structure but shows no other sign of pedogenic modification. As the 80–90 cm layer contains more carbon and less CaCO$_3$ then the overlying layers, it is identified as a buried A horizon and is almost certainly coeval with a buried soil described at the same depth in the nearby Devil's Kneading Trough by Kerney et al. (1964). They showed that the overlying colluvium was emplaced during the Sub-Atlantic period (Zone VIII), presumably as a result of intensive arable farming on the slopes above the coombe, and attributed fragments of iron slag resembling those found in the bAH horizon of profile 17 to Iron Age smelting in the vicinity. Typical rendzinas with prominent A horizons occur locally in valley bottoms where the ground has been less disturbed by recent cultivation.

Profile 17

Location : Wye, Kent (TR 079448)
Climatic regime: subhumid temperate
Altitude : 84 m   Slope: 1–2°
Land use : long-term grassland, formerly orchard

Horizons (cm)
0–23 Ah   Greyish brown (10YR 5/2) extremely calcareous silty clay loam (USDA silty clay loam) containing very few small chalk and shell fragments; moderate fine subangular blocky; friable; many fine roots; gradual boundary.
23–43 BC1   Greyish brown (2.5Y 5/2), very slightly stony, extremely calcareous silty clay loam (USDA silty clay loam); few chalk and shell fragments; moderate fine and medium subangular blocky; friable; common fine roots; gradual boundary.
43–80 BC2   Greyish brown to dark greyish brown (10YR 5/2–4/2), slightly stony, extremely calcareous silty clay loam (USDA silty clay loam); common chalk and few shell fragments; moderate medium and fine prismatic breaking to blocky; friable; few fine roots; gradual boundary.
80–90 bAh   Dark brown (10YR 3/3), very slightly stony, very calcareous silty clay loam (USDA silty clay loam); few medium subangular flints; few charcoal and iron-slag fragments; moderate fine and medium blocky to prismatic; firm; less porous than above; roots rare.

Analytical data

Horizon	Ah	BC1	BC2	bAh
Depth (cm)	0–23	23–43	43–80	80–90
sand 200 μm–2 mm %	5	6	6	6
60–200 μm %	8	7	8	9
silt 2–60 μm %	58	56	53	48
clay <2 μm %	29	31	33	37
organic carbon %	1.9	1.0	1.0	1.3
CaCO$_3$ equiv. <2 mm %	67	57	55	30
<2 μm %	11	10	7	4
pH (H$_2$O)	8.3	8.2	8.3	8.3
(0.01M CaCl$_2$)	7.8	7.8	7.8	7.8

Classification
England and Wales (1984): colluvial rendzina; carbonatic-loamy colluvium (Gore series).
USDA (1975): Typic Udifluvent; fine-loamy, carbonatic, mesic.
FAO (1974): Calcaric Fluvisol; medium textured.

4.4.5 Brown Rendzinas

As defined by Avery (1973, 1980), these are well drained soils with a thin brown or reddish B (normally Bw) horizon, a brownish A, or both, and more or less fragmented limestone, chalk or extremely calcareous unconsolidated material at or within 30 cm depth. The surface horizons are required to contain less than 40 per cent CaCO$_3$ (<2 mm fraction) and may be non-calcareous, but finely divided carbonates occur in or immediately beneath the solum. A thin, very dark, strongly aggregated Ah (mull) and a brighter coloured B (or AB) horizon are usually recognizable in woodland or old grassland profiles. The Ap horizons of cultivated phases (the anthropic brown rendzinas of Duchaufour, 1978, 1982) are typically brown or reddish brown, with value/chroma ratings of 4/3 or 4/4 in 10YR or redder hues.

Soils of this kind on limestone were first called brown rendzinas by Kubiena (1953), who attributed the development of a brownish Bw (or AB) horizon to advanced decalcification accompanied by liberation of ferric hydroxides from the solution residue of the rock, though in many places the 'browned' horizons are clearly derived mainly from loess or other extraneous material. In the FAO (1974) and US systems, variants with a Bw horizon extending to 25 cm or more below the surface are considered to have a cambic horizon (Chapter 3, p. 109) and are consequently classed as Cambisols or Inceptisols (Eutrochrepts) respectively. Others, including those in which a brown Ap rests directly on fragmented limestone with little fine earth, are grouped with very weakly developed soils as Regosols (US Orthents) despite evidence that the parent materials of the surface horizons have been considerably altered by weathering. In many English chalk and limestone landscapes, as in their French counterparts (e.g. Ducloux, 1971; Duchaufour, 1978), the solum depth fluctuates around 25 cm and reflects parent material irregularities produced by Devensian cryoturbation, which has resulted in irregular admixture of fragmented rock with thin superficial deposits or pre-Devensian solution residues. A variability study by Courtney and Webster (1973) of the extensive Sherborne map unit (Findlay, 1976) in the southern Cotswolds established a modal depth of 22 cm (mean 24 cm) to rock or rock rubble and showed that some 85 per cent of the included soils were between 15 and 35 cm deep. The 'stonebrash' soils characteristic of this and other Jurassic limestone cuestas to the north and south are thus partly brown rendzinas and partly deeper calcaric brown soils (p. 181) in proportions that vary from area to area.

Judging from the 1:250,000 soil-association maps (Mackney et al., 1983), brown rendzinas are considerably more extensive in England and Wales than those of the typical (humic) and grey subgroups. Though typically represented on Jurassic and Permian (Magnesian) limestones (Plate I.3), they are also widespread on chalk and chalky drift in the southern and eastern counties, and occur sporadically on older limestones to the north and west. The few rendzinas recorded in Scotland mostly conform to this subgroup, as do some of the soils identified as brown earths (Ballincurra series) or rendzinas (Kilcolgan series) on shallow calcareous drift in Western Ireland (Finch and Ryan, 1966; Finch et al., 1971).

Data on three representative profiles are given on pp. 140–2. Profile 18, described and sampled by F.M. Courtney, typifies the cultivated brown rendzinas on more or less disturbed and fragmented Jurassic limestone, clayey, fine loamy and coarse loamy variants of which are currently identified by the Soil Survey as Sherborne, Elmton and Marcham series respectively (Clayden and Hollis, 1984). These soils contain significant amounts of iron oxide, chiefly as goethite, which is derived from iron-bearing minerals in the parent rock and evidently promotes strong micro-aggregation, since particle-size analyses preceded by citrate-dithionite treatment to remove free iron yield significantly more clay than the conventional procedure using sodium hexametaphosphate for dispersion. Though not recorded in the profile described, another common feature is the occurrence, especially in subhumid areas, of a BCk or Ck horizon marked by incrustations of redeposited (secondary) calcium carbonate on the under sides of limestone fragments at depths between 25 and 60 cm. The available water capacities of these soils are often small but vary considerably from place to place depending on the depth to hard unbroken rock and the proportion, size and porosity of rock fragments in overlying horizons.

Profile 19 on periglacially disturbed Upper Chalk was sampled on a nearly level ridge top in the same area as profiles 12 and 15 above. Particle-size and mineralogical analyses (Cope, 1976) indicate that the fine earth is largely derived from loess, which has been mixed with fragmentary chalk and frost-shattered flints to a depth of 50–60 cm. The Ap horizon is only slightly calcareous and contains 5 per cent organic carbon, suggesting that the soil has not been cultivated for very long. Similar soils of the Andover series are widely distributed on the chalk downs south of the Thames and also on the Yorkshire Wolds (R.A. Jarvis et al., 1984), where the chalk is generally harder. They are often intricately interspersed with grey rendzinas and deeper brown soils in more or less regular

Figure 4.8: Chalk downland in Dorset, showing a pattern of small ancient fields within the larger modern enclosures (photo by D.C. Findlay).

patterns which may be periglacial in origin or reflect old field boundaries (Figure 4.8). The Andover soils mapped in Wessex are mostly more calcareous than the example described and are often difficult to distinguish consistently in the field from the grey rendzinas (Upton series) into which they grade as the $CaCO_3$ content increases to 40 per cent and more. Analogous coarse and fine loamy soils (Newmarket and Rudham series) in which the extraneous material contains more sand and less silt occur in eastern England (Hodge et al., 1984) and over smaller areas elsewhere.

Profile 20, recorded under old grassland on a drumlin in County Clare, is developed in medium textured highly calcareous glacial drift of Devensian (Midlandian) age. Counts of stones 0.6–2.5 cm in diameter showed that 92 per cent were of Carboniferous limestone, but mineralogical analyses of fine sand and clay fractions suggest strongly that much of the fine earth is derived from other sources, including igneous rocks and micaceous schists.

Profile 18

Location   : Temple Guiting, Gloucestershire (SP 121288)
Climatic regime: subhumid temperate
Altitude   : 237 m   Slope: 1°
Land use   : ley grass
Horizons (cm)

0–5 Ap1	Dark yellowish brown (10YR 3/4), slightly stony, very calcareous clay (USDA clay); common small to medium subangular limestone fragments; strong fine and medium subangular blocky; friable; many fine roots; clear smooth boundary.
5–24 Ap2	Dark brown to brown (7.5YR 4/4), moderately stony very calcareous clay (USDA clay); many limestone fragments as above; strong fine angular blocky; friable; common fine roots; clear irregular boundary.

24–30+ Medium to large, subangular oolitic
Cu   limestone fragments with a little
       brown earthy material in interstices.

Analytical data

Horizon	Ap1	Ap2
Depth (cm)	0–5	5–24
sand 200 µm–2 mm %	13	20
60–200 µm %	5	4
silt 2–60 µm %	34	36
clay <2 µm %	48	40
organic carbon %	4.8	4.0
$CaCO_3$ equiv. %	16	16
dithionite ext. Fe %	5.2	5.4
pH (H_2O)	8.0	8.1
(0.01M $CaCl_2$)	7.6	7.7

Classification
   England and Wales (1984): brown rendzina; clayey, lithoskeletal limestone (Sherborne series).
   USDA (1975): Typic (or Lithic) Udorthent; clayey-skeletal; carbonatic or oxidic, mesic.
   FAO (1974): Calcaric Regosol; fine textured.

Profile 19 (Cope, 1976)

Location	: Stapleford, Wiltshire (SU 087387)
Climatic regime:	subhumid temperate
Altitude	: 137 m Slope: 1°
Land use	: arable (wheat stubble)

Horizons (cm)

0–25   Dark brown (10YR 4/3), moderately
Ap   stony (26%), calcareous silty clay
        loam (USDA silt loam); common
        medium to large and many small and
        very small angular flint fragments;
        moderate fine and very fine subangu-
        lar blocky; friable; common fine and
        medium rhizomatous roots and straw
        fragments; abrupt smooth boundary.

25–40   Small to medium angular and suban-
BC   gular chalk fragments (50–70 per cent)
        and few medium angular flints in a
        massive brownish yellow (10YR 6/6)
        and locally brown (10YR 4/3) silty
        chalky matrix; little or no stratifica-
        tion; few roots; clear wavy boundary.

40–55   Weakly bedded, medium and large
2CuI   tabular chalk fragments (>70 per
        cent) with few, medium to large sub-
        rounded (nodular) flints, very pale
        brown (10YR 7/3-4) extremely calca-
        reous earthy material in interstices
        and few pockets containing small
        chalk fragments in a brownish yellow
        matrix; few roots; clear wavy bound-
        ary.

55+   Bedded fragmented chalk (90 per
2Cu2   cent) and flints as above, with white
        chalky material in interstices.

Analytical data

Horizon	Ap
Depth (cm)	0–25
sand 200 µm–2 mm %	8
60–200 µm %	3
silt 2–60 µm %	66
clay <2 µm %	23
organic carbon %	5.0
$CaCO_3$ equiv. %	4.5
pH (H_2O)	7.7
(0.01M $CaCl_2$)	7.5

Classification
   England and Wales (1984): brown rendzina; silty, lithoskeletal chalk (Andover series).
   USDA (1975): Typic Udorthent intergrading to Rendollic Eutrochrept; loamy-skeletal, carbonatic, mesic.
   FAO (1974): Calcaric Regosol intergrading to Calcic Cambisol; medium textured.

Profile 20 (Finch *et al.*, 1971)

Location	: Ballyvaghan, County Clare
Climatic regime:	perhumid temperate
Altitude	: 189 m Slope: 5–8°
Land use	: old grassland

Horizons (cm)

0–13   Dark brown (10YR 3/3) calcareous
Ah   clay loam (USDA clay loam) with
        some gravel; coarse granular; friable;
        abundant roots (root mat); gradual
        smooth boundary.

13–23 AB	Dark yellowish brown (10YR 3/4) calcareous clay loam (USDA clay loam) with some gravel; moderate coarse and fine granular; friable; plentiful roots; abrupt smooth boundary.	
23–38 Cu	Light grey (10YR 7/1) calcareous gravelly clay loam (USDA clay loam); massive; very few roots.	

Analytical data

Horizon	Ah	A/B	Cu
Depth (cm)	0–13	13–23	23–38
sand 200 μm–2 mm %	13	13	10
50–200 μm %	22	23	19
silt 2–50 μm %	37	34	38
clay <2 μm %	28	30	33
organic carbon %	4.1	2.0	0.5
nitrogen %	0.50	0.27	
C/N ratio	8.2	7.4	
citrate-dithionite ext. Fe %	1.8	1.6	0.4
pH (H$_2$O)	6.9	7.5	7.8

Classification
  Ireland (1980): rendzina (Kilcolgan series).
  England and Wales (1984): brown rendzina; fine loamy, drift with limestone (unnamed series).
  USDA (1975): Typic Udorthent; fine-loamy or loamy-skeletal, carbonatic, mesic.
  FAO (1974): Calcaric Regosol; medium textured.

4.4.6 Pararendzinas

Following Kubiena (1953), the term *pararendzina* is used in the England and Wales classification (Avery, 1980) to denote loamy or clayey rendzina-like soils that have bedrock or little altered unconsolidated material containing up to 40 per cent CaCO$_3$ directly below the A horizon or within 30 cm depth. Shallow or weakly developed soils meeting this requirement have been recorded in various parts of Europe on loess and moderately calcareous consolidated rocks but are generally confined to localized erosional sites and are evidently uncommon in the British Isles. Those so far identified as significant components of soil map units are on grey-green and red Triassic mudstones in south-west England (Avery, 1955; Findlay, 1965; Harrod, 1971), South Wales (Crampton, 1972) and the Midlands (Thomasson, 1971; Whitfield, 1974). Occurring in close association with calcaric brown soils (p. 181) or gley soils (p. 328) on convex slopes and eroded knolls, these soils resemble ochric rankers on similar but non-calcareous materials in showing a tendency to seasonal wetness coupled with marked susceptibility to drought, a combination that makes them difficult to manage productively for either grazing or arable cropping.

The following representative profile on greenish Triassic mudstone was described and sampled under a ryegrass ley which was severely poached at the time. Deeper variants intergrading to pelocalcaric gley soils (p. 333) have a thin mottled Bg horizon with prismatic structure immediately below the A.

Profile 21 (Cope, 1986)

Location	:	Hartpury, Gloucestershire (SO 788237)
Climatic regime:		subhumid temperate
Altitude	:	28 m Slope: 2° convex
Land use	:	ley grass

Horizons (cm)

0–23 Ap	Dark greyish brown (10YR 4/2), very slightly stony, slightly calcareous clay (USDA clay); few rounded quartzite pebbles at base; few very small charcoal fragments and patches of soft greenish grey marl; strong coarse angular blocky breaking to fine with some coarse platy in upper 10 cm; firm; abundant (0–10 cm) to many fine and few medium fleshy roots; abrupt smooth boundary.
23–27 AC(g)	Greenish grey (5GY 6/1) and reddish brown (5YR 4/4), weakly bedded and fragmented, soft, slightly calcareous mudstone breaking to fine angular blocky fragments, with fissures and channels containing greyish and yellowish brown earthy material; few small light grey (10Y 7/1) tabular calcareous sandstone fragments; few

	fine roots in channels; common manganiferous coats on fragment surfaces; abrupt smooth boundary.
27–95 Cr	Greenish grey (7.5Y GY 6/1), weakly bedded, soft, slightly calcareous mudstone breaking to angular blocky fragments as above; few reddish brown laminae and lenses of calcareous sandstone; no roots or earthworm channels.

Analytical data

Horizon	Ap
Depth (cm)	0–23
sand 200 μm –2 mm %	7
60–200 μm %	8
silt 2–60 μm %	39
clay <2 μm %	46
organic carbon %	3.3
$CaCO_3$ equiv. %	4.2
pH (H_2O)	7.4
(0.01M $CaCl_2$)	7.2

Classification
 England and Wales (1984): stagnogleyic pararendzina; clayey material passing to clay or soft mudstone (Hartpury series).
 USDA (1975): Typic (or Aquic) Udorthent; clayey, illitic (calcareous), mesic.
 FAO (1974): Calcaric Regosol; fine textured.

4.5 Sandy Regosols

4.5.1 General Characteristics, Classification and Distribution

Regosols were originally described by Thorp and Smith (1949) as 'an azonal group of soils consisting of deep unconsolidated rock (soft mineral deposits) in which few or no clearly expressed soil characteristics have developed; largely confined to recent sand dunes and to loess or glacial drift of steeply sloping land'. The term has since been used with various more precise but differing meanings and is applied in the FAO (1974) system to non-hydromorphic soils, including Orthents and Psamments in the US Taxonomy, that have formed in deposits other than recent alluvium and have a thin or weakly expressed (ochric) A horizon only. In the classification adopted here it is restricted in accordance with current usage by the Soil Survey of Scotland (1984) to non-alluvial sandy soils, including those classed as sand-rankers and sand-pararendzinas in England and Wales (Avery, 1980) and Albic Arenosols as defined by FAO.

Sandy regosols so defined are widely distributed in coastal dunes and in relatively flat areas of blown sand, called links or machairs in Scotland, which adjoin them in places on the landward side. They also occur very locally inland. The sandy subsurface horizons are often of considerable thickness but can rest on bedrock or a lithologically contrasting sediment such as beach shingle or till within 80 cm depth. Variants transitional to gley soils in which ochreous mottles indicative of periodic wetness appear between 40 and 80 cm occur in relatively low lying sites bordering dune slacks or swales and in places where the sandy C horizon overlies an impervious substratum. Few of these soils are cultivated; some in the Hebrides and elsewhere provide useful grazing and others have been afforested to arrest sand blow. Where the water table remains within root range for some part of the growing season, the land is less droughty than at higher levels and is consequently more amenable to agricultural use.

The maritime sandy regosols result from progressive colonization of blown sea sand by drought-resistant perennial plants whose growth can keep pace with sand accretion. Chief among these dune-building plants are sand couch grass (*Elymus farctus*), marram grass (*Ammophila arenaria*) and sea lyme grass (*Leymus arenarius*). As described by Salisbury (1952), Gimingham (1964) and Birse and Robertson (1976), the natural succession generally starts on or immediately behind the foreshore, where embryonic dunes are usually initiated by sea couch grass. At this stage the surface between the scattered salt-tolerant pioneer plants is bare and unstable, though the sand usually contains some organic matter derived from marine organisms and vegetative shoreline debris. In the succeeding main phase of dune building, the vegetation is typically dominated by vigorously growing marram grass, or less commonly sea lyme grass, but the surface between the grass tussocks remains mobile, erosion by wind in some places being accompanied by deposition in others. The annual production and decay of roots leads to irregular addition of organic residues (the raw soil humus of Kubiena, 1953), but any surface accumulations are slight and discontinuous, and are frequently buried by addition of fresh sand, so that exposed profiles present a speckled and banded appearance (Plate I.4). With the gradual fixation of the dunes, more species are able to establish themselves and there is a corresponding decrease in the vigour of *Ammophila*. Plants which here and there attain periods of dominance include mosses and lichens (grey dunes), sand sedge (*Carex arenaria*), finer grasses such as red fescue (*Festuca rubra*) and various herbaceous and woody species. It is at this stage that a more or less continuous Ah horizon becomes apparent. Further profile development takes different forms depending on the composition of the sand and the nature of the vegetation.

Sands along exposed western coasts in the Hebrides, Cornwall and Ireland are composed largely of shell fragments. The Blakeney (Norfolk) and Southport (Lancashire) dune systems studied by Salisbury (1922, 1925) are only slightly calcareous, whereas other coastal sands, particularly in eastern Scotland and beside the English Channel, contain little or no calcium carbonate. The sands on exposed coasts are also generally the coarsest, with median grain size exceeding 200 μm and even 600 μm in places. The vegetational succession on highly calcareous blown sand typically culminates in species-rich dune pasture, the soils of which are characterized by very friable to loose, densely through-rooted Ah horizons of moder type. On non-calcareous sands, and even on those that are slightly calcareous initially, the dunes are in places invaded by heathers (*Calluna vulgaris* and *Erica cinerea*) at the fixation stage. Under the resulting dune heath, a mor layer with F and thin H horizons forms at the surface; a thin bleached Ea horizon appears immediately beneath it and is succeeded locally by a brownish seam representing an incipient podzolic B horizon. A similar superficial horizon sequence develops where dunes are planted with conifers. Young soils under both types of cover were described as micro-podzols by Hall and Folland (1970).

Changes in dune soils with increasing age have been investigated by Salisbury (1922, 1925, 1952) and more recent workers (Wilson, 1960; Gimingham, 1964; Ball and Williams, 1974). Although the dates at which older dune systems were stabilized are difficult to determine precisely, and the results obtained by different authors are not directly comparable because of differences in sampling depths and procedures, the average organic matter and pH trends listed in Table 4.1 illustrate the variations found. The organic matter data show that contents increased faster at Southport than in the drier eastern and southern sites. The pH trends reflect the influence of initial $CaCO_3$ content and vegetation, the plant cover of the older Blakeney and Southport dunes consisting mainly of herbs, grasses and brambles (*Rubus* spp.), whereas those at the originally non-calcareous Studland site bore dune heath and those at Holkham a more or less closed cover of planted or self-sown pine.

The water and temperature regimes of such soils were also studied by Salisbury (1952) in several localities and by Ovington (1950) and Wright (1955) in the extensive Culbin Sands on the Moray Firth coast, where pine plantations were established from the 1920s onwards on non-calcareous dunes fixed artificially by 'thatching' with brushwood.

Table 4.1: Organic matter and pH trends in dune soils at four British sites

Quantity	Site	Dune age (years)		
		0	50	100
Organic matter content (loss on ignition %, corrected for loss of $CaCO_3$ from carbonates)	Blakeney, Norfolk (0–10 cm)[a]	0.2	0.4	0.7
	Southport, Merseyside (0–10 cm)[a]	0.5	0.9	4.0
	Studland, Dorset (0–5 cm)[b]	0.2	0.6	2.5
	Holkham, Norfolk (0–15 cm, excluding any superficial organic layer)[c]	0.2	0.6	1.7
pH (H_2O)	Blakeney, Norfolk (0–10 cm)[a] (average initial $CaCO_3$ content 0.3%	7.2	7.0	6.4
	Southport, Merseyside (0–10 cm)[a] (average initial $CaCO_3$ content 2%)	8.2	7.8	7.2
	Studland, Dorset (0–5 cm)[b] (average initial $CaCO_3$ content <0.1%)	7.0	5.2	4.3
	Holkham, Norfolk (0–15 cm)[c] (average initial $CaCO_3$ content c. 2%	8.8	6.5	5.5

[a] Salisbury (1922, 1925, 1952)
[b] Wilson (1960)
[c] Ball and Williams (1974)

Particularly in subhumid environments, the raw sands above the reach of the water table behave as 'edaphic deserts' on account of their very small available water capacity. Nearly bare dune surfaces also undergo very large diurnal temperature fluctuations in dry sunny weather, but the temperature range decreases markedly with depth because of the small thermal conductivity of the sand. Similarly, the water content of deeper layers remains relatively constant because the coarse texture severely limits upward movement, though once plants occupy the ground they quickly exhaust the small reserves of retained water in the subsoil. The capacity of the pioneer species to grow during rainless periods is attributable partly to their extensive root systems and their inbuilt ability to restrict day-time transpiration, and partly to night-time augmentation of the reserves by internal dew formation resulting from the marked temperature gradient between surface and subsoil that develops during sunny days. Salisbury (1952) showed that this occurs by measuring water contents in the morning and again at night. Following stabilization of the surface beneath a closed vegetative cover, diurnal temperature fluctuations decrease and the total water holding capacity of the upper soil layer increases in close accordance with increases in organic matter content, but much of the extra water is retained at suctions too great for it to be available to plants. Wright (1955) showed how the growth of trees radically alters the moisture regime by abstraction of water from their rooting zones, which were dried to wilting point for varying periods during the growing season. Under 45-year-old Corsican pine and birch at Culbin, the development of an Ah horizon was sufficient to maintain the upper layers of sand above wilting point throughout the summer when measurements were made. Under 80-year-old Scots pine, where the litter was less incorporated and adsorbed much of the precipitation, even these became dry.

The sandy regosols that occur sporadically inland are essentially similar to those in coastal sands but are nearly all non-calcareous. Variants in recent aeolian deposits resulting from wind erosion that followed clearance of sandy land for cultivation in historic or prehistoric times have been recorded in the East Anglian Breckland (Corbett, 1973) and a few other localities in the English lowlands (e.g. Hollis and Hodgson, 1974). Others of aeolian origin are associated with oro-arctic rankers (p. 121) and podzols on or just below the severely exposed summits of certain Scottish mountains (Birse, 1980); although composed mainly of bleached sand, these contain varying amounts of organic matter derived from pre-existing organic surface horizons nearby. Soils qualifying as sandy regosols also occur locally on pre-Holocene quartzose sands in the London and Hampshire basins, in colluvium resulting from erosion of sandy podzols, and in man-made sandy waste deposits.

Key to Subgroups

1. Sandy regosols with a calcareous C horizon
   Calcaric sandy regosols
2. Other sandy regosols
   Non-calcaric sandy regosols

4.5.2 Non-calcaric Sandy Regosols

These are the regosols in non-calcareous sands, including waste materials and pre-Holocene beds as well as those of recent aeolian origin. The latter are chiefly in dunes and links which flank parts of the coastline in eastern Scotland, southern England and elsewhere. The C horizons of these soils consist of uncoated or thinly coated grains and are typically greyish to light brown in colour. Depending on vegetation and land use history, some have a darkened A horizon up to 30 cm thick at the surface and others an unincorporated superficial organic layer which may overlie a more or less continuous Ea horizon. Those with an albic E 5 cm or more thick are set apart as podzolic sand-rankers in the England and Wales classification, but are difficult to distinguish consistently from those in originally iron-deficient sandy deposits exemplified by profile 24 (p. 149).

The most extensive waste materials in which such soils have developed are those resulting from the exploitation of china clay deposits in Cornwall and Devon (Bradshaw and Chadwick, 1980). Produced by hydrothermal alteration of granite, these consist of kaolin, mica, 'stent' (undecomposed rock) and quartz in proportions of approximately 1:1:1:6. As there is generally little overburden, they have been worked in large open pits, the quartz and stent being separated by washing and dumped on surrounding land. The dumped material consists mainly of coarse sand and fine quartz gravel with only about 5 per cent silt and clay. When freshly deposited it is subject to wind and water erosion, but the spoil heaps (Figure 4.9) are slowly colonized over periods ranging from 10 to 55 years by various acid-tolerant grasses, herbs and shrubs. In exposed situations the succession goes no further than heathland. More protected heaps are invaded by leguminous shrubs, including gorse (*Ulex europaeus*), broom (*Cytisus scoparius*) and tree lupin (*Lupinus arboreus*), which enhance the nitrogen status of the developing soil and play a central role in preparing the ground for establishment of sallow (*Salix cinerea*), rhododendron, birch and oak, the last two occurring only on sites more than 100 years old. Roberts *et al.* (1981) estimated the rates of accumulation of organic matter and nitrogen in the upper 21 cm at sites they sampled as 1,973 and 10.3 kg/ha/year respectively. On this basis a 100-year-old regosol on china-clay waste may be expected to contain about 0.5 per cent organic matter averaged over that depth.

The three profiles recorded on pp. 147–9 are in naturally occurring parent materials. Profile 22 under *Calluna*-lichen heath is located north of Aberdeen in links sand (Glentworth and Muir, 1963) which overlies a buried soil in till at 86 cm depth. It shows signs of podzolization and was originally

Figure 4.9: Heaps of china-clay waste near St Austell, Cornwall (photo by S.J. Staines).

classed as an iron podzol but has neither a clear albic E nor a B horizon that is sufficiently well expressed to qualify as Bs (p. 101) or spodic (Soil Survey Staff, 1975).

Profile 23 is in reddish blown sand which partly mantles Devensian river-terrace deposits near Stourport, Worcestershire. The area has a hummocky relief resulting from exploitation of the sand in shallow pits and drifting occurs where the ground surface is exposed. Dune bedding within 25 cm depth, coupled with the very small silt and clay contents, distinguish this soil from associated sandy brown soils of the Newport series (Clayden and Hollis, 1984).

Profile 24 was described and sampled in droughty arable land with added lime on pale coloured Paleogene (Reading Beds) sand in Berkshire. The absence of a distinct B horizon in this and similar sites may result from either accelerated erosion or the virtual absence of weatherable iron bearing minerals in the parent material. Humus podzols (Chapter 6, p. 254) occur under heathland in similar materials but there is no evidence that the soil described ever had a podzolic B (Bh) horizon.

Profile 22 (Glentworth and Muir, 1963)

Location	: Sands of Forvie, Aberdeenshire (NKO16276)
Climatic regime:	humid boreal
Altitude	: 30 m Slope: 2–3°
Vegetation	: lowland heath; heather (*Calluna vulgaris*) dominant, with sand sedge (*Carex arenaria*), crowberry (*Empetrum nigrum*), lichen (*Cladonia sylvatica*) abundant, and mosses (*Hylocomium splendens, Pleurozium schreberi*)

Horizons (cm)

L	Litter of mosses, lichens and *Calluna* twigs, c. 1 cm thick.
0–2 F	Dark brown, fibrous, partly decomposed organic material.
2–6 H	Very dark brown (5YR 2/1) to black, well decomposed organic material, mealy when crushed; weak granular; abundant fine roots; abrupt boundary.

6–9 Ah/Ea	Dark greyish brown (10YR 4/2) sand (USDA sand) with uncoated grains; weak subangular blocky; abundant roots; abrupt boundary.
9–50 BC	Brown (7.5Y 5/4) to light yellowish brown (10YR 6/4) sand (USDA sand); massive (very slightly compacted); frequent roots; gradual boundary.
50–86 Cu	Light yellowish brown (10YR 6/4) sand (USDA sand); single grain; loose; occasional roots; abrupt boundary.
86–93 2bAh	Dark reddish brown (5YR 3/2) humose sandy loam (USDA sandy loam) weak fine subangular blocky; occasional roots; sharp boundary.
93 Bf	Ironpan c. 1 mm thick; sharp boundary.
93–118 2bB?	Dark reddish brown (5YR 3/2) sandy loam (USDA sandy loam); medium subangular blocky; roots rare.

Analytical data *see* below.

Classification
   Scotland (1984): non-calcareous regosol.
   England and Wales (1984): typical sand-ranker; sandy stoneless drift (Beckfoot series).
   USDA (1975): Typic Undorthent; sandy over loamy; siliceous, mesic.
   FAO (1974): Dystric Regosol or Cambic Arenosol; coarse textured.

Profile 23 (Hollis and Hodgson, 1974)

Location	:	Hartlebury Common, Hereford and Worcester (SO 822704)
Climatic regime:		subhumid temperate
Altitude	:	30 m Slope: 2°
Vegetation	:	lowland heath (ungrazed common land); heather (*Calluna vulgaris*) dominant, with sparse ground layer of mosses and lichens; occasional tufts of wavy hair-grass (*Deschampsia flexuosa*), clumps of gorse (*Ulex europaeus*) and rare oak seedlings

Horizons (cm)

0–5 Ah1	Dark reddish brown (5YR 2/2) stoneless medium sand (USDA sand or fine sand); weak granular; loose; abundant fine and common medium woody (heather) roots; clear irregular boundary.
5–8 Ah2	Dark reddish brown (5YR 3/4) stoneless medium sand (USDA sand); single grain; loose, abundant roots; few medium (c. 4 mm) pores formed by ants; clear irregular boundary.
8–100 Cu	Yellowish red (5YR 5/7) stoneless medium sand (USDA sand); dune bedded, the bedding picked out by more ferruginous (5YR 4/5) sub-parallel bands (up to 10 mm thick) which are slightly hard; many roots down to 40 cm.

Analytical data for Profile 22

Horizon	Ah/Ea	BC	BC	Cu	2bAh	2bB?
Depth (cm)	6–9	9–16	27–41	60–70	86–93	108–118
silt[a] 2–50 μm %	3	2	2	2	21	19
clay[a] <2 μm %	<1	<1	<1	<1	12	13
loss on ignition %	3.9	1.3	0.5	0.4	7.6	3.7
organic carbon %	2.5	0.59	0.19			
nitrogen %	0.10	0.03	0.01			
C/N ratio	25	20	19			
TEB (me/100 g)	1.01	0.18		0.08	9.6	7.1
CEC (me/100 g)	6.7	1.3		0.98	18.6	10.4
% base saturation	15	14		8	52	68
pH (H$_2$O)	3.9	4.7	5.1	5.5	6.5	6.4

[a] by hydrometer method.

Analytical data

Horizon	Ah1	Ah2	Cu
Depth (cm)	0–5	5–8	8–100
sand 600 μm–2 mm %	5	5	6
200–600 μm %	46	53	59
60–200 μm %	40	35	31
silt 2–50 μm %	6	4	3
clay <2 μm %	3	3	1
organic carbon %	2.5	0.8	0.1
pH (H$_2$O)	4.3	4.4	5.2
(0.01M CaCl$_2$)	3.6	3.7	4.4

Classification
   England and Wales (1984): typical sand-ranker; sandy stoneless drift (Beckfoot series).
   USDA (1975): Typic Quartzipsamment; mesic, uncoated.
   FAO (1974): Dystric Regosol; coarse textured.

Analytical data

Horizon	Ap1	Ap2	Cu
Depth (cm)	0–9	9–18	18–90
sand 200 μm–2 mm %	40	36	38
60–200 μm %	36	40	53
silt 2–60 μm %	17	17	6
clay <2 μm %	7	7	3
organic carbon %	2.1	0.9	
CaCO$_3$ equiv. %	0	0.4	0
pH (H$_2$O)	5.8	7.1	6.6
(0.01M CaCl$_2$)	5.2	6.6	6.0

Classification
   England and Wales (1984): typical sand-ranker; medium sandy material passing to sand or sandstone (variant of Woking series).
   USDA (1975): Typic Quartzipsamment; mesic, uncoated.
   FAO (1974): Albic Arenosol; coarse textured.

Profile 24 (Jarvis, 1968)

Location	: Yattendon, Berkshire (SU 551741)
Climatic regime:	subhumid temperate
Altitude	: 109 m Slope: 4°
Land use	: arable (rape and grass seeds)

Horizons (cm)

0–9
Ap1   Greyish brown (2.5Y 6/2, dry), very slightly stony, medium sandy loam (USDA fine sandy loam); few, small to medium, angular flints and quartzite pebbles; weak fine subangular blocky; very friable; many roots; clear boundary.

9–18
Ap2   Pale brown (10YR 6/3, slightly moist), very slightly stony (as above) medium sandy loam (USDA fine sandy loam) faintly mottled with light yellowish brown (10YR 6/4); includes two horizontal bands (plough pans) with massive structure; few roots; added chalk; clear boundary.

18–90
Cu   Light grey (2.5Y 7/2, dry to slightly moist) medium sand (USDA sand) containing very few small flint fragments; single grain; loose; roots rare.

4.5.3 Calcaric Sandy Regosols

Soils of this subgroup have formed in 'fixed' dunes composed of shelly sand in numerous widely scattered coastal localities ranging from northern Scotland and the Hebrides to Cornwall and southern Ireland. Particularly in the Hebrides and other areas where the sand has blown inland to considerable distances, they also occur in shallower deposits over bedrock or Pleistocene sediments and are associated with calcaric sandy brown soils (p. 165) and sandy gley soils (p. 299) in level to gently undulating *machairs* (Figure 4.10) overlying raised beaches (Ritchie, 1976; Hudson *et al.*, 1982; Bibby *et al.*, 1982).

The Hebridean dunes and machairs support natural and improved pastures which provide valuable grazing in striking contrast to that available on acid peaty soils nearby, and a few relatively sheltered areas in Tiree, Uist and Benbecula, which lie wholly or partly in the humid temperate climatic zone (Table 1.12, p. 18) are cropped with cereals, chiefly oats, and potatoes. Wind erosion is a constant

Figure 4.10: Landscapes on windblown shelly sand at Balmartin, North Uist. The foreground is gently undulating machair with calcaric sandy regosols and sandy brown soils; beyond the road there are strongly undulating dunes with calcaric regosols and bare eroded patches (copyright Aerofilms Ltd.).

hazard, however, and minor element deficiencies affecting stock (cobalt, copper) and crops (manganese) have been reported. Seaweed and peat have been added as manures in addition to dung; some soils on the sands consequently have A horizons thick enough to qualify as man-made (Chapter 3, p. 107), but regosols with dark coloured A horizons no more than 15 cm thick are commonest, particularly on the dunes. In most areas of occurrence further south, where moisture deficiency is more severe, the soils are chiefly in fixed dunes used mainly for amenity purposes, with some afforestation locally. Data on two representative profiles are given below.

Profile 25 was described and sampled by C.C. Rudeforth under dune pasture on the west coast of Dyfed. The well sorted fine sand at this site is only slightly calcareous and the thin Ah horizon is almost completely decalcified. Profile 26 from Cornwall is in extremely calcareous shell sand. This soil was reclaimed for agriculture within the last 200 years (Staines, 1979) and resembles those of similar origin and composition in the Hebrides. It has a very dark coloured, appreciably loamy A horizon which is thick enough to qualify as mollic according to the FAO (but not the US) criteria and evidently incorporates manurial additions comprising both mineral and organic materials.

Profile 25

Location	: Ynyslas, Dyfed (SN 608940)
Climatic regime:	humid temperate
Altitude	: 5 m Slope: 10°
Land use and vegetation	: semi-natural dune pasture, 100 per cent cover; red fescue (*Festuca rubra*) dominant, with rest harrow (*Ononis spinosa*), buttercup (*Ranunculus* spp.), ribwort (*Plantago lanceolata*), violet (*Viola* spp.), moss and lichen

Horizons (cm)

L		Grass litter, c. 1 cm thick.
0–7 Ah		Dark brown (10YR 3/2; 10YR 5/3 dry) slightly calcareous fine sand (USDA fine sand) with paler areas reflecting irregular incorporation of more or less humified organic matter; weak granular to single grain; very friable to loose; abundant fine roots; clear smooth boundary.
7–150 Cu		Light brownish grey (2.5Y 6/2; 10YR 6/3 dry) slightly calcareous fine sand (USDA fine sand), becoming somewhat coarser below 50 cm; single grain; loose; few fine roots to 50 cm.

Analytical data

Horizon		Ah	Cu1	Cu2
Depth (cm)		0–7	7–50	50–90
sand	600 μm–2 mm %	<1	<1	<1
	200–600 μm %	9	9	21
	60–200 μm %	88	91	79
silt	2–60 μm %	2	<1	<1
clay	<2 μm %	1	<1	<1
organic carbon %		3.9	0.1	0.1
$CaCO_3$ equiv. %		<1	2.5	3.4
pH (H_2O)		7.1	7.9	8.2
(0.01M $CaCl_2$)		6.7	7.4	7.6

Classification
England and Wales (1984): typical sand-pararendzina; sandy stoneless drift (Sandwich series).
USDA (1975): Typic Udipsamment; mixed (calcareous), mesic, uncoated.
FAO (1974): Calcaric Regosol; coarse textured.

Profile 26 (Staines, 1979)

Location	: Phillack, Cornwall (SW 559383)
Climatic regime:	humid temperate
Altitude	: 45 m Slope: 2°
Land use	: grass ley

Horizons (cm)

0–22 Ap	Very dark brown (10YR 3/2–3) extremely calcareous loamy medium sand (USDA loamy sand) containing very few small slate fragments; weak medium and fine subangular blocky; very friable; many fine roots; clear smooth boundary.
22–90 Cu	Pale yellow (2.5Y 7/4), stoneless, extremely calcareous medium sand (USDA sand) composed mainly of shell fragments; single grain; loose; few earthworm channels infilled with soil similar to above; loose; few fine roots mainly in channels.

Analytical data

Horizon		Ap	Cu
Depth (cm)		2–20	25–90
sand	600 μm–2 mm %	5	1
	200–600 μm %	56	85
	60–200 μm %	17	12
silt	2–60 μm %	15	1
clay	<2 μm %	7	1
organic carbon %		3.4	
$CaCO_3$ equiv. %		50	75
pH (H_2O)		7.6	8.6
(0.01M $CaCl_2$)		7.4	8.1

Classification
England and Wales (1984): typical sand-pararendzina; carbonatic-sandy stoneless drift (Towans series).
USDA (1975): Typic Udipsamment; carbonatic, mesic, uncoated.
FAO (1974): Rendzina; coarse textured.

4.6 Rego-alluvial Soils

4.6.1 General Characteristics, Classification and Distribution

The term *rego-alluvial* is used here to describe weakly developed more or less well drained soils, including the Paternias and Borovinas of Kubiena (1953) and ranker-like and some rendzina-like alluvial soils of Avery (1980), that have developed in Holocene water-laid deposits and lack both a diagnostic B horizon and a prominent A more than 30 cm thick. Typical profiles have a thin or weakly expressed (ochric) A horizon passing downwards into a greyish or brownish C horizon that is finely stratified or has a massive or single-grain structure. Organic matter contents may decrease gradually or irregularly with depth where sedimentation has been slow or intermittent but are normally small (<0.35 per cent) in coarse textured subsurface horizons. Weakly developed soils of this kind are grouped with other alluvial soils as Fluvisols or Regosols in the FAO (1974) system and as Orthents, Psamments or Fluvents in the US Taxonomy, depending on the texture and stoniness of the materials between 25 and 100 cm depth.

Though widely distributed, soils of this group are nowhere extensive in the British Isles, and those recorded are nearly all non-calcareous. They are mainly gravelly or sandy (Orthents or Psamments), as most soils in finer textured alluvium are either gleyed within 40 cm of the surface (alluvial gley soils, p. 300) or have a brownish Bw horizon differentiated by structure, colour or both (alluvial brown soils, p. 169). Those of fluviatile origin are chiefly in hilly or mountainous areas where streams flow swiftly and there are marked changes in water level over short periods. Even in these situations, rego-alluvial soils are commonly confined to levees, gravel bars, low terraces and sloping fans, often of very limited extent, in which the water table seldom rises to the surface for more than a few days. Local variations in the thickness and organic matter content of surface and subsurface horizons are related to current or former sedimentation regimes. The least developed soils (raw alluvial soils of Avery, 1980) are associated with recently or sparsely vegetated banks of sand and gravel in stream beds and actively accumulating fans. These have thin A horizons which are discontinuous or scarcely perceptible. Other variants of limited extent are developed in shingle beaches of marine origin and in coarse textured downwash resulting from recent erosion of cultivated land or man-made dumps of unconsolidated material.

The use range and productive capacity of these soils depend on their texture and stoniness and the frequency of damaging floods. Coarse loamy and sandy types with few stones in the uppermost horizon are the most valuable, particularly in the Scottish Highlands, where in some localities they form the only ploughable land and the climate is too humid for substantial moisture deficits to occur at all frequently. In less humid areas, however, variants with very stony layers immediately below the surface are very drought susceptible.

Key to subgroups

1. Rego-alluvial soils with a calcareous C horizon
   Calcaric rego-alluvial soils
2. Other rego-alluvial soils
   Non-calcaric rego-alluvial soils

4.6.2 Non-calcaric Rego-alluvial Soils

Termed ranker-like alluvial soils by Avery (1980), these soils are developed in recent alluvial deposits derived from non-calcareous rocks and show no evidence of gleying in the upper 40 cm. Data on three representative profiles, all with sandy or sandy-gravelly subsurface horizons, are reproduced below.

Profile 27 is in riverine alluvium bordering the upper reaches of a small stream near Newton Stewart, Galloway, where the source rocks are greywacke and granite. According to Bown and Heslop (1979), flooding and deposition of silt are probably still active at this site. The brown colour of the loamy Ah horizon reflects incorporation of humified organic matter rather than weathering of the mineral material *in situ*.

Profile 28 is developed in similar sandy-gravelly alluvium in the upper Don valley, Aberdeenshire. Although the water table was 91 cm below the surface at the time of sampling and presumably fluctuates above and below this level, there is no clear evidence of gleying. As the estimated mean annual soil temperature at 50 cm depth is less than 8°C (Meteorological Office, 1968; Ragg and Clayden, 1973), this soil has a cryic temperature regime according to the US Taxonomy (Soil Survey Staff, 1975, p. 62). The land is nevertheless farmed on a ley-arable basis, suggesting that the specified soil temperature limit is inappropriate for application in boreal oceanic areas such as north-east Scotland (cp. Birse, 1980).

Profile 29 from Dungeness, Kent, is in marine shingle which had probably ceased to accumulate by the sixteenth century (Burnham, 1983b). The old beaches in this locality are composed almost entirely of water-rounded flint and form subparallel ridges ('fulls') separated by depressions ('lows'). The lows remain almost entirely bare but the fulls, which have somewhat smaller pebbles, support vegetation ranging from lichens, drought resistant grasses and herbs to gorse (*Ulex* spp.) scrub and wind-dwarfed holly (*Ilex aquifolium*) thickets. According to Scott (1965), an important part of the natural succession is played by nitrogen-fixing prostrate broom (*Cytisus scoparius*). This colonizes the shingle at an early stage, rooting to depths of 2 m or more in search of water, and its remains, together with humus from lichens and any entrapped dust, enrich the upper few centimetres in organic matter. With time, the broom clumps spread but eventually die, giving place to floristically rich heath-like vegetation which may in turn die out, leaving 'dirty shingle' open to fresh invasion by broom. Burnham (1983b) gives data on organic matter content and pH of surface soils (0–10 and 0–20 cm) typifying each vegetation stage. These show that the fine earth is almost entirely organic, with pH (in H_2O) values ranging from 3.6 to 4.4. Profile 29 may therefore have to be classed as a Histosol (Folist) in the US system.

Profile 27 (Bown and Heslop, 1979)

Location	: Auchinleck, Kirkcudbrightshire (NX 448703)
Climatic regime:	humid temperate
Altitude	: 88 m Slope: <1°
Land use and vegetation	: long-term grass with smooth stalked meadow grass (*Poa pratensis*), sweet vernal grass (*Anthoxanthum odoratum*), crested dogstail (*Cynosurus cristatus*), white clover (*Trifolium repens*) and chickweed (*Cerastium* spp.)

Horizons (cm)

0–25 Ah	Brown to dark brown (10YR 4/3) stony sandy loam (USDA sandy loam); moderate medium subangular blocky and granular; friable; many roots; gradual boundary.
25–45 2AC	Dark greyish brown (10YR 4/2) stony loamy sand (USDA loamy sand); friable; frequent roots; gradual boundary.
45–75+ 3Cu	Coarse sandy gravel; abundant greywacke and granite fragments with some large boulders; single grain; loose; few roots.

Analytical data

Horizon	Ah	2AC	3Cu
Depth (cm)	2–10	30–37	65–75
silt^a 2–50 μm %	17	8	5
clay^a <2 μm %	14	9	4
organic carbon %	3.9	1.8	
nitrogen %	0.29	0.14	
C/N ratio	13	13	
TEB (me/100 g)	7.4	0.8	0.8
CEC (me/100 g)	15.0	5.0	5.0
% base saturation	50	16	16
pH (H$_2$O)	6.1	5.7	5.7

^{a} by hydrometer method

Classification
  Scotland (1984): mineral alluvial soil.
  England and Wales (1984): typical ranker-like alluvial soil; loamy alluvium over non-calcareous gravel (Teign series).
  USDA (1975): Typic Udorthent; sandy-skeletal, mixed, mesic.
  FAO (1974): Dystric Fluvisol; medium to coarse textured.

Profile 28 (Heslop and Bown, 1969)

Location	:	Strathdon, Aberdeenshire (NJ 349119)
Climatic regime:		humid oroboreal
Altitude	:	290 m Slope: <1°
Land use	:	ley grass

Horizons (cm)

0–23 Ap — Dark greyish brown (10YR 4/2) stony sandy loam (USDA sandy loam); small subrounded stones; fine to medium subangular blocky to medium granular; friable; common to abundant roots; earthworms active; clear boundary.

23–41 2AC — Sandy gravel with pockets and lenses of dark greyish brown (10YR 4/2) loamy sand, single grain; loose; fine roots common to abundant; gradual boundary.

41–91 2Cu — Brown (10YR 5/3, general colour) coarse sandy gravel with layers of gravelly coarse sand (USDA sand or coarse sand); subrounded to subangular, small to large, fragments of granite, quartzite, epidiorite, hornblende schist and other stones; roots few to common in upper 20 cm; abrupt boundary.

91+ As above but below the water table.

Analytical data *see* below.

Classification
  Scotland (1984): mineral alluvial soil.
  England and Wales (1984): typical ranker-like alluvial soil; non-calcareous gravelly, acid crystalline rock (unnamed series).
  USDA (1975): Typic Cryorthent; sandy-skeletal, mixed.
  FAO (1974): Eutric Fluvisol; medium to coarse textured.

Analytical data for Profile 28

Horizon	Ap	2AC	2Cu	2Cu(g)
Depth (cm)	3–13	25–38	53–64	91–102
silt^a 2–50 μm %	28	3	6	5
clay^a <2 μm %	13	5	3	2
loss on ignition %	5.8	1.7	1.5	1.3
organic carbon %	2.5	0.8		
nitrogen %	0.19	0.04		
C/N ratio	13	20		
TEB (me/100 g)	6.1	1.7	1.5	1.9
CEC (me/100 g)	11.9	2.7	2.8	2.4
% base saturation	51	63	54	79
pH (H$_2$O)	5.5	5.9	6.0	6.0

^{a} by hydrometer method

Profile 29 (Green, 1968)

Location	:	Denge Beach, Dungeness, Kent (TR 087169)
Climatic regime:		subhumid temperate
Altitude	:	6 m Slope: <1°
Vegetation	:	semi-natural grassland with sweet vernal grass (*Anthoxanthum odoratum*), English stonecrop (*Sedum anglicum*) and other herbs

Horizons (cm)

0–10 H	Very dark greyish brown gravelly organic layer with abundant roots, mainly fine.
10–30 AC	Rather loose flint gravel and larger stones (pebbles) with many roots and some fine organic material.
30+ Cu	Loose flint gravel and pebbles; fine roots decrease with depth but some medium roots extend below 60 cm.

Classification
    England and Wales (1984): typical ranker-like alluvial soil; gravelly, very hard siliceous stones (Dungeness series).
    USDA (1975): Typic Udorthent or Folist; fragmental, siliceous, mesic.
    FAO (1974): Dystric Regosol? (stony phase).

4.6.3 Calcaric Rego-alluvial Soils

These are rego-alluvial soils in deposits derived wholly or partly from calcareous rocks. They are called typical rendzina-like alluvial soils in the England and Wales system and correspond to those described as *Borovina* and *Kalk-Paternia* by Kubiena (1953) and Mückenhausen *et al.* (1977) respectively. Well drained calcareous 'A–C soils' of this kind are common in the Rhine valley (e.g. profile 111 in Duchaufour, 1978, p. 23) and in the valleys of the Alpine Foreland in southern Germany and Austria, but are rare in the British Isles. The few that have been noted, for example in small areas mapped as Ribble series near Preston, Lancashire (Crompton, 1966), are associated with alluvial brown soils as minor inclusions in the soil-landscape patterns of floodplains with limestone catchments. None have been characterized in detail.

5
Brown Soils

5.1 General Characteristics, Classification and Extent

The general term *brown soil* is used here to denote more or less well drained (non-hydromorphic) soils with altered subsurface horizons, usually brown or reddish, in which iron oxides are bonded to silicate clays. Some have a colour/structure B horizon (Bw) and others an argillic (Bt) or podzolic B (Bs) as defined in Chapter 3; a few lack a distinct B horizon but have a prominent cumulic A (or A/B) horizon in which the colour of the mineral fraction is masked by deeply incorporated organic matter. Other properties such as texture and base status also vary greatly, depending on the parent material and the kind and degree of alteration it has undergone.

The major group so conceived includes calcareous soils with colour/structure B horizons as well as the originally non-calcareous or decalcified soils traditionally grouped in Britain as brown earths (Clarke, 1940; Robinson, 1949). According to current usage in the three national surveys, it embraces the brown soils and most pelosols and brown podzolic soils of Avery (1973, 1980); the brown calcareous, brown magnesian and brown forest soils as distinguished by the Soil Survey of Scotland (1984); and most of those classed as brown earths, grey-brown podzolic and brown podzolic in Ireland (Gardiner and Radford, 1980). Those with argillic B horizons are mainly Alfisols (Udalfs) in the US Taxonomy and Luvisols or Acrisols in the FAO (1974) system. Those without include Inceptisols (mostly Ochrepts) and Entisols (Psamments), corresponding to Cambisols and Arenosols respectively. Comparable divisions are made at soil group level in the classification used here. Each group is typified by freely draining profiles showing no evidence of periodic wetness and variants transitional to gley soils are placed in gleyic, stagnogleyic and pelo-(stagnogleyic clayey) subgroups. In contrast to the US and FAO systems, however, variants with prominent (mollic or umbric) A horizons are not set apart for the reasons mentioned in Chapter 2 (p. 70).

Two main factors, in addition to ground-surface stability, have evidently conditioned the development of these soils. The first is an adequate depth of unconsolidated parent material that is neither extremely calcareous nor extremely siliceous and is sufficiently permeable and well drained to permit free aeration, at least in the upper part of the profile. The second is a bioclimatic regime, typical of temperate regions where the natural vegetation is broadleaf forest, which promotes rapid decomposition of plant residues and consequent recycling of nitrogen and other nutrient elements. When these conditions are fully met, humus accumulates in limited amounts and is incorporated by earthworms to form a mull-like A horizon;

organic complexing agents derived from the litter are quickly biodegraded or immobilized at or near the surface of the soil; and hydrous iron oxides inherited from the parent material or formed by weathering remain closely attached to the silicate clay and help to bond clay and humus within aggregates.

In recent alluvial and colluvial deposits and in sands, the diagnostic B horizon (normally Bw) is often weakly expressed and differs little in composition from the parent material. Other brown soils show evidence of more pronounced alteration involving solution and removal of carbonates, chemical weathering of silicate minerals, accumulation of clay by illuviation or breakdown of coarser particles, or some combination of these processes. Where the parent material has been in place since the Devensian period or earlier and the upper horizons are non-calcareous, either originally or as a result of leaching, illuviation argillans (clay coats) can usually be detected at some depth in the profile, indicating that clay-size particles have been dispersed and translocated. On initially calcareous deposits and 'pre-weathered' sediments, this has generally resulted in differentiation of an argillic B horizon containing appreciably more clay than an overlying eluvial (Eb) horizon (Figure 5.1a). In other brown soils showing micromorphological evidence of clay illuviation, however, clay contents decrease with depth or remain about the same, either because the parent material is stratified and the uppermost layer originally contained more clay, as in many fluviatile deposits, or because loss of clay from the upper horizons by eluviation has been offset by gain due to weathering of coarser particles (Figure 5.1b).

In cool humid areas and on base deficient materials elsewhere, a mull-like Ah horizon persists naturally only so long as an active organic cycle is maintained. Progressive impoverishment of the upper horizons, accentuated where the ground is colonized by acidifying vegetation, causes it to be replaced by moder or mor and leads to corresponding

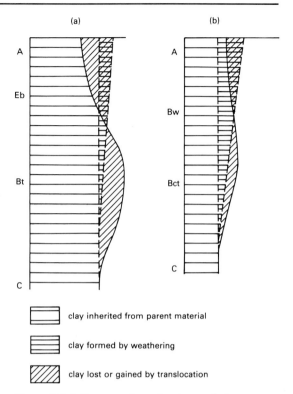

Figure 5.1: Influence of weathering and clay translocation on horizon development in brown soils: (a) with argillic B horizon; (b) without argillic B horizon (after Duchaufour, 1982).

changes in the underlying mineral horizons which mark the onset of podzolization (Figure 1.3, p. 9). These changes take different forms depending on the biotic regime and the composition of the mineral material. Where a superficial mor layer has formed under semi-natural vegetation or planted conifers, a thin eluvial (Ea or Eag) horizon resembling those of podzols commonly develops immediately beneath it, though the morphology and composition of underlying horizons may be little affected. Other strongly leached variants have a more or less ochreous Bs horizon (p. 101) directly beneath an Ah or Ap. These have been classed as brown podzolic soils in England and Wales and in Ireland but are grouped here as brown soils in approximate accordance with the Soil Survey of Scotland (Table 2.7, p.

61) and Forestry Commission (Table 2.8, p. 62) classifications.

Judging from the published soil association maps, brown soils cover some 45 per cent of England and Wales, about 40 per cent of Ireland, and between 10 and 15 per cent of Scotland. As shown in Figure 2.5 (p. 78), they occur mainly but not exclusively in lowland areas with subhumid or humid temperate climatic regimes. Except in certain recent alluvial deposits and rare oro-arctic phases, all bore deciduous forest at some stage in their development, but much of the ground has since supported arable crops or artificially maintained grassland for lengthy periods which in which some places began in Neolithic times. As a consequence all but the least labile soil properties such as texture, mineralogy and subsurface horizonation have been modified by cultural practices. In brown soils that were originally non-calcareous, local variations in pH and percentage base saturation resulting from past liming are such that a division into eutric (high base status) and dystric (low base status) classes seems impracticable for general use in soil surveys, though it is clearly important ecologically in areas that remain under semi-natural vegetation.

The six soil groups distinguished are defined as follows. Corresponding groups in the England and Wales classification are named in parentheses.

Key to Soil Groups

1. Brown soils with a podzolic B (Bs) horizon (p. 108)
    Podzolic brown soils (brown podzolic soils, excluding variants with a Bh horizon)
2. Other brown soils with an argillic B horizon (p. 109)
    Luvic brown soils (argillic brown earths, paleoargillic brown earths and argillic pelosols)
3. Other brown soils in which at least half of the upper 80 cm, or of the whole soil if bedrock supervenes, is sandy or sandy-skeletal (p. 84), and there is no loamy layer extending below 30 cm depth
    Sandy brown soils (brown sands and brown calcareous sands)
4. Other brown soils in recent alluvium (p. 113) more than 30 cm thick
    Alluvial brown soils (brown alluvial and brown calcareous alluvial soils)
5. Other brown soils that are calcareous within 50 cm depth, in and/or immediately below a colour/structure B horizon (p. 109)
    Calcaric brown soils (brown calcareous earths and calcareous pelosols without a gleyed subsurface horizon)
6. Other brown soils
    Orthic brown soils (brown earths and non-calcareous pelosols)

5.2 Sandy Brown Soils

5.2.1 General Characteristics, Classification and Distribution

These are coarse textured brown soils formed in sandy or sandy-gravelly superficial deposits and weathering products of quartzose sandstones. According to the definition adopted here, which is derived from the Netherlands classification of de Bakker and Schelling (1966), more than half of the upper 80 cm, or of the whole depth of soil if bedrock supervenes, consists of sandy or sandy-skeletal material and there is no loamy surface layer extending below 30 cm depth. The profiles are distinguished from those of sandy regosols (p. 143) and rego-alluvial soils (p. 152) by the occurrence of an apedal or weakly pedal Bw horizon, or occasionally a thick cumulic A horizon, that extends below 30 cm depth, lacks fine stratification, and has a brownish or reddish colour that differs to some extent from that of the underlying horizon and reflects the presence of free iron oxide closely associated with small but variable amounts of silicate clay and organic matter. One or more thin bands or irregular patches in which strongly orientated clay coats and bridges the sand grains generally appear at greater depths where the soil has formed in a Pleistocene or older deposit, but are insufficiently thick and continuous to constitute an argillic B horizon. Morphological and chemical properties characteristic of podzolic brown soils or podzols are also too weakly expressed to be diagnostic or are restricted to the upper few centimetres of soil under semi-natural vegetation or planted conifers. The sandy brown soils that lack a prominent (mollic or umbric) A are mostly Psamments in the US Taxonomy and Arenosols in the FAO (1974) scheme. Those with a prominent A are Udolls (FAO Phaeozems) or Umbrepts (FAO Humic Cambisols).

In England and Wales, where these soils have been divided into brown sands and brown calcareous sands (Avery, 1973, 1980), they cover about 2 per cent of the total surface area, chiefly on Pleistocene sands and sandy gravels in East Anglia, on Permo-Triassic sandstones and sandy glaciofluvial deposits in the Midlands, and less extensively on various other Quaternary and pre-Quaternary (e.g. Lower Greensand) formations. Except in thick well sorted aeolian and water-laid sands, they are usually interspersed with somewhat finer textured orthic or luvic brown soils; they also give place locally to podzols, particularly under existing or former heathland.

Soils of this group are less extensive in Scotland and Ireland and have not been specifically distinguished from other brown soils in either country. As classified and mapped by the Soil Survey of Scotland (1984), they include brown calcareous soils in shelly coastal sands (Fraserburgh association); some of the lowland brown forest soils in raised beach deposits (Carpour, Panbride, Dreghorn and Kirkcolm associations) and glaciofluvial sands and gravels (Darvel, Eckford/Innerwick, Gleneagles, Auchenblae, Collieston, Darnaway and Yarrow associations), and a few of the well drained soils in older sandy alluvium and aeolian sands (Links association). However, the series grouped as brown forest soils also include coarse loamy orthic brown soils, podzolic brown soils and humic man-made soils (Cultosols) as defined here. On the national soil map of Ireland (Gardiner and Radford, 1980), sandy brown soils are recorded as a significant component of only one association (no. 16), mainly located on coarse textured morainic and outwash deposits of Midlandian (Devensian) age in County Wexford.

Most sandy brown soils are or have been cultivated and some with prominent A horizons appear to have developed from heathland podzols through the combined effects of deep ploughing and the heavy manuring which these 'hungry' soils customarily received in pre-fertilizer days. The main factor limiting their productivity under present day farming conditions is inadequate moisture storage capacity. Those in the subhumid lowlands range from slightly to very droughty (Thomasson, 1979), depending on the difference between the average maximum potential soil moisture deficit (MD) and the effective available water capacity (A) of the profile as conditioned by the texture, stoniness, organic matter content and bulk density of horizons within the rooting zone. On very droughty phases (MD-A $>$ 50 mm under grass), typified by more or less stony soils of the Newport series (Clayden and Hollis, 1984) in eastern England and very stony or shallow variants elsewhere, yields of all the common crops are significantly limited by water stress in most years where no irrigation is provided. Dry springs in particular can seriously hinder germination and growth of seedlings. Other potential limitations are liability to wind or water erosion (Figure 5.2), and slaking and resultant surface or subsurface compaction in loamy sand types depleted of organic matter. On the other hand, the soils are easily worked at all seasons; those with few stones are consequently well suited to vegetable and nursery crops, especially in the more humid areas of occurrence.

Figure 5.2: Erosion on typical sandy brown soils (Newport series) near Claverley, Shropshire. Exposed fields are subject to wind erosion during spring when crop cover is thin (a) and compacted sloping land is at risk from water erosion (b) Copyright A.H. Reed).

Key to Subgroups

1. Sandy brown soils with a calcareous subsurface horizon (Bw, BC or C) starting within 50 cm depth
   1.1 with gleyic features (p. 112)
   Gleyic-calcaric sandy brown soils
   1.2 without gleyic features
   Calcaric sandy brown soils
2. Other sandy brown soils
   2.1 with gleyic features (p. 112)
   Gleyic sandy brown soils
   2.2 without gleyic features
   Typical sandy brown soils

5.2.2 Typical Sandy Brown Soils

These are freely draining non-calcareous sandy brown soils, including typical and some argillic brown sands according to the England and Wales system. They are more extensive and more widely distributed than those of the other subgroups. Profiles under semi-natural vegetation are moderately or strongly acid throughout and usually show signs of podzolization below a superficial organic layer, but lack both a diagnostic Bh and an ochreous B horizon containing enough pyrophosphate-extractable aluminium and iron to qualify as Bs. Cultivated phases are mostly near neutral due to liming and commonly have a prominent A horizon 25–40 cm thick. The topsoil may be loamy, either naturally or as a result of the formerly widespread practice of marling. It normally overlies a brown or reddish Bw which passes gradually into a sandy or sandy-gravelly Cu horizon or rests directly on sandstone bedrock (Plate I.5). Bands or lamellae in which silicate clay has accumulated as coats on the sand grains may occur in the lower part of the profile but are too thin and/or too deep to qualify as an argillic B horizon (Chapter 3, p. 109). Data on three representative profiles are reproduced on pp. 162–4.

Profile 30, with a prominent A and a distinct Bw horizon, is developed in fine sandy drift, probably late Devensian coversand (Matthews, 1970) in Humberside. The average maximum potential soil moisture deficit (MD) at this site is slightly greater than the profile available water capacity (135 mm), implying that the soil can be considered slightly to moderately droughty depending on the crop (Thomasson, 1979).

Profile 31 was described and sampled by A. J. Thomasson on Gleadthorpe Experimental Husbandry Farm in the Sherwood Forest district of Nottinghamshire. The lower horizons have weathered more or less in place from reddish Triassic sandstone but the prominent A contains a relatively high proportion of silt which may be partly of loessial origin. As is normal on reddish parent materials, the Bw horizon is yellower in hue than the less altered material below, presumably because it contains newly formed hydrous iron oxide which accompanies or replaces inherited haematite (Schwertmann and Taylor, 1977). A thin section at 80–85 cm depth confirmed that parts of the BC horizon have argillans coating and bridging the sand grains, but there is no distinctly more clayey horizon or series of clay-enriched bands thick enough to qualify as an argillic B. The annual rainfall (618 mm) and the potential soil moisture deficit (MD for grass c. 150 mm) are about the same as at the site of profile 30, but the soil has a smaller profile available water capacity of approximately 95 mm (Thomasson, 1979) and is therefore more droughty. Results of long-term ley-arable experiments reported by Harvey (1963) show that yields of both crops and grass were significantly lower at Gleadthorpe than at other experimental sites in England and Wales and there seems little doubt that water shortage was the main cause.

Profile 32 is located in Thetford Forest, Norfolk, which was planted from the 1920s onwards in parts of the East Anglian Breckland previously used for rough grazing or arable cropping. The 'chalk-sand drift' (Corbett, 1973) forming the C horizon is believed to consist of Devensian coversand which was irregularly mixed with the subjacent chalk or older chalky drift by cryoturbation. The soil is non-calcareous to 104 cm depth and has a thin clay enriched Bt or 'beta horizon' (Bartelli and Odell, 1960) lining the irregular lower boundary of the decalcified layer. Uncoated sand grains in the strongly acid A horizon, coupled with the slightly darker colour of the underlying horizon (23–41 cm), suggest that some mobilization of iron and organic matter has occurred, possibly since the forest was planted, but amounts of iron and carbon extracted by pyrophosphate from an adjacent profile afford no clear evidence of translocation and are too small to merit recognition of a podzolic B horizon (p. 108).

Profile 30 (Furness, 1985)

Location	: Great Hatfield, Humberside (TA 185433)
Climatic regime	: subhumid temperate
Altitude	: 18 m Slope: < 1°
Land use	: arable (cereals)

Horizons (cm)

0–30 Ap	Dark brown (7.5YR 3/2 broken and rubbed; 7.5YR 5/2 dry) loamy fine sand (USDA loamy fine sand) containing rare very small subangular flints; weak medium subangular blocky; friable; common fine roots; abrupt smooth boundary.
30–65 Bw	Strong brown (7.5YR 5/6 broken and rubbed; 7.5YR 6/4 dry) loamy fine sand (USDA loamy fine sand); very weak fine subangular blocky; very friable; few fine roots; gradual wavy boundary.
65–93 Cu	Reddish yellow (7.5YR 6/6 broken and rubbed; 7.5YR 7/4 dry) fine sand (USDA fine sand); single grain; loose; few fine roots.

Analytical data

Horizon	Ap	Bw	Cu
Depth (cm)	0–30	30–65	65–93
sand 600 μm–2 mm %	4	3	1
200–600 μm %	22	21	13
60–200 μm %	57	60	81
silt 2–60 μm %	11	10	3
clay < 2 μm %	6	6	2
organic carbon %	2.3		
$CaCO_3$ equiv. %	0.1	0	0
pyro. ext. Fe %	0.3	0.2	0.1
pyro. ext. Al %	0.1	0.1	0.1
total ext. Fe %	1.5	1.9	1.3
pH (H_2O)	7.3	5.6	5.1
(0.01M $CaCl_2$)	6.6	4.9	4.3
bulk density (g cm^{-3})	1.38	1.23	1.29
packing density (g cm^{-3})	1.44	1.28	1.31
retained water capacity (% vol.)	30	16	16
available water capacity (% vol.)	18	10	13
air capacity (% vol.)	17	38	35

Classification

England and Wales (1984): typical brown sand; fine sandy stoneless drift (Naburn series).

USDA (1975): Entic Hapludoll or Entic Haplumbrept; sandy, siliceous or mixed, mesic, coated.

FAO (1974): Haplic Phaeozem; coarse textured.

Profile 31

Location	: Gleadthorpe Experimental Husbandry Farm, Nottinghamshire (SK 591705)
Climatic regime	: subhumid temperate
Altitude	: 67 m Slope: 2°
Land use	: arable (disced after spring barley)

Horizons (cm)

0–30 Ap	Dark brown (7.5YR 3/2; 7.5YR 5/3 dry) slightly stony loamy medium sand (USDA loamy sand); common, mainly medium, quartzite pebbles; weak medium subangular blocky to single grain; very friable, many fine roots; sharp wavy boundary.
30–55 Bw	Brown (7.5YR 4/4; 7.5YR 6/4 dry) slightly stony medium sand (USDA sand); stones as above; single grain; loose; common fine roots; few partly infilled earthworm channels; clear wavy boundary.
55–100 BCt/Cu	Reddish brown (5YR 4/4) slightly stony medium sand (USDA sand) with patchy clay coats on sand grains and numerous paler, weakly layered inclusions with no clay coats; stones, structure and consistence as above; few fine roots; occasional earthworm channels.

Analytical data

Horizon	Ap	Bw	BCt/Cu
Depth (cm)	0–30	30–55	55–100
sand 600 μm–2 mm %	4	11	6
200–600 μm %	55	66	73
60–200 μm %	18	11	13
silt 2–60 μm %	16	9	6
clay < 2 μm %	7	3	2
organic carbon %	1.2	0.4	0.1
pyro. ext. Fe %	0.1	0.1	< 0.05
pyro. ext. Al %	0.1	0.1	< 0.05
total ext. Fe %	0.4	0.5	0.4
pH (H_2O)	6.2	6.6	5.6
(0.01M $CaCl_2$)	5.8	6.1	4.7

Analytical data for Profile 32

Horizon	Ah1	Ah2	AB(h)	Bw1	Bw2	Eb	Bt	Cu
Depth (cm)	0–4	4–23	23–41	41–61	61–81	81–97	97–104	104–120
sand 500 μm–2 mm %	4	4	5	4	2	2	3	<1[a]
200–500 μm %	41	43	45	45	51	50	50	46[a]
100–200 μm %	33	32	32	32	33	33	16	30[a]
50–200 μm %	13	14	12	13	10	11	14	13[a]
silt 2–50 μm %	5	4	3	4	3	3	4	5[a]
clay <2 μm %	4	3	3	2	1	1	13	6[a]
loss on ignition %	5.0	1.8	1.8	1.1	0.8	0.7	2.2	
$CaCO_3$ equiv. %	0	0	0	0	0	0	0	29
dithionite-ext. Fe %	1.1	1.0	0.9	0.8	0.8	0.7	1.8	
TEB (me/100 g)	0.91	0.10	0.10	0.10	0.11	0.10	4.53	
CEC (me/100 g)	6.5	3.3	3.2	1.9	1.5	1.2	7.3	
% base saturation	14	3	3	5	7	8	62	
pH (H_2O)	4.2	4.4	4.5	4.6	4.8	5.3	5.7	8.5
(0.01M $CaCl_2$)	3.3	3.9	3.9	4.1	4.3	4.7	5.1	7.7

[a] In decalcified mineral fraction < 2 mm

Classification
  England and Wales (1984): typical brown sand; medium or coarse sandy material passing to sand or soft sandstone (Cuckney series).
  USDA (1975): Entic Haplumbrept; sandy, siliceous or mixed, mesic, coated.
  FAO (1974): Haplic Phaeozem or Humic Cambisol; coarse textured.

Profile 32 (Corbett, 1973)

Location	: Santon Downham, Norfolk (TL 824840)
Climatic regime	: subhumid temperate
Altitude	: 38 m Slope: <1°
Land use and vegetation	: Scots pine plantation, 25 years old, with bracken (*Pteridium aquilinum*) understorey

Horizons (cm)

L/F — Intermittent layer of brown pine needles and moss over brown partly decomposed pine and bracken remains, c. 8 cm thick.

0–4 Ah1 — Black, very slightly stony, humose medium sand (USDA sand) with many uncoated grains; few angular and subangular flint fragments; weak granular to single grain; loose; common fine roots; clear wavy boundary.

4–23 Ah2 — Brown to dark brown (10YR 4/3; 10YR 5/3 dry) very slightly stony medium sand (USDA sand) with many uncoated grains; stones as above; massive; soft when dry, many fine and medium roots; gradual smooth boundary.

23–41 AB(h) — Dark brown (10YR 3/3; 10YR 4/2 dry) very slightly stony medium sand (USDA sand); stones, structure and consistence as above; many fine and medium roots; gradual smooth boundary.

41–61 Bw1 — Brown to dark brown (10YR 4/3; 10YR 4–5/3 dry) slightly to very slightly stony medium sand (USDA sand); angular and rounded flints; massive; soft when dry; many fine to coarse woody and fibrous roots; gradual smooth boundary.

61–81 Bw2 — Brown to dark brown (10YR 4/3; 10YR 6/4 dry) medium sand (USDA

	sand) containing occasional flints; single grain; loose; common fine to coarse, woody and fibrous roots; gradual boundary.
81–97 Eb	Yellowish brown (10YR 5/4; 10YR 6/3 dry) medium sand (USDA sand) containing occasional flints; massive; soft when dry; few medium woody roots; abrupt irregular boundary ranging downwards to 112 cm.
97–104 Bt	Strong brown (7.5YR 5/6) loamy medium sand to sandy loam (USDA loamy sand to sandy loam) containing occasional flints; massive; slightly hard when dry; few medium woody roots; clay coats on sand grains; sharp irregular boundary, ranging downwards in places to 120 cm.
104–120+ Cu	Light yellowish brown (10YR 6/4), slightly stony, very calcareous loamy sand to sandy loam; common chalk fragments and occasional subangular flints; massive; few medium woody roots.

Classification
  England and Wales (1984): argillic brown sand; sandy chalky drift (Worlington series).
  USDA (1975): Alfic Udipsamment; siliceous, mesic, coated.
  FAO (1974): Luvic Arenosol; coarse textured.

5.2.3 Gleyic Sandy Brown Soils

These are intergrades between typical sandy brown soils and sandy gley soils (p. 293), as indicated by ochreous mottles and/or greyish colours attributable to gleying between 40 and 80 cm depth. They are common components of catenary associations in areas where a seasonally fluctuating ground-water table occurs at variable depths in sandy Devensian deposits overlying impervious substrata such as till or glaciolacustrine clay. Those in drained agricultural land are unlikely to be wet within 70 cm depth for more than a few days at a time in most years (Robson and Thomasson, 1977) and can be moderately droughty.

The following representative profile from the Cheshire Plain is in glaciofluvial sands (Stockport Formation) containing igneous pebbles of Lake District origin. It has an Ap horizon of sandy loam texture, probably as a result of marling. This was widely practised in the area until the end of the nineteenth century, nearly every field containing a marl pit from which slightly calcareous till was dug and spread on surrounding land to alleviate acidity and give 'body' to the coarse textured surface soils.

Profile 33 (King, 1977)

Location	: Beeston, Cheshire (SJ 529583)
Climatic regime	: humid temperate
Altitude	: 61 m Slope: < 1°
Land use	: arable (barley)

Horizons (cm)

0–24 Ap	Dark greyish brown (10YR 4/2) very slightly stony medium sandy loam (USDA sandy loam); few small subangular fragments of igneous rock; moderate medium subangular blocky; very friable; many fine roots; clear smooth boundary.
24–42 Bw(g)	Dark yellowish brown (10YR 4/4) very slightly stony loamy medium sand (USDA loamy sand) with few fine light brownish grey (10YR 6/2) mottles; stones as above; very weak fine and medium blocky; very friable; common fine roots; gradual boundary.
42–90 Cg	Light brown (7.5YR 6/4) and light yellowish brown (10YR 6/4) very slightly stony medium sand (USDA sand) with common medium reddish yellow (7.5YR 6/8) mottles increasing in number and size with depth; stones as above; single grain; loose; few fine roots.

Analytical data

Horizon	Ap	Bw(g)	Cg
Depth (cm)	0–24	24–42	42–90
sand 600 µm–2 mm %	3	4	3
200–600 µm %	38	51	66
60–200 µm %	32	30	27
silt 2–60 µm %	17	11	3
clay < 2 µm %	10	4	1
organic carbon %	2.4	0.4	0.2
pyro. ext. Fe %	0.2	0.1	< 0.05
pyro. ext. Al %	0.1	0.1	0.05
total ext. Fe %	0.7	0.8	0.2
pH (H$_2$O)	5.1	5.8	5.4
(0.01M CaCl$_2$)	4.6	5.5	5.4

Classification
  England and Wales (1984): gleyic brown sand; sandy drift with siliceous stones (Ollerton series).
  USDA (1975): Aquic Quartzipsamment; mesic, uncoated.
  FAO (1974): Cambic Arenosol, coarse textured.

5.2.4 Calcaric Sandy Brown Soils

These are well drained sandy soils with a brownish Bw horizon which retains residual calcium carbonate or overlies a calcareous C horizon within 50 cm depth. They occupy numerous small areas on chalk-sand drift (p. 161) in the East Anglian Breckland (Corbett, 1973) and also occur in shelly coastal sands and pre-Quaternary sediments of similar composition, but are much less extensive than those of the typical subgroup. Formation of a 'colour B' horizon, which distinguishes them from calcareous sandy regosols (p. 149), appears to constitute the second stage of an evolutionary sequence on calcareous sand in which progressive decalcification leads to liberation of free iron and eventually to the development of more deeply leached profiles showing evidence of clay translocation, podzolization, or both (cf. Figure 1.3, p. 9).

Data on two examples are given on pp. 166–8. Profile 34 is in wind-blown shelly sand close to the coast of East Lothian. It has a well defined colour B horizon with ferruginous coats on the sand grains but otherwise resembles associated sandy regosols (e.g. profile 25, p. 151) in being little leached and containing little silt and clay. Cereals, principally barley, and market-garden crops are grown on similar soils nearby, though drought and wind erosion are serious limitations. Similar soils on stabilized coastal sands in western Scotland and the Hebrides commonly have very dark humose A horizons (Bibby et al., 1982; Hudson et al., 1982).

Profile 35 is on chalk-sand drift in Thetford Forest, Norfolk. By contrast with profile 32, which is located 5 km away on similar material, this soil is calcareous throughout and overlies the compact chalky C horizon at around 50 cm depth. Particle-size data for the profile were obtained on a carbonate-free basis and show that the non-carbonate fractions of the A and Bw horizons resemble that of the C horizon. This indicates that they result from decalcification of similar material *in situ*, though there is no reason to suppose that all horizons were equally calcareous originally. Placement of the soil in the US system depends on the presence or absence of a loamy or skeletal (> 35 per cent stones) layer within the upper metre and is therefore uncertain on the basis of the data provided.

In parts of the Breckland, as depicted in Figure 5.3, soils typified by profiles 32 and 35 occur in close association with rendzinas as components of polygonal or striped patterns (Figures 1.17a and 5.4) reflecting rhythmic lateral variations in the chalk content of the drift which Watt et al. (1966) attributed to frost heaving under periglacial conditions during the Devensian period. Although yields of un-irrigated farm crops can be severely limited in dry years on these sandy Breckland soils, there is no evidence that growth of pines is significantly affected by water shortage, apparently because they are able to draw moisture from the considerable reserves held in the chalky substrate.

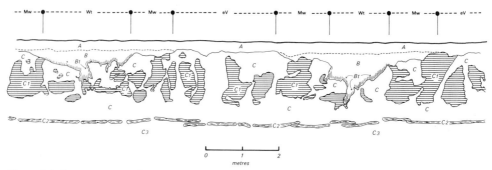

Figure 5.3: Section across slope at Grimes Graves, Norfolk (TL 811903) (after Corbett, 1973). A topsoil; B yellow-brown sand; Bt strong brown clay-enriched horizon; C1 chalk-sand mixture; C2 tabular flints almost *in situ*; C3 bedded chalk. Mw Methwold series (calcaric sandy brown soil); Wt Worlington series (typical sandy brown soil); eV Elveden series (brown rendzina).

Profile 34 (Ragg and Futty, 1967; Ragg and Clayden, 1973)

Location : Aberlady, East Lothian (NT 468807)
Climatic regime : humid temperate
Altitude : 2 m Slope: < 1°

Land use and vegetation : rough grazing with oat-grass (*Arrhenatherum elatius*), sheeps fescue (*Festuca ovina*), red fescue (*Festuca rubra*), marsh horsetail (*Equisetum palustre*) and marram grass (*Ammophila arenaria*)

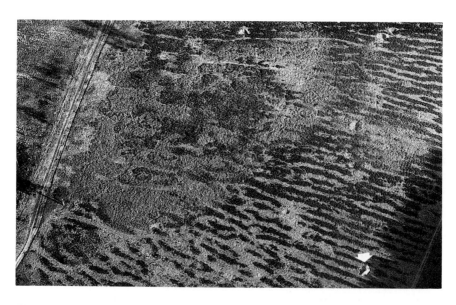

Figure 5.4: Striped pattern near Grimes Graves, Norfolk. Heather (dark tones) and grass stripes reflect cyclic soil variations as seen in Figure 5.3 at the site arrowed (Cambridge University Collection: copyright reserved).

Horizons (cm)

0–10 Ah
Dark greyish brown (10YR 4/2) stoneless calcareous medium sand (USDA fine sand); occasional small shell fragments; weak fine granular; loose; abundant roots; clear boundary.

10–23 AB
Brown (10YR 4/3) stoneless calcareous medium sand (USDA fine sand); few shell fragments; single grain; loose; abundant roots; clear boundary.

23–43 Bw
Strong brown (7.5YR 5/5) stoneless calcareous medium sand (USDA fine sand); single grain; frequent roots; shell fragments present; clear boundary.

43–132 Cu
Light greyish brown (7.5YR 6/3) stoneless calcareous medium sand (USDA fine sand), becoming coarser below 120 cm; single grain; loose; occasional roots to 120 cm; very small shell fragments present and layer of shells at 50–55 cm.

Analytical data *see* below

Classification

Scotland (1984): brown calcareous soil (Fraserburgh series).

England and Wales (1984): typical brown calcareous sand; sandy stoneless drift (unnamed series).

USDA (1975): Typic Udipsamment; mixed (calcareous), uncoated.

FAO (1974): Cambic Arenosol; coarse textured.

Profile 35 (Corbett, 1973)

Location : Bromehill, Norfolk (TL 797882)

Climatic regime : subhumid temperate
Altitude : 23 m   Slope: 4°
Land use and vegetation : Scots pine plantation, 30 years old, interplanted with beech; field layer includes rosebay willow herb (*Chamaenerion angustifolium*), creeping soft grass (*Holcus mollis*) and Cocksfoot (*Dactylis glomerata*)

Horizons (cm)

L/F
Plant litter and living roots, c. 3 cm thick.

0–3 Ah1
Very dark greyish brown (10YR 3/2), stoneless, slightly calcareous humose sand (USDA sand) with few uncoated grains; weak fine granular; loose; abundant fine roots forming mat; sharp smooth boundary.

3–13 Ah2
Dark greyish brown to dark brown (7.5YR 4/2.5), very slightly stony, slightly calcareous medium sand (USDA sand); few small chalk and flint fragments; weak fine granular and single grain; loose; many fine roots; abrupt smooth boundary.

13–51 Bw
Yellowish brown (10YR 5/4), very slightly stony, slightly calcareous medium sand (USDA sand); stones as above; single grain to massive; slightly hard when dry; few fine roots; abrupt wavy boundary.

51–70+ Cu
Pale grey (10YR 7/2) 'chalk-sand drift' consisting of subangular chalk fragments in a matrix of sand and disintegrated chalk; massive; firm; hard when dry.

Analytical data for Profile 34

Horizon	Ah	AB	Bw	Cu	2Cu
Depth (cm)	0–10	13–23	28–38	78–86	122–132
sand[a] 200 μm–2 mm %	44	45	46	39	59
50–200 μm %	52	52	52	58	39
silt[a] 2–50 μm %	2	1	1	1	1
clay[a] < 2 μm %	2	2	1	2	1
organic carbon %	2.3	0.4			
CaCO$_3$ equiv. %	6.6	9.1	8.3	9.6	10.9
pH (H$_2$O)	7.8	8.0	8.0	8.4	8.8

[a]By hydrometer method

Analytical data

Horizon	Ah2	Bw	Cu
Depth (cm)	3–13	13–51	51–70
sand[a] 500 μm–2 mm %	10	8	7
200–500 μm %	46	43	48
100–200 μm %	27	29	22
50–100 μm %	12	15	18
silt[a] 2–50 μm %	2	2	2
clay[a] < 2 μm %	3	3	3
organic carbon %	0.7	0.2	
$CaCO_3$ equiv. %	2.9	3.7	25
dithionite-ext. Fe %	0.9	0.7	0.5
pH (H_2O)	8.1	8.3	8.5
(0.01M $CaCl_2$)	7.5	7.5	7.7

[a]In decalcified mineral fraction < 2 mm

Classification
  England and Wales (1984): typical brown calcareous sand; sandy chalky drift (Methwold series).
  USDA (1975): Typic Udipsamment (or Udorthent); sandy (or sandy-skeletal), carbonatic or mixed (calcareous), mesic.
  FAO (1974): Cambic Arenosol; coarse textured.

5.2.5 Gleyic-calcaric Sandy Brown Soils

These are closely related soils occurring locally in calcareous sandy or sandy-gravelly deposits affected by seasonal rising ground water. Typical profiles have ochreous mottles between 40 and 80 cm depth, indicating that they are intergrades between calcaric sandy brown soils and sandy gley soils (p. 299). The following example is located in the same area as profile 34 (p. 166) above and is developed in similar shelly sand. The water table rises to within 50 cm of the surface during the winter months and pits nearby revealed a 2 Cg horizon of silty clay at about 1 m depth.

Profile 36 (Ragg, 1974)

Location	: Dirleton, East Lothian (NT 505845)
Climatic regime	: humid temperate
Altitude	: 15 m Slope: < 1°
Land use	: long-term grass

Horizons (cm)

0–31 Ap	Dark yellowish brown (10YR 3/4) calcareous loamy medium sand (USDA loamy fine sand or loamy sand) with common fine dark reddish brown (5YR 3/4) mottles, especially near base; nearly massive; very friable; abundant fine roots; abrupt smooth boundary.
31–50 Bw(g)	Dark brown to brown (7.5YR 4/4) slightly calcareous (shelly) medium sand (USDA fine sand or sand) with common fine dark reddish brown (5YR 3/4) mottles; single grain; loose; common fine roots; gradual smooth boundary.
50–80 BC(g)	Brown (7.5YR 5/4) calcareous (shelly) medium sand (USDA fine sand or sand) with common medium yellowish red (5YR 5/8) mottles and streaks; single grain; loose; few roots; clear smooth boundary.
80–100 Cg	Dark greyish brown (2.5Y 4/2) calcareous (shelly) medium sand (USDA fine sand or sand); wet; single grain; loose; no roots.

Analytical data

Horizon	Ap	Bw(g)	BC(g)	Cg
Depth (cm)	0–25	31–50	50–80	80–100
sand[a] 200 μm–2 mm %	76	55	88	87
50–200 μm %	10	40	7	8
silt[a] 2–50 μm %	7	1	2	3
clay[a] < 2 μm %	7	4	3	2
organic carbon %	1.7	0.5		
$CaCO_3$ equiv. %	5.7	4.4	6.0	6.5
pH (H_2O)	8.2	8.3	8.5	8.6

[a]By hydrometer method

Classification
  Scotland (1984): brown calcareous soil intergrading to calcareous ground-water gley (Luffness series).
  England and Wales (1984): gleyic brown calcareous sand; stoneless drift (unnamed series).
  USDA (1975): Aquic Udipsamment; mixed (calcareous), mesic, uncoated.
  FAO (1974): Cambic Arenosol; coarse textured.

5.3 Alluvial Brown Soils

5.3.1 General Characteristics, Classification and Distribution

These soils have formed in medium to fine textured alluvial deposits accumulated during the last few thousand years in river floodplains (Figure 5.5) and intertidal flats. They are distinguished from rego-alluvial soils (p. 152) by the presence of a colour/structure B (Bw) horizon, a thick cumulic A horizon, or both, and from alluvial gley soils (p. 300) by the absence of a gleyed or hydrocalcic subsurface horizon that starts within 40 cm depth. The latter requirement implies that the parent deposit is at least moderately permeable and that the water table is nearly always more than about 50 cm below the surface, either naturally or as a result of artificial drainage. In addition, the parent material must have been in place long enough for biological homogenization (Hoeksema, 1953) and the effects of seasonal wetting and drying to have destroyed any fine stratification to depths exceeding 30 cm, and to have produced a subsurface horizon with pedal structure.

The degree of horizon development and the extent to which structural reorganization is accompanied by chemical alteration vary in accordance with the age and composition of the alluvium. As Green (1968) demonstrated in Romney Marsh, variants in calcareous marine sediments embanked in mediaeval times or earlier show clear evidence of carbonate leaching and 'browning' attributable to transformation of iron-bearing minerals, giving Bw horizons which qualify as cambic according to the criteria used in the US and FAO systems. The alluvial brown soils that meet these requirements and lack a prominent A horizon are therefore identifiable as Inceptisols (mostly Fluventic or Fluvaquentic Eutrochrepts) or Cambisols respectively; those with a prominent A horizon are Hapludolls (Phaeozems) or Haplumbrepts (Humic Cambisols) depending on base status. Other variants, corresponding to the allochthonous brown warp soils of Kubiena (1953) have subsurface horizons that qualify as Bw on a structural basis but inherit their brown or reddish colour from the parent deposits and are not true cambic horizons inasmuch as they show little evidence of leaching or weathering *in situ*. On this basis the soils are Entisols (mostly Fluvents) in the US Taxonomy or Fluvisols according to FAO. It is often difficult, however, to distinguish acquired from inherited characteristics in such soils, as the parent deposits usually become finer upwards and lower horizons frequently have colours of lower chroma due to gleying. Hence the occurrence of a brownish subsurface horizon with more clay and a higher chroma or redder hue than the underlying horizon does not necessarily indicate that silicate clay and iron oxide have accumulated as a result of weathering *in situ* (cp. Duchaufour, 1982, p. 187).

Soils of this group are widely distributed and locally important in England and Wales and also occur in Scotland and Ireland but have not been specifically distinguished from other alluvial soils in either country. Those in Scotland have mostly been grouped with rego-alluvial and alluvial gley soils as mineral alluvial soils (Soil Survey of Scotland, 1984), and Larney *et al*. (1981) described examples which they studied in County Monaghan as brown earths. The main areas of occurrence are in floodplains of major rivers and on artificially embanked and drained marine or estuarine alluvium adjoining the lower tidal reaches of the Trent, Yorkshire Ouse, Humber and Severn, bordering the Wash, and in Romney Marsh beside the English Channel. Much of the Trent, Ouse and Humber 'warpland' was itself created artificially during the nineteenth and earlier centuries by allowing silt-laden water to flood through sluice gates at high tide and deposit its sediment on land embanked for the purpose (Heathcote, 1951). When a metre or more of 'warp' had accumu-

Figure 5.5: The Ribble floodplain near Baldeston, Lancashire (Photo by H.S. Bentley).

lated, the water supply was withdrawn and the area left to dry until the surface was firm enough for a pioneer crop, usually grass, to be sown; then, after a further period of consolidation (ripening), the land was divided into fields, drained, and brought into regular arable use.

Wherever these soils occur, they are generally associated with somewhat lower lying and usually finer textured alluvial gley soils in patterns related to current or former depositional regimes; ungleyed rego-alluvial soils in which no B horizon is discernible also appear as minor components in places, especially in the youngest and coarsest deposits. In some coastal marshes, brown soils are mainly located in low ridges corresponding to former tidal creeks and gley soils in intervening 'pool' areas. As in similar Dutch landscapes (Edelman, 1950), the creek ridges (called Roddons or Rodhams in the Fens) are believed to result from differential shrinkage following drainage, the sands and silts beneath the ridges having shrunk less than the clay or clay-over-peat layers which underlie the pools (Figure 5.6). In Romney Marsh (Green, 1968), creek ridge and pool patterns are confined to 'old marshland' where most of the soils are superficially decalcified; in younger 'innings' enclosed since the fourteenth century, the soils remain calcareous to the surface and relics of creeks and old river courses survive in places as narrow swampy strips bordered by higher levees. River floodplains show similar patterns, with well drained brown soils on levees and finer textured gley soils in neighbouring backswamps (Figure 5.7). Soil patterns in the Humber warplands are determined in part by the boundaries of areas artificially raised by warping, but brown soils are again mainly situated close to the rivers and gley soils further away.

The alluvial brown soils which are regularly flooded are unsuitable for arable cultivation but often bear very productive grassland. Others, particularly silty variants in marine and estuarine deposits, support a wide range of agricultural and horticultural crops and form some of the most valuable arable land in Britain. They have large available water

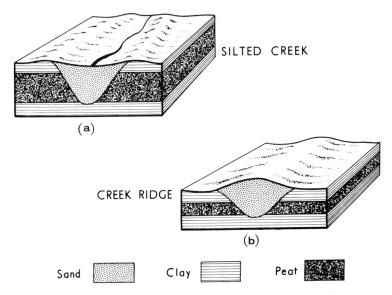

Figure 5.6: Block diagrams showing creek-ridge development (from Green, 1968).

capacities and the presence of a water table at moderate depths further ensures that crops rarely suffer from drought. Capping, compaction and workability problems arise under intensive cultivation but can be avoided or minimized by careful management so long as adequate under drainage is provided.

Key to Subgroups

1. Alluvial brown soils with pelo-characteristics (p. 112)
   1.1 with a calcareous subsurface horizon starting no more than 50 cm below the surface
       Pelocalcaric alluvial brown soils
   1.2 with non-calcareous subsurface horizons extending to more than 50 cm depth
       Pelo-alluvial brown soils
2. Other alluvial brown soils with a calcareous subsurface horizon starting no more than 50 cm below the surface
   2.1 with gleyic features (p. 112)
       Gleyic-calcaric alluvial brown soils
   2.2 without gleyic features
       Calcaric alluvial brown soils

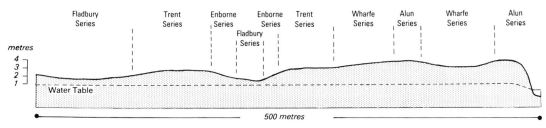

Figure 5.7: Soil series in a typical cross-section of the Trent floodplain in Nottinghamshire (after Ragg *et al.*, 1984). *Alun series* coarse loamy typical alluvial brown soil *Wharfe series* fine loamy typical alluvial brown soil *Trent series* fine loamy gleyic alluvial brown soil *Enborne series* fine loamy typical alluvial gley soil *Fladbury series* clayey pelo-alluvial gley soil

3. Other alluvial brown soils
   3.1 with gleyic features (p. 112)
          Gleyic alluvial brown soils
   3.2 without gleyic features
          Typical (loamy) alluvial brown soils

5.3.2 Typical (Loamy) Alluvial Brown Soils

Non-calcareous alluvial brown soils that are predominantly loamy in texture and show no evidence of subsurface gleying to a depth of at least 60 cm are treated as the typical subgroup. These soils are mainly confined to river floodplains with large seasonal variations in water level, such as those of the Severn and Wye in the Welsh Borderland. They are usually in levees bordering existing or former stream channels and are commonly flooded for short periods, but most of them receive little fresh sediment at present. Horizons within the rooting zone are permeable and do not contain stagnant water for periods long enough to cause appreciable gleying except in topsoils where poaching has occurred. Many of the soils are uniformly loamy, friable and porous to depths of a metre or more, making them exceptionally valuable for agriculture when protected from floods. Others have gravelly or sandy layers within 80 cm depth. These grade into non-calcareous rego-alluvial soils and are more susceptible to drought.

Data on two representative profiles under grass are reproduced below. Profile 37 is in brownish coarse loamy alluvium derived mainly from Lower Palaeozoic rocks in the Tywi valley, Dyfed. It shows evidence of deep reaching earthworm activity and has a Bw horizon with moderately developed subangular blocky structure merging downwards into a nearly massive C horizon which is clearly stratified. The B horizon can be considered cambic on the basis that it contains slightly more clay than the C and the profile is classified accordingly in the US and FAO systems, though there is a strong possibility that the upward clay increase is inherited. The physical data confirm that this deep stone-free soil has a large available water capacity and that the subsurface horizons are moderately porous to at least 80 cm depth.

Profile 38 is in alluvium derived mainly from granite and shale (Conry *et al.*, 1970) in the valley of the river Liffey in County Kildare. It differs from the preceding profile in having a thick cumulic A horizon which appears to incorporate a weakly expressed Bw. The soil is almost uniformly dark coloured (Munsell value 3 or less) to a depth of 122 cm, below which the substratum, waterlogged at the time of sampling, is partly cemented by ferrimanganiferous deposits. The horizon notation has been inserted tentatively by the writer; particle-size analyses show marked lithologic discontinuities at 64 and 122 cm and the horizons between these depths are interpreted as buried A horizons on the basis of their slightly darker colours and enhanced organic carbon contents, compared with those above and below, though both features are possibly inherited. It is noteworthy in this connection that the Liffey rises in the granitic Wicklow mountains, where most of the higher lying soils have dark humose or peaty surface horizons. Similar riverine alluvial soils with thick dark A or A/B horizons have been recorded by Clayden (1971) in the Teign valley, Devon, and by Jarvis (1977) in the Tyne valley, Northumberland.

Profile 37 (Wright, 1980)

Location	: Llandeilo, Dyfed (SN 621217)
Climatic regime	: humid temperate
Altitude	: 23m Slope: < 1°
Land use	: long-term grass

Horizons (cm)

0–15 Ah   Dark brown to brown (10YR 4/3) stoneless sandy silt loam (USDA loam) with common linear strong brown (7.5YR 5/6) mottles along root channels (poached surface); moderate fine subangular blocky; friable; many fine roots; earthworms present; clear smooth boundary.

15–45 Bw	Brown (7.5YR 4/4) stoneless sandy silt loam (USDA loam); moderate medium and fine subangular blocky; friable; many to common fine roots; earthworms present; gradual smooth boundary.
45–70 BC	Brown (7.5YR 4.5/4) stoneless sandy silt loam (USDA loam to sandy loam) with red (2.5YR 4/6) patches (material derived from Devonian red beds) and lenses of coarse loamy sand 0.5–2 cm in diameter (5% of horizon); weak medium and coarse subangular blocky; very friable; common fine roots; earthworms present; gradual smooth boundary.
70–120 Cu(g)	Brown (7.5YR 4.5/4) stoneless sandy loam with few red patches, few fine pinkish grey (7.5YR 6/2) mottles and bands or lenses of coarse loamy sand up to 1 cm wide and 10 cm apart; massive; very friable; few roots.

Analytical data

Horizon	Ah	Bw	BC
Depth (cm)	0–15	15–45	45–70
sand 200 μm–2 mm %	7	8	12
60–200 μm %	27	29	34
silt 2–60 μm %	50	49	44
clay < 2 μm %	16	14	10
organic carbon %	2.6	1.0	0.5
pH (H_2O)	5.6	5.9	6.0
(0.01M $CaCl_2$)	4.8	5.0	5.2
bulk density (g cm^{-3})	1.18	1.44	1.42
packing density (g cm^{-3})	1.32	1.57	1.51
retained water capacity (% vol.)	48	37	35
available water capacity (% vol.)	24	23	25
air capacity (% vol.)	6	9	12

Classification
England and Wales (1984): typical brown alluvial soil; coarse loamy river alluvium (Alun series).
USDA (1975): Fluventic Dystrochrept (or Eutrochrept); coarse-loamy, mixed, mesic.
FAO (1974): Dystric (or Eutric) Cambisol; medium textured.

Profile 38 (Conry et al., 1970)

Location : Celbridge, County Kildare
Climatic regime : subhumid temperate
Altitude : 59 m   Slope: < 1°
Land use : long-term grass

Horizons (cm)

0–25 Ah	Dark brown (10YR 3/3) clay loam (USDA loam); moderate medium and fine subangular blocky; friable; plentiful roots; gradual boundary.
25–64 ABw	Dark brown (10YR 3/3) sandy silt loam (USDA sandy silt loam); moderate medium granular; friable; plentiful roots; gradual boundary.
64–89 2bAh1	Very dark to dark greyish brown (10YR 3–4/2) silty clay loam (USDA silty clay loam); weak medium subangular blocky; fair root supply; gradual boundary.
89–122 2bAh2	Very dark greyish brown (10YR 3/2) silty clay loam (USDA silty clay loam); weak medium subangular blocky; wet; sparse roots; gradual boundary.
122–147 3Cg(f)	Light olive brown (2.5Y 5/4) sandy loam (USDA sandy loam) with very prominent manganese-iron concretions in upper part; wet; no roots; abrupt boundary.
147+ 3Cgfm	Hard manganese–iron pan

Analytical data *see* p. 174

Classification
Ireland (1980): Regosol (Alluvial soil) (Liffey series).
England and Wales (1984): typical brown alluvial soil; fine loamy river alluvium (Trent series).
USDA (1975): Cumulic Haplumbrept (intergrading to Hapludoll); fine-silty, mixed, mesic.
FAO (1974): Humic Cambisol; medium textured.

5.3.3 Gleyic Alluvial Brown Soils

These are similar soils with mottles attributable to gleying within 60 cm of the surface. They are transitional to typical alluvial gley soils (p. 303) in morphology and are found in

Analytical data for Profile 38

Horizon	Ah	ABw	2bAh1	2bAh2	3Cg(f)
Depth (cm)	0–25	25–64	64–89	89–122	122–147
sand 200 μm–2 mm %	9	9	1	2	57
50–200 μm %	27	34	10	4	17
silt 2–50 μm %	45	41	61	61	14
clay < 2 μm %	19	16	28	33	12
organic carbon %	3.2	0.8	1.3	1.3	0.5
nitrogen %	0.36	0.11	0.17	0.18	
C/N ratio	9.0	7.3	8.0	7.2	
citrate-dithionite ext. Fe %	1.1	1.1	1.4	1.9	0.8
TEB (me/100 g)	13.4	9.8	19.8	23.4	9.0
CEC (me/100 g)	33.6	17.8	25.0	28.4	9.2
% base saturation	40	55	79	83	98
pH (H_2O)	5.8	6.2	6.5	6.6	6.7

correspondingly intermediate floodplain sites where the water table rises into the lower part of the profile for significant periods in winter and spring. Soils of this subgroup also occur locally in loamy marine or estuarine alluvium which was originally non-calcareous, as in the Solway Firth area (Longtown series in Kilgour, 1979), or which has been deeply decalcified, as in the old marshland of Green (1968). The most gleyed variants have a grey and brown mottled horizon (Bg or BCg) starting between 40 and 50 cm depth; others have no strongly gleyed horizon above 80 cm, the matrix colours remaining brownish or reddish throughout. Most of these soils are at least moderately permeable, and as ground-water levels have often been lowered artificially, the morphology may not reflect the current water regime.

The following representative profile, located in the Vyrnwy valley, Powys, is developed in rather more than a metre of river alluvium overlying glacial drift with fragipan features in the upper part. Although subject to occasional floods, the land had been ploughed and carried a grass ley when sampled. Micromorphological studies of this and other alluvial soils in the area by Thompson (1983) revealed the common occurrence in lower horizons of poorly sorted ped and channel cutans (matrans) that consist of translocated clay, silt and humus and resemble those found in seasonally flooded Bangladesh soils by Brammer (1971).

Profile 39 (Thompson, 1982)

Location	: Llandysilio, Powys (SJ 268197)
Climatic regime	: subhumid temperate
Altitude	: 65 m Slope: < 1°
Land use	: ley grass

Horizons (cm)

0–18 Ap	Dark brown (10YR 3/3) silty clay loam (USDA silt loam) containing occasional small subrounded siltstone fragments, moderate medium subangular blocky with dark greyish brown (10YR 4/2) faces; firm; common fine roots; few sesquioxidic coats, sharp smooth boundary.
18–52 Bw	Dark yellowish brown (10YR 4/4) stoneless silty clay loam (USDA silty clay loam); weak medium prismatic with brown to dark brown (10YR 4/3) ped faces; moderately porous; firm; common fine roots; many dark coats on ped faces and in large vertical worm channels; gradual smooth boundary.

Analytical data for Profile 39

Horizon	Ap	Bw	Bw(g)	BCg1	BCg2	2BCgx	3Cu
Depth (cm)	0–18	18–52	52–63	63–76	76–107	107–119	132–150
sand 200 μm–2 mm %	5	5	4	5	3	16	64
60–200 μm %	7	6	6	6	3	6	14
silt 2–60 μm %	62	60	60	59	65	57	11
clay < 2 μm %	26	29	30	30	29	21	11
organic carbon %	3.4	1.0	0.7	0.6	0.5	0.3	0.4
CaCO$_3$ equiv. %	0.2	0	0	0	0	0	0
pH (H$_2$O)	6.5	6.7	6.5	6.4	6.4	6.3	6.3
(0.01M CaCl$_2$)	6.2	6.0	5.8	5.7	5.7	5.7	5.7

52–63 Bw(g) — Olive brown (2.5Y 4/4) stoneless silty clay loam (USDA silty clay loam) with few fine yellowish brown (10YR 5/6) mottles and 1 cm areas of light brownish grey (2.5Y 6/2); moderate medium prismatic with brown to dark brown (10YR 4/3) faces; firm; common fine roots; few soft rounded ferri-manganiferous concentrations; organic coats and earthworm channels as above; clear wavy boundary.

63–76 BCg1 — Light olive brown (2.5Y 5/4) stoneless silty clay loam (USDA silty clay loam) with common fine strong brown (7.5YR 4/6) mottles and red (2.5YR 4/6) root traces; moderate coarse prismatic with greyish brown (10YR 5/2) faces; firm; few fine roots; ferri-manganiferous concentrations, coats and worm channels as above; abrupt wavy boundary.

76–107 BCg2 — Light brownish grey (2.5Y 6/2) stoneless silty clay loam (USDA silty clay loam); moderate coarse prismatic with light brownish grey (2.5Y 6/3) faces; firm; few fine roots; common rounded ferri-manganiferous soft concentrations and rare soft 'iron pipes'; common silt coats; abrupt irregular boundary.

107–119 2BCgx — Light brownish grey (2.5Y 6/3) slightly stony clay loam (USDA silt loam) with many medium strong brown (7.5YR 5/6) mottles; common small rounded and tabular sandstone fragments; massive; very firm; no roots; silt caps on larger stones; sharp wavy boundary.

119–150 2Cu(g)/3Cu — Light olive brown (2.5Y 5/4) very slightly stony clay loam with strong brown mottles, passing at 132 cm into olive brown (2.5Y 4/4) stoneless sandy loam.

Classification
England and Wales (1984): gleyic brown alluvial soil; fine silty river alluvium (Clwyd series).
USDA (1975): Fluvaquentic Eutrochrept; fine-silty, mixed, mesic.
FAO (1974): Gleyic Cambisol; medium textured.

5.3.4 Pelo-alluvial Brown Soils

These are non-calcareous clayey alluvial soils with relatively good natural drainage. Though nowhere extensive, they occur in association with pelo-alluvial gley soils (Chapter 7, p. 306) in the floodplains of rivers with clayland catchments such as the Stratford and Bristol Avons, and have also been identified in older marine alluvium. Typical profiles have a brown or reddish Bw horizon with blocky to prismatic structure, passing between 40 and 80 cm depth into a mottled Bg or BCg. The following example from Northamptonshire was described and sampled in recently drained arable land bordering a tributary of the river Nene which drains an area of Jurassic clays and ironstones, mantled on higher ground by Chalky Boulder Clay. As the main source rocks are calcareous in the un-weathered state, the alluvium presumably consists of pre-weathered and transported soil material. The soil cracks strongly in dry summers and coefficient of linear extensibility (C.O.L.E.) values reported by Reeve et al. (1980) markedly exceed the threshold value of 0.09 chosen to define vertic subgroups in the US Taxonomy. Analytical data on < 2 μm separates (CEC and K$_2$O content) suggest that the clay fractions are smectitic (Avery and Bullock, 1977) and this was confirmed by X-ray examination. The medium bulk density values (Table 3.7, p. 98) are characteristic of clayey soils in recent alluvial deposits (Hall et al., 1977). Air capacity decreases with depth as bulk density increases, but remains above 5 per cent to 80 cm depth, suggesting that the subsurface horizons are appreciably permeable at moisture states approximating to field capacity.

Profile 40 (Reeve, 1978; Reeve and Hall, 1978; Reeve et al., 1980)

Location	: Brockhall, Northamptonshire (SP 630619)
Climatic regime	: subhumid temperate
Altitude	: 83 m Slope: < 1°
Land use	: arable (bare ploughland)

Horizons (cm)

0–23 Ap	Dark brown (10YR 3/3) stoneless clay (USDA clay); weak cloddy (angular blocky) with strong fine angular blocky (frost tilth) at surface; firm; few fine roots and inclusions of cereal stubble; sharp smooth boundary.
23–45 Bw(g)	Yellowish brown (10YR 5/4) stoneless clay (USDA clay) with faint ochreous mottling below 35 cm; strong fine angular blocky; firm; few fine roots; gradual smooth boundary.
45–80 Bg	Greyish brown (10YR 5/2) and strong brown (7.5YR 5/6), prominently mottled stoneless clay (USDA clay); moderate coarse prismatic breaking to coarse angular blocky with light yellowish brown to light brownish grey (2.5Y 6/3) faces; firm; roots rare; gradual smooth boundary.
80–120 BCg	Light brownish grey (2.5Y 6/2) and strong brown, prominently mottled stoneless clay (USDA silty clay to clay); structure as above but weaker; wet; water table at 110 cm (March, 1974).

Analytical data *see* below.

Analytical data for Profile 40

Horizon	Ap	Bw(g)	Bg	BCg
Depth (cm)	0–23	30–40	60–70	100–110
sand 200 μm–2 mm %	5	1	2	2
60–200 μm %	8	6	3	6
silt 2–60 μm %	35	35	31	42
clay < 2 μm %	52	58	64	50
< 0.2 μm %	22	24	25	22
fine clay/total clay	0.42	0.41	0.39	0.44
organic carbon %	4.4	2.7	1.2	1.1
dithionite ext. Fe %	4.9	5.2	4.9	4.5
pH (H$_2$O)	6.4	6.6	7.1	7.4
(0.01M CaCl$_2$)	5.9	6.2	6.7	6.9
clay < 2 μm				
CEC (me/100 g)	57	57	49	49
K$_2$O %	2.0	2.0	2.1	1.9
bulk density (g cm^{-3})	0.88	1.08	1.13	
packing density (g cm^{-3})	1.35	1.60	1.71	
retained water capacity (% vol.)	57	51	52	
available water capacity (% vol.)	25	18	13	
air capacity (% vol.)	10	8	6	

Classification
England and Wales (1984): pelogleyic brown alluvial soil; clayey river alluvium (Wyre series).
USDA (1975): Vertic Fluvaquent intergrading to Fluvaquentic (or Vertic) Eutrochrept; very-fine, montmorillonitic, mesic.
FAO (1974): Eutric Fluvisol intergrading to Gleyic (or Vertic) Cambisol; fine textured.

5.3.5 Calcaric Alluvial Brown Soils

These are freely draining soils with a brownish Bw horizon, a cumulic A, or both, in recent alluvium derived wholly or partly from calcareous rocks. Except for the presence of calcium carbonate, which may occur partly as fragmentary mollusc shells, they resemble those of the typical (non-calcareous) subgroup and present similar variations in morphology, but are much less extensive.

The following representative profile in coarse loamy river alluvium was described and sampled in the floodplain of the river Swale, which has its headwaters on the Carboniferous Limestone series of the northern Pennines. The alluvium at this site is clearly stratified, every horizon boundary corresponding to a more or less well marked lithostratigraphic discontinuity. A young surface deposit (0–13 cm), within which Ah and Bw horizons can be distinguished, overlies a very dark buried topsoil that qualifies as a prominent (mollic) A horizon and rests at 60 cm on a yellowish brown sandy layer traversed by partly infilled worm channels; below this is a second finer textured dark horizon which could be a second buried A horizon, though it may also have originated as a subaqueous sediment in an abandoned meander.

Profile 41 (Allison and Hartnup, 1981)

Location	: Scruton, North Yorkshire (SE 313934)
Climatic regime	: subhumid temperate
Altitude	: 29 m Slope: < 1°
Land use	: long-term grass

Horizons (cm)

0–10 Ah — Very dark greyish brown (10YR 3/2; 10YR 4/2 dry) stoneless calcareous fine sandy silt loam (USDA loam); moderate medium granular; friable; abundant fine roots; earthworms, potworms and springtails common; sharp smooth boundary.

10–13 2Bw — Yellowish brown (10YR 5/4 10YR 5/2 dry), stoneless, slightly calcareous fine sandy loam (USDA fine or very fine sandy loam); moderate fine subangular blocky; many fine roots; potworms, insect larvae and earthworms present; sharp smooth boundary.

13–31 3bAh — Very dark greyish brown (10YR 3/2; 10YR 5/2 dry) with 20 per cent dark brown (10YR 3-4/3), slightly calcareous fine sandy loam (USDA fine or very fine sandy loam) containing few small sandstone fragments;

Analytical data for Profile 41

Horizon	Ah	2Bw	3bAH	4bAB	4bBw	5bAh(g)
Depth (cm)	0–10	10–13	13–31	31–60	60–94	94–120
sand 200 μm–2 mm %	6	3	12	22	31	3
60–200 μm %	42	64	39	49	51	15
silt 2–60 μm %	40	27	37	18	11	58
clay < 2 μm %	12	6	12	11	7	24
organic carbon %	4.9	1.9	3.2	2.6	1.3	2.2
CaCO$_3$ equiv. %	5.0	4.7	1.8	0	3.9	1.3
pH (H$_2$O)	7.4	7.8	7.5	7.5	7.9	7.8
(0.01M CaCl$_2$)	7.0	7.2	7.0	7.0	7.3	7.2

	moderate medium subangular blocky; friable; many fine roots, mainly in partly infilled worm channels; clear smooth boundary.
31–60 4bAB	Dark brown (10YR 3/3; 10YR 4/2 dry) stoneless fine sandy loam (USDA fine sandy loam); moderate coarse subangular blocky; friable; common fine roots as above; earthworms and insect larvae present; clear smooth boundary.
60–94 4bBw	Yellowish brown (10YR 5/4; 2.5Y dry), stoneless, slightly calcareous loamy medium sand (USDA loamy fine sand); weak very coarse blocky; very friable; common fine roots mainly in earthworm channels partly filled with darker material; abrupt smooth boundary.
94–120 5bAh(g)	Dark brown (10YR 3/3; 2.5Y 5/2 dry), stoneless slightly calcareous silty clay loam (USDA silt loam) with few fine yellowish red (5YR 4/6) mottles localized in small patches; moderate coarse prismatic; firm; few roots; earthworm channels.

Analytical data *see* p. 177.

Classification
England and Wales (1984): typical brown calcareous alluvial soil; coarse loamy river alluvium (Swale series).
USDA (1975): Fluventic Hapludoll; coarse-loamy, mixed, mesic.
FAO (1974): Calcaric Phaeozem, medium textured.

5.3.6 Gleyic-calcaric Alluvial Brown Soils

These are similar but more extensive soils with mottles attributable to gleying in and/or below a loamy Bw horizon, which may be partly or completely decalcified by leaching. They occur in association with alluvial gley soils in embanked marine or estuarine alluvium and occupy small areas in river floodplains, including those of the Thames (Usher series in Jarvis, 1968) and its major tributaries. Some in the Humber warplands (Plate I.6) and the silt-lands surrounding the Wash have been cultivated more or less continuously since they were embanked. Except where deep drainage has been provided, the water table normally rises into the lower horizons in winter and spring.

Data on two representative profiles are reproduced below. Profile 42 from Romney Marsh was described and sampled under old pasture in land enclosed between the fourteenth and sixteeenth centuries (Green, 1968). The partially decalcified fine silty upper horizons (Ah, AB and Bw) show clear evidence of mixing by earthworms and pass at 46 cm into a mottled Cg horizon retaining traces of the original stratification. Profile 43 is in deeply drained and intensively cultivated warpland adjoining the river Trent in north Lincolnshire; this deeply ploughed fertile soil is coarser in texture (coarse silty), apparently little leached, and predominantly brownish to a depth of 150 cm, with few signs of fine stratification. As in very silty soils generally, much of the very large available water capacity is held at low suctions, and the air capacity values indicate that the profile is slightly porous throughout. Assuming that the thick dark Ap horizon qualifies as a mollic rather than an anthropic epipedon (p. 71), the soil is a Hapludoll according to the US Taxonomy and a Phaeozem in the FAO system.

Profile 42 (Green, 1968)

Location	: Old Romney, Kent (TR 007212)
Climatic regime	: subhumid temperate
Altitude	: 3 m Slope: < 1°
Land use	: long-term grass
Horizons (cm)	
0–10 Ah	Very dark greyish brown (10YR 3/2), stoneless, slightly calcareous silty clay loam (USDA loam to clay loam); moderate fine to medium granular; very friable; abundant fine roots; clear irregular boundary.
10–25 AB	Dark greyish brown (10YR 4/2) and yellowish brown (10YR 5/4), stone-

	less calcareous silty clay loam (USDA loam), the darker colours associated with ped faces and infilled channels; very weak coarse prismatic breaking easily to moderate medium subangular blocky; friable; many fine roots; earthworms active; clear boundary.
25–46 2Bw	Brown (10YR 4/3), stoneless, very calcareous clay loam (USDA loam) with some yellowish brown (10YR 5/4–6) patches; moderate fine to medium subangular blocky; very friable; few channels with dark greyish infillings; common fine roots; earthworms active; clear boundary.
46–102 2Cg	Finely stratified, pale brown (10YR 6/3) silty clay loam to silty clay and light grey to pale brown (2.5Y 7/2–10YR 6/3) very fine sandy loam and loamy very fine sand, the latter with prominent ochreous mottling; few partly infilled earthworm channels as above; few roots.
102–183+ 3Cg	Finely stratified, light grey (5Y 7/2) very calcareous very fine sandy loam to loamy very fine sand with yellowish brown (10YR 5/4) clayey layers and prominent vertical ferruginous streaks partly associated with pores; passing to dark grey fine sand (CG) at 183 cm; wet.

Analytical data *see* below

Classification
England and Wales (1984): gleyic brown calcareous alluvial soil; fine silty marine alluvium (Blacktoft series).
USDA (1975): Fluvaquentic Eutrochrept; fine-silty, mixed, mesic.
FAO (1974): Gleyic Cambisol; medium textured.

Profile 43 (Reeve and Thomasson, 1981)

Location	: East Stockwith, Lincolnshire (SK 785944)
Climatic regime	: subhumid temperate
Altitude	: 4 m Slope: < 1°
Land use	: arable (cereals)

Horizons (cm)

0–35 Ap	Very dark greyish brown (10YR 3/2), stoneless, slightly calcareous silt loam (USDA silt loam); moderate fine subangular blocky; friable; many fine roots; sharp wavy boundary.
35–44 Bw(g)	Brown to dark brown (7.5YR 4/4), stoneless, calcareous silty loam (USDA silt loam) with few extremely fine strong brown (7.5YR 5/6) mottles; weak fine subangular blocky; very friable; common fine roots; clear smooth boundary.
44–150 BC(g)	Brown to dark brown (10YR 4/3), stoneless, very calcareous silt loam (USDA loam) with common fine dark yellowish brown (10YR 4/6) and fine brown (10YR 5/3) mottles; weak angular blocky; very friable; few fine roots.

Analytical data for Profile 42

Horizon	Ah	AB	2Bw	2Cg
Depth (cm)	0–10	10–25	25–46	46–102
sand 200 μm–2 mm %	< 1	< 1	< 1	< 1
60–200 μm %	15	12	35	37
silt 2–60 μm %	58	63	43	42
clay < 2 μm %	27	25	22	21
organic carbon %	2.9	1.1	0.5	0.3
$CaCO_3$ equiv. %	3.0	8.2	12	11
dithionite ext. Fe %	1.3	1.2	1.2	0.8
pH (H_2O)	7.6	8.0	8.2	8.2
(0.01M $CaCl_2$)	7.2	7.5	7.6	7.6

Analytical data for Profile 43

Horizon	Ap	Bw(g)	BC(g)	BC(g)
Depth (cm)	10–20	35–40	75–85	105–115
sand 200 µm–2 mm %	1	1	1	2
60–200 µm %	2	2	2	3
silt 2–60 µm %	80	82	83	83
clay < 2 µm %	17	15	14	12
organic carbon %	2.0	1.4	0.9	0.9
$CaCO_3$ equiv. %	8.7	11	12	11
pH (H_2O)	8.0	8.3	8.3	8.3
(0.01M $CaCl_2$)	7.5	7.7	7.8	7.8
bulk density (g cm^{-3})	1.38	1.33	1.31	1.32
packing density (g cm^{-3})	1.53	1.47	1.44	1.43
retained water capacity (% vol.)	39	42	41	41
available water capacity (% vol.)	25	26	28	28
air capacity (% vol.)	8	8	9	9

Classification
  England and Wales (1984): Gleyic brown calcareous alluvial soil; coarse silty marine alluvium (Romney series).
  USDA (1975): Fluventic (or Fluvaquentic) Hapludoll; coarse-silty, mixed, mesic.
  FAO (1974): Calcaric Phaeozem; medium textured.

5.3.7 Pelocalcaric Alluvial Brown Soils

These are calcareous or partially decalcified clayey alluvial soils with brownish Bw horizons. They are located in slightly elevated sites, generally of small extent, in drained coastal marshes and beside sluggish streams draining areas in which most of the soils are derived from calcareous argillaceous rocks. The following example was recorded by C.A.H. Hodge in arable land in the floodplain of the Great Ouse, about 30 metres from the river bank. This site was subject to occasional short-period floods at the time of sampling. The parent material is alluvium derived ultimately from Jurassic clays and chalky till (Chalky Boulder Clay). Carbonate contents of successive horizons show no evidence of decalcification *in situ*. The clay fractions are richer in smectite than in profile 40 (p. 176) and the soil cracks widely and deeply in dry seasons.

Profile 44

Location : Buckden, Cambridgeshire (TL 2136559)
Climatic regime : subhumid temperate
Altitude : 12 m   Slope: < 1°
Land use : arable (barley stubble)

Horizons (cm)

0–23
Ap
  Very dark greyish brown (10YR 3/2), very calcareous clay (USDA silty clay to clay); compound, weak medium prismatic and moderate very fine subangular blocky; friable, common fine roots; earthworm casts present; clear smooth boundary.

23–48
Bw(g)
  Brown (10YR 4/3) and dark yellowish brown (10YR 4/4), very faintly mottled, very calcareous clay (USDA clay) with darker colours on ped faces; strong coarse angular to subangular blocky; firm; few roots; few small (2 mm) freshwater shells; gradual smooth boundary.

48–71
Bg
  Distinctly and finely mottled, greyish brown (10YR 5/2) and yellowish brown (10YR 5/6) very calcareous clay (USDA clay); moderate medium prismatic with mainly greyish brown faces; firm; few channels with dark greyish coats or infillings; roots and shell fragments as above; gradual smooth boundary.

Analytical data for Profile 44

Horizon	Ap	Bw(g)	Bg	BCg
Depth (cm)	5–20	30–45	55–70	75–90
sand 200 μm–2 mm %	2	3	2	3
60–200 μm %	6	5	6	7
silt 2–60 μm %	42	35	35	38
clay < 2 μm %	50	57	57	52
organic carbon %	5.3	3.1	1.6	1.6
$CaCO_3$ equiv. %	13	14	13	16
dithionite ext. Fe %	2.5	2.7	3.0	2.9
pH (H_2O)	7.6	7.9	8.1	8.2
(0.01M $CaCl_2$)	7.3	7.4	7.7	7.6
clay < 2 μm				
CEC (me/100 g)		72	74	81
K_2O %		1.8	1.8	2.0

71–90 BCg Prominently but finely mottled, light grey (10YR 6/1) and yellowish brown (10YR 5/6) very calcareous clay (USDA clay); weak medium to coarse prismatic; firm; roots, shell fragments and channels as above; few white mycelium-like bodies, probably secondary $CaCO_3$.

Analytical data *see* above

Classification
   England and Wales (1984): pelogleyic brown calcareous alluvial soil; clayey river alluvium (Uffington series).
   USDA (1975): Mollic (or Vertic) Fluvaquent intergrading to Fluvaquentic (or vertic) Eutrochrept; fine, montmorillonitic, mesic.
   FAO (1974): Gleyic (or Vertic) Cambisol; fine textured.

5.4 Calcaric Brown Soils

5.4.1 General Characteristics, Classification and Distribution

This group consists of loamy and clayey brown soils with colour/structure B (Bw) horizons in or over calcareous materials other than recent alluvial deposits. It includes the brown calcareous earths and some calcareous pelosols of Avery (1973, 1980) and similar soils classed as brown calcareous by the Soil Survey of Scotland (1984) and as brown earths of high base status in Ireland (Gardiner and Radford, 1980). They are mostly Eutrochrepts according to the US Taxonomy and Calcic Cambisols in the FAO (1974) scheme, but some have a mollic A horizon and are therefore Hapludolls (FAO Phaeozems).

In England and Wales, where soils of this group are estimated to cover about 6 per cent of the total surface area, the parent materials include Devensian head, till and water-sorted drift containing fragmentary chalk or limestone, Flandrian colluvium of similar composition, and older calcareous deposits in sites where any deeply decalcified horizons formed by pre-Devensian weathering have been removed by erosion or mixed with subjacent unweathered materials by cryoturbation. The equivalent soils in Ireland are developed in glacial or glaciofluvial drift containing comminuted Carboniferous limestone. Those in Scotland are associated with sporadically occurring limestone outcrops of very small extent.

Except in recent colluvial materials, which show little sign of post-depositional decalcification, these soils exhibit stages of leaching and horizon differentiation intermediate between rendzinas and luvic brown soils (Figure 1.3, p. 9.) and normally occur in close association with one or both of these groups, their distribution being governed by variations in initial carbonate content and the extent to which leaching has been counteracted by bioturbation, recent erosion or soil creep. The colour/structure B horizon, the presence of which distinguishes the calcaric brown soils from related rendzinas, may result from alteration of the parent material in place or consist of weathered material that has moved downslope. It is typically brown but can be olive or greyish, particularly in materials derived from certain Cretaceous rocks (Lower Chalk and Upper Greensand) which weather without yielding appreciable amounts of free iron. In this case the soil is considered to be a brown soil rather than a rendzina only if the subsurface horizon contains less than 40 per cent calcium carbonate (< 2 mm) and is significantly less calcareous than the underlying horizon.

Although both A and B horizons may be completely decalcified, most of these soils remain calcareous throughout, partly through the action of earthworms and other burrowing animals in bringing finely divided carbonates to the surface. Those decalcified to depths greater than about 40 cm usually show some evidence of clay translocation and grade into luvic brown soils in which a distinct argillic B horizon can be identified. A BCk or Ck horizon containing redeposited (secondary) calcium carbonate is discernible in places, especially in subhumid areas, and variants intergrading to gley soils may have a grey and brown mottled Bg or BCg horizon below the Bw.

Calcaric brown soils resemble rendzinas in their chemical properties and ecological relationships and have often been grouped with them, particularly in Britain (Avery, 1965; Soil Survey of Scotland, 1984) and France (CPCS, 1967; Duchaufour, 1978, 1982). Generally, however, the upper horizons contain less 'active' calcium carbonate and variants overlying bedrock are freely penetrable by roots to greater depths. The effective capacity of the soil to retain available moisture and nutrients is correspondingly greater and this is commonly reflected in more vigorous growth of semi-natural vegetation or crop plants. In the subhumid lowlands where most of these soils occur, a very large proportion of the land they occupy is cultivated. Physical limitations, usually minor, affecting land use on particular soil series or phases include moderately steep slopes, stoniness, droughtiness, periodic subsurface wetness and workability problems.

Key to Subgroups

1. Calcaric brown soils with pelo- characteristics (p. 112)
 Pelocalcaric brown soils
2. Other calcaric brown soils with gleyic features (p. 112)
 2.1 with a slowly permeable subsurface horizon (p. 110)
 Stagnogleyic calcaric brown soils
 2.2 without a slowly permeable subsurface horizon
 Gleyic calcaric brown soils
3. Other calcaric brown soils in recent colluvium (p. 114) more than 40 cm thick
 Colluvial calcaric brown soils
4. Other calcaric brown soils
 Typical calcaric brown soils

5.4.2 Typical Calcaric Brown Soils

These are freely draining calcaric brown soils in which the characteristic Bw horizon passes downwards into a less altered (more calcareous) substratum other than recent colluvium. They are typically developed in head deposits overlying and partly derived from chalk or harder limestone; others are in calcareous glacial or river drift and in materials weathered more or less in place from soft pre-Quaternary rocks. Most of them are loamy and those that are clayey generally have fissured bedrock at moderate depths, as in the Moreton series on Jurassic limestone (Ragg *et al.*, 1984) and the Blewbury series on Lower Chalk (Findlay *et al.*, 1984). Data on three representative profiles in agricultural land are reproduced on pp. 183–5.

Profile 45 was described and sampled by D. Mackney on the Grassland Research Institute farm at Hurley, Berkshire. The parent material is loamy head in which sand and clay derived from nearby Tertiary beds has been mixed with frost-shattered chalk and flint derived from the subjacent Upper Chalk. Cryoturbation structures were clearly evident in the BC and C horizons when the profile was exposed. Similar but silty soils of the Panholes (Plate I.7) and Coombe series (Clayden and Hollis, 1984), in which loess has been mixed with locally derived materials, are also widely distributed on the English Chalk, particularly in counties south of the Thames.

Profile 46 is in coarse loamy drift over Jurassic limestone in Lincolnshire; this soil has been more deeply ploughed but otherwise resembles the preceding profile in morphology and mode of formation. The sand fraction is probably of aeolian origin in part. Analogous fine loamy soils (Aberford series) are widely distributed on the Permian (Magnesian) limestone cuesta which extends from Nottingham northwards into Yorkshire (R.A. Jarvis *et al.*, 1984).

Profile 47 from County Clare is in morainic (glaciofluvial) drift of Midlandian (Devensian) age derived from limestone, sandstone, shale and granite. Located in old meadowland, it has a well developed mollic A horizon which is slightly acid. The succeeding Bw horizon is also leached of carbonates but remains near neutral in reaction.

Amounts of silicate clay and free iron oxide decrease with depth in all three profiles, reflecting residual accumulation due to removal of carbonates and slight weathering of the non-calcareous fraction.

Profile 45

Location : Hurley, Berkshire (SU 818824)
Climatic regime : subhumid temperate
Altitude : 94 m   Slope: 3° convex
Land use : ley grass
Horizons (cm)

0–21
Ap
Dark brown (10YR 3/3; 10YR 4/2.5 rubbed), very slightly stony, **slightly calcareous sandy clay loam** (USDA sandy clay loam); angular flint fragments; moderate medium subangular blocky; friable; many fine roots; earthworms present; abrupt smooth boundary.

21–42
Bw
Brown to dark brown (7.5YR 4/4), slightly stony, calcareous sandy clay loam (USDA sandy loam to sandy clay loam); common angular flint and few subrounded chalk fragments; weak fine to medium subangular blocky; friable; common fine roots; common earthworm channels with dark brown infillings; clear boundary.

42–56/78
BC
Light yellowish brown (10YR 6/4), very stony very calcareous medium sandy loam (USDA fine sandy loam) with irregular brown (7.5YR 4/4) inclusions; abundant subrounded chalk fragments and few subangular and subrounded (nodular) flints; weak blocky to massive; friable; few roots; earthworm channels as above; abrupt wavy boundary.

56/78+
2Cu
Subangular chalk fragments and few large flints set in a very pale brown (10YR 7/3) compact chalky matrix.

Analytical data *see* p. 184

Classification
England and Wales (1984): typical brown calcareous earth; fine loamy material over lithoskeletal chalk (Soham series).
USDA (1975): Rendollic Eutrochrept; loamy-skeletal, mixed, mesic.
FAO (1974): Calcic Cambisol; medium textured.

Analytical data for Profile 45

Horizon	Ap	Bw	BC
Depth (cm)	0–21	21–42	42–56
sand 600 μm–2 mm %	4	1	3
200–600 μm %	26	26	23
60–200 μm %	25	28	24
silt 2–60 μm %	22	25	31
clay < 2 μm %	23	20	19
organic carbon %	2.2	0.8	
CaCO$_3$ equiv.			
< 2 mm %	2.0	8.2	35
< 2 μm %	nil	0.4	5.4
dithionite ext. Fe %	1.9	1.9	1.4
pH (H$_2$O)	7.5	8.1	8.3
(0.01M CaCl$_2$)	7.3	7.6	8.0

Profile 46 (George and Robson, 1978

Location : Rauceby, Lincolnshire (TF 005482)
Climatic regime : subhumid temperate
Altitude : 62 m   Slope: < 1°
Land use : arable (winter barley)
Horizons (cm)

0–35 Ap
Dark brown to brown (10YR 4/3), very slightly stony, slightly calcareous medium sandy loam (USDA fine sandy loam); few limestone fragments; moderate medium subangular blocky; very friable; many fine roots; sharp smooth boundary.

35–64 Bw
Brown to dark brown (7.5YR 4/4), slightly stony, very calcareous medium sandy loam (USDA sandy loam); common limestone fragments; weak medium to coarse subangular blocky; very friable; few fine roots; common worm channels partly infilled with brown soil from above; gradual boundary.

64+ 2Cr
Weathering oolitic limestone with yellowish brown (10YR 5/8), fine earth in cracks.

Analytical data

Horizon	Ap	Bw
Depth (cm)	0–35	35–64
sand 600 μm–2 mm %	2	13
200–600 μm %	32	26
60–200 μm %	35	30
silt 2–60 μm %	17	19
clay < 2 μm %	14	12
organic carbon %	1.4	
CaCO$_3$ equiv.		
< 2 mm %	2.4	31
< 2 μm %	0	2.7
dithionite ext. Fe %	1.8	1.5
pH (H$_2$O)	7.6	8.2
(0.01M CaCl$_2$)	7.1	7.7

Classification
England and Wales (1984): typical brown calcareous earth; coarse loamy material over lithoskeletal limestone (Cranwell series).
USDA (1975): Rendollic Eutrochrept; coarse-loamy or loamy-skeletal, mixed, mesic.
FAO (1974): Calcic Cambisol; coarse textured.

Profile 47 (Finch *et al.*, 1971)

Location	: Kilkishen, County Clare
Climatic regime	: humid temperate
Altitude	: 36 m Slope: < 2–3°
Land use and vegetation	: long-term grass; common bent grass (*Agrostis capillaris*) dominant, with tall oat (*Arrhenatherum avenaceum*), cocksfoot (*Dactylis glomerata*), red fescue (*Festuca rubra*), plantain (*Plantago lanceolata*) and red clover (*Trifolium pratense*).

Horizons (cm)

0–26 Ah	Very dark greyish brown (10YR 3/2) medium sandy loam (USDA fine sandy loam); moderate fine granular; friable; root mat in top 10 cm; clear smooth boundary.
26–48 Bw	Dark yellowish brown (10YR 4/4) medium sandy loam (USDA fine sandy loam); weak fine subangular blocky; friable; plentiful roots; clear irregular boundary.
48–69 Cu	Very pale brown (10YR 7/4), gravelly calcareous medium sandy loam (USDA fine sandy loam); massive; few roots.

Analytical data *see* below.

Classification
- Ireland (1980): brown earth of high base status (Baggotstown series).
- England and Wales (1984): typical brown calcareous earth; coarse loamy drift with limestone (Ruskington series).
- USDA (1975): Typic Hapludoll; coarse-loamy or loamy-skeletal, mixed or carbonatic, mesic.
- FAO (1974): Haplic Phaeozem; medium textured.

5.4.3 Colluvial Calcaric Brown Soils

Well drained soils of this subgroup occur in dry valley bottoms (Figure 5.8) and at the base of slopes in cultivated or formerly cultivated land overlying Chalk, Jurassic and Permian limestones, or calcareous drift. Compared with colluvial rendzinas (p. 137) in similar situations, these soils are less calcareous and are evidently derived mainly from pre-existing brown soils upslope. As a consequence the profiles have brownish Bw (or ABw) horizons which normally contain less than 40 per cent calcium carbonate. In contrast to those of the typical subgroup, carbonate contents usually decrease with depth within the colluvial layer, reflecting exposure of progressively deeper

---

Analytical data for Profile 47

Horizon	Ah	Bw	Cu
Depth (cm)	0–26	26–48	48–69
sand 200 µm–2 mm %	27	26	27
50–200 µm %	34	40	41
silt 2–50 µm %	29	23	26
clay < 2 µm %	10	11	6
organic carbon %	2.3	0.9	0.2
citrate-dithionite ext. Fe %	0.6	0.5	0.3
TEB (me/100 g)	14.9	8.3	
CEC (me/100 g)	19.0	9.6	4.0
% base saturation	78	87	sat.
pH (H_2O)	6.1	6.8	8.3

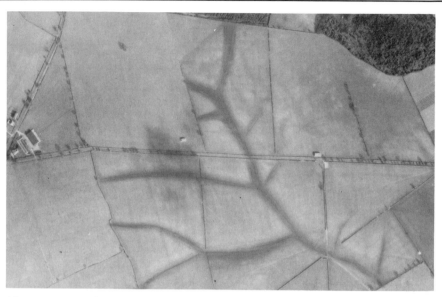

Figure 5.8: Aerial view of arable land over Jurassic limestone between Empingham and Tickencote, Leicestershire, with brown rendzinas (light tones) and typical or pelo-calcaric brown soils on the interfluves and colluvial calcaric brown soils in dry valleys (Copyright Soil Survey and Land Research Centre).

and more calcareous horizons in the source areas subject to erosion. This feature affords a useful means of identification in addition to organic matter distribution and other indications of recent colluvial origin. It also implies that the subsurface horizons designated Bw are not strictly cambic horizons as conceived in the US system, in which case the soils have to be classed as Entisols (Udifluvents) rather than as Inceptisols (Eutrochrepts).

The following representative profile was described and sampled by D. Mackney in a Chiltern dry valley near Marlow, Buckinghamshire. The chalky colluvium at this site rests at 112 cm depth on a darker more flinty layer interpreted as the A horizon of a buried luvic brown soil (p. 209) with an A/Eb/Bt horizon sequence over chalky head. A similar but shallower colluvial soil sampled in a nearby valley (grid ref. SU 813853) had a more pronounced buried A contianing 5 per cent organic carbon. Charcoal from this horizon gave a radiocarbon date of 1865 ± 80 years BP, suggesting that colluvial deposition was initiated by forest clearance during or before the Roman occupation. A noteworthy feature at both sites is that the colluvium contains numerous flint fragments up to 10 cm in diameter, suggesting that soil creep rather than sheetwash was the major transporting process. Similar stony colluvial soils have been recorded in the South Downs (Bell, 1983) and in dry valleys entrenching Jurassic limestone dipslopes in Lincolnshire (George and Robson, 1978) and Oxfordshire (Jarvis and Hazleden, 1982).

Profile 48

Location	: Marlow, Buckinghamshire (SP 630619)
Climatic regime	: subhumid temperate
Altitude	: 56 m Slope: 1°
Land use	: long-term grass

Horizons (cm)

0–18 Dark greyish brown (10YR 4/2) slightly
Ap   stony, very calcareous clay loam (USDA silt loam); common medium and large angular and subangular flints and

18–56 Bw	few subrounded chalk fragments; strong fine and medium subangular blocky; friable to firm; abundant fine roots; earthworms present; clear wavy boundary. Brown (10–7.5YR 4/4), slightly stony very calcareous clay loam (USDA silt loam); stones as above; moderate medium and coarse subangular blocky; friable; many fine roots; earthworms present; gradual boundary.
56–112 BC	Dark brown (7.5YR 3/4), moderately stony, slightly calcareous, sandy silt loam to clay loam (USDA silt loam to loam); stones as above; weak subangular blocky; friable; charcoal fragments; few fine roots; earthworms present; abrupt boundary.
112–122 2bAh	Very dark brown to dark brown (10YR 3/2–3), moderately stony, very slightly calcareous sandy silt loam (USDA loam); many angular flints and few rounded flint pebbles; common charcoal fragments; no roots; abrupt boundary.
122–165 2bEb	Dark brown (7.5YR 3/4), moderately flinty, very slightly calcareous sandy silt loam (USDA loam); stones as above, mainly small flint fragments <1 cm; weak angular blocky; gradual boundary.
165–181 2bBt	Reddish brown (5YR 4/4) flinty sandy clay (USDA sandy clay loam to sandy clay).
181–200 3Cu	Soft white chalky Head.

Classification
England and Wales (1984): colluvial brown calcareous earth; fine loamy calcareous colluvium (Dullingham series).
USDA (1975): Typic Udifluvent; fine-loamy, mixed (calcareous), mesic.
FAO (1974): Calcaric Fluvisol; medium textured.

5.4.4 Gleyic Calcaric Brown Soils

These are intergrades between typical or colluvial calcaric brown soils and calcaric (p. 328) or calcaric alluvial (p. 311) gley soils, as shown by the appearance of ochreous mottles or other evidence of current or former wetness within 60 cm of the surface. They occur in permeable calcareous deposits which overlie impervious substrata at depths below 80 cm and occupy level or depressional sites where the water table rises seasonally into the lower part of the profile. The parent materials include late Pleistocene river drift and head, usually loamy or loamy over sandy or gravelly, and thin partly weathered limestone beds over clay or clay shale, as in the Aswardby series described by George and Robson (1978) in Lincolnshire. Soils developed wholly or partly in recent colluvium are also included, and some in springline sites have a substratum of white calcareous tufa precipitated from seepage water charged with calcium bicarbonate. In other variants, typified by the Grove series (Jarvis, 1973) on

Analytical data for Profile 48

Horizon	Ap	Bw	BC	2bAh	2bEb	2bBt
Depth (cm)	0–18	18–56	56–112	112–122	122–155	165–181
sand 600 μm–2 mm %	4	4	5	7	11	12
200–600 μm %	10	10	14	20	20	27
60–200 μm %	8	8	7	9	9	7
silt 2–60 μm %	56	57	56	48	47	19
clay <2 μm %	22	21	18	16	13	35
organic carbon %	3.9	1.1	1.1	0.6	0.3	0.4
$CaCO_3$ equiv. %	31	25	4.6	0.8	0.8	tr.
pH (H_2O)	7.4	7.5	7.4	7.3	7.2	7.2
(0.01M $CaCl_2$)	7.2	7.3	7.2	7.2	7.1	6.9

gravelly drift derived from Upper Greensand and Lower Chalk, the B horizon retains an olive or greyish colour which cannot be attributed to strong gleying because related soils with no evidence of wetness have similar colours.

The first representative profile recorded below is marginal to the typical subgroup. It is sited in the Upper Thames valley on a very gentle slope separating the Northmoor (Floodplain) terrace from the higher Summertown-Radley terrace (Briggs, 1976), and is developed in stratified loamy and gravelly drift regarded as head by Jarvis and Hazleden (1982). The large silt content of the Bw horizon is attributable to inclusion of loess. Ochreous mottles appear at about 56 cm in the lower part of this horizon, and the grey and brown mottled Cgk horizon contains coats (calcitans), nodules and pore fillings of redeposited (secondary) calcium carbonate.

Profile 50, which was not sampled for analysis, is located in a valley-side slope north of Chelmsford, Essex, where London Clay is overlain successively by fluvial gravels and chalky till (Chalky Boulder Clay), and springs emerge at the junction between it and the gravels. The Ap and Bw horizons, which show no evidence of gleying, pass downwards into pale brown to white very calcareous tufaceous material with ferruginous and manganiferous segregations on the faces of prismatic structural units. Radiocarbon dates ranging from 7,500 to 10,000 years BP were obtained by Kerney et al. (1980) from organic remains in tufa covered by varying thicknesses of chalky 'hillwash' at the foot of the North Downs near Folkestone, and the tufa in the profile described is probably of similar age, implying that the A and Bw horizons are in recent colluvium.

Profile 49 (Jarvis and Hazleden, 1982)

Location   : Cote, Oxfordshire (SP 359020)
Climatic regime: subhumid temperate
Altitude   : 72 m   Slope: 1°
Land use   : ley grass
Horizons (cm)

0–19
Ap
Dark brown (9YR 3/3), very slightly stony, slightly calcareous clay loam (USDA loam); few very small to medium rounded limestone, quartz and subangular flint fragments; moderate fine and medium subangular blocky; friable to firm; many fine roots; clear smooth boundary.

19–37
Bw
Brown (7.5YR 4/4; 10YR 5/6 rubbed), very slightly stony, slightly calcareous clay loam (USDA loam to clay loam) with dark brown (10YR 4/3) ped faces and infilled worm channels; stones as above; moderate medium subangular blocky; friable; common fine roots; gradual boundary.

37–61
Bw(g)
Strong brown (7.5YR 5/6), stoneless, slightly calcareous clay loam (USDA loam to silt loam) with few very fine yellowish red (5YR 5/6) mottles and common soft fine manganiferous nodules in lower 5 cm; brown (7.5YR 4/4) ped faces and pore fillings; weak medium prismatic breaking to medium subangular blocky; friable; few roots; few partly filled earthworm channels.

61–78
2BC(g)
Yellowish brown (10YR 5/6), stoneless, very calcareous medium sandy loam (USDA fine sandy loam) with common fine and medium strong brown (7.5YR 5/6) and light yellowish brown (10YR 6/4) mottles; weak medium subangular blocky; very friable; few roots and earthworm channels as above; few very weakly cemented ferri-manganiferous concentrations; abrupt irregular boundary.

78–92
3BC(g)
Yellowish brown (10YR 5/6), very gravelly, very calcareous sandy loam (variable clay content) with common fine and medium strong brown (7.5YR 5/6) and greyish brown (2.5Y 5/2) mottles; stones as above, mainly very small and small limestone pebbles; massive; very friable; earthworm channels and manganiferous concentrations as above; some

Analytical data for Profile 49

Horizon	Ap	Bw	Bw(g)	2BC(g)	4Cgk
Depth (cm)	0–19	19–37	37–61	61–78	92–120
sand 600 μm–2 mm %	6	2	2	4	2
200–600 μm %	13	8	6	17	5
60–200 μm %	17	15	16	37	20
silt 2–60 μm %	39	48	54	28	46
clay <2 μm %	25	27	22	14	27
organic carbon %	2.0	1.0			
CaCO$_3$ equiv.					
<2 mm %	4.2	2.2	3.7	36	29
<2 μm %	0.4	tr.	0.4	0.7	1.1
dithionite ext. Fe %	3.0	2.6	2.5	2.3	1.8
pH (H$_2$O)	7.5	7.7	7.8	8.0	8.0
(0.01M CaCl$_2$)	7.2	7.4	7.5	7.6	7.6

pores in lower part lined with secondary CaCO$_3$; abrupt irregular boundary.

92–120 4Cgk Greyish brown (10YR 5/2), stoneless, very calcareous clay loam (USDA loam to clay loam) with many fine and medium strong brown (7.5YR 5/6) mottles; massive with platy tendency and few light grey (10YR 6/1) to pale brown (10YR 6/3) vertical faces partially coated with secondary CaCO$_3$; friable; rare roots; common soft calcareous nodules <3 mm and secondary CaCO$_3$ infilling many fine and medium pores; common soft manganiferous concentrations.

Analytical data *see* above.

Classification
England and Wales (1984): gleyic brown calcareous earth; fine loamy drift with limestone (Astrop series).
USDA (1975): Typic Eutrochrept; fine-loamy, mixed, mesic.
FAO (1974): Gleyic Cambisol; medium textured.

Profile 50 (Allen and Sturdy, 1980)

Location : Faulkbourne, Essex (TL 795175)
Climatic regime: subhumid temperate
Altitude : 30 m Slope: 3°
Land use : long-term grass

Horizons (cm)

0–25 Ap Brown (10YR 4/3), slightly stony, very calcareous clay loam; common very small and small flint fragments; fine and medium subangular blocky; friable; abundant fine roots; earthworms present; sharp boundary.

25–50 Bw Yellowish brown (10YR 5/4), stoneless, very calcareous silty clay loam; very fine and fine subangular blocky; friable; many fine roots; gradual boundary.

50–70 BCk(g) Pale brown (10YR 6/3), stoneless, extremely calcareous silt loam with much secondary calcium carbonate; fine and medium prismatic; few fine roots; abundant dendritic manganiferous segregations on ped faces; gradual boundary.

70–96 Ck(g) White, stoneless, extremely calcareous material (tufa) with some ferruginous and manganiferous staining on faces of coarse prismatic peds; few small fossil bivalve shells.

Classification
England and Wales (1984): Gleyic brown calcareous earth; fine silty calcareous colluvium (unnamed series).
USDA (1975): Rendollic Eutrochrept?; fine-(or coarse-) silty, carbonatic, mesic.
FAO (1974): Calcic Cambisol; medium textured.

5.4.5 Stagnogleyic Calcaric Brown Soils

These are loamy calcaric brown soils that show signs of gleying in or above a slowly permeable substratum such as chalky till or slightly weathered argillaceous rock, the occurrence of which within 80 cm depth distinguishes them from those of the gleyic subgroup. The impeding layer is denser and more coarsely structured than the overlying Bw horizon, and is usually finer in texture. Slightly gleyed calcareous or superficially decalcified soils with these characteristics are nowhere extensive in the British Isles and have so far only been identified and mapped, either separately or as components of associations, in a few localities in northern and eastern England, including the Maltby area of Yorkshire (Fordoles and Maltby series in Hartnup, 1977) and parts of Humberside (Arnold series in Furness, 1985) and Lincolnshire (Cannamore association in Hodge et al., 1984).

The following representative profile is in 'coverloam' (Corbett, 1979) over chalky drift (till or solifucted till) with a clay loam matrix. The former deposit, which forms the surface soil over much of central East Anglia, is believed to consist of Devensian coversand which has been irregularly mixed with finer textured subjacent materials by cryoturbation. Its thickness and composition vary considerably over short distances, giving rise to complex soil-landscape units in each of which several soil series are represented. On level interfluves underlain by thick Anglian till (Chalky Boulder Clay), stagnoluvic gley soils exemplified by profile 153 (p. 366) are generally dominant. Profile 51 is located in the sideslope of a shallow valley where the soils are naturally better drained and in places less leached. The particle-size analyses show a marked lithologic discontinuity at 55 cm between the cover deposit in which the Ap and Bw(g) horizons are developed and the compact chalky substratum.

Profile 51 (Corbett, 1979)

Location	:	Pulham St Mary, Norfolk (TM 207848)
Climatic regime:		subhumid temperate
Altitude	:	33 m Slope: gentle
Land use	:	long-term grass

Horizons (cm)

0–24 Ap	Very dark grey (10YR 3/1), very slightly stony, slightly calcareous clay loam (USDA sandy clay loam); few very small and small angular flint and rounded chalk fragments and quartzose pebbles; moderate medium and coarse granular grading to fine subangular blocky; abundant fine and few medium fleshy roots; abrupt boundary.
24–55 Bw(g)	Yellowish brown (10YR 5/4–6) very slightly calcareous sandy clay loam

Analytical data for Profile 51

Horizon	Ap	Bw(g)	2BC(g)1	2BC(g)2
Depth (cm)	0–24	24–55	55–80	80–90
sand 600 μm –2 mm %	6	5	3	4
200–600 μm %	24	26	9	7
60–200 μm %	20	19	9	9
silt 2–60 μm %	26	25	46	51
clay <2 μm %	24	25	33	29
organic carbon %	2.5	0.6		
CaCO$_3$ equiv. %	2.7	tr.	34	38
pH (H$_2$O)	7.7	8.0	8.4	8.4
(0.01M CaCl$_2$)	7.3	7.5	7.6	7.5

55–80 2BC(g)1	(USDA sandy clay loam) with common medium strong brown (7.5YR 5/6) mottles; rare small angular flint fragments; moderate medium prismatic; clear boundary.
Yellowish brown (10YR 5/6), slightly, stony very calcareous clay loam (USDA clay loam) with common medium yellowish brown (10YR 5/8) and few fine brown (10YR 5/3) and greyish brown (10YR 5/2) mottles, the latter mainly around stones; common chalk pebbles, mainly small, and rare medium and large flint fragments; moderate coarse prismatic; gradual boundary.
80–90+ 2BC(g)2	Yellowish brown (10YR 5/6), slightly stony, very calcareous clay loam (USDA clay loam) with common, fine and medium, grey and light grey (10YR 5–6/1) mottles, particularly around stones; stones as above; weak coarse prismatic.

Analytical data *see* p. 190

Classification
England and Wales (1984): stagnogleyic brown calcareous earth; fine loamy chalky drift (Cannamore series).
USDA (1975): Typic Hapludoll; fine-loamy, mixed, mesic.
FAO (1974): Haplic (or Gleyic) Phaeozem; medium textured.

5.4.6 Pelocalcaric Brown Soils

These are similar soils in which textures are predominantly clayey. They are grouped with associated pelocalcaric gley soils (p. 333) as calcareous pelosols in the current England and Wales classification (Avery, 1980) and are much more extensive than those of the preceding subgroup, particularly on Chalky Boulder Clay (Hanslope series; Plate I.8) and calcareous Jurassic rocks (Evesham and Haselor series) in eastern and south-western England and the Midlands (Figure 2.5, p. 78). The upper horizons crack in dry seasons and ped faces in the Bw commonly have a glazed or polished appearance resulting from pressures caused by swelling during the winter period when the cracks tend to close. As a rule, however, fissures are more closely spaced and interconnected than at corresponding depths in the related gley soils. The subsurface horizons are also generally wet for shorter periods, either because surplus water is removed naturally by slow percolation or downslope seepage or because the soils are more responsive to drainage measures (Robson and Thomasson, 1977). Mole drainage is particularly effective in these soils, presumably because their high calcium status promotes flocculation of the clay and there is therefore less blockage of the channels by dispersed material than in non-calcareous soils of similar texture (Childs, 1943). Data on two representative profiles in agricultural land are given on pp. 192–4.

Profile 52 is on interbedded calcareous clay shale and argillaceous limestone which form the lower part of the Lower Lias formation along most of its outcrop. Although the Bw and BC horizons are both slightly mottled, indicating that drainage is impeded to some extent, they are slightly to moderately porous (Table 3.7, p. 98) as judged by the air capacity values and the broken limestone band below 67 cm depth further aids disposal of excess surface water. As is common in clay soils, less than one-third of the retained water is available to plants.

Profile 53, on Chalky Boulder Clay in north Buckinghamshire, was described and sampled by D. Mackney in an arable field that had carried a succession of spring cereal crops. Probably as a consequence of the necessarily ill-timed cultivations involved, the Ap horizon has a high bulk density. It also contains significantly less organic matter than the A horizon of profile 52 and the subsurface horizons are distinctly less porous. Fine clay/total clay ratios and free iron oxide contents decrease steadily from the surface downwards, reflecting the effects of weathering unaccompanied by appreciable eluviation of fine clay.

X-ray analysis of <2 μm separates, along with the CEC values and potassium contents quoted, show that the clay fractions of both these soils consist mainly of mica and interstratified mica-smectite in varying proportions. Judging from the figures obtained when CEC is recalculated on a <2 mm soil basis, neither is likely to have a shrinkage potential large enough for it to qualify as Vertic according to the criteria used in the US system (Avery and Bullock, 1977; Reeve et al., 1980).

Profile 52 (Ragg et al., 1984)

Location	:	Compton Verney, Warwickshire (SP 321508)
Climatic regime:		subhumid temperate
Altitude	:	91 m Slope: 2° straight
Land use	:	arable (cereals)

Horizons (cm)

0–21 Ap
: Dark greyish brown (10YR 4/2; 2/5Y 4/2 rubbed, 10YR 5/2 dry) very calcareous clay (USDA clay) containing occasional quartzose pebbles and limestone fragments; moderate medium subangular blocky; firm; many fine roots; sharp smooth boundary.

21–53 Bw(g)
: Olive brown (2.5Y 4/4 broken and rubbed; 10YR 5/3 dry) grading in places to dark yellowish brown (10YR 4/4), very calcareous clay (USDA clay) with few fine brown to dark brown (7.5YR 4/3) mottles; occasional angular limestone fragments; strong medium angular blocky; firm; common fine roots; abrupt wavy boundary.

53–67 BC(g)
: Olive brown (2.5Y 4/4 broken and rubbed; 2.5Y 6/4 dry) stoneless very calcareous clay (USDA silty clay) with common fine olive yellow (2.5Y 6/6) and few fine dark greyish brown (2.5Y 4/2) mottles; strong medium prismatic with dark greyish brown faces; very firm; few fine roots, mainly in fissures; few worm channels with greyish coatings; sharp boundary.

67+ Cu/Cr
: Fragmented hard limestone (Blue Lias) overlying slightly weathered calcareous clay-shale.

Analytical data

Horizon	Ap	Bw(g)	BC(g)
Depth (cm)	0–21	21–53	53–67
sand 200 μm–2 mm %	10	5	5
60–200 μm %	4	5	6
silt 2–60 μm %	39	37	44
clay <2 μm %	47	53	45
organic carbon %	3.2	1.7	
CaCO$_3$ equiv.			
<2 mm %	14	13	31
<2 μm %	1.3	0	8.5
pH (H$_2$O)	8.0	8.0	8.1
(0.01M CaCl$_2$)	7.4	7.4	7.5
clay <2 μm			
CEC (me/100 g)	41	46	
K$_2$O %	3.5	3.5	
bulk density (g cm^{-3})	1.32	1.44	1.45
packing density (g cm^{-3})	1.74	1.91	1.85
retained water capacity (% vol.)	44	35	38
available water capacity (% vol.)	11	9	12
air capacity (% vol.)	5	11	8

Classification

England and Wales (1984): typical calcareous pelosol; clayey material passing to clay with interbedded limestone (Haselor series).

USDA (1975): Aquic Eutrochrept; fine, mixed or illitic, mesic.

FAO (1974): Eutric or Gleyic Cambisol; fine textured.

Profile 53

Location	:	Hardmead, Buckinghamshire (SP 945464)
Climatic regime:		subhumid temperate
Altitude	:	91 m Slope: 2°
Land use	:	arable (wheat stubble)

Horizons (cm)

0–24 Ap
: Brown to dark brown (10YR 4/3) slightly stony calcareous clay (USDA

	clay) with olive brown (2.5Y 4/4) inclusions; common small and medium subangular flint and limestone fragments and few quartzose pebbles; weak coarse angular blocky with fine angular blocky in upper 1 cm; firm; common fine roots; earthworms present; abrupt smooth boundary.		many fine to medium yellowish brown to strong brown (10YR–7.5YR 5/6) and greyish brown (2.5Y 5/2) mottles; stones as above, but more of chalk; weak medium prismatic; very firm; few fine roots; gradual boundary.
24–48 Bw(g)1	Olive brown (2.5Y 4/4) slightly stony calcareous clay (USDA clay) with very slightly greyer ped faces; stones as above, together with few subrounded chalk fragments; medium prismatic breaking to moderate medium and fine angular blocky; some smooth glazed ped faces; very firm; common fine roots; earthworms present; gradual wavy boundary.	62–110 BCgk	Coarsely mottled grey (N 6/0), brown to dark brown (10YR 4/3) and strong brown (7.5YR 5/6), slightly stony, very calcareous silty clay to clay (USDA silty clay); stones as above, mainly subrounded chalk fragments; massive, with common fine pores, apparently old root channels and mostly outlined in grey; grey stress-oriented coats around stones; very firm; roots rare; soft powdery white deposits of secondary calcium carbonate.
48–62 Bw(g)2	Olive brown (2.5Y 4/4), slightly stony, very calcareous clay (USDA clay) with		

Analytical data for Profile 53

Horizon	Ap	Bw(g)1	Bw(g)2	BCgk
Depth (cm)	0–24	24–48	48–62	61–110
sand 500 μm–2 mm %	4	4	3	4
200–500 μm %	2	4	4	3
50–200 μm %	13	10	8	7
silt 2–50 μm %	36	37	38	44
clay <2 μm %	45	45	47	42
<0.2 μm %	25	24	21	12
fine clay/total clay	0.56	0.53	0.45	0.29
organic carbon %	1.9	1.0		
CaCO$_3$ equiv.				
<2 mm %	8.1	11	18	38
<2 μm %	0	0	4.9	7.5
dithionite ext. Fe %	2.4	2.3	2.3	1.8
pH (H$_2$O)	8.4	8.4	8.5	8.4
(0.01M CaCl$_2$)	7.5	7.8	7.7	7.9
silicate clay <2 μm				
CEC (me/100 g)	57	52		37
K$_2$O%	2.4	2.6		3.0
bulk density (g cm^{-3})	1.39	1.43		1.69
packing density (g cm^{-3})	1.80	1.83		1.97
retained water capacity (% vol.)	43	42		35
available water capacity (% vol.)	17	14		10
air capacity (% vol.)	4	4		2

Classification
England and Wales (1984): typical calcareous pelosol; clayey chalky drift (Hanslope series).
USDA (1975): Aquic Eutrochrept; fine, mixed, mesic.
FAO (1974): Gleyic Cambisol, fine textured.

5.5 Orthic Brown Soils

5.5.1 General Characteristics, Classification and Distribution

These soils also have a loamy or clayey Bw (or ABw) horizon formed in materials other than recent alluvium but contain no finely divided calcium carbonate of natural origin in or immediately beneath the solum and are consequently more or less acid in the absence of added lime. As the prefix *orthic* implies, they embody the central concept of brown soils identified in west European countries by the terms brown earth, *Braunerde* (Kubiena, 1953; Mückenhausen et al., 1977) and *sol brun* (CPCS, 1967; Duchaufour, 1982). According to the definition adopted here, the group includes the brown earths (*sensu stricto*) and non-calcareous pelosols of Avery (1973, 1980), some of the soils called brown forest and brown magnesian in Scotland (Ragg et al., 1978; Soil Survey of Scotland, 1984) and most of those classed as acid brown earths in Ireland (Gardiner and Radford, 1980). Those without a prominent mollic or umbric A horizon are nearly all Ochrepts (Dystrochrepts; Dystric Eutrochrepts; Fragiochrepts) in the US Taxonomy and Dystric or Eutric Cambisols in the FAO system. Variants with prominent A horizons are much less common, occurring chiefly in heavily manured agricultural land, in recent colluvium, and in materials derived from basic igneous rocks. These can be Hapludolls (FAO Phaeozems) or Umbrepts (FAO Humic Cambisols) depending on percentage base saturation in and below the topsoil.

The A and Bw horizons of these soils have similar clay contents and are differentiated, in some cases weakly, by organic matter content, colour and structure. A layer of more or less decomposed litter (F and/or H horizon) generally occurs at the surface in uncultivated profiles on base-deficient materials and may be underlain by a thin (<5 cm) Ea or Eag horizon depleted of iron, but there is no underlying illuvial horizon of significant thickness. Except in colluvial deposits which overlie pre-existing soil horizons, the Bw normally passes downwards into a less altered BC or C horizon or rests directly on bedrock. In subgroups intergrading to gley soils it overlies a mottled horizon which may qualify as Bg, BCg or Cg. As noted in the introduction to this chapter, the lower horizons commonly contain argillans as grain coatings or pore fillings; they may also exhibit fragipan features (Chapter 3, p. 106), especially in loamy late Pleistocene deposits.

Judging from available soil survey information, orthic brown soils cover about 13 per cent of England and Wales and a somewhat smaller proportion of Ireland, but are much less extensive in Scotland, the majority of the well drained soils mapped there as brown forest qualifying as podzolic according to the criteria used here (Ragg et al., 1978). The commonest parent materials are residual or locally transported weathering products of consolidated non-calcareous rocks; others include river and glacial drift of Devensian age and thin loessial deposits over massive Carboniferous limestone in the Pennines and elsewhere. In all the more humid areas of occurrence (Figure 2.5, p. 78), orthic brown soils grade locally into podzolic brown soils (p. 232) and are generally replaced by them as altitude and rainfall increase.

The colour, texture and mineralogy of these soils vary in accordance with the composition of the parent material and its weatherability. In unconsolidated sediments with sand and silt fractions composed dominantly of quartz, loss of bases by leaching is generally accompanied by slight weathering or transformation of accessory silicate minerals or 2:1 lattice clays leading to liberation of small amounts of hydrated iron oxide, but the particle-size distribution remains little affected. Where the coarser fractions contain higher proportions of silicate clay or clay-forming minerals, however, as in more or less stony materials derived from shales, slates and crystalline rocks, weathering in the soil zone has generally resulted in a considerable increase in the content of clay-size particles.

Parent rocks rich in iron-bearing minerals give rise to orthic brown soils containing unusually large amounts of free iron. Following Kubiena (1953), those in which the ratio of percentage dithionite-extractable iron to percentage clay exceeds 0.25 have been distinguished as *ferritic brown earths* in England and Wales (Avery, 1980). Soils meeting this requirement are common on Jurassic 'ironstones' in the English Midlands and have also been identified on basic igneous rocks (e.g. Loveland and Bullock, 1976). Other characteristic features of brown soils in basic igneous materials are strong aggregation, relatively high base status, large phosphorus-retention capacity, and high cation-exchange capacity in relation to conventionally measured clay content, particularly in the lower horizons. The latter property is at least partly attributable to the presence of undispersed aggregates of expansible clay minerals (vermiculite or smectite) in the silt and sand fractions (Mitchell and Muir, 1937; McAleese and Mitchell, 1958; Curtin and Smillie, 1981). Variants derived from ultrabasic rocks, chiefly serpentinite, are distinguished as brown magnesian soils by the Soil Survey of Scotland (1984) and also occupy small areas in the Lizard peninsula (Staines, 1984). The lower horizons of these soils are neutral or alkaline and magnesium is the dominant exchangeable cation, often exceeding calcium by a factor of five or more. The sand and silt fractions are composed largely of serpentine minerals (antigorite, chrysotile, fibrolite, talc) or aggregates of their weathering products (e.g. trioctahedral smectite) and contents of chromium, cobalt and nickel are abnormally high, reaching levels which are associated in places with distinctive semi-natural plant communities (Birse, 1982).

With few exceptions, soils of this group are suited to a wide range of uses where climatic and topographic conditions permit. Being more than 30 cm deep and at least moderately well drained by definition, inherent physical limitations are seldom severe. Although the soils are naturally acid and often phosphate deficient, they generally respond well to manurial amendments. In the areas where they are most extensive, the climate favours grass production rather than arable cropping and some of the land is too steep or too rocky for intensive cultivation. Variants with bedrock, fragipan or gravelly layers at moderate depths can be moderately droughty, however, especially in less humid areas.

Key to subgroups

1. Orthic brown soils with pelo-characteristics (p. 112)
   Pelo-orthic brown soils
2. Other orthic brown soils with gleyic features (p. 112)
   2.1 with a slowly permeable subsurface horizon (p. 110)
   Stagnogleyic orthic brown soils
   2.2 without a slowly permeable subsurface horizon
   Gleyic orthic brown soils
3. Other orthic brown soils in recent colluvium (p. 114) more than 40 cm thick
   Colluvial orthic brown soils
4. Other orthic brown soils
   Typical orthic brown soils

5.5.2 Typical Orthic Brown Soils

The typical subgroup, which is also the most extensive, includes freely draining brown

Figure 5.9: The Hopesay valley, Shropshire. Typical orthic brown soils (brown earths) over Silurian siltstones and fine-grained sandstones underlie nearly all the moderately to steeply sloping land (British Geological Survey: NERC copyright).

soils that have been classed as typical and ferritic brown earths in England and Wales, brown forest and brown magnesian soils in Scotland, and acid brown earths in Ireland. They are typically loamy and the few that are clayey are more friable and finely structured than other soils of similar texture. As in the typical calcaric brown soils (p. 182), the material in which the A and Bw horizons have developed may have moved downslope, but variants in dry valley deposits identifiable as recent colluvium (p. 114) are excluded. Data on four representative profiles are given on pp. 197–200, two under forest and two in agricultural land.

Profile 54, located under oak–hazel woodland in the Welsh Borderland (Figure 5.9), has an Ah horizon of acid mull over a yellowish brown fine silty Bw which grades downwards into shattered Silurian siltstone. Percentage base saturation values are low throughout and decrease with depth, reflecting the circulation of cations from lower horizons to the surface by way of the leaf fall. Materials derived directly or indirectly from Palaeozoic shales, slates, cleaved mudstones and fine-grained sandstones give rise to similar fine loamy (Denbigh series) or fine silty (Barton series) soils over large areas in Wales and south-west England (Plate II.1), and also in various parts of Ireland, where they are typified by the Clonroche and Ballindaggan series (Gardiner and Ryan, 1964; Conry and Ryan, 1967; Gardiner and Culleton, 1978). Some, like the profile described, have bedrock at moderate depths; others overlie stony head or glacial drift which is in places compacted (fragipan).

Profile 55 is in thin drift overlying and mainly derived from red sandstone (Upper Old Red Sandstone) in rolling country northwest of Berwick-on-Tweed. This coarse loamy cultivated soil is weakly structured and has a high base status resulting from regular applications of lime. Reddish loamy soils with similar horizons are widely distributed in England and Wales on Devonian and Permo-Triassic sandstones (Eardiston and Bromsgrove series) and conglomerates (Crediton series), and in drift derived from Devonian sandstones and siltstones (Milford series).

Profile 56, on dolerite in South Devon,

was chosen to represent the orthic brown soils derived from basic crystalline rocks. The weakly differentiated A and Bw horizons are strongly aggregated, percentage base saturation is high throughout and the cation-exchange capacity of the saprolite forming the BC horizon is very large in relation to the conventionally determined clay content. Only very small proportions of the total free iron contents are extracted by pyrophosphate and amounts of pyrophosphate-extractable aluminium are also small, though significantly more aluminium is removed from analogous horizons of similar soils by acid oxalate (e.g. Loveland and Bullock, 1976). This indicates that the inorganic part of the exchange complex contains appreciable amounts of poorly ordered (allophanic) aluminous material, as found in Scottish soils of similar provenance by Kirkman et al. (1966) and Tait et al. (1978). Larger amounts occur in the Bs horizons of (andic) podzolic brown soils in basic or intermediate igneous materials as exemplified by profile 79 (p. 238).

Profile 57, under beechwood with acid mull, typifies the iron-rich orthic brown soils distinguished as ferritic brown earths by Avery (1973, 1980). Sited on a nearly level dipslope plateau, it is developed in the cryoturbated weathering products of Middle Lias 'Marlstone' and resembles agricultural soils of the same (Banbury) series studied in detail by Storrier and Muir (1962). The unweathered parent rock is a greenish shelly ferruginous limestone containing siderite (ferrous carbonate) and chamosite (an iron aluminosilicate). Free iron oxides have accumulated in the weathered horizons along with layer-lattice clays (chiefly vermiculite and kaolin) as a result of solution and removal of carbonates and oxidation of the iron-bearing minerals. Total extractable iron contents range from 19 to 23 per cent (27 to 33 per cent as Fe_2O_3) but amounts of both iron and aluminium extracted by pyrophosphate are no greater than in other orthic brown soils of similar texture and base status. These trends again contrast with those characteristic of podzolic brown soils and are consistent with Storrier and Muir's finding that most of the iron oxide in the profiles they studied was well crystallized goethite. The large iron oxide contents appear to promote structural stability and to inhibit eluviation of clay in these soils, which have a high reputation agriculturally. Despite the stoniness of the subsurface horizons, crops seldom suffer severely from drought, possibly because the stones are porous and retain significant amounts of available water (Thomasson, 1971).

Profile 54 (Mackney and Burnham, 1966)

Location	: Edgton, Shropshire (SO 380865)
Climatic regime:	humid temperate
Altitude	: 290 m Slope: 3°
Land use and vegetation	: oak-hazel woodland; field layer dominated by brambles (*Rubus fruticosus*) and bracken (*Pteridium aquilinum*)

Horizons (cm)

L/F	Partially decomposed litter with active mesofauna, including Collembola and slugs, less than 1 cm thick.
0–0.6 H	Very dark brown (10YR 2/2) loamy organic material; granular; very friable; many faecal pellets and earthworm casts; very porous; abundant fine roots; abrupt irregular boundary.
0.6–12 Ah	Dark greyish brown (10YR 4/2) slightly stony silty clay loam (USDA silty clay loam); common siltstone fragments up to 10 cm; moderate fine subangular blocky and granular, with common earthworm casts; friable; many medium woody and fleshy roots; abrupt irregular boundary.
12–38 Bw1	Yellowish brown (10YR 5/4) moderately stony silty clay loam (USDA silt loam); stones as above, up to 15 cm; fine subangular blocky and granular; friable; common medium woody roots; common channels filled with dark greyish brown soil from horizon above; gradual boundary.
38–51	Yellowish brown (10YR 5/4) very

Analytical data for Profile 54

Horizon	H	Ah	Bw1	Bw2
Depth (cm)	0–0.6	2–12	15–35	40–51
sand 200 μm–2 mm %		8	9	8
50–200 μm %		2	2	1
silt 2–50 μm %		62	63	64
clay <2 μm %		28	26	27
loss on ignition %	37	13.3	5.6	4.6
organic carbon %	17.7	5.4		
nitrogen %	0.97	0.40		
C/N ratio	18	13		
pyro ext. Fe %		0.5	0.5	0.3
pyro ext. Al %		0.4	0.4	0.3
pyro. Fe+Al %/clay %		0.03	0.03	0.02
total ext. Fe %		1.4	1.5	1.5
TEB (me/100 g)		3.50	1.57	1.41
CEC (me/100 g)		24.2	11.6	14.8
% base saturation		14	14	9
pH (H₂O)	4.2	4.4	4.3	4.4

Bw2	stony silty clay loam (USDA silt loam to silty clay loam); abundant siltstone fragments; structure and consistence of interstitial fine earth as above; common medium to coarse woody roots.
51–64 Cu	Fragmented olive siltstone with interstitial cracks almost free of fine earth.

Analytical data *see* above.

Classification
    England and Wales (1984): typical brown earth; fine silty material over lithoskeletal siltstone (Barton series).
    USDA (1975): Typic Dystrochrept; loamy-skeletal, mixed, mesic.
    FAO (1974): Dystric Cambisol; medium textured.

Profile 55 (Ragg and Futty, 1967; Ragg and Clayden, 1973)

Location	: Chirnside, Berwickshire (NT 896575)
Climatic regime:	humid temperate
Altitude	: 120 m Slope: 3°
Land use	: arable

Horizons (cm)
    0–23   Dark brown to brown (7.5YR 4/2)

Ap	sandy loam (USDA sandy loam); occasional large subrounded grey-wacke fragments; weak coarse granular; very friable; frequent roots; sharp boundary.
23–38 Bw	Reddish brown (2.5YR 5/4) stony sandy loam (USDA fine sandy loam); frequent small subrounded and subangular stones; weak subangular blocky; firm and slightly compact; occasional roots; clear boundary.
38–90 BC(x)	Red (10R 5/6) stony sandy loam (USDA sandy loam); frequent angular sandstone fragments; massive; firm; few roots; clear boundary.
90+ Cr	Red (10R 4/6) sandstone.

Analytical data

Horizon	Ap	Bw	BC	Cr
Depth (cm)	4–15	29–35	51–60	68–76
silt^a 2–50 μm %	27	27	23	24
clay^a <2 μm %	12	6	9	9
organic carbon %	3.0	0.8		
TEB (me/100 g)	12.9	5.4	4.5	4.1
CEC (me/100 g)	12.9	6.6	4.5	4.1
% base saturation	100	83	100	100
pH (H₂O)	7.1	6.8	6.9	6.9

[a] by hydrometer method

Classification
  Scotland (1984): brown forest soil (Hobkirk series).
  England and Wales (1984): typical brown earth, reddish-coarse loamy material over lithoskeletal sandstone (Eardiston series).
  USDA (1975): Dystric Eutrochrept; coarse-loamy, siliceous or mixed, mesic.
  FAO (1974): Chromic Cambisol; coarse to medium textured.

Profile 56 (Hogan, 1978)

Location   : Dartington, Devon (SX 777619)
Climatic regime: humid temperate
Altitude   : 64 m   Slope: 8° convex
Land use   : ley grass
Horizons (cm)
  0–22   Brown to dark brown (7.5YR 4/4)
  Ap   moderately stony clay loam (USDA loam); many small subangular dolerite fragments; strong medium to fine subangular blocky; friable; abundant fine roots; clear smooth boundary.
  22–61 Brown to dark brown (7.5YR 4/4),
  Bw   very slightly stony, medium sandy loam (USDA sandy loam) with many coarse brown to strong brown (7.5YR 5/5) rotted dolerite fragments; few small harder fragments; strong fine subangular blocky; very friable; common fine roots; few soft irregular ferri-manganiferous concentrations; gradual smooth boundary.
  61–107 Brown to strong brown (7.5YR 5/5)
  BC   slightly stony sandy loam (USDA sandy loam) consisting of rotted dolerite (saprolite); with common medium subangular corestones; massive; very friable; few fine roots and many ferri-manganiferous concentrations as above.

Analytical data

Horizon	Ap	Bw	BC
Depth (cm)	0–22	22–61	61–107
sand 600 μm–2 mm %	14	16	19
200–600 μm %	11	24	27
60–200 μm %	17	19	23
silt 2–60 μm %	35	27	22
clay <2 μm %	23	14	9
organic carbon %	1.7	0.6	
pyro. ext. Fe %	0.2	0.1	0.1
pyro. ext. Al %	0.1	0.1	<0.05
total ext. Fe %	4.0	3.5	3.1
TEB (me/100 g)	21.6		34.1
CEC (me/100 g)	30.2		35.9
% base saturation	72		95
pH (H$_2$O)	5.9	6.8	7.0
(0.01M CaCl$_2$)	5.2	5.9	6.2
bulk density (g cm^{-3})	1.23	1.13	1.07
packing density (g cm^{-3})	1.43	1.25	1.15
retained water capacity (% vol.)	41	34	35
available water capacity (% vol.)	21	14	14
air capacity (% vol.)	13	23	25

Classification
  England and Wales (1984): typical brown earth intergrading to ferritic brown earth; coarse loamy passing to soft weathered basic crystalline rock (variant of Erisey series).
  USDA (1975): Dystric Eutrochrept; coarse-loamy, oxidic, mesic.
  FAO (1974): Eutric Cambisol; medium textured.

Profile 57 (Thomasson, 1971)

Location   : Croxton Kerrial, Leicestershire (SK 805323)
Climatic regime: subhumid temperate
Altitude   : 143 m   Slope: <1°
Land use and   broadleaf woodland dominated
  vegetation   : by beech (*Fagus sylvatica*) with field layer mainly of bramble (*Rubus fruticosus*)
Horizons (cm)
  L/F   Partially decomposed litter, c. 2 cm thick.
  0–8   Dark brown (7.5YR 4/2) very slightly
  Ah   stony clay loam (USDA clay loam); few small ironstone fragments; strong medium to fine subangular blocky and granular; friable; many fine roots; abrupt smooth boundary.
  8–20   Reddish brown (5YR 4/4) slightly
  AB   stony clay loam (USDA clay loam); common small and medium ironstone fragments; strong medium and fine subangular blocky; friable;

	common medium woody roots; dark brown ped faces and infilled worm channels; gradual boundary.
20–34 Bw1	Reddish brown (5YR 4/4) slightly stony clay loam (USDA loam); stones as above; strong medium subangular blocky; friable; common woody roots; gradual boundary.
34–60 Bw2	Strong brown to yellowish red (7.5–5YR 5/6) moderately stony clay loam (USDA loam); many, small to large, subangular ironstone fragments; moderate fine subangular blocky; friable; few woody roots; gradual boundary.
60–90 BC/Cr	Strong brown (7.5YR 5/8) very stony sandy loam (USDA sandy loam) passing into soft ironstone rock weathered *in situ*.

Analytical data *see* below.

Classification
England and Wales (1984): ferritic brown earth; fine loamy over lithoskeletal ironstone (Banbury series).
USDA (1975): Typic Dystrochrept; loamy-skeletal, oxidic, mesic.
FAO (1974): Chromic Cambisol; medium textured.

5.5.3 Colluvial Orthic Brown Soils

These are well drained brown soils in non-calcareous colluvial deposits that have accumulated in dry valley bottoms and on footslopes, mainly as a result of accelerated erosion since the land was deforested and brought into cultivation. The profiles are characterized by weakly differentiated A and Bw horizons and organic matter contents which decrease gradually or irregularly with increasing depth. Some have a brownish Bw underlain by a darker horizon which may be a buried A or a layer containing a higher proportion of transported organic matter; others have a very dark coloured horizon (ABw) extending below 30 cm depth. Where neither of these features is apparent, the soil may be identified as colluvial if it occupies a 'receiving' site and the organic carbon content exceeds 0.2 per cent to at least 1.25 m below the surface or if fragments of charcoal or artefacts such as earthenware occur in the lower part of the rooting zone. To be classed as an orthic brown soil rather than as a rego-alluvial soil (p. 152), the material below 30 cm depth must have been in place long enough

Analytical data for Profile 57

	Ah	AB	Bw1	Bw2	BC
Horizon					
Depth (cm)	0–8	8–20	20–34	34–60	60–90
sand 200 μ–2 mm %	21	18	24	24	28
50–200 μm %	16	19	19	24	27
silt 2–50 μm %	35	34	34	31	30
clay <2 μm %	28	29	23	21	15
organic carbon %	4.4	2.4			
nitrogen %	0.39	0.24			
C/N ratio	11	10			
pyro. ext. Fe %	0.40	0.25	0.13	0.04	0.03
pyro. ext. Al %	0.24	0.22	0.15	0.08	0.06
total ext. Fe %	19	20	22	23	23
TEB (me/100 g)	5.4	1.9	5.1		
CEC (me/100 g)	29.9	24.9	19.6		
% base saturation	18	8	26		
pH (H$_2$O)	4.6	4.5	5.1		
(0.01M CaCl$_2$)	4.1	4.0	5.6		

for structure forming agencies to have obliterated any fine stratification, though this may be evident in the surface horizon where recent sedimentation has occurred, as in profile 58. Deep colluvial soils of this type have been recorded in numerous localities. As they occupy limited areas, however, and are not always readily distinguishable from associated soils in older deposits, they have seldom been separated on soil maps and have only recently been considered to merit systematic recognition as soil series (Avery, 1980; Clayden and Hollis, 1984). Data on two examples are given below.

Profile 58 was described and sampled in west Gloucestershire. The coarse loamy colluvium in which it is developed is derived from higher lying brown soils on red Triassic sandstone and overlies stony head of similar origin. The upper 30 cm comprising the Ap horizon has a laminated structure and evidently accumulated very recently by downslope washing which was actively proceeding at the time of sampling (26 Feb 1979). Below this, a weakly expressed brownish Bw horizon extends to 85 cm and is succeeded by a darker and appreciably finer textured horizon interpreted in the field as a buried topsoil. Although the organic carbon data do not support this conclusion, possibly because the thickness sampled was too great, the presence of charcoal fragments coupled with the weak horizonation indicates that the upper 150 cm is recent colluvium.

Profile 59 is located in a lower slope near the southern margin of Strathmore, Perthshire, on coarse textured drift derived from Devonian (Old Red Sandstone) rocks. It has very dark surface and subsurface horizons giving the appearance of an unusually thick topsoil, which at this site apparently results from slow colluviation rather than from former manurial practices as in Plaggen soils (p. 386). Judging from the loss on ignition values, the organic matter content exceeds 2 per cent to at least 80 cm depth.

Profile 58 (Cope, 1986)

Location	: Newent, Gloucestershire (SO 727268)
Climatic regime:	subhumid temperate
Altitude	: 38 m Slope: 2° concave
Land use	: arable (roots)

Horizons (cm)

0–15 Ap1 Dark reddish brown (5YR 3/3; 5YR 4/4 rubbed; 7.5YR 4/4 dry) very slightly stony fine sandy loam (USDA fine sandy loam) with thin laminae of (loamy) fine sand; few small subangular quartzite fragments; weak very coarse platy (finely stratified); friable; common fine roots; abrupt boundary.

15–30 Ap2 Dark reddish brown (5YR 3/3; 5YR 4/4 rubbed; 7.5YR 4/4 dry) very slightly stony fine sandy loam (USDA fine sandy loam) with few thin sandy laminae and few glass and iron ore fragments; stones as above; moderate very coarse platy (stratified); firm; common fine roots; clear wavy boundary.

30–45 ABw Reddish brown (5YR 4/3; 5YR 4/4 rubbed; 7.5YR 4/4 dry) very slightly stony fine sandy loam (USDA fine sandy loam) with rare thin sandy laminae; stones as above; weak coarse angular blocky; friable; common fine roots; few small charcoal fragments; abrupt wavy boundary.

45–85 Bw Reddish brown (5YR 4/4 broken and rubbed; 7.5YR 5/4 dry) very slightly stony fine sandy loam (USDA fine sandy loam) with few thin sandy laminae and few glass and iron ore fragments; stones as above; moderate very coarse platy (stratified); firm; common fine roots; clear wavy boundary.

85–150 bAh Dark reddish brown (5YR 3/3; 5YR 4/4 rubbed; 7.5YR 5/4 dry) very slightly stony fine sandy loam (USDA fine sandy loam); stones as above; moderate medium angular blocky; firm; roots and charcoal fragments as above; clear wavy boundary.

Analytical data for Profile 58

Horizon	Ap1	Ap2	Ab	Bw	bAh	2BC
Depth (cm)	0–15	15–30	30–45	45–85	85–150	150–180
sand 600 μm–2 mm %	2	2	2	2	5	19
200–600 μm %	18	17	17	16	19	22
100–200 μm %	33	30	29	30	22	19
60–100 μm %	18	15	15	15	10	8
silt 2–60 μm %	19	24	25	26	29	23
clay <2 μm %	10	12	12	11	15	9
organic carbon %	1.1	1.3	1.0	0.4	0.3	0.2
pH (H$_2$O)	7.6	7.5	6.4	6.9	6.6	6.8
(0.01M CaCl$_2$)	6.9	6.7	5.9	6.1	5.9	6.1
bulk density (g cm^{-2})		1.39	1.35	1.56	1.55 [a]	
packing density (g cm^{-3})		1.49	1.45	1.65	1.68 [a]	
retained water capacity (% vol.)		30	26	23	25[a]	
available water capacity (% vol.)		18	16	14	12[a]	
air capacity (% vol.)		19	23	19	16[a]	

[a] samples at 95–100 cm

150–180 2BC	Reddish brown (5YR 4/4 broken and rubbed; 5YR 7/6 dry) slightly stony medium sandy loam (USDA sandy loam); common medium subangular quartzite fragments; weak fine subangular blocky; wet; few fine roots; clear boundary.
180–200 3Cu	Reddish brown (5YR 4/4) moderately stony medium sandy loam; many quartzite fragments; massive; wet.

Analytical data *see* above.

Classification
   England and Wales (1984): colluvial brown earth; reddish-coarse loamy non-calcareous colluvium (Newent series).
   USDA (1975): Dystric Fluventic Eutrochrept; coarse-loamy, mixed, mesic.
   FAO (1974): Eutric Fluvisol; coarse textured.

Profile 59 (Laing, 1976)

Location   : Newtyle, Angus (No 293416)
Climatic regime: humid boreal
Altitude   : 76 m   Slope: 3°
Land use   : ley grass

Horizons (cm)

0–33 Ap	Very dark greyish brown (10YR 3/2) sandy loam to sandy clay loam (USDA sandy loam); few stones; moderate coarse angular blocky, breaking to coarse and fine granular; friable; many grass roots; earthworms present; abrupt smooth boundary.
33–66 ABw	Dark brown (7.5YR 3/2) stony sandy loam to sandy clay loam (USDA sandy loam); weak coarse angular blocky, breaking to medium and fine granular; frequent roots; earthworms present; abrupt smooth boundary.
66–81 BC	Reddish brown (5YR 4/4) very stony loamy sand (USDA loamy sand) many angular to subrounded stones of varying size, including angular fragments of flaggy sandstone; massive; very friable to loose; few roots.

Analytical data *see* p. 203.

Classification
   Scotland (1984): Brown forest soil (Vinny series).
   England and Wales (1984): colluvial brown earth; reddish-coarse loamy non-calcareous colluvium (Newent series).
   USDA (1975): Pachic Haplumbrept; coarse-loamy, mixed, mesic.
   FAO (1974): Humic Cambisol; medium textured.

Analytical data for Profile 59

Horizon	Ap	Ap	ABw	ABw	BC
Depth (cm)	3–13	20–30	36–46	53–63	71–81
silt^a 2–50 μm %	16	18	16	13	8
clay^a <2 μm %	18	18	16	15	10
loss on ignition %	6.9	7.5	5.0	5.1	3.4
organic carbon %	2.6	2.6			
nitrogen %	0.24	0.25			
C/N ratio	11	10			
TEB (me/100 g)	8.1	8.0	3.6	3.4	2.3
CEC (me/100 g)	15.9	15.6	11.1	10.7	6.9
% base saturation	51	51	32	32	33
pH (H$_2$O)	6.1	6.1	6.1	6.0	6.0

^{a} by hydrometer method

5.5.4 Gleyic Orthic Brown Soils

These are intergrades between typical or colluvial orthic brown soils and orthic (p. 339) or alluvial (p. 300) gley soils which occur as members of catenary associations in permeable non-calcareous parent materials. Mottles attributable to gleying appear in or below the Bw horizon at 60 cm or less, indicating that the lower part of the rooting zone is or was periodically saturated by ground water rising from below. Soils of this subgroup are typically developed in medium to coarse textured head or river drift (Plate II.2), but some have bedrock at moderate depths and others consist wholly or partly of recent colluvium. Although the Bw horizons are normally loamy, lower horizons are frequently sandy or sandy-gravelly. The current or former zone of fluctuation of the water table is marked in some profiles by a distinct ferruginous horizon (BCgf or Cgf) in which iron mobilized by gleying has precipitated. Data on two representative profiles, both in agricultural land, are given on pp. 204–6.

Profile 60 is located near Worcester in the early Devensian Kidderminster terrace of the Severn (Mitchell et al., 1973). It is sandy from 67 cm downwards and the air capacity data confirm that the overlying coarse loamy Bw and Bw(g) horizons are very porous. Thin sections showed that the small amounts of clay in the BCg and 2Cgf horizons are irregularly distributed and occur mainly as argillans coating the sand grains, but there is no distinct, relatively fine textured horizon identifiable as an argillic B within the upper 120 cm. In agricultural soils of this type, the gleyic features commonly relate to an earlier undrained condition (Robson and Thomasson, 1977).

Profile 61 is sited in a shallow valley on the Woburn Experimental Farm in Bedfordshire. It has a gleyed horizon starting at 40 cm depth and is consequently marginal to the alluvial gley group (p. 300) but was selected for inclusion because the soils and superficial deposits in this area have been investigated in great detail by Catt et al. (1977). They showed that the Lower Greensand substratum, consisting of ferruginous sand and sand-rock, is mantled on lower slopes by loamy colluvium derived partly from the Greensand and partly from superincumbent glacial drift, relics of which exist nearby. Lacustrine sediments including a peat layer underlie the colluvium in places and a sample of the peat provided a radiocarbon date of 3,085 ± 85 years BP, proving that the colluvium was emplaced since that time. The profile described has more than 0.5 per cent organic carbon in all horizons to a depth of 126 cm and the 104–126 cm layer is very dark coloured, suggesting that it represents a former ground surface.

Profile 60 (Palmer, 1982)

Location	: North Claines, Hereford and Worcester (SO 862595)
Climatic regime:	subhumid temperate
Altitude	: 51 m Slope: 1°
Land use and vegetation	: long-term grass; perennial rye-grass (*Lolium perenne*) dominant

Horizons (cm)

0–14 Ah — Very dark greyish brown (10YR 3/2) stoneless medium sandy loam (USDA sandy loam) with few fine yellowish brown (10YR 5/4–6) mottles along root channels; strong medium and fine subangular blocky; friable; abundant fine and few medium fleshy roots; abrupt smooth boundary.

14–43 Bw — Brown to dark brown (9YR 4/3) slightly stony medium sandy loam (USDA sandy loam); common small and very small quartzose pebbles; moderate coarse angular blocky; friable; many fine roots; few coarse vertical earthworm channels with dark greyish brown coatings; clear smooth boundary.

43–67 Bw(g) — Brown to dark brown (8YR 4/3) slightly stony medium sandy loam (USDA sandy or fine sandy loam) with common brown (7.5YR 5/4) and few light brown (7.5YR 6/4) mottles, the latter only in the lowest 5 cm; stones as above but mainly medium and small; weak coarse and medium angular blocky; friable; many fine roots concentrated in infilled earthworm channels; clear wavy boundary.

67–90 BCg(t) — Brown (7.5YR 5/3) very slightly stony loamy medium sand (USDA loamy sand) with many medium and fine greyish brown (10YR 4.5/2), yellowish brown (10YR 5/6) and yellowish red (5YR 4/6) mottles; stones as above; weak coarse angular blocky with dark greyish brown faces; moist to wet; common fine roots and earthworm channels as above; clear smooth boundary.

90–120 2Cgf — Yellowish red (5YR 4/6) very slightly stony medium sand (USDA sand) with many medium and coarse reddish grey (5YR 5/2), light reddish brown (5YR 6/3) and strong brown (7.5YR 5/6) mottles; stones as above,

Analytical data for Profile 60

Horizon	Ah	Bw	Bw(g)	BCg(t)	2Cgf
Depth (cm)	0–14	14–43	43–67	67–90	90–120
sand 600 μm–2 mm %	3	4	3	3	5
200–600 μm %	39	40	37	38	44
60–200 μm %	29	27	32	39	42
silt 2–60 μm %	18	19	19	14	4
clay <2 μm %	11	10	9	6	5
<0.2 μm %	5	4	4	3	3
organic carbon %	2.9	0.8			
pH (H_2O)	6.3	6.2	5.4	5.6	5.8
(0.01M $CaCl_2$)	5.9	5.9	4.7	5.0	5.4
bulk density (g cm^{-3})	1.18	1.23	1.37		
packing density (g cm^{-3})	1.27	1.32	1.45		
retained water capacity (% vol.)	38	28	22		
available water capacity (% vol.)	21	17	12		
air capacity (% vol.)	17	24	25		

mainly small; single grain; wet (water entered pit from this horizon at one corner); few fine roots; many irregular very weakly cemented ferri-manganiferous concentrations.

Classification
England and Wales (1984): gleyic brown earth; coarse loamy drift with siliceous stones (Arrow series).
USDA (1975): Dystric Eutrochrept or Typic Dystrochrept; coarse-loamy, siliceous or mixed, mesic.
FAO (1974): Gleyic Cambisol; medium textured.

Profile 61 (Catt et al., 1977)

Location	: Husborne Crawley, Bedfordshire (SP 963359)
Climatic regime:	subhumid temperate
Altitude	: 88 m Slope: 1°
Land use	: arable

Horizons (cm)

0–21 Ap — Dark brown to very dark greyish brown (7.5–10YR 3/2) very slightly stony sandy loam (USDA fine sandy loam); few small subangular flint fragments; weak coarse to medium subangular blocky; friable; common fine roots; sharp smooth boundary.

21–40 Bw — Reddish brown (5YR 4/3) sandy loam (USDA fine sandy loam); weak coarse to medium subangular blocky; firm; few roots; clear smooth boundary.

40–67 2Bg — Dark reddish grey (5YR 4/2) sandy clay loam (USDA fine sandy loam to sandy clay loam) with very many fine dark reddish brown (5YR 3/4) and greyish mottles; structure and roots as above; firm; abrupt smooth boundary.

67–88 3BC(g) — Dark brown (10YR 4/3) sandy loam with many fine dark reddish brown (5YR 3/4) and greyish mottles; structure as above; friable; gradual boundary.

88–104 3Cg — Dark greyish brown (10YR 4/2) sandy loam with many fine mottles as above; massive; gradual boundary.

104–119 4bAhg1 — Very dark greyish brown (10YR 3/2) very slightly stony fine sandy loam (USDA fine sandy loam) with many greyish and ochreous mottles; few small ferruginous sandstone fragments and large subangular flints at base; ochreous sandy pockets and ferruginous coats on ped faces; moderate coarse to medium subangular blocky; friable; abrupt smooth boundary.

119–126 4bAhg2 — Very dark grey (10YR 3/1) very slightly stony clay loam (USDA loam to sandy clay loam) with few fine brown mottles; few quartzite pebbles; moderate medium subangular blocky; firm; abrupt irregular boundary.

126–147 5bBg — Greyish brown (10YR 5/2) sandy loam (USDA fine sandy loam) with common medium yellowish brown and grey mottles; weak medium blocky with some sandy coats on ped faces; firm; few fine woody roots.

Analytical data for Profile 61

Horizon	Ap	Ap	Bw	2Bg	4bAhg1	4bAhg2	5bBg
Depth (cm)	0–7	7–21	21–40	40–67	104–119	119–126	126–147
sand 500 µm–2 mm %	3	3	2	2	2	1	1
250–500 µm %	19	20	17	15	15	11	11
60–225 µm %	49	49	46	41	47	33	35
silt 2–60 µm %	18	16	21	22	26	34	41
clay <2 µm %	11	12	14	20	10	21	12
organic carbon %	2.1	1.6	0.8	0.7	0.6	0.6	0.5
$CaCO_3$ equiv. %	0.1	0.1	0.1	0	0	0	0
dithionite ext. Fe %	2.4	2.3	3.1	4.6	0.9	0.8	0.4
pH (H_2O)	6.8	7.1	7.4	7.0	7.0	6.9	6.8
(0.01M $CaCl_2$)	6.3	6.3	6.5	6.3	6.2	6.2	6.1

Classification
> England and Wales (1984): gleyic brown earth intergrading to typic alluvial gley soil; coarse loamy non-calcareous colluvium.
> USDA (1975): Mollic (Aeric) Fluvaquent?; coarse-loamy, mixed or siliceous, mesic.
> FAO (1974): Mollic Gleysol?; coarse textured.

5.5.5 Stagnogleyic Orthic Brown Soils

These soils also show signs of gleying in the lower part of the profile but differ from those of the preceding subgroup in having a relatively impermeable BC or C horizon within 80 cm depth which impedes downward percolation. The parent materials include medium textured tills, compact head deposits and weathering products of pre-Quaternary rocks such as siltstone. In most cases the impeding layer has some or all of the properties of a fragipan (Chapter 3, p. 106) and a clay content similar to or smaller than that of the Bw horizon. Data on two representative profiles are given below. The first is in originally grey parent material and the second in reddish (haematitic) material which resists reduction.

Profile 62, sited under deciduous woodland in the High Weald of Kent, is developed in thin locally derived head overlying soft siltstone (Tunbridge Wells Sand). It is strongly acid throughout and there are common greyish mottles in the lower part of the Bw horizon, implying that it is marginal morphologically to stagno-orthic gley soils (p. 344) in similar material. The dense substratum has well developed fragipan characteristics, including brittle consistence when moist and roughly vertical bleached veins which appear as a polygonal network when cut horizontally. Thin sections at 54–58 and 85–90 cm showed few fine argillans, mainly in the bleached veins.

Profile 63 is in reddish loamy till overlying and mostly derived from fine grained Lower Old Red Sandstone sediments (Laurencekirk association) in the Howe of the Mearns, Kincardineshire. It is an example of the imperfectly drained medium to fine textured soils grouped by the Soil Survey of Scotland (1984) as brown forest soils with gleying, most of which qualify as gley soils according to the morphological criteria adopted in both this account and in the US (Ragg and Clayden, 1973) and FAO systems. The thick Ap incorporates the upper part of the B horizon as found in analogous soils under semi-natural vegetation and the remaining part is only differentiated from the very dense C horizon by weak coarse blocky to prismatic structure and light brown mottling associated with ped faces and stone surfaces. Base saturation is more than 90 per cent throughout, probably as a result of liming. Soils of this kind form some of the most productive agricultural land in north-east Scotland though their adaptability is clearly limited to some extent by seasonal wetness due to slow permeability.

Profile 62 (Fordham and Green, 1980)

Location : Pembury, Kent (TQ 623432)
Climatic regime: subhumid temperate
Altitude : 104 m   Slope: 2°
Land use and
  vegetation : broadleaf woodland; chestnut coppice with oak standards; field layer dominated by bramble (*Rubus fruticosus*).

Horizons (cm)
L/F   More or less decomposed litter, mainly oak and chestnut leaves.
0–7
Ah   Dark brown (10YR 3/3) silt loam (USDA silt loam) containing rare very small and small siltstone fragments; moderate fine and medium subangular blocky; common medium woody roots; clear boundary.
7–30
Bw   Yellowish brown (10YR 5/6) silt loam (USDA silt loam) with rare siltstone fragments as above; moderate to weak, medium and fine subangular blocky; friable; common woody roots; gradual boundary.
30–42
Bw(g)   Yellowish brown (10YR 5/5–6) and brown (7.5YR 5/4) very slightly stony

	silt loam (USDA silt loam) with common fine light olive-grey (5Y 6/2) mottles; few siltstone fragments; moderate to weak medium prismatic breaking to fine subangular blocks; friable to firm; common woody roots; gradual boundary.
42–70 2BCg(x)	Light grey (5Y 7/2) silt loam (USDA silt) with many sharp yellowish brown (10YR 5/6) mottles; rare siltstone fragments; weak coarse prismatic; firm; roots rare; gradual boundary.
70–90+ 2BCgx	Light grey (5Y 7/2) and yellowish brown (10YR 5/6), very slightly stony, prominently mottled silt loam (USDA silt) penetrated by a network of vertical grey (5Y 7/1) veins with strong brown (7.5YR 5/8) fringes which bound very weak coarse prisms or polyhedra; very firm and brittle when moist and very hard when dry; grey veins slightly softer; roots rare.

Analytical data *see* below.

Classification
  England and Wales (1984): stagnogleyic brown earth; coarse silty material passing to siltstone (unnamed series, formerly Curtisden).
  USDA (1975): Typic Fragiochrept or Aeric Fragiaquept; coarse-silty, siliceous or mixed, mesic.
  FAO (1974): Gleyic Cambisol?; medium textured (fragipan phase).

Profile 63 (Grant and Heslop, 1981)

Location : Laurencekirk, Kincardineshire (NO 722724)
Climatic regime: humid boreal
Altitude : 60 m   Slope: 2°
Land use : ley grass
Horizons (cm)

0–36 Ap	Dark reddish grey (5YR 4/2) very slightly stony clay loam (USDA loam); few very small to large subangular and subrounded stones; moderate medium subangular blocky; friable; many fine roots; earthworms present; clear wavy boundary.
36–80 Bw(g)	Reddish brown (2.5YR 4/4) slightly stony clay loam or sandy clay loam (USDA loam) with common fine to coarse light brown (7.5YR 6/4) mottles, some associated with weathering stones; common stones as above; weak coarse subangular blocky tending to prismatic; firm; few fine roots; gradual boundary.
80–130 Cu(gx)	Reddish brown (2.5YR 4/4) slightly stony sandy loam (USDA sandy loam to loam) with common very fine and fine yellowish red (5YR 5/6) mottles; stones as above; massive; firm; no roots.

Analytical data for Profile 62

Horizon	Ah	Bw	Bw(g)	2BCg(x)	2BCgx
Depth (cm)	0–7	7–30	30–42	42–70	70–90
sand 200 μm–2 mm %	5	4	3	<1	<1
60–200 μm %	14	15	14	2	2
silt 2–60 μm %	68	67	70	92	91
clay <2 μm %	13	14	13	6	7
organic carbon %	1.9	0.7			
pyro. ext. Fe %	0.29	0.21	0.14	0.03	
pyro. ext. Al %	0.06	0.06	0.06	0.03	
pyro. Fe+Al %/clay %	0.03	0.02	0.02	0.01	
total ext. Fe %	1.0	1.2	1.2	0.5	
TEB (me/100 g)	0.86	0.55	0.55	0.30	
CEC (me/100 g)	11.9	8.0	7.0	3.3	
% base saturation	7	7	8	8	
pH (H$_2$O)	4.1	4.2	3.8	4.5	4.8
(0.01M CaCl$_2$)	3.6	3.9	3.1	4.0	4.0

Analytical data for Profile 63

Horizon	Ap	Bw(g)	Bw(g)	Cu(gx)
Depth (cm)	13–23	40–50	60–70	100–110
silt[a] 2–50 μm %	30	29	29	30
clay[a] <2 μm %	20	19	20	17
organic carbon %	2.4	0.20	0.10	0.04
TEB (me/100 g)	14.6	13.6	11.8	15.2
CEC (me/100 g)	15.0	13.8	12.4	15.7
% base saturation	97	99	95	97
pH (H_2O)	6.5	6.4	6.8	6.9
(0.01M $CaCl_2$)	6.0	6.0	6.0	6.2
bulk density (g cm^{-3})	1.52		1.71	1.76
packing density (g cm^{-3})	1.70		1.89	1.91

[a] by hydrometer method

Classification
   Scotland (1984) brown forest soil with gleying (Laurencekirk series).
   England and Wales (1984): stagnogleyic brown earth; reddish-fine loamy drift with siliceous stones (Llangendeirne series).
   USDA (1975): Dystric Eutrochrept; fine-loamy, mixed, mesic.
   FAO (1974): Eutric Cambisol; medium textured.

5.5.6 Pelo-orthic Brown Soils

These are the clayey brown soils with slowly permeable subsurface horizons which are set apart as non-calcareous pelosols in the England and Wales classification (Avery, 1973, 1980). They occur in relatively well drained sites on non-calcareous argillaceous rocks and fine textured Pleistocene sediments, and are distinguished from the more extensive pelo-orthic gley soils (p. 348) on similar materials by the presence of a brownish or reddish blocky or prismatic Bw horizon which is at most slightly mottled to a depth of at least 40 cm. This horizon normally has a sepic plasmic fabric (Brewer, 1964) with plasma separations (Table 3.6, p. 91) in which the clay-size material is preferentially oriented; it may contain slightly more clay than the A when the latter is developed in a lithologically distinct surface layer, but evidence of clay translocation is otherwise insufficient for it to qualify as argillic (Bt). On originally greyish materials, as in profile 64 below, the Bw generally passes downwards into a more coarsely structured Bg or BCg horizon with prominent ochreous mottles. In reddish (haematitic) materials, however, the lower horizons retain their original colour and gleyic features are weakly expressed. The non-calcareous sedimentary rocks from which most of these soils are derived give rise to clay fractions of micaceous or mixed composition (Avery and Bullock, 1977), with relatively small proportions of fine (<0.2 μm) clay. Their shrinkage potential is correspondingly small and the soil horizons often feel more silty and less plastic than would be expected from the particle-size analyses.

The following representative profile is sited in moderately sloping old grassland on Jurassic (Middle Lias) clay-shale in north Gloucestershire. X-ray diffraction data on clay separates from this soil, together with the CEC and K_2O values quoted, show that they consist mainly of mica and interstratified mica-vermiculite in varying proportions, with smaller amounts of kaolinite and chlorite.

Profile 64 (Courtney and Findlay, 1978)

Location   :   Maugersbury, Gloucestershire (SP 175233)

Climatic regime: subhumid temperate
Altitude : 170 m Slope: 5°
Land use : long-term grass (ridge and furrow microrelief)
Horizons (cm)
0–16 Ah — Dark brown (10YR 3/3) stoneless clay (USDA clay); strong fine and very fine subangular blocky; friable; abundant fine roots; abrupt smooth boundary.
16–42 Bw — Yellowish brown to light olive brown (10YR–2.5Y 5/4) clay (USDA clay) containing occasional very small ironstone fragments; strong coarse, breaking to moderate medium, subangular blocky; friable to firm; common fine roots; clear wavy boundary.
42–57 Bg — Olive (5Y 5/3–4) clay (USDA clay) with common yellowish brown (10YR 5/6) and strong brown (7.5YR 5/6) mottles; occasional ironstone fragments as above; moderate medium prismatic breaking to coarse blocky; firm; few ferri-manganiferous (2–5 mm) nodules; few fine roots; clear smooth boundary.
57–90+ BCg — Olive grey (5Y 5/2) clay (USDA clay) with common yellowish brown (10YR 5/6) and light olive brown (2.5Y 5/6) mottles; ironstone fragments as above; weak coarse blocky becoming platy below 70 cm; very firm; common 5–20 mm ferri-manganiferous nodules and soft concentrations; roots rare.

Analytical data

Horizon	Ah	Bw	Bg	BCg
Depth (cm)	0–16	16–42	42–57	57–90
sand 200 µm–2 mm %	3	5	13	11
50–200 µm %	14	16	13	9
silt 2–50 µm %	23	23	21	29
clay <2 µm %	60	56	53	51
organic carbon %	4.7	1.7		
dithionite ext. Fe %	3.2	4.5	4.3	3.8
pH (H$_2$O)	6.3	7.3	6.6	6.7
(0.01M CaCl$_2$)	5.7	6.9	5.9	6.1
clay <2 µm				
CEC (me/100 g)	46	45	44	42
K$_2$O %	1.9	1.9	2.2	2.6

Classification
England and Wales (1984): typical non-calcareous pelosol; clayey material passing to clay or soft mudstone (Stow series).
USDA (1975): Aquic Dystric Eutrochrept; fine, mixed, mesic.
FAO (1974): Gleyic Cambisol; fine textured.

5.6 Luvic Brown Soils

5.6.1 General Characteristics, Classification and Distribution

These soils have been called argillic brown earths, paleo-argillic brown earths and argillic pelosols in England and Wales (Avery, 1973, 1980) and grey-brown podzolic soils in Ireland (Gardiner and Radford, 1980). In earlier English publications (e.g. Avery, 1964, 1965; Mackney and Burnham, 1964) they were identified by the French term *lessivé*, and *luvic* (from Greek *Louo* to wash) has the same connotation, implying the presence of an argillic (clay-illuvial) B horizon. Most British soils of this group are Alfisols (Udalfs) in the US Taxonomy; others are Ultisols (Udults) or Mollisols (Argiudolls) depending on percentage base saturation at specified depths and the presence or absence of a mollic epipedon, neither of which is differentiating in the classification used here. According to the FAO system, in which similar but not identical separations are made, Luvisols, Podzoluvisols, Acrisols and Luvic Phaeozems are included.

Soils of this group cover about the same proportionate area as the orthic brown soils (brown earths *sensu stricto*) in England and Wales (Mackney *et al.*, 1983). The main areas of occurrence (Figure 2.5, p. 78) are in southern and eastern England and the Midlands, where argillic B horizons have formed in a wide range of late Pleistocene (Devensian) and older deposits. Luvic brown soils are also extensive on glacial and glaciofluvial drift in the central plain of Ireland. In Scotland they are relatively uncommon and have not been distinguished from other 'brown forest soils' by the Soil Survey.

The typical horizon sequence in these soils is A/Eb/Bt, with modifications related to parent material, degree of leaching and land use history. Many of them have formed in or over originally calcareous material. Where this underlies the solum at depths ranging from 40 to 120 cm, the lower boundary of the Bt normally coincides with the depth of decalcification and is generally abrupt but often irregular. In originally non-calcareous or more deeply decalcified deposits, argillans may line widely spaced fissures to depths of several metres and the lower boundary of the solum is correspondingly ill defined. An eluvial subsurface horizon is nearly always discernible in undisturbed profiles but is commonly absent in cultivated land, particularly where accelerated erosion has occurred, in which case the Ap may incorporate part of the original Bt. As in the orthic brown soils, surface horizons under semi-natural vegetation range in character from mull to mor, the most leached variants having a mor layer with signs of superficial podzolization immediately beneath it. The Eb horizon, where present, is normally differentiated from the Bt by a weaker structure and paler brownish colour, especially when dry, as well as by a smaller clay content. In many places, however, the profile is composite (Figure 1.9, p. 15), the Bt horizon having formed in an originally finer textured layer, so that the overall clay increase with depth is partly inherited and partly pedogenic in origin. Intergrades to gley soils occur where the Bt or the immediately underlying layer is slowly permeable or where ground water rises seasonally from below.

Where the E horizon is strongly leached and depleted of clay, it commonly has a pale coloured lower part which penetrates the Bt as sandy or silty coats (skeletans) on ped faces or as deep vertically oriented tongues. Variants characterized by tonguing of 'albic' material into or through an argillic B horizon are placed in glossic (from Greek *glossa* = tongue) groups (e.g. Glossudalfs, Fraglossudalfs) in the US Taxonomy and are similarly set apart as Podzoluvisols in the FAO (1974) scheme. On the basis of micromorphological and mineralogical studies, French and Belgian pedologists (e.g. Jamagne, 1972; De Coninck *et al.*, 1976; Jamagne *et al.*, 1984) have attributed glossic features to a secondary phase of clay translocation which ensues when the Bt horizon becomes dense and relatively impermeable (fragipan-like) and drainage is preferentially restricted to widely spaced fissures or channels, which in some cases appear to have originated under periglacial conditions as frost cracks or infilled ice wedges (Figure 1.6, p. 12). According to this explanation, periodic stagnation of water above the E/Bt boundary causes localized removal and segregation of iron, and as the deferrated clay so produced is readily dispersed and eluviated, it tends to be removed from the iron depleted parts and to accumulate as thick layered argillans at greater depths.

The luvic brown soils in late Pleistocene (Devensian) sediments such as loess or till and in erosion surfaces of comparable age bevelling older formations generally have brown or reddish brown Bt horizons with dominant chromas of four or less (Plate II.3). Others, including those grouped as paleoargillic brown earths in England and Wales, are characterized by chromic argillic B horizons (Chapter 3, p. 109) that are strong brown to red or red-mottled, at least in the lower part (Plate II.4). Although some of these soils appear to have formed in deposits of Devensian age or to have inherited their strong colours from pre-Quaternary rocks, the great majority are on 'older drift' or quasi-residual deposits such as Clay-with-flints and have complex or occasionally polycyclic profiles (Figure 1.9, p. 15) showing distinctive morphological features that can be attributed to weathering and pedological reorganization of the subsoil materials during one or more pre-Devensian interglacial periods. In most places, however, the upper horizons (A, A and E, or A, E and part of the Bt) have evidently formed in thin superficial layers consisting of loess or coversand which has been irregularly mixed with material derived from the pre-existing soil by cryoturbation or solifluxion, or by later pedoturbation under temperate conditions. The profile as a whole may show features attributable to periglacial disturbance, such as ice-wedge casts or preferentially oriented stones, and the 'paleo-argillic' horizon may represent either the truncated remains of a pre-existing profile, or consist of old material which has been transported over a short distance but retains characteristic features of its original fabric. In some places a Bt horizon has formed in the cover deposit and evidently results from a post-depositional phase of clay illuviation which has also affected the older soil material below.

Many of these soils overlie chalk or harder limestone at moderate depths. Others are non-calcareous to depths well below the rooting zone and some are less than 35 per cent base saturated at more than 1.50 m below the upper boundary of the argillic B, implying that they are Ultisols (mostly Paleudults) rather than Alfisols (Paleudalfs) in the US Taxonomy.

As the luvic brown soils are mainly located in level to moderately sloping lowland areas with moderate rainfall, they are mostly well suited to arable cultivation, though the inherently weak structure of the upper horizons leads to slaking, surface capping and ploughpan formation, particularly where organic matter contents have been depleted under continuous cropping. Freely draining variants in more or less re-worked loess (brickearth) and other permeable loamy materials with few stones and large available water capacities are the most valuable agriculturally. Others are less adaptable or less consistently productive because of droughtiness, stoniness, unfavourable working qualities, seasonal wetness, or some combination of these limitations. Where the soils extend into more humid areas, as in parts of central Ireland, they are mainly used for pasture, which is highly productive when well managed.

Key to Subgroups

1. Luvic brown soils with gleyic features (p. 112) but no slowly permeable subsurface horizon (p. 110)
   Gleyic luvic brown soils
2. Other luvic brown soils with a chromic argillic B horizon (p. 109)
   2.1 with a slowly permeable subsurface horizon and gleyic features
       Stagnogleyic chromoluvic brown soils
   2.2 without a slowly permeable subsurface horizon
       Chromoluvic brown soils
3. Other luvic brown soils with a slowly permeable subsurface horizon
   3.1 with pelo- characteristics (p. 112)
       Peloluvic brown soils
   3.2 without pelo- characteristics
       Stagnogleyic luvic brown soils
4. Other luvic brown soils
   Typical luvic brown soils

5.6.2 Typical Luvic Brown Soils

The group is typified by freely draining soils with loamy or occasionally sandy upper horizons and brownish argillic B horizons that fail to qualify as chromic (Chapter 3, p. 109). Called typical argillic brown earths and argillic brown sands in the England and Wales classification, these soils have formed in permeable Devensian sediments, including more or less reworked loess (brickearth), river-terrace deposits, glaciofluvial drift and head, and to a lesser extent in older drifts and materials weathered more or less in place from pre-Quaternary rocks such as siltstone and fine grained sandstone. Most, but not all, of the parent materials were calcareous when deposited.

The Bt horizon normally has a blocky or prismatic structure with clay coats (argillans) on ped faces, particularly in the lower part, but can be massive if it is coarse loamy in texture. It is reddish brown (5YR or redder hue) in materials derived directly or indirectly from red beds and olive or even greenish coloured in variants derived from rocks such as Upper Greensand malmstone (Harwell and Selborne series) and associated glauconitic beds (Ardington series), which weather without liberating appreciable amounts of free iron (Loveland and Findlay, 1982). Eluvial and illuvial horizons are most clearly expressed on calcareous substrata, particularly under semi-natural vegetation. In other situations they are often less well defined and intergrades to orthic or sandy brown soils occur in which the B horizons fail to meet either the particle-size or the micromorphological criteria for recognition as argillic. Where the land has been in agricultural use for long periods, there is often much mixing by worms. Horizon boundaries are consequently merging, pre-existing argillans are disrupted, and agricutans (Jongerius, 1970) composed of clay, silt and organic matter which have been washed down channels and cracks from the ploughed layer are commonly more conspicuous. Data on four representative profiles, one under deciduous woodland and three in agricultural land, are reproduced on pp. 213–17.

Profile 65 was described and sampled in a Chiltern beechwood with no record of previous cultivation. The C horizon is a head deposit composed of loess and fragmentary chalk, together with shattered flints and clay derived from Clay-with-flints which caps the highest ground nearby. Particle-size and mineralogical data reported by Avery *et al.* (1959) indicate that the overlying horizons result from decalcification of similar material, as there is hardly any vertical change in the proportionate amounts and composition of the non-calcareous sand and silt fractions. Percentage base saturation is lowest in the Eb horizon and increases in the Ah, presumably as a result of biological cycling. Thin sections of the strongly expressed Bt horizon (Bullock, 1974) showed up to 8 per cent illuvial clay, partly as argillans lining voids and partly as elongated strands interpreted as infilled channels.

Profile 66 is in long cultivated land on brickearth in the West Sussex coastal plain, where deep silty soils of this kind support a wide range of agricultural and horticultural crops (Figure 5.10). The parent deposit consists of wind-borne loess which has been reworked by solifluxion or fluvial action, as evidenced by the presence of occasional small flints, fragmentary chalk in the C horizon, and the relatively large sand contents of horizons below 137 cm. The proportion of loess-derived material is much larger than in profile 65, however, and this is reflected in the yellower hue and smaller clay content of the thick argillic B, which is barely differentiated by colour from the overlying eluvial horizon.

Profile 67 is located in the Wye valley west of Hereford, where the underlying Devonian (Old Red Sandstone) rocks are mantled by Devensian glacial drift containing Lower Palaeozoic siltstone and sandstone fragments in a reddish loamy matrix which is often slightly calcareous at depth. Although horizons are only weakly differentiated by colour and texture in this soil, thin sections confirmed that argillans are common in both the Bt and the underlying BCt. The small amount of calcium carbonate in the upper horizons probably result from liming.

Profile 68 is on a highly calcareous gravelly glaciofluvial deposit of Devensian (Midlandian) age in County Carlow. Similar materials underlie numerous widely scattered areas in south-central Ireland (Association 30 in Gardiner and Radford, 1980). Some have a hummocky surface relief, with shallower and more droughty soils on hummocks and poorly drained calcaric gley soils in depressions, but well drained luvic brown (grey-brown podzolic) soils predominate. The stratified parent material at the gently sloping site described contains numerous rounded Carboniferous limestone fragments along with

Figure 5.10: The West Sussex Coastal Plain near Angmering (British Geological Survey: NERC copyright).

smaller amounts of sandstone and granite. It is decalcified to depths ranging from 70 to 97 cm and the well defined Bt horizon has correspondingly undulating lower and upper boundaries. Under the relatively dry climatic conditions prevailing in this area, such soils are largely devoted to tillage crops, including malting barley and sugar beet.

Profile 65 (Avery et al., 1959; Avery, 1964; Bullock, 1974)

Location	:	Monks Risborough, Buckinghamshire (SP 833040)
Climatic regime:		subhumid temperate
Altitude	:	198 m Slope: 2° straight
Land use and vegetation	:	broadleaf woodland; beech high forest with seedling ash in spaces left by felling; field layer, sparse in dense shade, includes bramble (*Rubus fruticosus*) and scattered herbs, chiefly woodruff (*Galium odoratum*), enchanters nightshade (*Circaea lutetiana*) and wood sorrel (*Oxalis acetosella*)

Horizons (cm)

L/F   Litter, mainly beech leaves, up to 3 cm thick; partially decomposed F layer thin and discontinuous.

0–8 Ah   Dark grey to greyish brown (10YR 4/1–2), moderately stony, humose silt loam (USDA silt loam), lighter in colour when dry; many, mainly medium, angular and subangular flint fragments; moderate granular with numerous cavities; friable; soft when dry; many woody roots; small earthworms present; abrupt irregular boundary.

8–28 Eb   Brown (7.5YR 4–5/4; 10YR 6/4 dry) moderately flinty silt loam (USDA silt loam); flint fragments as above; few channels infilled with darker soil; weak granular; friable; soft when dry; many woody roots; clear wavy boundary.

28–43 EB   Brown to dark brown (7.5YR 4/4) moderately flinty silty clay loam (USDA silt loam); flint fragments as above; weak medium subangular blocky with clay coats on some ped faces; friable; common woody roots; infilled channels as above; clear wavy boundary.

Analytical data for Profile 65

Horizon	Ah	Eb	EB	Bt	Bt	Cu
Depth (cm)	0–8	8–28	28–43	43–69	69–84	84–100
sand 200 μm–2 mm %	5	5	5	3	3	4[a]
60–200 μm %	9	8	7	5	4	5[a]
silt 2–60 μm %	70	75	67	48	47	53[a]
clay <2 μm %	16	13	21	44	46	38[a]
organic carbon %	5.3	1.6				
nitrogen %	0.35	0.18				
C/N ratio	15	8.9				
CaCO$_3$ equiv. %	0	0	0	0	0	33
dithionite ext. Fe %	1.2	1.1	1.4	2.3	2.3	
TEB (me/100 g)	6.6	1.7	4.5	24	26	
CEC (me/100 g)	24	11.1	12.9	33	35	
% base saturation	27	15	35	71	76	
pH (H$_2$O)	4.6	4.5	5.0	5.3	5.8	8.1
(0.01M CaCl$_2$)	4.0	3.8	4.2	4.7	5.4	7.5

[a] In decalcified mineral fraction <2 mm

43–84 Bt   Reddish brown (5YR 4/4) moderately flinty silty clay (USDA silty clay), flint fragments as above, becoming less numerous with depth; moderate medium to coarse subangular blocky, becoming more angular with increasing depth, with slightly darker coloured clay coats on ped faces; firm; few roots and infilled channels; common fine manganiferous nodules and soft concentrations; horizon 30–43 cm thick, with sharp wavy lower boundary.

84–100+ Cu   Brown to reddish yellow (7.5YR 5/4–6/6) moderately stony very calcareous silty clay loam (USDA silty clay loam); many subrounded chalk and common angular flint fragments; patches of soft disintegrated chalk; massive; firm; few roots.

Analytical date see above.

Classification
  England and Wales (1984): typical argillic brown earth; coarse silty over clayey, drift with siliceous stones (unnamed series, formerly Charity).
  USDA (1975): Typic Hapludalf: fine, mixed, mesic.
  FAO (1974): Chromic Luvisol; medium textured.

Profile 66 (Hodgson, 1967)

Location : Barnham, Sussex (SU 960037)
Climatic regime: subhumid temperate
Altitude : 7 m   Slope: <1°
Land use : ley grass
Horizons (cm)

0–20 Ap   Dark greyish brown (10YR 4/2) stoneless silt loam (USDA silt loam) with few fine and medium yellowish brown (10YR 5/6) mottles; weak fine and medium subangular blocky with some fine granular; friable; abundant (in upper 7.5–10 cm) to many fine roots; abrupt smooth boundary.

20–38 Eb1   Dark yellowish brown to yellowish brown (10YR 4–5/4) stoneless silt loam (USDA silt loam) with few very faint fine mottles; structure as above; friable; common large vertical earthworm channels and dark greyish brown ped faces and channel linings; many fine roots; gradual boundary.

38–58 Eb2   Dark yellowish brown (10YR 4/4) stoneless silty clay loam (USDA silt loam); weak fine platy, friable; earthworm channels as above; common fine roots; gradual boundary.

58–71 Bt1	Dark yellowish brown (10YR 4/4) stoneless silty clay loam (USDA silt loam); moderate fine and medium prismatic; dark brown (10YR 4/3) clay coats on some ped faces; firm; hard when dry; earthworm channels and roots as above; gradual boundary.
71–132 Bt2	Dark yellowish brown (10YR 4/4) stoneless silty clay loam (USDA silt loam to silty clay loam); moderate fine to medium prismatic with dark brown clay coats on ped faces; friable to firm; few very faint fine mottles and fine manganiferous nodules below 95 cm in places; earthworm channels as above and few fine roots concentrated in them; gradual boundary.
132–157 2Bt3	Yellowish brown (10YR 5/4) stoneless clay loam (USDA silty clay loam to clay loam); moderate medium and coarse prismatic with very faint fine mottles and dark brown clay coats on ped faces and lining many pores; friable; roots rare; gradual boundary.
157–246 2BCt(g)	Yellowish brown (10YR 5/4) stoneless clay loam (USDA clay loam) with many fine pale brown (10YR 6/3) and strong brown (7.5YR 5/8) mottles; moderate medium and coarse prismatic with clay coats as above; friable; abrupt boundary.
246–274 2BC(g)	Yellowish brown (10YR 5/4), pale brown (10YR 6/3) and strong brown (7.5YR 5/6) mottled, very calcareous clay loam (USDA loam) containing large nodular flints coated with brown clay, many rounded chalk fragments and patches of finely divided chalk.

Analytical data *see* below.

Classification
England and Wales (1984): typical argillic brown earth; silty stoneless drift (Hamble series).
USDA (1975): Typic Hapludalf; fine silty, mixed, mesic.
FAO (1974): Orthic Luvisol; medium textured.

Profile 67 (Ragg et al., 1984)

Location   : Webton, Hereford and Worcester (SO 410361)
Climatic regime: subhumid temperate
Altitude   : 99 m   Slope: 1°
Land use   : arable (fallow)
Horizons (cm)

| 0–28 Ap | Reddish brown (5YR 4/3), slightly stony, slightly calcareous sandy silt loam (USDA silt loam); common medium rounded micaceous siltstone fragments; moderate medium subangular blocky; friable to firm; common fine roots; abrupt smooth boundary. |

Analytical data for Profile 66

Horizon	Ap	Eb1	Eb2	Bt1	Bt2	2Bt3	2BCt(g)	2BC(g)
Depth (cm)	0–15	23–36	43–53	61–69	76–86	137–150	163–178	259–274
sand 200 µm–2 mm %	2	2	1	<1	<1	4	3	5
60–200 µm %	10	10	6	4	5	19	25	20
silt 2–60 µm %	71	71	74	71	71	49	48	53
clay <2 µm %	17	17	19	25	24	28	24	22
organic carbon %	2.1	0.9						
$CaCO_3$ equiv. %	0	0	0	0	0	0	0	24
dithionite ext. Fe %	1.2	1.2	1.3	1.7	1.8	1.7	1.4	0.8
TEB (me/100 g)	12.3	8.6	7.8	12.0	14.6	14.8	13.4	
CEC (me/100 g)	17.4	13.0	11.9	14.8	16.8	16.8	15.1	
% base saturation	71	66	66	81	87	88	89	
pH (H_2O)	6.0	6.0	6.1	6.1	6.6	6.7	6.7	7.7
(0.01M $CaCl_2$)	5.2	5.2	5.2	5.3	5.8	6.0	6.1	7.0

Analytical data for Profile 67

Horizon	Ap	Eb	Bt	BCt
Depth (cm)	0–28	28–51	51–77	77–110
sand 200 μm–2 mm %	12	13	10	18
60–200 μm %	15	15	17	15
silt 2–60 μm %	60	59	55	52
clay <2 μm %	13	13	18	15
<0.2 μm %	5	5	7	7
fine clay/total clay	0.38	0.38	0.39	0.47
organic carbon %	0.9	0.9		
$CaCO_3$ equiv. %	2	2	<1	<1
pH (H_2O)	7.7	8.0	7.8	7.9
(0.01M $CaCl_2$)	7.1	7.3	7.1	7.1
bulk density (g cm^{-3})	1.19	1.49	1.56	1.54
packing density (g cm^{-3})	1.36	1.60	1.72	1.67
retained water capacity (% vol.)	26	31	30	31
available water capacity (% vol.)	14	15	13	13
air capacity (% vol.)	29	12	10	9

28–31 Eb — Reddish brown (5YR 4/3), slightly stony, slightly calcareous sandy silt loam (USDA silt loam); stones as above; moderate fine subangular blocky; firm; common fine roots; abrupt smooth boundary.

51–77 Bt — Reddish brown (2.5YR 4/4), slightly stony, very slighty calcareous clay loam (USDA silt loam to loam); stones as above; moderate medium prismatic; firm; few fine roots; common clay coats; clear smooth boundary.

77–110 BCt — Dark reddish brown (2.5YR 3/4), moderately stony, very slighty calcareous sandy silt loam (USDA loam to silt loam); stones as above; weak medium angular blocky; firm; common clay coats.

Analytical data *see* above

Classification

England and Wales (1984): typical argillic brown earth; reddish-coarse loamy drift with siliceous stones (Escrick series).

USDA (1975): Typic Hapludalf; coarse-loamy, mixed, mesic.

FAO (1974): Chromic Luvisol; medium textured.

Profile 68 (Conry and Ryan, 1967; Gardiner and Radford, 1980)

Location : Agricultural Institute, Oakpark, County Carlow

Climatic regime: subhumid temperate

Altitude : 60 m   Slope: 2–3°

Land use and vegetation : reseeded pasture, mainly of rough stalked meadow grass (*Poa trivialis*) and white clover (*Trifolium repens*), with about 10 per cent perennial ryegrass (*Lolium perenne*)

Horizons (cm)

0–9 Ap1 — Very dark greyish brown (10YR 3/2) gravelly sandy loam (USDA sandy loam); moderate fine and very fine granular; very friable; uncoated quartz grains; abundant roots; clear smooth boundary.

9–36 Ap2 — Dark brown (10YR 3/3) gravelly sandy loam (USDA sandy loam); moderate fine and medium granular; friable; uncoated quartz grains; plentiful roots; clear smooth boundary.

36–57/70 — Brown to dark yellowish brown (10YR 4/3–4) sandy loam (USDA

Analytical data for Profile 68

Horizon	Ap1	Ap2	Eb	Bt	BC	2Ck
Depth (cm)	0–9	9–36	36–57	70–87	87–97	97–110
sand 200 μm–2 mm %	34	34	46	38	41	
50–200 μm %	23	24	20	15	31	
silt 2–50 μm %	28	27	24	23	13	
clay <2 μm %	15	15	10	24	15	
organic carbon %	4.8	1.5	0.4	0.6	0.1	<0.1
citrate-dithionite ext. Fe %	1.5	1.1	0.8	1.4	0.6	0.3
TEB (me/100 g)	21.4	15.7		13.5		
CEC (me/100 g)	24.9	16.0	7.3	13.8	8.0	1.9
% base saturation	86	98	sat.	98	sat.	sat.
pH (H$_2$O)	6.9	7.5	7.4	7.5	7.4	8.5

Eb		sandy loam) with little gravel; weak granular; very friable; many worm channels lined with A horizon material; plentiful roots; abrupt wavy boundary.
57/70–67/87	Bt	Dark greyish brown to brown (10YR 4/2–3) gritty sandy clay loam to clay loam (USDA sandy clay loam); moderate medium subangular blocky with clay coats on ped faces; friable to firm; plentiful roots; clear wavy boundary.
67/87–70–97	BC	Brown to dark brown (10YR 4/3) slightly calcareous gritty sandy loam (USDA sandy loam); weak granular; very friable; sparse roots; abrupt irregular (tonguing) boundary.
70/97+	2Ck	Grey (5Y 5/1) very calcareous gravelly coarse sand; single grain; loose; no roots; secondary calcium carbonate in bands.

Analytical data *see* above.

Classification
  Ireland (1980): grey-brown podzolic (Athy complex, moderately deep component).
  England and Wales (1984): typical argillic brown earth; coarse loamy material over calcareous gravel (Rougemont series).
  USDA (1975): Mollic (or Glossic) Hapludalf; fine-loamy over sandy or sandy-skeletal; mixed, mesic.
  FAO (1974): Orthic (or Calcic) Luvisol; medium textured.

5.6.3 Gleyic Luvic Brown Soils

These are luvic brown soils with gleyic features in the lower part of the profile that can be attributed to the presence of deep seated ground water for all or part of the year. Medium to coarse textured deposits such as brickearth and river drift are the main parent materials and the argillic B horizons are either moderately permeable or overlie permeable layers. The ground-water body may be widespread and perennial, as in certain low lying coastal localities and river valleys in chalkland areas, or it may be 'perched' above an impervious substratum and disappear in dry periods. Some variants have a relatively coarse textured horizon below the Bt with prominent iron concentrations (BCgf or Cgf) marking current or former water table levels. Others, including soils of the Hook series (Hodgson, 1967) in thick brickearth, resemble those of the stagnogleyic subgroup in morphology but are distinguishable by the absence of a slowly permeable horizon (Chapter 3, p. 110) within 80 cm depth or by the presence of free ground water within or below the solum for some part of the year. Data on two representative profiles in agricultural land are reproduced below.

Profile 69 is in a coarse loamy fluvial

deposit of Devensian age in the northern part of the Vale of York. The strongly gleyed argillic B is weakly expressed texturally but micromorphological examination showed that it contains common to many argillans, particularly in the lower part. Judging from the packing density and air capacity data, this horizon is at least slightly porous throughout, and the presence of ferruginous 'pipes' around old root channels in the underlying grey Cg horizon affords clear evidence of current or former saturation with ground water (Chapter 7, p. 291).

Profile 70 was described and sampled by R.G. Sturdy in the valley of the river Stour south of Sudbury, Suffolk. It is in loamy head which occupies a broad shallow depression in the valley side and overlies impervious London Clay at 1.3 m below the surface. The upper metre of head in which A, E and B horizons have formed has a substantial loess component but the basal layer is sandier and acts as an aquifer conveying water laterally above the clay substratum. Although all the subsurface horizons are distinctly to prominently mottled, indicating that they were periodically wet before artificial drainage was provided, those above 104 cm depth are predominantly brown and only those below 88 cm qualify as strongly gleyed (suffix g) according to the prescribed criteria. None contained free water when the profile was sampled in September 1978.

Profile 69 (Allison and Hartnup, 1981)

Location   : Morton upon Swale, North Yorkshire (SE 330910)
Climatic regime: subhumid temperate
Altitude   : 27 m   Slope: 1°
Land use   : ley grass
Horizons (cm)

0–24 Ap	Dark greyish brown (10YR 4/2) medium sandy loam (USDA fine sandy loam) containing rare small limestone and sandstone fragments; weak coarse angular blocky; friable; many fine and few medium fleshy roots; earthworms present; abrupt smooth boundary.
24–44 Eb	Dark yellowish brown (10YR 4/4; 10YR 6/3 dry) stoneless fine to medium sandy loam (USDA fine sandy loam); moderate fine to coarse subangular blocky; friable; common roots; common infilled worm channels; abrupt wavy boundary.
44–55 Btg1	Light brownish grey (2.5Y 6/2) stoneless sandy loam to clay loam (USDA

Analytical data for Profile 69

Horizon	Ap	Eb	Btg1	Btg2	2Cg
Depth (cm)	0–24	24–44	44–55	55–96	96–120
sand 600 µm–2 mm %	2	2	1	1	1
200–600 µm %	17	15	16	16	11
60–200 µm %	35	34	33	30	45
silt 2–60 µm %	30	34	32	34	26
clay <2 µm %	16	15	18	19	17
organic carbon %	2.3	1.8	0.6	0.4	0.5
pH (H$_2$O)	6.7	6.8	5.4	5.4	5.4
(0.01M CaCl$_2$)	6.2	6.2	4.7	4.6	4.6
bulk density (g cm^{-3})	1.39	1.26	1.42	1.59	
packing density (g cm^{-3})	1.53	1.39	1.58	1.76	
retained water capacity (% vol.)	35	34	33	30	
available water capacity (% vol.)	19	16	15	13	
air capacity (% vol.)	13	16	13	10	

55–96 Btg2	fine sandy loam) with many very fine yellowish brown (10YR 5/6) mottles; moderate coarse angular blocky; firm; common fine roots mainly in worm channels lined with topsoil material; common rounded ferri-manganiferous nodules; clear wavy boundary. Light grey to grey (10YR 6/1) very slightly stony clay loam (USDA fine sandy loam to loam) with common fine strong brown (7.5YR 5/8) mottles; few small sandstone fragments; moderate coarse to medium prismatic with grey (10YR 5/1) faces; firm; few fine roots; earthworm channels fewer than above; abrupt smooth boundary.
96–120 2Cg	Grey (5Y 5/1) stoneless fine sandy loam (USDA fine sandy loam) with very many coarse yellowish brown (10YR 5/8) mottles; massive, but slightly to moderately porous; firm; few fine roots; few irregular soft ferruginous concentrations and 'iron pipes' around old root channels.

Analytical date *see* p. 218.

Classification
  England and Wales (1984): gleyic argillic brown earth; coarse loamy stoneless drift (unnamed series, variant of Wighill).
  USDA (1975): Aquic (or Aquultic) Hapludalf; fine-loamy, mixed, mesic.
  FAO (1974): Gleyic Luvisol; medium textured.

Profile 70

Location	: Great Henny, Essex (TL 876373)
Climatic regime:	subhumid temperate
Altitude	: 46 m Slope: 2°
Land use	: arable (cereals)

Horizons (cm)

0–31 Ap	Dark greyish brown (10YR 4/2 broken and rubbed) very slightly stony sandy silt loam (USDA fine sandy loam to loam) with common very fine brown (10YR 5/3) mottles (silt concentrations); few medium subangular flint fragments; moderate medium subangular blocky; friable; common fine roots; earthworms present; sharp smooth boundary.
31–58 Eb(g)	Yellowish brown (10YR 5/4; 10YR 4/3 rubbed, 10YR 6/4 dry) sandy silt loam (USDA loam to silt loam) with common fine pale brown (10YR 6/3) mottles; rare small flint fragments; moderate coarse subangular blocky with yellowish brown ped faces; friable; common fine roots and earthworm channels; many irregular ferri-manganiferous nodules; clear wavy boundary.
58–88 Bt(g)	Brown (7.5YR 5/4, broken and rubbed) silty clay loam (USDA silt loam) with many fine greyish brown (10YR 5/2) and strong brown (7.5YR 5/6) mottles; rare small flint fragments; moderate medium prismatic with brown (7.5YR 5/4) faces and many distinct continuous clay coats; firm; few fine roots, mainly in earthworm channels; common soft irregular ferri-manganiferous concentrations; gradual smooth boundary.
88–104 Btg	Brown (7.5YR 5/4, broken and rubbed) sandy silt loam to clay loam (USDA loam to silt loam) with many medium pale brown (10YR 6/3) and strong brown (7.5YR 5/6) mottles; rare small flint fragments; weak medium prismatic breaking to medium subangular blocky, with brown (10YR 5/3) faces; firm; few fine roots and earthworm channels; many distinct continuous clay coats and ferri-manganiferous concentrations as above; clear wavy boundary.
104–132 2BCtg	Greyish brown (10YR 5/2) slightly stony medium sandy loam (USDA fine sandy loam) with many coarse strong brown (7.5YR 5/6) mottles; common flint fragments, mainly medium angular; weak very coarse blocky with brown (7.5YR 5/2) faces; firm; few roots; common clay coats surrounding and bridging sand grains; common irregular ferri-manganiferous nodules; sharp wavy boundary.
132–140 3BCg	Brown (7.5YR 5/2) stoneless silty clay to clay (USDA silty clay) with

Analytical data for Profile 70

Horizon	Ap	Eb(g)	Bt(g)	Btg	2BCtg	3BCg
Depth (cm)	0–21	31–58	58–88	88–104	104–132	132–140
sand 600 μm–2 mm %	7	6	2	2	8	<1
200–600 μm %	21	15	7	11	26	<1
60–200 μm %	18	12	9	17	34	13
silt 2–60 μm %	43	53	61	52	19	45
clay <2 μm %	11	14	21	18	13	42
<0.2 μm %	5	7	12	11	10	
fine clay/total clay	0.45	0.50	0.57	0.61	0.77	
organic carbon %	1.2	0.4	0.3	0.2		
dithionite ext. Fe %	1.1	1.3	1.5	1.5	1.5	1.2
pH (H$_2$O)	6.3	7.2	7.5	7.8	7.6	7.7
(0.01M CaCl$_2$)	5.5	6.3	6.6	6.8	6.8	6.7

common fine greenish grey (5GY 6/1) and strong brown mottles; weak very coarse angular blocky with brown (7.5YR 5/2) faces; firm; no roots.

Analytical date see above.

Classification
England and Wales (1984): gleyic argillic brown earth; coarse loamy drift with siliceous stones (Wix series).
USDA (1975): Aquic Hapludalf; fine-loamy, mixed, mesic.
FAO (1974): Orthic Luvisol; medium textured.

5.6.4 Stagnogleyic Luvic Brown Soils

These are intergrades between typical luvic brown soils and stagnoluvic gley soils (p. 364) in which drainage is impeded to some extent by a relatively impermeable Bt horizon or an immediately underlying slowly permeable layer. They are more extensive than those of the typical subgroup in England and Wales and also occur in the drier lowlands of Ireland and southern Scotland but have not been differentiated from related kinds of soil in either country. The substrata include medium to fine textured tills, head deposits, and weathering products of pre-Quaternary rocks such as mudstone and siltstone. Frequently, however, the eluvial horizons are developed in an initially coarser textured surface layer, implying that the increase in clay content with depth is partly inherited. There are normally signs of gleying in or above, the Bt, and prominent greyish and ochreous mottling appears in the most gleyed variants at around 50 cm below the surface. Others, particularly on originally reddish (haematitic) materials, have no gleyed horizon identifiable by the suffix g in or immediately beneath the solum, though water regime studies reported by Robson and Thomasson (1977) have shown that some part of the rooting zone is wet for substantial periods in most years. The first of the following two representative profiles from the English Midlands is of the latter type.

Profile 71 is under oak–ash–hazel woodland with a thin acid mull surface horizon. It is on soft, red, slightly calcareous Devonian siltstone which, together with interbedded fine grained sandstone, forms the soil over considerable areas in the Welsh Borderland. Base saturation is very low in the Eb horizon but increases to more than 60 per cent in the lower part of the well defined Bt. Thin sections from this horizon showed common argillans and intrapedal clay concentrations interpreted as disrupted argillans. The BC horizon, which also contains argillans, consists largely of partially weathered rock fragments (lithorelicts) showing few signs of pedological reorganization. Although this soil was originally classed as freely drained by Mackney and Burnham (1966), ped faces in the Bt horizon are slightly paler than the interiors and bore films of moisture when the profile was sampled, indicating that water stagnates periodically in the intraped fissures. The greenish veins and blotches within the peds are inherited from the parent material.

Profile 72, sampled under old grass in Northamptonshire, is developed in originally greyish chalky till (Chalky Boulder Clay) overlain by a thin loamy superficial deposit which probably incorporates sand of aeolian origin (Perrin et al., 1974). The mottled argillic B (Btg) horizon and the underlying BC are in the till, the uppermost part of which has been decalcified and further enriched in clay by illuviation, as evidenced by the common occurrence of argillans in thin sections of these horizons. Judging from the air capacity and packing density data, the Btg is the least permeable horizon in the profile.

Profile 71 (Mackney and Burnham, 1966; Ragg and Clayden, 1973; Avery and Bullock, 1977)

Location	:	Stottesdon, Shropshire (SO 663845)
Climatic regime:		humid temperate
Altitude	:	165 m Slope: 2°
Land use and vegetation	:	broadleaf woodland; oak (*Quercus robur*) and ash (*Fraxinus excelsior*) with hazel (*Corylus avellana*) coppice and field layer dominated by bramble (*Rubus fruticosus*); ridge and furrow microrelief indicates that the land was once cultivated.

Horizons (cm)

L/F — Very thin litter layer, mainly of oak leaves, over partly decomposed leaves and twigs, less than 1 cm thick; active mesofauna.

0–3 H/Ah — Dark brown (7.5YR 3/2) stoneless humose silty clay loam (USDA loam); upper 5 mm granular (insect droppings) following by very fine subangular blocky; friable; abundant fine to coarse roots; earthworms active; abrupt irregular boundary tonguing downwards along root and earthworm channels.

3–20 Eb — Reddish brown (5YR 4/4) stoneless silty clay loam (USDA silt loam) medium to coarse subangular blocky, breaking very easily to fine and very fine; friable; many fine to coarse roots; common root and earthworm channels; manganiferous nodules 1–5 mm; gradual boundary.

20–28 EB — Reddish brown (5YR 4/4) stoneless silty clay loam (USDA silty clay loam); coarse blocky breaking less easily to

Analytical data for Profile 71

Horizon	H/Ah	Eb	EB	Bt(g)	Bt(g)	Bt(g)	BCt
Depth (cm)	0–3	3–20	20–28	28–33	33–38	38–75	75–117
sand 200 μm–2 mm %	4	3	3	1	3	1	1
50–200 μm %	14	12	8	2	4	3	29
silt 2–50 μm %	59	61	60	58	55	58	41
clay <2 μm %	23	24	29	39	38	38	29
organic carbon %	6.3	2.4					
dithionite ext. Fe %	1.2	1.4	2.0	2.0	1.8	1.4	1.8
TEB (me/100 g)	4.2	1.0	2.7			9.3	11.3
CEC (me/100 g)	20.0	20.0	11.1			15.1	12.4
% base saturation	17	5	24			62	91
pH (H$_2$O)	4.2	4.4	4.8	4.8	5.1	5.2	5.9
(0.01M CaCl$_2$)	3.7	3.7	4.0	4.0	4.3	4.5	5.6

| 28–75 Bt(g) | smaller peds; friable; many roots; earthworm channels and casts; few soft manganiferous concentrations; gradual boundary.

Dark reddish brown (2.5–5YR 3/4) stoneless silty clay (USDA silty clay loam) finely veined and blotched with greenish grey; prismatic breaking to coarse blocky with reddish brown (5YR 5/4) faces; firm to very firm; ped faces wet, moist internally; common roots; few earthworm channels; all ped faces have clay coats, thickest on vertical faces; common manganiferous coats but no hard nodules. |
| 75–117 BCt | Reddish brown micaceous clay loam (USDA clay loam), calcareous below 117 cm. |

Analytical data see p. 221.

Classification
England and Wales (1984): stagnogleyic argillic brown earth; reddish-fine silty over clayey material passing to siltstone (unnamed series, formerly Bromyard).
USDA (1975): Typic Hapludalf; fine, illitic, mesic.
FAO (1974): Chromic Luvisol; medium textured.

Profile 72 (Reeve, 1978)

Location   : Holdenby, Northamptonshire (SP 689679)
Climatic regime: subhumid temperate
Altitude   : 126 m   Slope: 3°
Land use   : long-term grass
Horizons (cm)

0–9 Ah1	Dark brown (10YR 3/3 broken and rubbed; 2.5Y 5/2 dry) stoneless humose clay loam (USDA clay loam) with common very fine reddish brown (5YR 4/4) mottles associated with root channels; strong fine granular; friable; abundant fine roots; abrupt smooth boundary.
9–28 Ah2	Brown to dark brown (10YR 4/3; 10YR 4/4 rubbed; 10YR 5/2 dry) very slightly stony clay loam (USDA loam); few small rounded quartzite and angular flint fragments; strong fine subangular blocky; friable; abundant fine roots; gradual boundary.
28–54 Eb	Dark yellowish brown (10YR 4/4 broken and rubbed; 10YR 5/4 dry) very slightly stony clay loam (USDA loam); stones as above, with a few large quartzite pebbles; strong fine angular blocky; friable; many roots; few earthworm channels with dark linings; gradual boundary.
54–85 2Bt(g)	Dark yellowish brown to yellowish brown (10YR 4–5/4) very slightly stony clay (USDA clay) with common fine light brownish grey (10YR 6/2)

Analytical data for Profile 72

Horizon	Ah1	Ah2	Eb	2Bt(g)	2BC(g)
Depth (cm)	0–5	10–25	30–45	60–75	100–110
sand 200 µm–2 mm %	11	13	13	8	11
60–200 µm %	28	29	30	16	17
silt 2–60 µm %	33	33	33	33	42
clay <2 µm %	28	25	24	43	30
organic carbon %	6.4	3.2	1.3		
$CaCO_3$ equiv. %	0	0	0	0	15
dithionite ext. Fe %	3.7	4.1	4.6	4.0	4.1
pH (H_2O)	5.4	5.4	6.7	7.8	8.5
(0.01M $CaCl_2$)	4.6	4.5	5.7	7.1	7.8
bulk density (g cm^{-3})		1.08	1.43	1.47	1.51
packing density (g cm^{-3})		1.33	1.64	1.85	1.78
retained water capacity (% vol.)		45	32	39	33
available water capacity (% vol.)		24	15	10	13
air capacity (% vol.)		14	14	5	10

	and yellowish brown (10YR 5/6) mottles; stones include small to large angular flint, subangular ironstone and rounded quartzite fragments; moderate medium prismatic with brown (10YR 5/3) faces; very firm; common fine roots; few earthworm channels; abrupt smooth boundary.
85–110+ 2BC(g)	Yellowish brown (10YR 5/6) and greyish brown (10YR 5/2) distinctly mottled, slightly to moderately stony, very calcareous clay loam (USDA clay loam); stones as above, together with common small to medium sub-rounded chalk fragments; weak coarse angular blocky with yellowish brown (10YR 5/4) faces; very firm; few roots.

Analytical data *see* p. 222

Classification
   England and Wales (1984): stagnogleyic argillic brown earth; fine loamy over clayey, chalky drift (Ashley series).
   USDA (1975): Aquic Hapludalf; fine, mixed, mesic.
   FAO (1974): Orthic Luvisol; medium textured.

5.6.5 Peloluvic Brown Soils

These are the clayey luvic brown soils set apart as argillic pelosols in the England and Wales classification. They are common on moderately calcareous mudstones such as Keuper Marl (Mercia Mudstone) where there is little surface drift and also occur on Chalky Boulder Clay and other fine textured Quaternary deposits, often in close association with less leached pelocalcaric brown soils (p. 191). Although they have only been positively identified in England and Wales, similar soils are probably represented in small, relatively well drained areas on clayey glacial or glaciolacustrine drift in eastern Scotland (Peterhead, Tipperty and Whitsome associations) and central Ireland (e.g. Rathcannon series in Finch and Ryan, 1966). They are distinguished from the pelo-calcaric brown soils in similar materials by the presence of a decalcified blocky to prismatic Bt horizon with significantly more clay than horizons above and below, and from the stagnogleyic luvic brown soils by their predominantly clayey texture and consequent propensity to seasonal swelling and shrinking. The topsoil may be loamy but normally contains more than 30 per cent clay and a clearly expressed Eb horizon is generally absent, particularly in cultivated land.

The following representative profile is of the Worcester series on red Triassic mudstone in the Needwood Forest district of Staffordshire. It has a well marked Bt horizon with macro- and micro-morphological evidence of clay translocation in and below it, but the changes in clay content with depth cannot be attributed entirely to this cause as the smaller amount in the BC is likely to reflect incomplete dispersion of partly weathered mudstone fragments (Dumbleton and West, 1966) and the surface horizon contains appreciable amounts of sand and occasional stones which are clearly of extraneous origin. The clay fractions are dominantly micaceous and X-ray diffraction data on a sample from the Bt2 horizon showed that small amounts of chlorite and interstratified mica-smectite are also present. As in profile 71 on more silty red beds, gleying is very weakly expressed, but water retention and hydrologic studies of Worcester soils reported by Thomasson and Robson (1971), Hall *et al.* (1977), Robson and Thomasson (1977) and McGowan (1984) have confirmed that the B horizons transmit water very slowly during the winter half-year. As they are also denser than most clayey subsurface horizons, their available water capacities are relatively small. Crops, particularly grass, consequently suffer from drought on these soils, especially where hard slightly weathered mudstone is encountered at shallower depths than in the profile described. McGowan (1984) showed that this layer, which commonly has an angular blocky ('starchy') structure, often remains unsaturated in winter, even when the surface is waterlogged. Rewetting is apparently restricted because the dense Bt horizon has a small shrink-swell capacity in relation to clay content (Reeve et al., 1980) and the small volume of desiccation cracks formed during the summer soon disappears on reswelling. Where the relatively permeable substratum is within about 60 cm depth, deep subsoiling is effective in facilitating downward passage of surface water and improving the ability of the soil to supply water to crops during rainless periods.

Profile 73 (Jones, 1983)

Location	:	Tatenhill, Staffordshire (SK 187224)
Climatic regime:		subhumid temperate
Altitude	:	117 m Slope: 12°
Land use and vegetation	:	long-term grass with Yorkshire fog (*Holcus lanatus*), common bent-grass (*Agrostis capillaris*), perennial ryegrass (*Lolium perenne*) and white clover (*Trifolium repens*) abundant or frequent

Horizons (cm)

0–18   Ah   Dark brown (7.5YR 3/3 broken and rubbed; 5YR 4/3 dry), very slightly stony, very slightly calcareous silty clay (USDA silty clay loam); few medium quartzose pebbles and angular sandstone fragments; moderate medium and coarse subangular blocky; firm; many fine roots; earthworms active; abrupt wavy boundary.

18–36   Bt(g)1   Dark reddish brown (5YR 3/4; 5YR 4/4 rubbed; 5YR 5/3 dry), very slightly stony, very slightly calcareous silty clay (USDA silty clay); stones as above; also rare angular flint and black shale fragments; moderate medium and coarse angular blocky with reddish brown (5YR 4/3) faces; firm; common fine roots; few earthworm channels infilled with dark brown casts; common rounded black ferri-manganiferous nodules; abrupt irregular boundary.

36–64   Bt(g)2   Dark reddish brown (2.5–5YR 3/4; 2/5YR 4/3 dry), stoneless very slightly calcareous clay (USDA silty clay to clay) with common greenish grey (5GY 5/1) speckles; moderate medium

Analytical data for Profile 73

Horizon	Ah	Bt(g)1	Bt(g)2	BCt
Depth (cm)	0–18	18–36	36–64	64–100
sand 200 μm–2 mm %	4	2	<1	<1
60–200 μm %	7	5	4	8
silt 2–60 μm %	50	46	43	53
clay <2 μm %	39	47	53	39
<0.2 μm %	16	18	21	13
fine clay/total clay	0.41	0.38	0.40	0.33
organic carbon %	2.5	0.7		
$CaCO_3$ equiv. %	<1	<1	<1	14
pH (H_2O)	6.9	7.7	8.0	8.4
(0.01M $CaCl_2$)	6.5	7.1	7.2	7.8
clay <2 μm				
CEC (me/100 g)	41	33	29	24
K_2O %	5.6	5.6	6.4	6.4
bulk density (g cm^{-3})	1.32	1.50	1.53	
packing density (g cm^{-3})	1.67	1.92	2.00	
retained water capacity (% vol.)	44	35	37	
available water capacity (% vol.)	18	13	10	
air capacity (% vol.)	16	8	5	

64– 100+ BCt/Cr	and coarse prismatic with reddish brown (5YR 4/3) faces; very firm; few fine roots; discontinuous clay coats on prism faces; few soft manganiferous concentrations; clear irregular boundary. Dark reddish brown (2.5YR 3/4; 2.5YR 4/3 dry), stoneless, very calcareous silty clay (USDA silty clay loam); weak (adherent) angular blocky; very firm, becoming extremely firm and difficult to dig with spade below 100 cm; few roots; reddish brown (5YR 4/3) clay coats on faces of blocks.

Analytical data *see* p. 224.

Classification
  England and Wales (1984): argillic pelosol; reddish-clayey material passing to clay or soft mudstone.
  USDA (1975): Mollic Hapludalf; fine, illitic, mesic.
  FAO (1974): Chromic Luvisol; fine textured.

5.6.6 Chromoluvic Brown Soils

These are the freely draining luvic brown soils with strong brown to red chromic argillic B horizons (Chapter 3, p. 109). Together with those of the following subgroup, they are mainly but not exclusively confined to parts of England and Wales that lay outside the Devensian ice limits and are typically associated with dissected plateaux, terraces and pediments which presumably represent the more or less disturbed remnants of pre-Devensian ground surfaces. They are commonest over chalk and other limestones, the resistance to denudation of which has favoured preservation of such surfaces and associated weathered horizons. Others are in or over Wolstonian (Table 1.1, p. 14) or older fluvial or head deposits and in weathering products of non-calcareous rocks, including pre-Cambrian gneisses and schists in south Devon (Findlay *et al.*, 1984). None have been positively identified in Scotland or Ireland and they are probably rare in both countries, but small patches of red soil over limestone which may meet the prescribed requirements have been described in south Wexford by Gardiner and Ryan (1962) and in north-west Scotland by Lawson (1983).

Like the typical luvic brown soils with 'ordinary argillic B horizons', soils of this subgroup are mainly loamy or loamy over clayey. A few in erosional sites, typified by the Winchester series (Clayden and Hollis, 1984) in thin Clay-with-flints over chalk, are clayey and some have sandy or sandy gravelly horizons above or below the diagnostic B. Data on two examples are reproduced on pp. 226–8.

Profile 74 is in ancient woodland on the interfluve between the Wylye and Nadder valleys west of Salisbury, Wiltshire, where the underlying Upper Chalk is capped by a thin layer of Clay-with-flints *sensu stricto* (Loveday, 1962). This distinctive horizon, which constitutes the chromic argillic B, is overlain by a flinty silty surface layer in which the acid mull Ah and Eb horizons are developed. Micromorphological and mineralogical studies of similar Clay-with-flint layers indicate that they consist partly of chalk solution residues, chiefly flint, and partly of illuvial clay derived from remains of Tertiary beds or other formerly overlying materials (Avery *et al.*, 1959; Loveday, 1962; Hodgson *et al.*, 1967; Bullock, 1974). As in associated rendzinas (e.g. profile 19, p. 141) and typical luvic brown soils (e.g. profile 65, p. 213), the silty layer is considered on the basis of particle-size and mineralogical analyses (Cope, 1976) to have originated from a cover of loess. This was apparently mixed by cryoturbation with frost-shattered flint and fine material from the pre-existing Clay-with-flint soil, and subsequent clay eluviation served to accentuate the contrast in texture between the two layers. Clay separates from the 2Bt horizon in a nearby cultivated soil were found to be smectitic (Avery and Bullock, 1977), whereas samples from the A and Eb horizons contained very little smectite and more mica, together with substantial amounts of more or less interstratified vermiculite and chlorite. Essentially similar soils derived from loess over 'decalcification clay' occur on Carboniferous limestone in Somerset (Findlay, 1965) and Derbyshire (Piggott, 1962; Ragg et al., 1984).

Profile 75 from Essex is in deeply weathered and cryoturbated loamy and gravelly material, commonly known as 'hoggin', which overlies pre-Anglian fluvial sands and gravels (Rose and Allen, 1977). The particle-size analyses show a marked increase in the sand/silt ratio at around 56 cm depth, indicating that the upper horizons are again developed in a lithologically distinct surface layer consisting largely of loess. In this soil, however, the upper boundary of the composite Bt horizon is within the cover deposit and only the lower part has chromic characteristics. A thin section of the brown Bt1 horizon showed common argillans lining voids, confirming that substantial clay translocation has taken place since the loess was deposited. Similar argillans were identified in sections of the more brightly coloured and ferruginous Bt2 (with rare stones) and Bt3 (very stony) subhorizons, but in these nearly all the clay size material (plasma) shows preferred orientation and appears to consist largely of fragments of earlier formed argillans which have been disrupted and incorporated in the matrix (Bullock, 1974, 1985). In coarser textured soils of the St Albans series (Clayden and Hollis, 1984) on similar deposits, the subsurface horizons are predominantly sandy-gravelly and the argillic B horizon is of sandy loam or loamy sand texture, with yellowish red clay coats on sand grains and stones.

Profile 74 (Cope, 1976)

Location	: Barford St Martin, Wiltshire (SU 042336)
Climatic regime:	subhumid temperate
Altitude	: 160 m Slope: 1°
Land use and vegetation	: broadleaf (oak–beech) woodland with hazel coppice.

Horizons (cm)

L	Litter of hazel, beech and oak leaves and twigs, 1–2 cm thick.
0–5 Ah1	Dark brown (10YR 3/3) stoneless silty clay loam (USDA silt loam); strong fine subangular blocky and granular; friable; abundant fine to medium and few coarse woody roots; earthworms active; abrupt smooth boundary.
5–10 Ah2	Brown to dark brown (10YR 4/3) stoneless silty clay loam (USDA silt loam); moderate fine subangular blocky; friable; roots as above but few fine; abrupt smooth boundary.
10–35/40 Eb	Dark yellowish brown (10YR 4/4; 10YR 5/4 dry) moderately stony silty clay loam (USDA silt loam); many medium and large angular flint fragments; weak fine subangular blocky; friable; common medium and few large woody roots; abrupt wavy boundary.
35/40– 70/95 2Bt	Yellowish red (5YR 5/6) very stony clay (USDA clay); abundant large and very large subrounded (nodular) and subangular (broken) flints; moderate coarse subangular blocky (strong

Analytical data for Profile 74					
Horizon		Ah1	Ah2	Eb	2Bt
Depth (cm)		0–5	5–10	10–35	35–90
sand 200 μm–2 mm %		5	3	6	4
60–200 μm %		2	2	2	2
silt 2–60 μm %		70	71	68	12
clay <2 μm %		23	24	24	82
organic carbon %		4.0	3.1		
pH (H$_2$O)		6.2	6.1	5.4	6.1
(0.01M CaCl$_2$)		6.0	5.9	4.8	5.7

when dry); firm; few woody roots; few fine manganiferous nodules; sharp wavy boundary.

70/95–120 Cu Bedded, medium and large tabular chalk fragments with brownish yellow (10YR 6/6) coats and interstitial material and very few, very large unbroken flint nodules; few narrow fissures infilled with yellowish red clay; roots rare.

Analytical data *see* p. 226.

Classification
  England and Wales (1984): typical paleo-argillic brown earth; fine silty over clayey; drift with siliceous stones (Carstens series).
  USDA (1975): Typic Hapludalf; very-fine, montmorillonitic, mesic.
  FAO (1974): Chromic Luvisol; medium textured.

Profile 75 (Allen and Sturdy, 1980)

Location   : Terling, Essex (TL 779144)
Climatic regime: subhumid temperate
Altitude   : 40 m   Slope: 2°
Land use   : long-term grass
Horizons (cm)

0–12 Ah Brown to dark brown (10YR 4/3) very slightly stony sandy silt loam (USDA sandy loam to loam); few small to large flint fragments and rare quartzose pebbles; moderate medium and coarse subangular blocky; friable; many fine roots; earthworm channels and casts; abrupt smooth boundary.

12–33 Eb Yellowish brown (10YR 5/4) and dark yellowish brown (10YR 4/4) slightly stony medium sandy loam (USDA sandy loam); stones as above; moderate medium and fine subangular blocky; friable; common fine roots; earthworm channels and casts; faint patchy clay and manganiferous coats on stones; gradual smooth boundary.

33–56 Bt Brown to dark brown (7.5YR 4/4) moderately stony sandy clay loam (USDA sandy clay loam); stones as above; moderate fine subangular blocky; friable; few roots; earthworm channels as above; many distinct discontinuous clay coats and prominent patchy manganiferous coats on stones; clear wavy boundary.

56–78 2Bt Strong brown to yellowish red (7.5–5YR 5/6) sandy clay loam (USDA sandy clay loam) with rare stones in the upper part; moderate coarse subangular blocky with some dark brown (7.5YR 4/2) faces; firm; few roots following earthworm channels (1 per cent); clay and organic matter coats on ped faces; abrupt wavy boundary.

78–114 3Bt Strong brown to yellowish red (7.5–5YR 5/6) very stony sandy clay loam (USDA sandy clay loam); abundant small and medium rounded and subrounded flint pebbles; massive; very firm; roots rare; many entire prominent clay coats around stones and sand grains.

Analytical data for Profile 75

Horizon	Ah	Eb	Bt	2Bt	3Bt
Depth (cm)	0–12	12–33	33–56	56–78	78–114
sand 600 µm–2 mm %	11	14	12	4	17
200–600 µm %	26	27	25	25	42
60–200 µm %	11	11	15	33	8
silt 2–60 µm %	35	31	25	13	7
clay <2 µm %	17	17	23	25	26
organic carbon %	4.4				
dithionite ext. Fe %	2.5	2.4	4.9	5.6	6.1
pH (H$_2$O)	5.3	5.9	6.4	6.9	7.1
(0.01M CaCl$_2$)	4.8	5.3	5.8	6.4	6.6

Classification
> England and Wales (1984): typical paleo-argillic brown earth; fine loamy drift with siliceous stones (Sheldwich series, formerly Terling).
> USDA (1975): Typic Paleudalf (or Hapludalf); fine-loamy, oxidic, mesic.
> FAO (1974): Chromic Luvisol; medium textured.

5.6.7 Stagnogleyic Chromoluvic Brown Soils

These are genetically similar and often closely associated soils with slowly permeable subsurface horizons and mottles and/or ped-face colours attributable to gleying within 60 cm of the surface. They are extensive on Clay-with-flints and other deeply weathered plateau deposits in southern England and occur more sporadically further north on Anglian or Wolstonian tills (Table 1.1, p. 14) and pre-Ipswichian river drift. A loamy Eb horizon is usually but not always present and can be mottled, particularly in the lower part, which in places extends downwards as irregular projections or tongues (glossic features).

The deposits in which most of these soils have formed overlie pervious substrata such as chalk or gravel and lie well above the permanent water table, implying that gleying in the upper one to two metres penetrated by roots results exclusively from impeded drainage of surface water. In other situations, exemplified on Anglian or pre-Anglian fluvial deposits in parts of Essex (Allen and Sturdy, 1980; Sturdy and Allen, 1981), semi-hydromorphic brown soils with chromic argillic B horizons of loamy texture overlie coarser textured layers in which ground water is retained above impervious substrata at greater depths. These were set apart as gleyic paleo-argillic brown earths by Avery (1980) but are difficult to distinguish morphologically from stagnogleyic variants in texturally similar materials. The two representative profiles on which field and analytical data are given on pp. 229–32 are both in plateau deposits.

Profile 76 is under semi-natural beech–oak woodland on the Chiltern plateau about 2 km from profile 65 (p. 213). Like profile 74 above, it is developed in a silty surface layer over Clay-with-flints, but the latter deposit is more than three metres thick at this site and is evidently derived mainly from Tertiary (Reading Beds) clays and sands, the weathered and disturbed remains of which have been mixed with flints released by solution of the underlying Upper Chalk. Although all the horizons sampled are moderately to strongly acid, base saturation in the lower part of the Bt exceeds the limiting value of 35 per cent chosen to distinguish Alfisols from Ultisols in the US Taxonomy. Cultivated soils of the same series (Batcombe), which are widespread in the Chilterns and on similar plateau deposits south of the Thames, have generally received regular dressings of chalk and are often base saturated to at least 1.5 m depth. At Rothamsted, where selected profiles have been studied in detail by Avery and Bullock (1969), Weir *et al.* (1969) and Avery *et al.* (1972), the brightly coloured B horizon commonly starts directly below the Ap and eroded variants with clayey topsoils also occur.

Particle-size and mineralogical analyses and micromorphological studies of samples from profile 76 reported by Avery *et al.* (1959) indicate that the soil material to a depth of 74 cm is partly of loessial origin and, as in profile 75, the occurrence of identical void argillans both above and below this depth affords evidence that the downward increase in clay content results partly from clay translocation since the loess was deposited. In the thick chromic argillic B horizon below 41 cm, much of the clay-size material consists of plasma separations (sepic plasmic fabric) and concentrations (papules or embedded argillans) with varying degrees of preferred orientation which is masked in places by red ferruginous segregations. The clay fractions of this and similar soils contain expansible layer-silicates (vermiculite or smectite) accompanied by substantial but generally sub-

Plate I

1. Typical ranker (Typic Hapludoll) on weathered dolerite: Christow, Devon. 2. Typical rendzina (Typic Rendoll) on chalk: Whipsnade, Bedfordshire. 3. Brown rendzina (Typic Udorthent) on Permian Magnesian Limestone: Bramham, North Yorkshire. 4. Calcaric Sandy Regosol (Typic Udipsamment) in blown sand with buried A horizons: Skara Brae, Orkney. 5. Typical sandy brown soil (Typic Quartzipsamment) on red Permian sandstone: Exeter, Devon. 6. Gleyic-calcaric alluvial brown soil (Fluventic Eutrochrept) in silty estuarine alluvium (warp): Kilpin, Humberside. 7. Typical calcaric brown soil (Rendollic Eutrochrept) in flinty silty head over chalk: Nutley, Hampshire. 8. Pelocalcaric brown soil (Aquic Eutrochrept) in chalky till: Stony Stratford, Buckinghamshire.

Plate II

1. Typical orthic brown soil (Dystric Eutrochrept) in head derived from Carboniferous shale and sandstone: South Tawton, Devon. **2.** Gleyic orthic brown soil (Aquic Dystric Eutrochrept) in coarse loamy over sandy river drift: Apperley, Gloucestershire. **3.** Typical luvic brown soil (Typic Hapludalf) in loess: Pegwell Bay, Kent. **4.** Stagnogleyic chromoluvic brown soil (Aquic Paleudalf) in Plateau (Northern) Drift with superficial admixture of loess: Northleigh, Oxfordshire (Photo by P. Bullock). **5.** Typical podzolic brown soil (Typic Dystrochrept) in head derived from Silurian siltstone and shale: Llanddwyn, Powys. **6.** Humoferric podzol (Duric Haplorthod?) in glaciofluvial sand and gravel mainly derived from acid crystalline rocks: Glen Quoich, Aberdeenshire. **7.** Lowland humus podzol (Typic Haplohumod) in Eocene (Barton Beds) sand: Lyndhurst, Hampshire. **8.** Oro-arctic humus podzol (Typic Cryohumod or Cryorthod) in granitic drift: Beinn a'Bhuird, Aberdeenshire.

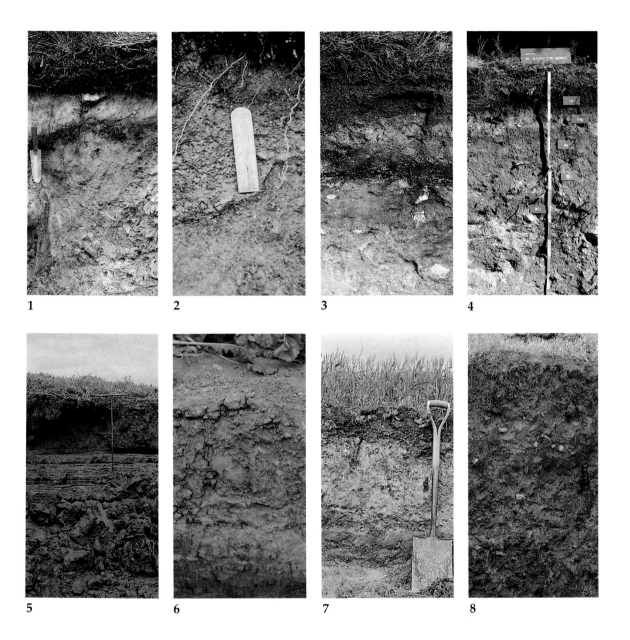

Plate III

1. Luvic gley-podzol (Ultic Fragiaquod) in head derived from Cretaceous siltstones and fine-grained sandstones: Ashdown Forest, East Sussex. **2.** Ironpan stagnopodzol (Histic Placaquept) in granitic drift: Cairn o'Mounth, Kincardineshire. **3.** Humus-ironpan stagnopodzol (Fragic Placohumod) in head derived from subjacent Carboniferous sandstone: Thrunton Forest, Northumberland. **4.** Ferric stagnopodzol (Histic Sideraquod) in granitic head: Postbridge, Devon. **5.** Saline alluvial gley soil (Typic or Sulfic Fluvaquent) in unreclaimed salt marsh: Scolt Head Island, Norfolk. **6.** Calcaric alluvial gley soil (Typic Fluvaquent) in coarse silty marine alluvium, reclaimed from salt marsh in 1948: Friskney, Lincolnshire. **7.** Ground-water calcaric gley soil (Typic Haplaquoll) in loamy chalky drift: Burwell, Cambridgeshire. **8.** Stagno-orthic gley soil (Aeric Fragiaquept) in loamy drift derived from Lower Palaeozoic mudstones and sandstones: Llangadog, Dyfed (Photo by P.S. Wright).

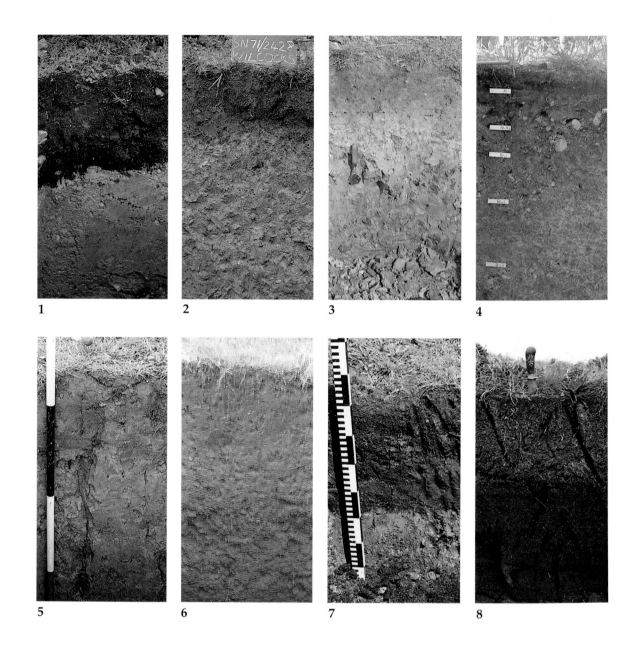

Plate IV

1. Drained humic orthic gley soil (Histic Humaquept) in loamy drift derived from acid and basic crystalline rocks: Tillycorthie, Aberdeenshire. 2. Humic stagno-orthic gley soil (Typic Humaquept) in loamy drift derived from Carboniferous sandstone and shale: Quarter Bach, Dyfed (Photo by P.S. Wright). 3. Stagnoluvic gley soil (Mollic Albaqualf) in brownish till mainly derived from Carboniferous shale and sandstone: Bradnop, Staffordshire. 4. Chromic stagnoluvic gley soil (Aeric Albaquult) in Pebbly Clay Drift: Epping Forest, Essex. 5. Peloluvic gley soil (Typic or Vertic Ochraqualf) on Jurassic (Upper Lias) clay-shale: Shutford, Oxfordshire (Photo by M.G. Jarvis). 6. Loamy cultosol (Plaggept) on pre-existing podzolic brown soil: Clonakilty, Co. Cork. 7. Acid-sulphate fen soil (Terric Sulfohemist) over Fen Clay: Mepal, Cambridgeshire. 8. Raw sapric bog soil (Terric Medisaprist) in blanket peat over till: Auckengill, Caithness.

sidiary amounts of kaolinite and mica. In the extremely clayey 3BCt(g) horizon, which immediately overlies chalk and resembles the Clay-with-flints *sensu stricto* in profile 74, the expansible component is predominantly smectite. Clays from the upper horizons contain more vermiculite, which is interstratified to some extent with chlorite layers and may either be inherited from the loess or result from transformation of pre-existing 2:1 layer minerals under acid conditions.

Profile 77 (Plate II.4), also in uncultivated land, is on Plateau (Northern) Drift of Anglian or pre-Anglian age (Shotton *et al.*, 1980; Hey, 1986) which caps hills adjoining the Thames and Evenlode valleys in the Oxford district. It is strongly acid to at least 2 m depth, there are indications of incipient podzolization in the upper few centimetres of mineral soil, and base saturation values for the lower horizons place it close to the limit between Alfisols and Ultisols in the US system. Silt/sand ratios reveal a clear lithologic discontinuity roughly coincident with the lower boundary of the Eb horizon, but the upper layer is less silty than in profile 76, indicating that it contains a smaller proportion of loess-derived material. The clayey 2Bt(g) horizon, constituting the uppermost part of the second lithological layer and of the chromic B, is penetrated by wedge-shaped loamy inclusions (tongues), the composition and micromorphology of which are consistent with admixture of materials from the upper and lower layers. That disturbance took place under periglacial conditions is indicated by the occurrence of vertically oriented stones both between and within the tongues, which probably originated during the Devensian stage as infilled frost cracks. The argillic B horizon is strong brown to reddish yellow in the upper part, with red mottles which increase in size and number with depth and become dominant in the slowly permeable BCt horizons below 1 m. As in the preceding profile, thin sections of successive subhorizons show complex microfabrics with discrete concentrations of preferentially oriented and more or less iron-stained clay occurring partly as argillans bounding natural surfaces, partly as linear or irregularly shaped bodies unassociated with voids, partly as regularly shaped papules (Table 3.6, p. 92) and partly as thick coats around sand grains embedded in the matrix (embedded grain argillans). The distribution patterns of these pedological features, coupled with the occurrence of sharply defined bodies of contrasting lithology in the lower horizons, support the conclusion of Bullock and Murphy (1979) that some originated as components of larger masses of soil material (fossil aggregates) which were transported from elsewhere during emplacement of the drift, whereas others originated at or near their present position. As some but not all of the papules are reddened (rubified), and other reddened areas consist of ferruginous segregations transgressing the boundaries between papules and matrix, it seems that at least two phases of rubification are represented, separated by one or more phases of disruption or transport. Wholesale or partial reddening of loamy or clayey soils is most likely to have taken place during either the Ipswichian or the Cromerian interglacial (Table 1.1, p. 14) when, judging from palaeo-entomological evidence (Coope, 1977), summers were significantly warmer than in the intervening Hoxnian. It is therefore probable that the parent deposit incorporates transported soil material reddened during the Cromerian and was subjected at the Ipswichian stage to a further phase of reddening, leading to the formation of red ferruginous segregations which have survived more or less *in situ*.

Profile 76 (Avery *et al.*, 1959; Avery, 1964; Ragg and Clayden, 1973)

Location : Hampden, Buckinghamshire (SP 832021)
Climatic regime: subhumid temperate

| Altitude | : | 229 m Slope: 1° straight |
| Land use and vegetation | : | broadleaf woodland; beech (*Fagus sylvatica*) standards with some oak (*Quercus robur*) and holly (*Ilex aquifolium*); field layer dominated by bramble (*Rubus fruticosus*) with occasional herbs, chiefly wood sorrel (*Oxalis acetosella*) and yellow archangel (*Lamiastrum galeobdolon*); tussock grass (*Deschampsia caespitosa*) common nearby in spaces left by felling. |

Horizons (cm)

L	Fresh litter, mainly beech leaves, up to 3 cm thick.
F/H	Partially decomposed litter, up to 3 cm thick, followed in places by traces of black finely granular humus (H).
0–4 Ah	Very dark grey to dark greyish brown (10YR 3/1–4/2) very slightly stony silt loam (USDA silt loam); few small angular flint fragments; weak granular; friable; abundant woody roots; small earthworms present; abrupt irregular boundary.
5–23 Eb	Yellowish brown (10YR 5/4–6); 10YR 6/3–4 dry) very slightly stony silt loam to silty clay loam (USDA silt loam); few, mainly small, flint fragments as above; very weak fine blocky and granular; friable; few channels infilled with darker granular soil; many woody roots; clear wavy boundary.
23–41 Bt	Strong brown (7.5YR 5/6) slightly stony silty clay loam (USDA silty clay loam) with few fine brown (7.5YR 5/4) and yellowish red (5YR 5/6) mottles; stones as above, small to medium; moderate fine and medium subangular blocky with paler coloured silty coats on ped faces; friable to firm; common woody roots; few fine ferri-manganiferous concretions; clear wavy boundary.
41–74 Bt(g)	Strong brown to yellowish red (7.5–5YR 5/6) moderately stony clay (USDA clay) faintly to distinctly mottled with colours ranging from brown (10YR 5/3) to red (2.5YR 5/4); stones as above but some larger; moderate medium to coarse subangular blocky with smooth ped faces paler in colour than interiors; firm; few woody roots; continuous clay coats on most ped faces; common fine rounded ferri-manganiferous concretions; gradual boundary.
74–152 2Bt(g)	Yellowish red (5YR 5/6) slightly stony clay (USDA clay) with common reddish yellow (7.5YR 6/6), red (2.5YR 5/6–8) and light brownish grey to light grey (10YR 6/2–7/1) mottles; common large and very large broken and

Analytical data for Profile 76

Horizon	Ah	Eb	Bt	Bt(g)	2Bt(g)	3BCt(g)
Depth (cm)	0–4	4–23	23–41	41–74	91–152	198–228
sand 200 μm–2 mm %	3	2	1	1	8	2
60–200 μm %	7	9	8	6	16	4
silt 2–60 μm %	76	71	62	39	18	8
clay <2 μm %	14	18	29	54	58	86
organic carbon %	4.5	1.5				
dithionite ext. Fe %	1.4	1.6	2.0	3.3	3.6	
TEB (me/100 g)	2.5	0.7	0.4	2.3	9.8	
CEC (me/100 g)	22.7	11.6	11.0	19.4	19.6	
% base saturation	11	6	5	12	50	
pH (H$_2$O)	4.4	4.5	4.5	4.7	5.0	4.7
(0.01M CaCl$_2$)	3.9	4.1	4.1	4.1	4.0	4.0

152–230+ 3BCt(g)	nodular flints with whitish rinds; weak coarse blocky; firm (red material relatively friable); few woody roots; grey plastic clay around stones and lining fissures and channels; gradual boundary.
Similar flinty clay, becoming more uniform in colour and finer in texture with increasing depth, and containing increasing proportions of very large little-broken nodular flints.

Analytical data *see* p. 230.

Classification
England and Wales (1984): stagnogleyic paleo-argillic brown earth; fine silty over clayey, drift with siliceous stones (Batcombe series).
USDA (1975): Aquic Paleudalf; fine, mixed or vermiculitic, mesic.
FAO (1974): Orthic Acrisol; medium textured.

Profile 77 (Bullock and Murphy, 1979)

Location	:	Northleigh, Oxfordshire (SP 401138)
Climatic regime:		subhumid temperate
Altitude	:	120 m Slope: 1°
Land use and vegetation	:	ungrazed common land; bracken (*Pteridium aquilinum*) dominant with grassy patches (mainly bent-grass (*Agrostis* spp.) and occasional hawthorn (*Crataegus monogyna*) and elder (*Sambucus nigra*) bushes.

Horizons (cm)

3–0 (L/F)	Bracken litter.
0–0.5/1 H	Black decomposed litter; fine granular; loose; many fleshy roots (bracken rhizomes); sharp irregular boundary.
0.5/1–1/3 Ah1	Dark to very dark greyish brown (10YR 4–3/2) humose clay loam (USDA silt loam) containing rare quartzose pebbles and many uncoated sand grains; moderate granular; very friable; many fleshy roots (bracken rhizomes); sharp irregular boundary.
1/3–12/17 Ah2	Brown to dark brown (10YR 4/3) very slightly stony silty clay loam (USDA silt loam) with darker (10YR 4/2) and paler (10YR 4/4) inclusions; few quartzite and vein quartz and rare sandstone pebbles; weak fine subangular blocky; friable; many fine and few fleshy (rhizomatous) roots; few channels with darker infillings; abrupt wavy boundary.
12/17– 45/50 Eb	Dark yellowish brown and yellowish brown (10YR 4–5/4) slightly stony clay loam (USDA silt loam) with few faint light yellowish brown (10YR 6/4) mottles; stones as above; weak medium and fine subangular blocky; consistence and roots as above; few medium ferri-manganiferous nodules; abrupt wavy boundary.
45/50– 50/78 Eb(g)/Bt(g)	(discontinuous and rather heterogeneous horizon occurring as wedge-shaped inclusions in the underlying horizon) Yellowish brown to light yellowish brown (10YR 5/4–6/4, paler when dry) moderately to very stony clay loam (USDA loam) with few light brownish grey to pale brown (10YR 6/2–7/3) mottles and strong brown (7.5YR 5/6–8) relatively fine textured inclusions; stones as above, some vertically oriented; very weak subangular blocky; friable; few roots, some dead (traces); very few fine manganiferous nodules; abrupt very irregular boundary.
45/50– 80/88 2Bt(g)1	Strong brown to reddish yellow (7.5YR 5/6–6/8) moderately to very stony clay (USDA clay) with many fine light yellowish brown (10YR 6/4), few light grey and common fine to medium red (2.5YR–10R 5/6–8) mottles, stones as above, often vertically oriented; weak medium and fine angular blocky with mainly brown (7.5YR 5/4) faces; friable to firm; few fine and rare woody roots; patchy clay coats on ped faces and around stones; very few nodules as above; clear wavy boundary.
80/88– 100/106 2Bt(g)2	Strong brown (7.5YR 5/6–8) and red (2.5YR–10R 5/6–8), coarsely mottled, moderately stony clay (USDA clay) with common, fine and medium light brownish grey to light grey (10YR

Analytical data for Profile 77

Horizon	Ah1	Ah2	Eb	Eb(g)/Bt(g)	2Bt(g)1	2Bt(g)2	2BCt(g)	3BCt(g)
Depth (cm)	0.5–3	3–12	15–44	44–58	50–80	80–100	110–170	180–200
sand 200 μm–2 mm %	10	9	13	20	16	12	14	5
60–200 μm %	12	11	10	17	12	11	14	10
silt 2–60 μm %	58	60	58	43	25	25	23	38
clay <2 μm %	20	20	19	20	47	52	49	47
organic carbon %	6.3	3.6		0.7				
pyro. ext. Fe %		0.6	0.5	0.3	0.04	0.02	0.02	
pyro. ext. Al %		0.2	0.2	0.2	0.2	0.2	0.2	
total ext. Fe %		1.5	1.9	2.3	5.3	4.7	4.8	
TEB (me/100 g)		1.7		3.1			5.5	7.6
CEC (me/100 g)		19.0		8.9			16.6	18.4
% base saturation		9		35			33	41
pH (H$_2$O)	4.3	3.9	4.1	5.0	4.9	4.7	4.6	4.7
(0.01M CaCl$_2$)	3.9	3.4	3.6	4.0	4.0	3.8	3.7	3.9
bulk density (g cm^{-3})		0.88	0.96	1.25	1.33	1.40		
packing density (g cm^{-3})		1.06	1.13	1.42	1.51	1.86		
retained water capacity (% vol.)		38	40	38	41	41		
available water capacity (% vol.)		21	25	21	11	10		
air capacity (% vol.)		27	22	15	9	6		

100/106–120/180 2BCt(g) 6–7/2) mottles, particularly around stones (as above); weak medium to coarse angular blocky with many light grey clay coats on faces, in channels and around stones; firm (compact in place but crumbles easily when removed); few roots; clear boundary.

120/180–230 3BCt(g) Red (as above), light grey and strong brown to reddish yellow, coarsely mottled moderately stony clay (USDA clay) with more red and less brown than above and light grey mottles occurring partly as subvertical streaks; stones as above; few gravelly clay loam inclusions c. 20 cm in diameter; massive; consistence as above; with red material relatively friable and greyish material moderately to very plastic; clay coats as above; roots rare; clear very wavy boundary.
Red (as above) very slightly stony clay (USDA clay) with common medium to coarse light grey and few reddish yellow mottles; lense-shaped gravelly pocket between 200 and 230 cm on one side of section.

Analytical data *see* above.

Classification
England and Wales (1984): stagnogleyic paleo-argillic brown earth; fine loamy over clayey; drift with siliceous stones (Hornbeam series).
USDA (1975): Albaquic (or Glossaquic) Paleudalf intergrading to Paleudult; fine, mixed(?), mesic.
FAO (1974): Orthic Acrisol, medium textured.

5.7 Podzolic Brown Soils

5.7.1 General Characteristics, Classification and Distribution

As indicated in the introduction to this chapter, these soils have a friable and more or less ochreous Bs horizon below an Ah or Ap and no intervening E or Bh horizon of significant thickness. Most of the well drained acid soils previously grouped as brown podzolic in England and Wales (Avery, 1980) and in Ireland (Conry et al., 1972; Gardiner and Radford, 1980), meet these requirements, as do the majority of those that have been called brown forest in Scotland (Ragg et al., 1978) and upland or podzolic brown earths by the Forestry Commission (Table 2.7, p. 62). Cultivated soils which may originally have had E, or E and Bh, horizons are included, together with uncultivated soils that can be equated with the ochric brown soils (*sols bruns ocreux*) of Duchaufour (1982) and correspondingly regarded as resulting from an early stage of podzolization in materials containing significant amounts of weatherable silicate minerals. Variants of both kinds are grouped here with the brown soils rather than the podzols for three reasons. In the first place, the amorphous or poorly ordered sesquioxidic materials accumulated in the Bs horizons of those that remain under semi-natural vegetation appear to have originated mainly as a result of strong leaching and weathering of silicates under the influence of acid organic matter rather than by translocation from above (cp. Crompton, 1960; Avery et al., 1977). Secondly, both cultivated and uncultivated variants closely resemble the related orthic brown soils (brown earths) in agronomic and silvicultural qualities. Finally, many if not most of them qualify as Inceptisols (Ochrepts or Andepts) rather than as Spodosols in the US Taxonomy and as Cambisols (or Andosols) in the FAO system because the Bs horizons fail to meet the criteria prescribed for a spodic horizon.

Like the orthic brown soils, those of this group are most commonly developed in stony superficial deposits derived from subjacent consolidated rocks. They are widely distributed at altitudes up to about 400 m in humid and perhumid areas of the British Isles but are only very sporadically represented in the subhumid lowlands (Figure 2.5, p. 78). Oro-arctic (cryic) phases called subalpine and alpine brown soils by Birse (1980) have been noted in a few places on readily weatherable crystalline rocks within the oro-arctic zone. In some areas of occurrence, including upland Wales, the Lake District and the Southern Uplands of Scotland, much of the ground they underlie consists of steep slopes which originally bore oak forest with acidophile ground vegetation and now support acidic grassland with bracken or coniferous plantations. Elsewhere, as in parts of north-east Scotland and southern Ireland, slopes are gentler and a high proportion of the land is, or has been, cultivated. The morphological and chemical properties of the horizons within ploughing depth are correspondingly variable, depending on local differences in land use history.

Under semi-natural vegetation there is normally a thin organic surface layer (F/H) overlying a dark coloured Ah horizon, a very narrow or discontinuous Ea or Eag, or both, and in some places a shallow or weakly expressed Bh (Ah/Bh) is perceptible. Intergrades to stagnopodzols (p. 270) are common on upper slopes in perhumid upland localities and include profile forms (Figure 6.5, p. 270) described as 'podzols with gleying' by Crampton (1963). In these a thin peaty Oh horizon (hydromor) is underlain by lenticular or cup-shaped greyish patches (Eag) which commonly occupy the centres of coarse prismatic structural units and are bordered by discontinuous ochreous seams (incipient thin ironpans). Other variants have prominent A horizons (umbric or occasionally mollic), some containing enough organic matter to qualify as humose topsoils (Chapter 3, p. 107). These occur in both cultivated and uncultivated land and commonly have higher C/N ratios than similar horizons in less strongly leached soils. The underlying Bs horizons are generally distinguishable from Bw horizons in similar material by stronger colours, weaker and more or less smeary consistence, finer and more granular (crumb-like) structure, and a positive reaction to the Fieldes and Perrott (1966) sodium fluoride test for reactive hydroxy-aluminium (Chapter 3, p. 102). They normally pass downwards into less brightly coloured and less weathered BC and C horizons, which are frequently indurated to some extent (fragipan or duric horizon) and can show signs of gleying. Bisequal profiles (Chapter 3, p. 99) with an argillic (Bt) horizon below the Bs may also occur. These were distinguished as paleo-argillic brown podzolic soils by Avery (1980) but few have been positively identified.

Some soils conforming to this group have significantly more pyrophosphate-extractable iron, aluminium, or both, in all or part of the Bs than in the overlying horizon, whilst others with B horizons of similar morphology and composition do not. The apparent absence of an illuvial horizon can be attributed in some cultivated phases to incorporation of part or all of the original B in the Ap. It may also result from disturbance caused by uprooting of trees, truncation by erosion as postulated by Robinson et al. (1949) and Adams and Raza (1978), or precipitation in the upper horizons of mobile sesquioxides transported from upslope (Schweikle, 1973). Generally, however, accumulation of clay-size material, including iron oxides, by weathering of coarser mineral particles in the upper part of the profile appears to have inhibited or at least masked downward translocation of aluminium and iron (cp. Duchaufour, 1982, p. 80).

Certain of these soils, particularly variants derived from basic or intermediate crystalline rocks such as the Malvern and Bowden series in England and Wales (Clayden and Hollis, 1984), many of the 'brown forest soils' mapped as members of the Darleith/ Kirktonmoor, Sourhope and Insch associations by the Soil Survey of Scotland (1984) and some 'acid brown soils' of basaltic origin in Ulster (Association 17 in Gardiner and Radford, 1980), contain significant amounts of aluminium in amorphous or paracrystalline inorganic combinations, including allophane or imogolite, which are extracted by acid oxalate (Chapter 3, p. 94) or 0.5N NaOH (Hashimoto and Jackson, 1960) but not by pyrophosphate (Loveland and Bullock, 1975, 1976; Tait et al., 1978; Loveland, 1984; Payton, 1988). These could well be classed as andic rather than podzolic (Duchaufour, 1982), or even in some cases as Andepts (FAO Andosols). They have not been separated in this account, partly because the various analytical differentiating criteria that have been proposed for use in the US taxonomy and elsewhere (e.g. Flach et al., 1980; Parfitt et al., 1983) have yet to be applied systematically to British soils.

Although the podzolic brown soils resemble lithologically similar orthic brown soils in agronomically important attributes such as drainage and workability, the proportionately larger amounts of amorphous 'sesquioxidic' materials in the exchange complex tend to give them a more porous and finely aggregated structure and larger pH-dependent cation exchange and phosphorus retention capacities (Evans and Smillie, 1976). Their inherently low nutrient status is fairly readily overcome by the use of lime and fertilizers, and moisture deficiency is not a problem in the main areas of occurrence except where the soil is unusually shallow over bedrock or indurated material. Climatic and topographic limitations, including high rainfall and steep, uneven or bouldery slopes, make much of the land unsuitable for tillage. For both pasture and forestry, however, the soils are generally more productive than the stagnopodzols and other peaty topped soils which usually replace them at higher altitudes.

Key to Subgroups

1. Podzolic brown soils with gleyic features (p. 112)
    1.1 with a slowly permeable subsurface horizon (p. 110)
        Stagnogleyic podzolic brown soils
    1.2 without a slowly permeable subsurface horizon
        Gleyic podzolic brown soils

2. Other brown podzolic soils
Typical podzolic brown soils

5.7.2 Typical Podzolic Brown Soils

The typical subgroup, which is by far the most extensive, comprises the freely draining podzolic brown soils in which the Bs horizon passes downwards into an unmottled BC or C or rests directly on bedrock. As indicated above, they are mostly loamy or loamy-skeletal but sandy and sandy-gravelly variants with B horizons of loamy sand texture also occur, chiefly in glaciofluvial deposits derived from rocks containing significant amounts of weatherable minerals. Data on four representative profiles in different parent materials are given on pp. 237–41. The first two are under acidic grassland with bracken on steep upland slopes and the others in cultivated land. All except the last have a more or less indurated layer below the Bs, though variants with loose rubbly C horizons can also be found in similar situations. None of them qualifies unequivocally as a Spodosol according to the criteria listed by Soil Survey Staff (1975), but profiles 80 and 81 are classed tentatively as Haplorthods (FAO Leptic Podzols) on the basis that the Bs horizons are coarse loamy and contain pellet-like micro-aggregates.

Profile 78 (Linhope series) was described and sampled in a forest ride near Innerleithen, Peebles-shire. The parent material is stony head derived from Lower Palaeozoic greywacke. Soils of the essentially identical Manod series (Plate II.5) are widely distributed in materials of similar origin in upland Wales (Rudeforth *et al.*, 1984) and south-west England (Findlay *et al.*, 1984). Similar soils also predominate in several parts of Ireland (Association 9 in Gardiner and Radford, 1980). In the example selected, stone and sand contents decrease and clay and silt contents increase from the C horizon upwards, showing that formation of fine earth by weathering of fragmentary rock has been a major developmental process. Compared with less strongly leached orthic brown soils from similar sedimentary rocks (e.g. profile 54, p. 197), the B horizon contains more organic matter with a higher C/N ratio and a relatively high proportion extractable by pyrophosphate. Amounts of pyrophosphate-extractable aluminium and iron are larger, both absolutely and in relation to clay content, and the slight maxima recorded in the ochreous Bs suggest some illuvial accumulation. Thin sections (Ragg *et al.*, 1978) showed that both A and B horizons have very porous fabrics composed largely of fine pellet-like aggregates (Bullock and Clayden, 1980), probably of biological origin, which in the Bs are mostly rewelded into coarser subangular blocky peds. Large (500–600 μm) earthworm casts are common in the A horizon and channels containing them were noted in the Bs. Sections of the dense BCx and Cx horizons showed a progressive decrease in macroporosity with depth; silt coats are common on the upper surfaces of stones in these horizons and some voids are lined by argillans with strong continuous orientation.

Profile 79 is an example of the 'andic podzolic brown soils' derived from crystalline rocks. Located in the Cheviot Hills, Northumberland (Figure 5.11), it is developed in very stony head or locally derived till overlying porphyritic andesite. As in the preceding profile, the measured clay contents decrease markedly with depth and the pyrophosphate extraction data afford no evidence that either organic-bound iron or aluminium has been translocated to a significant extent. There is, however, distinctly more oxalate-extractable aluminium in the Bs than in overlying and underlying horizons. Samples from both the humose Ah and the Bs give a strong positive reaction to the Fieldes and Perrott sodium fluoride test (p. 102) and have physical proper-ties characteristic of Andepts (Andosols), including low fine-earth bulk density, fine granular (crumb) structure and smeary (thixotropic) consistence. Compared with texturally

Figure 5.11: Harthope valley, Cheviot Hills, Northumberland. Podzolic brown soils in drift derived from andesitic lavas underlie the valley side slopes (Copyright Geoffrey N. Wright).

similar Andepts from volcanic ash, however, amounts of oxalate-extractable aluminium are considerably smaller in relation to clay content, and Payton (1988) reports further chemical and mineralogical data showing that clay fractions of the A and B horizons consist dominantly of layer silicates rather than allophane, with hydroxy-Al interlayered vermiculite as a major component. There is an abrupt increase in bulk density at the upper boundary of the underlying fragipan (BCx), parts of which are very weakly cemented.

Profile 80 is in enclosed farmland adjoining Bodmin Moor, Cornwall. Here the parent material is gravelly granitic head (growan) and the soil is correspondingly coarser in texture, though it resembles profiles 78 and 79 in other respects. As nearby uncultivated soils in similar material resemble profile 86 (p. 252) in having indistinct eluvial (Ah/Ea) and illuvial (Bh) horizons of podzol type within the upper 30 cm, it is likely that the prominent very dark Ap results from mixing of such horizons, so obscuring any evidence of sesquioxide translocation. Similar soils of granitic origin occur in numerous other British localities, including the margins of Dartmoor (Clayden, 1971), the Aberdeen area (Countesswells association) where they were mostly identified as cultivated podzols by Glentworth and Muir (1963), the lower slopes of the Leinster mountains south of Dublin, and the Mourne area of Down and Armagh (associations 8 and 12 in Gardiner and Radford, 1980).

Profile 81 is an example of the well drained cultivated soils classed as brown podzolic in south-west Ireland. Sited in improved pasture in the Dingle peninsula, County Kerry, it is formed on coarse grained sandstone (Annascaul Beds) and has a brownish Ap over a yellowish red coarse loamy B horizon with weak fine granular structure. As shown on the national soil association map (associations 6 and 15), similar soils in materials derived wholly or partly from Old Red Sandstone rocks predominate over considerable areas extending from Waterford westwards to Kerry. These and other Irish soils have been identified as brown podzolic by the presence of a very friable B horizon measurably enriched in free iron (dithionite-extractable) and cannot be equated directly with those called brown podzolic in England and Wales because few determinations of pyrophosphate-extractable iron and aluminium have been made. Comparison with morphologically and genetically similar English soils, however, suggests strongly that the B horizons general qualify as Bs. However, Conry et al. (1972) showed clearly that all four cultivated profiles they studied originally had more or less continuous albic E horizons and that the B horizons were enriched in translocated organic matter as well as in iron and aluminium. In some places deep cultivation by spade or other hand implements had produced Plaggen soils (Chapter 8, p. 386) with Ap horizons more than 50 cm thick.

Profile 78 (Ragg, 1974; Ragg et al., 1978)

Location	:	Elibank Forest, Peeblesshire (NT 342351)
Climatic regime:		humid oroboreal
Altitude	:	275 m Slope: 20°
Land use and vegetation	:	ride in Sitka spruce plantation: field layer mainly of common bent-grass (*Agrostis capillaris*), sheeps fescue (*Festuca ovina*) and bracken (*Pteridium aquilinum*).

Horizons (cm)

L	Mainly bracken litter, c. 2 cm thick.
0–1 F	Very dark brown (10YR 2/2) more or less decomposed litter, mainly of bracken, with some mineral grains and abundant fleshy roots (bracken rhizomes).
1–15 Ah	Dark brown (10YR 3/3) slightly stony humose clay loam (USDA loam – clay loam) with common uncoated quartz grains; common greywacke fragments; weak medium and fine granular; friable; abundant fine roots; many casts and faecal pellets; clear smooth boundary.
15–20 AB	Dark yellowish brown (10YR 3/4) and strong brown (7.5YR 5/6) moderately stony clay loam (USDA loam); stones as above; weak medium and coarse granular; friable; abundant fine roots; abrupt smooth boundary.
20–40 Bs	Strong brown (7.5YR 5/6) moderately stony clay loam (USDA loam) stones as above; weak granular; very friable; smeary feel; common fine roots; gradual wavy boundary.
40–60 BCx	Strong brown (7.5YR 5/6) very stony sandy loam (USDA sandy loam) stones as above; massive; very firm

Analytical data for Profile 78

Horizon	Ah	AB	Bs	BCx	Cx
Depth (cm)	1–15	15–20	20–40	40–60	60–80
silt[a] 2–50 µm %	46	49	48	43	40
clay[a] <2 µm %	18	22	19	6	4
organic carbon %	8.4	7.2	1.9	0.5	0.4
nitrogen %	0.62	0.51	0.12		
C/N ratio	14	14	16		
pyro. ext. Fe %	0.9	0.9	1.0	0.2	
pyro. ext. Al %	0.4	0.4	0.5	0.4	
pyro. ext. C %	2.3	1.9	0.7	0.2	
pyro. Fe+Al %/clay %	0.07	0.06	0.08	0.10	
total ext. Fe %	1.6	1.7	1.7	1.0	
TEB (me/100 g)	4.2	2.8	0.2	0.1	0.1
CEC (me/200 g)	25.5	23.8	9.0	4.8	4.1
% base saturation	17	12	2	2	2
pH (H$_2$O)	4.7	4.6	4.7	4.8	5.2

[a] by hydrometer method

60–100+ Cx	Yellowish brown (10YR 5/4) very stony sandy loam (USDA sandy loam); abundant angular stones, many with fine sandy or silty cappings; massive; very firm and brittle; rare roots to 85 cm; less compact below 100 cm.		

and brittle; few fine roots; gradual wavy boundary.

60–100+ Cx — Yellowish brown (10YR 5/4) very stony sandy loam (USDA sandy loam); abundant angular stones, many with fine sandy or silty cappings; massive; very firm and brittle; rare roots to 85 cm; less compact below 100 cm.

Analytical data *see* p. 237.

Classification
    Scotland (1984): brown forest soil (Linhope series).
    England and Wales (1984): typical brown podzolic soil; fine loamy material over lithoskeletal mudstone and sandstone or slate (Manod series).
    USDA (1975): Typic Fragiochrept (or Cryochrept); loamy-skeletal, mixed, mesic?
    FAO (1974): Dystric Cambisol; medium textured (fragipan phase).

Profile 79 (Payton, 1988)

Location : Langlee, Northumberland (NT 963236)

Climatic regime: humid oroboreal
Altitude : 240 m Slope: 15°
Land use and vegetation : rough grazing; acidic grassland, formerly oak woodland, with abundant bracken (*Pteridium aquilinum*) and frequent heather (*Calluna vulgaris*) and wood sorrel (*Oxalis acetosella*)

Horizons (cm)

L	Grass stems bracken and moss litter, c. 3 cm thick.
0–2 F	Very dark brown (10YR 2/2) very fibrous mat of partially decomposed moss, bracken and grass remains; abundant fine roots.
2–3 H	Black (10YR 2/1) organic material; weak fine granular; abundant roots; abrupt smooth boundary.
3–20 Ah	Dark brown (7.5YR 3/3) moderately stony (25 per cent) humose clay loam (USDA loam); many very small to medium angular and subangular andesite fragments; weak fine granular; very friable; many fine fibrous and medium fleshy roots (bracken rhizomes); gradual smooth boundary.

Analytical data for Profile 79

Horizon	Ah	Ah	ABs	Bs	Bs	BCx	BCx
Depth (cm)	3–8	8–20	20–35	35–45	45–60	60–70	70–100
sand 500 µm–2 mm %	13	17	16	22	29	30	28
250–500 µm %	6	8	7	10	11	10	12
50–250 µm %	14	13	15	16	17	21	20
silt 2–50 µm %	42	38	45	37	30	32	33
clay <2 µm %	25	24	17	15	13	7	7
organic carbon %	8.0	5.8	4.0	2.7	2.6	0.1	0.1
pyro. ext. Fe %	0.8	0.8	0.8	0.7	0.3	0.1	0.1
pyro. ext. Al %	0.7	0.4	0.6	0.5	0.3	0.1	0.1
pyro. ext. C %	2.8	2.3	1.9	1.6	1.4	0.1	0.1
pyro. Fe+Al %/clay %	0.06	0.05	0.08	0.08	0.05		
oxalate ext. Fe %	1.2	1.2	1.4	1.5	1.1	0.3	0.2
oxalate ext. Al %	0.7	0.8	1.0	1.1	1,4	0.4	0.4
citrate-dithionite ext. Fe %	1.7	1.7	1.7	1.8	1.5	0.8	0.8
pH (H$_2$O)	4.7	4.8	4.8	4.9	4.9	5.5	5.5
(M NaF)[a]		10.4	11.4	11.3	11.2	9.3	
bulk density (g cm^{-3})		0.51		0.67		1.74	1.64

[a] 1 g soil (<2 mm) in 50 cm^3 M NaF after 2 minutes

20–35 ABs	Dark brown to strong brown (7.5YR 3/6) moderately stony sandy silt loam (USDA loam); stones as above, very small to large; weak fine granular; very friable; releasing water and becoming smeary on manipulation; many roots; gradual wavy boundary.
35–60 Bs	Strong brown (7.5YR 4/6) moderately stony (33 per cent) coarse sandy loam (USDA coarse sandy loam); weak medium granular to fine subangular blocky; consistence as above; common fine and few coarse woody roots; sharp wavy boundary marked by mat of living and dead roots.
60–110 BCx	Pale brown to brown (7.5YR 6–5/3) very stony (45 per cent) coarse sandy loam (USDA coarse sandy loam); stones as above; massive to weak coarse platy (lenticular); very firm and brittle; discontinuous very weak cementation; irregularly shaped medium macropores often with distinct silty coats; prominent light greyish brown silt coats on lenticular ped surfaces and capping many stones; rare roots; pan discontinuous vertically and laterally, with looser and more porous inclusions containing little fine earth; clear irregular boundary.
110+ Cu/Cr	Angular andesite rubble with little fine earth, passing to andesite *in situ*.

Analytical data *see* p. 238.

Classification
England and Wales (1984): humic brown podzolic soil, fine loamy material over lithoskeletal basic crystalline rock (Bowden series).
USDA (1975): Andic Fragiumbrept; loamy-skeletal, mixed, mesic.
FAO (1974): Dystric Cambisol; medium textured (fragipan phase).

Profile 80 (Staines, 1976)

Location	:	Camelford, Cornwall (SX 123835)
Climatic regime:		perhumid temperate
Altitude	:	241 m Slope: 6°
Land use	:	ley grass (first year)

Horizons (cm)

0–24 Ap	Very dark grey to greyish brown (10YR 3/1–2) slightly stony coarse sandy silt loam (USDA sandy loam); common, very small to medium, subangular granite fragments; weak coarse angular blocky breaking to fine subangular blocky and granular; very friable; many fine roots; earthworms present; clear boundary.
24–45 Bs	Strong brown (7.5YR 4/6) slightly stony coarse sandy loam (USDA coarse sandy loam) with medium and coarse tonguing inclusions of dark soil similar to above; stones as above; moderate fine granular; very friable; common fine roots; earthworms present; gradual boundary.
45–75 BC(x)	Brown to dark yellowish brown (10–7.5YR 4/4) very stony coarse sandy loam, becoming extremely stony at base; abundant small to large subangular granite fragments; weak fine blocky tending to platy; friable to firm becoming more compact (hard to dig) with depth; common fine pores; few roots.

Analytical data

Horizon	Ap	Bs
Depth (cm)	2–20	26–42
sand 600 μm–2 mm %	19	33
200–600 μm %	16	13
60–200 μm %	11	10
silt 2–60 μm %	38	37
clay <2 μm %	16	7
organic carbon %	3.0	0.8
pyro. ext. Fe %	0.8	0.3
pyro. ext. Al %	0.3	0.3
pyro. ext. C %	1.3	0.6
pyro. Fe+Al %/clay %	0.07	0.08
total ext. Fe %	1.7	0.8
pH (H_2O)	6.2	6.2
(0.01M $CaCl_2$)	5.9	5.3

Classification
England and Wales (1984: typical brown podzolic soil; coarse loamy material over lithoskeletal acid crystalline rock (Moretonhampstead series).

USDA (1975): Entic Haplorthod?; loamy-skeletal, mixed, mesic.
FAO (1974): Leptic Podzol?; medium textured.

Profile 81 (Conry and O'Shea, 1973)

Location : Annascaul, Co. Kerry
Climatic regime: perhumid temperate
Altitude : 61 m   Slope: 2–3°
Land use and vegetation : long-term grass with timothy (*Phleum pratense*), perennial ryegrass (*Lolium perenne*), Yorkshire fog (*Holcus lanatus*) and white clover (*Trifolium repens*)

Horizons (cm)

0–25/33 Ap	Brown to dark brown (7.5YR 4/2) stony sandy loam (USDA sandy loam); moderate fine and medium granular; friable; abundant roots; clear wavy boundary.
25–56/64 Bs	Yellowish red (5YR 4/6) stony sandy loam (USDA sandy loam); weak fine granular; friable; abundant roots; clear wavy boundary.
56–100+ Cu	Greyish brown (2.5Y 5/2) very stony sandy loam (USDA sandy loam); massive; no roots.

Analytical data

Horizon	Ap	Bs	Cu
Depth (cm)	0–25	33–56	56–100
sand 200 µm–2 mm %	35	43	39
50–200 µm %	23	23	24
silt 2–50 µm %	26	27	30
clay <2 µm %	16	7	7
organic carbon %	3.6	1.6	0.4
nitrogen %	0.35	0.09	
C/N ratio	10	18	
citrate-dithionite ext. Fe %	1.7	2.7	0.6
TEB (me/100 g)	9.6	8.6	1.5
CEC (me/100 g)	27.2	24.0	6.2
% base saturation	35	36	25
pH (H$_2$O)	5.7	6.1	6.0

Classification
  Ireland (1980): brown podzolic soil (Gurteen series).
  England and Wales (1984): typical brown podzolic soil; coarse loamy material over lithoskeletal sandstone (Batch series).
  USDA (1975): Entic Haplorthod?; coarse loamy or loamy-skeletal, mixed, mesic.
  FAO (1974): Leptic Podzol?; medium textured.

5.7.3 Gleyic Podzolic Brown Soils

Podzolic brown soils of this subgroup occur locally in places where the lower horizons are subject to periodic saturation by ground water. The profiles show ochreous mottles and/or greyish colours attributable to gleying below the Bs but there is no slowly permeable subsurface horizon within 80 cm depth. The following representative profile is located at the base of a slope on stony glaciofluvial drift overlying till in the Eden valley, Cumbria. The parent deposit derives its reddish colour from the neighbouring Permo-Triassic sandstone but the stones it contains are mainly of andesitic rock (Borrowdale Volcanic series) transported from the Lake District. Drainage is impeded below 90 cm by compact and locally indurated finer textured material. Like profiles 80 and 81, this cultivated soil may once have been a podzol.

Profile 82 (Matthews, 1977)

Location : Hunsonby, Cumbria (NY 597368)
Climatic regime: humid temperate
Altitude : 130 m   Slope: 2° concave
Land use : arable

Horizons (cm)

0–27 Ap	Very dark brown (7.5YR 2/2) very stony fine sandy loam (USDA fine sandy loam); abundant, small to large, rounded and subrounded igneous pebbles and cobbles; weak fine subangular blocky; friable; common fine roots; earthworms present; sharp wavy boundary.
27–46 Bs	Yellowish red (5YR 4/8) very stony loamy medium sand (USDA loamy coarse sand); stones as above; weak granular; very friable; common fine

	roots; earthworm channels; gradual boundary.
46–80 2BC(g)	Yellowish red (5YR 4/8) and brown (7.5YR 5/4) moderately stony fine sandy loam to loamy fine sand (USDA loamy fine sand) with common, faint to distinct brownish mottles and few black (2/0) manganiferous segregations towards base; stones as above, mainly small and medium; structure and consistence as above; few roots; clear irregular boundary.
80–90+ 3Cu(g)	Reddish brown (5YR 4/4) very stony loamy sand to sandy loam with brown (7.5YR 5/4) mottles and common fine ferri-manganiferous segregations; stones as above; massive to single grain; very friable; roots rare.

Analytical data

Horizon	Ap	Bs	2BC(g)
Depth (cm)	0–27	27–46	46–80
sand 600 µm–2 mm %	17	28	5
200–600 µm %	14	21	8
60–200 µm %	38	34	61
silt 2–60 µm %	19	10	21
clay <2 µm %	12	7	5
organic carbon %	4.1	2.5	0.4
pyro. ext. Fe %	0.5	0.3	0.2
pyro. ext. Al %	0.2	0.3	0.1
pyro. ext. C %	1.1	0.7	0.4
pyro. Fe+Al %/clay %	0.06	0.09	0.06
total ext. Fe %	1.6	1.4	0.9
pH (H$_2$O)	6.2	6.3	6.0

Classification
England and Wales (1984): gleyic brown podzolic soil; sandy-gravelly, basic crystalline rock (Glassonby series).
USDA (1975): Entic Haplumbrept or Haplorthod; sandy-skeletal, oxidic, mesic.
FAO (1974): Humic Cambisol or Leptic Podzol; coarse textured.

5.7.4 Stagnogleyic Podzolic Brown Soils

These are podzolic brown soils with signs of gleying in or above a slowly permeable subsurface horizon that starts within 80 cm depth. The impeding layer usually has fragipan characteristics in some part, as in the following example located in gently sloping ploughed and reseeded grassland near Llandeilo, Dyfed. This soil is developed in stony drift (till or head) derived from Lower Palaeozoic mudstones and sandstones. Though finer in texture than most podzolic brown soils, it is differentiated from associated stagnogleyic orthic brown soils of the Sannan series (Clayden and Hollis, 1984) by the presence of a weakly expressed Bs containing considerably more extractable iron and appreciably more pyrophosphate-extractable aluminium than the overlying horizon.

Profile 83 (Wright, 1980)

Location : Llandeilo, Dyfed (SN 644279)
Climatic regime: humid temperate
Altitude : 140 m   Slope: 2°
Land use : long-term grass
Horizons (cm)

0–25 Ap(g)	Brown to dark brown (10YR 4/3) very slightly stony silty clay loam (USDA clay loam to silty clay loam) with common very fine linear yellowish red (5YR 5/8) mottles along root channels; few, mainly small, mudstone and sandstone fragments; moderate fine subangular blocky; friable; many fine roots; earthworms present; clear wavy boundary.
25–48 Bs(g)	Yellowish brown (10YR 5/6) slightly stony silty clay (USDA silty clay loam) with 30 per cent strong brown (7.5YR 4/6) inclusions and common fine pale brown (10YR 6/3) mottles; common, mainly medium, subangular sandstone and mudstone fragments; moderate medium to fine subangular blocky, becoming finer and more granular in strong brown inclusions; friable; common fine and few coarse

48–60 Bg1	woody roots; earthworms present; clear wavy boundary. Light yellowish brown (2.5Y 6/4) slightly stony clay loam with common fine and medium light brownish grey (2.5Y 6/2) and common very fine and fine yellowish red (5YR 4/8) mottles; stones as above; moderate medium prismatic; friable to firm; common roots; earthworms present; clear wavy boundary.	
60–80 Bg2	Light brownish grey (2.5Y 6/2) moderately stony silty clay loam to clay loam (USDA silt loam to loam) with common coarse strong brown (7.5YR 5/8) mottles, stones as above; moderate medium and coarse prismatic; firm; few fine roots; clear wavy boundary.	
80–100 BCx(g)	Yellowish brown (10YR 5/4) moderately stony clay loam with many medium light grey to grey (10YR 6/1) and few fine strong brown (7.5YR 5/6) mottles; stones as above; massive; vesicular macropores; very firm and brittle; no roots.	

Analytical data

Horizon	Ap(g)	Bs(g)	Bg2
Depth (cm)	0–25	25–48	60–80
sand 200 μm–2 mm %	8	6	9
60–200 μm %	8	5	10
silt 2–60 μm %	50	51	59
clay <2 μm %	34	38	22
organic carbon %	3.9		
pyro. ext. Fe %	0.8	1.8	0.2
pyro. ext. Al %	0.3	0.5	0.2
pyro. Fe+Al %/clay %	0.03	0.06	0.02
total ext. Fe %	1.8	3.2	1.0
pH (H_2O)	5.5	5.8	6.2
(0.01M $CaCl_2$)	4.8	5.2	5.6

Classification

England and Wales (1984): stagnogleyic brown podzolic soil; fine silty drift with siliceous stones (variant of Sannan series).

USDA (1975): Aquic Fragiochrept; fine-loamy, mixed or illitic, mesic.

FAO (1974): Gleyic Cambisol; medium textured (fragipan phase).

6
Podzols

6.1 General Characteristics, Classification and Extent

As originally used in Russia (Muir, 1961b) and subsequently in various other countries (e.g. Baldwin *et al.*, 1938; Stace *et al.*, 1968), the term *podzolic* referred to leached acid soils, including podzols, with light coloured eluvial horizons and illuvial horizons enriched in silicate clay, sesquioxides, or both. In western Europe, however, its meaning has been restricted to soils in which the characteristic accumulation products are sesquioxides and organic matter. Those with argillic B horizons only have been grouped with the brown soils (e.g. Avery, 1973, 1980; Mückenhausen *et al.*, 1977; Duchaufour, 1982) or set apart at the highest categorical level (e.g. de Bakker and Schelling, 1966), and the same distinction is drawn in the US and FAO systems.

According to the classification adopted here, podzols are required to have a podzolic B horizon (Bh and/or Bs), a thin ironpan (Bf), or both, as defined in Chapter 3 (p.108). In addition, if the B horizon consists only of an uncemented (friable) Bs, there must be an overlying albic E (Ea or Eag) at least 5 cm thick. Thus, whilst many of these soils have a superficial organic layer and an underlying eluvial horizon, both can be absent, particularly in cultivated land. They are divided at group level into non-hydromorphic podzols, gley-podzols and stagnopodzols. Those of the first group are typically pervious and well aerated; gley-podzols have a gleyed horizon below a podzolic B, whereas stagnopodzols show evidence of anaerobic conditions above a thin ironpan or Bs horizon but few or no signs of gleying in deeper horizons. The ten subgroups recognized are distinguished by further variations in the kind and sequence of subsurface horizons (Figure 6.1) which in turn reflect differences in parent material or other factors conditioning profile development.

The non-hydromorphic podzols and gley-podzols are nearly all Spodosols according to the US Taxonomy and Podzols in the FAO system, but some of the stagnopodzols qualify as Inceptisols (Placaquepts) because they lack a spodic B horizon as currently defined and the presence of a thin ironpan (placic horizon) is not alone sufficient to place a soil in the Spodosol order.

Judging from the national soil association maps, podzols of one kind or another cover rather less than 5 per cent of England and Wales, around 25 per cent of Scotland and some 8 per cent of Ireland. Their distribution (Figure 2.5, p.78) reflects the influence of climate, parent material and biotic factors. In the subhumid lowlands they occur only in coarse textured siliceous sediments and on initially finer, base-deficient deposits where clay contents of the uppermost horizons have been reduced by prior eluviation. In cooler

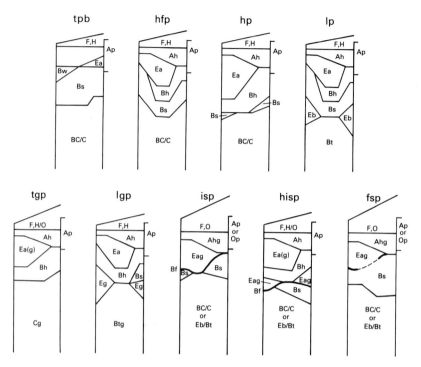

Figure 6.1: Horizon patterns in podzolic brown soils and podzols: *tbp* typical podzolic brown soils; *hfp* humoferric podzols; *hp* humus podzols; *lp* luvic podzols; *tgp* typical gley-podzols; *lgp* luvic gley-podzols; *isp* ironpan stagnopodzols; *hisp* humus-ironpan stagnopodzols; *fsp* ferric stagnopodzols.

and more humid areas they have formed in a progressively wider range of materials but are mainly restricted to initially well drained sites with pervious substrata. Although some of the podzols are, or have been, cultivated, and considerable areas have been planted with conifers during the last half century, much of the ground underlain by these soils remains under semi-natural vegetation, particularly at higher altitudes.

As indicated in Chapter 1 (p.8), podzol development can generally be attributed to the establishment on originally acid or strongly leached mineral materials of mor-forming biotic regimes which promote mobilization and downward movement of organic matter, aluminium and iron. However, by contrast with boreal regions of continental Europe and North America, where coniferous or mixed forests cover vast areas, the commonest vegetation types on undisturbed podzols in the British Isles are dwarf shrub heaths and other low growing acidophile plant communities dominated by grasses, bracken or mosses. In the oro-arctic (montane) zone and certain extremely exposed coastal localities, one or other of these have almost certainly occupied the ground since the close of the last glacial period. Elsewhere, as shown in Figure 1.15 (p.23) a forest cover developed and reached its maximum extent during the post-glacial climatic optimum (c. 6,000–7,000 years BP) but later gave place to heath or moorland, a change that was increasingly promoted by human activities from Neolithic times onwards (Gimingham, 1972; Pennington, 1974; Godwin, 1975). That podzols formed under forest in places is evident from their general occurrence in surviving remnants of the once widespread Caledonian pinewoods in eastern Scotland (Steven and Carlisle, 1959; Gauld, 1982) and their more sporadic appearance in ancient oak or beech woodland in south-west Ireland (Turner and Watt, 1939) and several parts of England (Dimbleby and Gill, 1955; Avery, 1958; Mackney, 1961). In areas where the primaeval forest was deciduous, however, it seems that development of soils with well marked podzol characters generally followed invasion by heathers or other undemanding plants which are unable to circulate nutrients as effectively as trees. This conclusion is supported by the nature of soil horizons preserved beneath archaeological structures (e.g. Evans, 1975), the widespread occurrence of ericoid pollen in podzol profiles irrespective of previous or existing vegetation (Dimbleby, 1962, 1965), and detailed studies of profile morphology in areas with historic evidence of deforestation (e.g. Romans and Robertson, 1975).

The three soil groups into which the podzols are divided are defined as follows. Corresponding classes in the England and Wales classification are named in parentheses.

Key to Soil Groups

1. Podzols with a gleyed subsurface horizon (p.109) below a podzolic B (p.108) and no thin ironpan (p.108)

   Gley-podzols (Gley-podzols)
2. Other podzols with either
   (a) a thin ironpan (p.108) below a peaty topsoil (p.107), a gleyed albic E (p.108) or a Bh horizon; or
   (b) a gleyed albic E (Eag or Eg) or albic densipan (p.108) over a Bs horizon (usually mottled in the upper part)

   Stagnopodzols (Stagnopodzols)
3. Other podzols
   Non-hydromorphic podzols (podzols *sensu stricto*, humic cryptopodzols and some humic brown podzolic soils)

Figure 6.2: Holt Heath, Dorset, on sandy Eocene beds (Nature Conservancy Council: copyright reserved).

6.2 Non-hydromorphic Podzols

6.2.1 General Characteristics, Classification and Distribution

These are the more or less well drained podzols having a bleached (albic) E horizon, a distinct dark Bh, a cemented Bs, or some combination of these features. Including Orthods and Humods in the US Taxonomy, and Orthic, Humic and Leptic Podzols in the FAO scheme, they occur in numerous widely scattered localities ranging from existing or former heathland in the English lowlands (Figure 6.2) to the summits of Scottish mountains (Figure 6.3), but are far less common than stagnopodzols in the humid and perhumid uplands at intermediate altitudes. In Ireland, where they are relatively inextensive, they have not been separated systematically from stagnopodzols in soil surveys. Most of the ground underlain by these soils carries semi-natural vegetation or planted conifers, as only those in which one or more of the diagnostic horizons extends below ploughing depth retain their identity under cultivation.

The group is typified by *humoferric* (humus-iron) podzols (Plate II.6) with a non-peaty surface horizon followed by an Ea/Bh/Bs horizon sequence betokening marked translocation of organic matter and iron. It also includes a variety of genetically related and often closely associated profile forms (Figure 6.1) in which one or other of these horizons is absent or weakly expressed. Those without an albic E horizon, either naturally or through cultivation, are distinguished from podzolic brown soils and humic rankers (p.122) by the presence of a subsurface Bh. Others previously distinguished as iron podzols or ferric podzols have an albic E over a friable or cemented Bs; and some, with or without an albic E, have a dark coloured Bh only. These include lowland *humus podzols* (Humods) in iron-deficient sands (Plate II.7) and frost-affected montane soils (Plate II.8) termed alpine podzols by the Soil Survey of Scotland (1984).

Argillans can often be identified below the podzolic B, either in a distinct finer textured argillic (Bt) horizon as in *luvic podzols* (Figure 6.1), or in a BC horizon of similar or coarser texture, which is commonly indurated to some extent (fragipan or duric horizon). The clay content may also reach a maximum in the podzolic B itself, either as a direct result of eluviation prior to podzolization or because silicate clay has been decomposed in the E horizon and been reconstituted in the B from the mobile decomposition products (Guillet et al., 1975).

Of the variants occurring at lower altitudes, those with E, or E and Bh horizons thick enough to remain evident in cultivated land are largely confined to siliceous sandy and sandy-gravelly materials, including Tertiary and Lower Cretaceous sands in southern England (M.G. Jarvis et al., 1984), Permo-Triassic sandstones and glaciofluvial deposits in the Midlands (Ragg et al., 1984), glaciofluvial and raised-beach sands adjoining the Moray Firth (Grant, 1960), and lithologically similar deposits in the Screen area of Wexford (Gardiner and Ryan, 1964). The illuvial accumulation products in these soils occur chiefly as cutans coating and bridging the sand grains, giving B horizons which are massive and frequently cemented. In cooler and more humid upland areas with an oroboreal climatic regime, variants with Ea horizons 10 cm or more thick are restricted to exceptionally well drained sites on stony weathering products of hard quartz-rich rocks. The Anglezarke association mapped on Carboniferous and Jurassic sandstones in northern England (R.A. Jarvis et al., 1984) includes podzols of this type, some of which have peaty surface horizons and can therefore be regarded as intergrading to stagnopodzols.

In parent materials which are loamy or richer in weatherable minerals, the eluvial horizons are generally thinner and can be masked by incorporated organic matter; the

Figure 6.3: The dissected summit plateau of the Cairngorms from south-east of Loch Avon, Banffshire. Oro-arctic humus podzols in mountain-top detritus are the dominant soils over much of this landscape (copyright Aerofilms Ltd).

illuvial horizons are typically friable, with granular (crumb) structure and pellety microfabric, and more or less indurated BC horizons are common. Non-hydromorphic podzols with these features are particularly widespread under heather moor and coniferous woodland in north-east Scotland, where they have developed in drifts derived from acid and intermediate igneous rocks, schists, slates and sandstones. Many of the associated freely draining cultivated soils appear to have had similar profiles originally but are now podzolic brown soils according to the criteria adopted here.

Although parent material is clearly one factor causing differences in profile morphology, several authors (e.g. Scheys et al., 1954; Mackney, 1961; Guillet, 1975) have cited evidence showing that vegetational history has also played a role, 'iron podzols' with friable Bs horizons commonly reflecting an initial stage of podzolization under forest and humus-iron podzols with compact Bh horizons resulting from replacement of forest by heath. Anderson et al. (1982) also consider that the Bs horizon normally forms first, primarily through downward translocation of inorganic sols followed by illuvial accumulation of aluminium as allophane or imogolite and iron as ferrihydrite. As the Bs horizons of most iron podzols, including those studied by Anderson *et al.*, contain more alkali-extractable carbon than the Ea horizons, however, it seems likely that both elements are transported at least partly as light coloured metal-organic complexes (cf. Buurman and Van Reeuwijk, 1984).

In their natural (unimproved) state, all these soils are strongly acid and deficient in available nutrients. Reserves of nitrogen and bases are mainly located in the organic surface layer, which has a high C/N ratio and mineralizes very slowly, whilst phosphorus, manganese, cobalt and copper are transported and immobilized in the B horizon along with aluminium and iron. In the sandy and sandy-gravelly variants that retain their identity under cultivation, these chemical defects have usually been alleviated by the use of lime and fertilizers, and in some places their physical and chemical properties have been substantially modified by heavy organic manuring, marling, or both. As a rule, however, inadequate moisture storage capacity remains a serious limitation to crop production in subhumid areas, especially where rooting is restricted by a compact or cemented subsurface horizon which has not been disrupted by deep ploughing or subsoiling. Deficiencies of magnesium, manganese, boron or copper can also be induced or accentuated by overliming, and the weak structure of the eluvial horizons consititutes a considerable erosion hazard in newly cultivated land.

Well drained podzols present fewer limitations for forestry than for agricultural or horticultural cropping. Of the major coniferous species used for afforestation, Scots and Corsican pine are well adapted to these soils, and providing that rootable depth is adequate and cultural measures are directed to ensuring that the trees are well established, neither low nutrient status nor drought appears to limit their performance significantly.

Key to Subgroups

1. Non-hydromorphic podzols with an argillic B horizon (p.109)
   Luvic podzols
2. Other non-hydromorphic podzols with a Bs horizon 5 cm or more thick
   Humoferric podzols
3. Other non-hydromorphic podzols
   Humus podzols

6.2.2 Humoferric Podzols

According to the definition adopted here, these are non-hydromorphic podzols having a Bh horizon over a Bs, or a Bs only, and no underlying argillic (Bt) horizon. The variants without a distinct dark Bh have been called iron podzols by the Soil Survey of Scotland (e.g. Glentworth, 1962; Glentworth and Muir, 1963; Romans, 1970) and ferric podzols in England and Wales (Avery, 1980); they are usually intricately interspersed with profiles conforming to the central concept and have only occasionally been mapped separately in soil surveys. Nearly all these soils are Orthods in the US Taxonomy and Orthic Podzols in the FAO (1974) systems. A few have an iron-deficient Bh and therefore qualify as Humods (FAO Humic Podzols) but are not easily distinguished in the field.

Though widely distributed in the British Isles, soils of this subgroup are restricted to parent materials containing appreciable amounts of free iron or iron-bearing minerals and occur mainly but not exclusively in areas with temperate or boreal climatic regimes. Variants located within or just below the oro-arctic zone are distinguished as subalpine podzols by the Soil Survey of Scotland (1984). Under semi-natural vegetation or planted conifers there is usually a light coloured Ea horizon above the podzolic B but this can be absent in cultivated land and in uncultivated variants intergrading to podzolic brown soils, as exemplified by profile 86 (p.252). Intergrades to gley-podzols (p.261) and humus-ironpan stagnopodzols (p.280) are also locally common, the former having mottles indicative of gleying at depths below 40 cm and the latter a very thin or discontinuous ironpan at the Bh/Bs interface. Data on four representative profiles are reproduced on pp. 250–4.

Profile 84 is in thin sandy drift over red Permian sandstone in Cumbria. It is sited in former heathland afforested with conifers and has a well developed Bh horizon in which pyrophosphate-extractable iron, aluminium and carbon all attain maximal values. Profile 85, described and sampled by D. Mackney in the north Shropshire plain, represents a cultivated phase developed in sandy glaciofluvial drift. The illuvial horizons resemble those of the preceding profile except that the Bs is more clearly differentiated by colour and contains as much extractable iron and aluminium as the Bh. Although the B horizons of both profiles are appreciably cemented, the cementation is insufficiently strong or continuous to prevent penetration by roots, especially of pine. Earthworm channels were also noted below the Bh in the cultivated soil, which has a near-neutral reaction throughout, presumably as a result of liming or marling.

Profile 86 is located in steeply sloping heather moor near the upper altitudinal limit of the oroboreal zone in the eastern Grampians. The parent material is drift derived from fine grained schists and phyllites and textures are correspondingly loamy. The eluvial (Ah) and Bh horizons are both much thinner than in profile 84; because of their large organic matter contents they are only slightly differentiated by colour when moist, but the Ah contains numerous uncoated sand grains and the illuvial character of the Bh is clearly evident from the analytical data, which show large increases in pyrophosphate-extractable iron and aluminium coupled with a smaller increase in extractable carbon. The underlying friable Bs has the largest extractable aluminium content and passes downwards into a much more compact BC horizon. This may be very weakly cemented (duric horizon) but is less strongly indurated than corresponding horizons in granitic drift at lower altitudes in north-east Scotland (Romans, 1970). Morphologically similar soils of the Moor Gate and Bowden series (Clayden and Hollis, 1984) in drift from acid and basic igneous rocks respectively have been classed as humic brown podzolic soils in England and Wales (Avery *et al.*, 1977). Admixture of the upper horizons by ploughing gives rise to a dark humose Ap and the Bh is no longer discernible.

Profile 87 was described and sampled by A.D. Walker in a Scots pine plantation in Strathspey. It is developed in sandy glaciofluvial gravel derived mainly from Dalradian schists and is included as an example of the iron podzols lacking a Bh horizon. As in similar soils under nearby remnants of indigenous pine forest, there is a sharply defined mor layer at the surface, directly underlain by a strongly bleached Ea horizon. The friable Bs is enriched in organic carbon as well as in sesquioxides and merges downwards into a weakly cemented less ochreous horizon containing insufficient pyrophosphate-extractable aluminium and iron to qualify as Bs. This rests abruptly on an extremely stony cemented layer (duric horizon) which evidently acts as a barrier to root penetration. Iron podzols with or without an indurated subsoil layer are common in north-east Scotland (Glentworth, 1962; Romans, 1970) and have also been identified in several other British localities, including Exmoor (Hogan and Harrod, 1982) and south-west Ireland, where variants modified by cultivation were described as brown podzolic soils by Conry *et al.* (1972). According to the criteria used here, cultivated analogues are considered as podzols rather than podzolic brown soils only if part of the E horizon remains, of if the Bs is cemented.

Profile 84 (Matthews, 1977)

Location	: Dolphenby, Cumbria (NY 570309)
Climatic regime	: humid temperate
Altitude	: 123 m Slope: < 1°
Land use and vegetation	: Scots pine (*Pinus sylvestris*) and larch (*Larix decidua*) plantation with field layer of grasses and bracken (*Pteridium aquilinum*)

Horizons (cm)

- L/F — More or less decomposed grass and pine litter, c. 10 cm thick.
- 0–8 Ah — Dark reddish brown (5YR 2/2), very slightly stony, humose loamy medium sand (USDA sand to loamy sand) with many uncoated grains; few, medium and large, subangular to rounded fragments of igneous rock and sandstone; weak medium to fine granular; many fine and medium to coarse fleshy and woody roots; abrupt smooth boundary.
- 8–18 Ea — Grey (5YR 5/1) very slightly stony medium sand (USDA sand); stones as above; single grain; very friable to loose; few fine and common medium and coarse woody roots; abrupt wavy boundary.
- 18–30 Bh(m) — Dark reddish brown (5YR 2/2) stoneless loamy medium sand (USDA sand); weak coarse platy; firm; very weakly cemented in parts; few fine roots in cracks and common coarse woody roots; gradual wavy boundary.
- 30–48 Bs(m) — Dark reddish brown (2.5YR 3/4) very slightly stony loamy medium sand (USDA loamy sand to sand); few small to medium angular sandstone fragments; very weak coarse blocky; firm; very weakly cemented in parts; common coarse woody roots; gradual boundary.
- 48–90 BC — Yellowish red (5YR 4/6) very slightly stony medium sand (USDA sand); few weathering sandstone fragments; single grain; loose; few coarse woody roots.
- 90+ Cr — Yellowish red (5YR 4/6) soft medium grained sandstone.

Analytical data *see* below.

Classification

England and Wales (1984): humo-ferric podzol; medium to coarse sandy material passing to sand or soft sandstone (Shirrell Heath series).

USDA (1975: Typic Haplorthod; sandy, siliceous or mixed, mesic, cemented.

FAO (1974): Orthic Podzol; coarse textured.

Analytical data for Profile 84

Horizon	Ah	Ea	Bh(m)	Bs(m)	BC
Depth (cm)	0–8	8–18	18–30	30–48	48–90
sand 600 µm–2 m %	11	7	6	4	3
200–600 µm %	48	55	52	55	55
60–200 µm %	28	28	30	29	34
silt 2–60 µm %	8	7	5	3	3
clay <2 µm %	5	3	7	9	5
organic carbon %	12	0.9	2.9	1.2	0.2
pyro. ext. Fe %	0.2	0.05	0.8	0.3	0.05
pyro. ext. Al %	0.1	0.02	0.3	0.2	0.1
pyro. ext. C %	2.7	0.4	2.1	0.7	0.2
pyro. Fe+Al %/clay %			0.16	0.07	
total ext. Fe %	0.4	0.2	1.2	0.9	0.7
pH (H$_2$O)	4.0	4.1	4.0	4.1	4.2
(0.01M CaCl$_2$)	3.4	3.5	3.6	4.0	4.2

Profile 85

Location	: Cheswardine, Shropshire (SJ 696296)
Climatic regime	: subhumid temperate
Altitude	: 122 m Slope: 1°
Land use	: ley grass

Horizons (cm)

0–22
Ap
Very dark brown (7.5YR 2/2) slightly stony loamy medium sand (USDA loamy sand) containing many uncoated grains; common quartzose pebbles; weak fine and medium subangular blocky; very friable; many roots; earthworms present; abrupt smooth boundary.

22–30
Bh
Black (5YR 2/1) to dark reddish brown (5YR 2/2), very slightly stony, humose loamy medium sand (USDA loamy sand); stones as above; coarse platy; friable to firm; very weakly cemented in parts; few roots, mainly in fissures; abrupt irregular boundary, tonguing downwards to 60 cm.

30–48
Bs
Yellowish red (5YR 4/8) very slightly stony loamy medium sand (USDA loamy fine sand); massive; firm and very weakly cemented in upper part, friable below; few fine roots; few earthworm channels; gradual boundary.

48–90
2BC
Brown (7.5YR 4/4) and light brown (7.5YR 6/4) stoneless medium sand (USDA sand); single grain; loose; rare roots and earthworm channels.

Analytical data *see* below.

Classification

England and Wales (1984): humo-ferric podzol; sandy drift with siliceous stones (Redlodge series).

USDA (1975): Typic Haplorthod; sandy, mixed or siliceous, mesic, cemented.

FAO (1974): Orthic Podzol; coarse textured.

Analytical data for Profile 85

Horizon	Ap	Bh	Bs	2BC
Depth (cm)	2–22	22–30	30–48	48–90
sand 600 μm–2 mm %	4	3	2	3
200–600 μm %	37	31	27	46
60–200 μm %	38	46	49	44
silt 2–60 μm %	15	14	16	6
clay <2 μm %	6	6	6	1
organic carbon %	2.9	4.3	1.1	
pyro. ext. Fe %	0.1	0.4	0.4	<0.05
pyro. ext. Al %	0.1	0.3	0.3	0.1
pyro. ext. C %	0.9	3.0	0.9	0.3
pyro. Fe+Al %/clay %		0.12	0.12	
total ext. Fe %	0.5	0.9	1.2	0.3
pH (H$_2$O)	6.7	6.9	6.7	6.6
(0.01M CaCl$_2$)	6.4	6.5	6.5	6.4

Profile 86 (Heslop and Bown, 1969; Ragg and Clayden, 1973)

Location : Ladylea, Strathdon, Aberdeenshire (NJ 338170)
Climatic regime: humid oroboreal
Altitude : 490 m Slope: 18°
Land use and vegetation : rough grazing; moorland with heather (*Calluna vulgaris*), bell-heather (*Erica cinerea*), crowberry (*Empetrum nigrum*) and cowberry (*Vaccinium vitis-idaea*)

Horizons (cm)

0–4 F/H Black (5YR 2/1) mainly well decomposed organic material containing some uncoated mineral grains; abundant fine roots; abrupt boundary.

4–10 AH Very dark grey to black humose sandy loam (USDA fine sandy loam) containing uncoated sand grains; massive; frequent roots; abrupt boundary.

10–13 Bh Dark reddish brown (5YR 2.5/2) humose sandy loam (USDA fine sandy loam); coarse blocky; friable; abrupt irregular boundary.

13–38 Brown (7.5YR 4.5/4.5) sandy loam Bs (USDA fine sandy loam) medium to coarse subangular blocky; friable; frequent fine roots; gradual boundary.

38–70 BCx(m) Yellowish brown (10YR 5/4) sandy loam (USDA sandy loam); coarse blocky to platy; firm to very firm (moderately indurated); occasional roots; some humus and iron staining on old root channels; gradual boundary.

70–122+ Cu(x) Brown (10YR 5/2.5) sandy loam (USDA sandy loam); massive; frequent mica schist fragments; silt cappings on upper surfaces of stones.

Analytical data *see* below.

Classification

Scotland (1984): humus-iron podzol (Foudland series).
England and Wales (1984): humic brown podzolic soil; coarse loamy material over lithoskeletal acid schist (unnamed series).
USDA (1975): Cryic Fragiorthod (or Duric Cryorthod); coarse loamy or loamy-skeletal, mixed, mesic, uncemented.
FAO (1974): Orthic (or Leptic) Podzol; coarse to medium textured (fragipan phase?).

Analytical data for Profile 86

Horizon	F/H	Ah	Bhs	Bs	BCx(m)	Cu(x)	Cu(x)
Depth (cm)	0–4	4–10	10–13	18–28	44–54	84–94	114–122
silt[a] 2–50 µm %		20	29	31	34	27	21
clay[a] <2 µm %		7	7	9	7	4	6
organic carbon %	49	12	9.0	3.6	1.3	0.4	0.2
nitrogen %	1.54	0.61	0.34				
C/N ratio	32	19	25				
pyro. ext. Fe %		0.1	0.6	0.6	0.08	0.05	0.04
pyro. ext. Al %		0.3	0.7	0.9	0.4	0.2	0.1
pyro. ext. C %		3.8	4.7	2.7	0.8	0.1	0.1
pyro. Fe+Al %/clay %		0.06	0.19	0.17	0.07		
citrate-dithionite ext. Fe %		0.6	2.8	2.3	1.0	0.9	0.9
TEB (me/100 g)	19.7	2.2	1.4	0.2	0.2		
CEC (me/100 g)	84	35	36	17	5.7		
% base saturation	23	6	4	1	3		
pH (H$_2$O)	4.1	4.3	4.6	4.9	5.1	5.5	5.5
(0.01M CaCl$_2$)	3.1	3.1	3.4	3.9	4.4	4.5	4.6

[a] by hydrometer method

Profile 87

Location	:	Rothiemurchus, Inverness-shire (NH 880070)
Climatic regime:		humid oroboreal
Altitude	:	290 m Slope: 12°
Land use and vegetation	:	Scots pine plantation about 25 years old; field layer of heather (*Calluna vulgaris*); bilberry (*Vaccinium myrtillus*), cowberry (*Vaccinium vitis-idaea*) and feather mosses.

Horizons (cm)

L — Fresh litter, ca. 2 cm thick.

0–5 F1 — Dark reddish brown (5YR 3-2/2) partially decomposed litter; felty.

5–10 F2/H — Dark reddish brown (5YR 2/2) more decomposed litter; moderately humified but layering still visible; abrupt wavy boundary with discontinuous H horizon less than 1 cm thick at base.

10–18 Ea — Pinkish to reddish grey (5YR 6–5/2) stony gravelly loamy sand (USDA loamy sand); common subrounded and rounded stones up to 10 cm, mainly schist; weak medium blocky; very friable; common woody and fibrous roots; abrupt slightly wavy boundary.

18–30 Bs — Yellowish red (5YR 5/8) stony loamy sand (USDA loamy sand); stones as above; weak medium blocky; very friable; common woody and fibrous roots; gradual smooth boundary.

30–46 BC(m) — Reddish yellow (5YR 6/8) stony loamy sand (USDA loamy sand); stones as above; weak medium angular blocky; very weakly cemented in parts; common woody and fibrous roots; abrupt wavy boundary.

46–94 BCm — Brown (7.5YR 5/4) very or extremely stony sand (USDA sand); abundant rounded and subrounded stones, mainly schist less than 5 cm, with occasional granite fragments up to 30 cm; massive; weakly and locally very weakly cemented; roots rare; common diffuse ferruginous staining in stone cavities; gradual boundary.

94–160 Cu — Light yellowish brown (10YR 6/4) very or extremely stony sand (USDA sand); stones as above; single grain; mainly loose but locally very weakly cemented in upper part; roots rare; iron staining as above.

Analytical data for Profile 87

Horizon	F1	F2/H	Ea	Bs	BC(m)	BCm	Cu
Depth (cm)	0–5	5–9	11–17	22–28	35–42	55–65	97–107
silt[a] 2–50 μm %			19	21	21	<1	1
clay[a] <2 μm %			4	4	2	2	2
loss on ignition %	74	94	3.4	7.8	4.7	1.6	
organic carbon %	43	50	2.0	3.6	0.8	0.1	
nitrogen %	1.38	1.35	0.13	0.07	0.03		
C/N ratio	31	37	15	51	28		
pyro. ext. Fe %			0.09	0.2	0.03	0.03	
pyro. ext. Al %			0.06	0.6	0.2	0.1	
pyro. ext. C %			0.6	0.7	<0.05	<0.05	
pyro. Fe+Al % clay %				0.2	0.05		
total ext. Fe %			0.2	1.5	1.2	0.2	
TEB (me/100 g)	33	19.5	0.3	0.5	0.3		
CEC (me/100 g)	176	195	8.5	12.1	5.0		
% base saturation	19	10	4	4	6		
pH (H$_2$O)	4.1	3.6	4.0	4.9	5.2	5.6	5.4
(0.01M CaCl$_2$)	3.3	2.9	3.3	4.5	4.8	5.2	5.3

[a] by hydrometer method

Classification
  Scotland (1984): iron podzol (Forres series).
  England and Wales (1984): ferric podzol; sandy-gravelly, non-calcareous gravel (unnamed series).
  USDA (1975): (Duric) Cryorthod; sandy-skeletal, mixed.
  FAO (1974): Orthic Podzol; coarse textured (fragipan phase?).

6.2.3 Humus Podzols

These are non-hydromorphic podzols with a black to dark reddish brown Bh horizon and no underlying Bs or Bt of significant thickness. As originally described by Frosterus (1914), and later by Kubiena (1953), they are typically developed in iron-deficient sandy materials. In the England and Wales classification the Bh horizon is accordingly required to contain less than 0.4 per cent free iron and similar but not identical differentiating criteria are used to distinguish Humods from Orthods in the US Taxonomy and Humic Podzols from Orthic Podzols in the FAO (1974) system. Thus a Humod must have a Bh in which the ratio of free iron to carbon is less than 0.2 and a Humic Podzol one containing insufficient free iron (<0.5 per cent) to redden on ignition. In the classification adopted here, however, no limit is placed on iron content because experience has shown that the previously proposed criteria are difficult to apply consistently in the field. More or less well drained podzols conforming to this less restrictive definition occur in existing or former heathland on siliceous sandy or sandy-gravelly deposits, including certain Tertiary and Lower Cretaceous sands in southern England and windblown sands bordering the Moray Firth, and in weakly weathered mountain top detritus at higher altitudes, but are considerably less extensive in the British Isles as a whole than the humoferric podzols into which they frequently grade.

The lowland humus podzols commonly have thick Ea horizons, extending in places to depths of a metre or more, and the Ea/Bh boundary is often convoluted, with downward projections apparently following former root channels. Another characteristic feature is the common occurrence of thin wavy humus-enriched laminae both above and below the main B horizon. These extremely impoverished soils are especially prone to erosion by wind or water when the vegetative cover and associated organic surface layer are disrupted. This can result in exposure of the compacted Bh horizon and concomitant accumulation of aeolian or colluvial deposits of bleached sand in which sandy regosols (p. 143) subsequently develop.

The humus podzols in montane situations are set apart as alpine podzols by the Soil Survey of Scotland (1984) but are considered here as oro-arctic (cryic) phases. They normally show features attributable to freeze-thaw processes (Ragg and Bibby, 1966), including upper horizons with a relatively loose and porous fabric and the frequent absence of a superficial organic layer devoid of mineral grains. In places, particularly on quartz-rich rocks, there is a well defined Ea horizon above the Bh (e.g. Plate II.8 and profile 91 on p. 258), but weakly differentiated profiles (e.g. profile 90) called cryptopodzolic rankers by Duchaufour (1982) and humic cryptopodzols by Avery (1973, 1980) are more common. These have humose upper horizons (Ah/Ea) that contain less pyrophosphate-extractable aluminium and iron than the underlying Bh and are distinguishable from it in the field by the absence of organic coatings on sand grains and stones. Both eluvial and illuvial horizons are typically developed in a very or extremely stony layer, often with a stone pavement at the surface, which may rest directly on bedrock but commonly overlies a denser and less stony layer in which the stones have silty cappings.

These soils were first distinguished and characterized in Britain at altitudes above about 700 m in the Grampian Highlands by Romans *et al.* (1966), who described them as 'alpine soils' and interpreted them as A/C profiles. They were subsequently identified and mapped in various parts of Scotland by Heslop and Bown (1969), Bown (1973), Futty and Dry (1977) and Bown and Heslop (1979), and their overall distribution is depicted on the 1:250,000 soil association maps published in 1982. Ragg and Clayden (1973) reported analytical data on two representative profiles from the Southern Uplands, which confirmed that the characteristic Bh horizons qualify as spodic according to the US system. Judging from these and later analyses, the Scottish alpine podzols comprise both Humods (Cryohumods or Cryic Fragihumods) and Orthods (Cryothods or Cryic Fragiorthods), depending on the ratio of free iron to carbon in the Bh. The landscape units in which they appear as dominant or subdominant components (Figures 6.3 and 4.6) normally include rankers, lithosols, and bare rock or scree as well as podzols and are characterized by various low growing plant communities, including alpine lichen heath, stiff sedge–fescue (*Carex bigelowii – Festuca vivipera*) grassland and fescue–woolly fringe moss (*Rhacomitrum lanuginosum*) heath (Soil Survey of Scotland, 1984), which provide scanty pasturage for sheep and deer. Variants with peaty surface horizons also occur, particularly in saddles and hollows where snow accumulates and the soil is periodically saturated with water. Similar terrain has been noted at high altitudes in the Lake District and Snowdonia and on the summit of the Cross Fell range in the northern Pennines but no representative profiles have been fully characterized in any of these areas.

Field and analytical data on two representative lowland profiles, both qualifying as Humods, and two oro-arctic (alpine) humus podzols are reproduced on pp.256–9. Profile 88 was described and sampled by D.J. Eldridge in an abandoned sand pit in west Norfolk. As the Lower Cretaceous (Sandringham) sands at this site contain little silt and clay and less than 0.1 per cent extractable iron, the illuvial accumulation products in the thick Bh horizon consist essentially of organic matter and aluminium. Profile 89 (Plate III.7) is located in dry heathland on fine, slightly glauconitic Tertiary (Barton Beds) sand in the New Forest, Hampshire. Here the Bh horizon is shallower and thinner and both it and the succeeding horizons between 42 and 91 cm depth contain more free iron. The massive brownish yellow horizon directly below the Bh is designated BCt rather than Bs because it has less than 0.4 per cent pyrophosphate-extractable Fe + Al and a thin section showed that the sand grains are thinly coated and in places bridged by illuvial clay.

The following montane profiles are both in granitic materials. Profile 90, located near the summit of Cairnsmore of Fleet in Galloway, has bedrock at 28 cm depth and there is only a very thin C horizon above it. As in profile 86 (p.252), the Ah and Bh horizons are both humose and the former has more organic matter, but amounts of pyrophosphate-extractable iron and aluminium and the proportion of the total carbon extracted by pyrophosphate are much larger in the Bh. As the fine earth fraction of this horizon contains 0.8 per cent free iron, the soil is an Orthic Podzol according to the FAO system, but it is a Cryohumod in the US Taxonomy because the ratio of free iron to organic carbon is only 0.08. Profile 91 was described and sampled by E.L. Birse, R.E.F. Heslop and J.S. Robertson at an altitude of 1,265 m in the Cairngorms. Here the ground surface consists largely of stones and boulders and podzols like that described, with a grey Ea horizon in the extremely stony surface layer, are confined to relatively stable sites on gentle slopes. The total extractable iron and organic carbon contents of the Bh1 horizon are 0.7 and 3.4 per cent respectively, so that this soil is again an Orthic Podzol according to FAO but just fails to qualify as a Humod.

Profile 88

Location	:	Leziate, Norfolk (TF 685184)
Climatic regime:		subhumid temperate
Altitude	:	20 m Slope: 3°
Vegetation	:	open birch woodland with field layer dominated by bracken (*Pteridium aquilinum*)

Horizons (cm)

L/F	More or less decomposed litter, c. 1 cm thick.
0–30 Ah	Black (2.0) medium sand (USDA fine sand) containing occasional quartzose sandstone fragments, sand grains mainly uncoated; single grain; very friable; common medium to large woody roots; sharp wavy boundary.
30–76 Ea	Pinkish grey (7.5YR 7/2; 7.5YR 6/4 rubbed) medium sand (USDA fine sand) containing occasional sandstone fragments as above; single grain; loose; about twelve dark coloured subhorizontal laminae c. 1 mm thick and rare red to yellowish red soft plate-like concentrations; few woody roots; sharp wavy boundary (72–85 cm).
76–108 Bh1m	Black (2/0) stoneless humose medium sand (USDA sand) with coated grains; massive; firm, very weakly cemented; roots rare; abrupt irregular boundary (100–120 cm).
108–113 Bh2	Dark reddish brown (5YR 3/3) stoneless medium sand (USDA sand) less compact than above; horizon discontinuous, with sharp irregular lower boundary (110–120 cm); many rounded soft concentrations as above.
113–200 Cu	Pinkish grey (7.5YR 7/2; 7.5YR 5/2 rubbed) stoneless fine sand (USDA fine sand); single grain; loose; many black subhorizontal laminae 1–5 mm thick.

Analytical data *see below*.

Classification

England and Wales (1984): humus podzol; medium or coarse sandy material passing to sand or soft sandstone (Leziate series).
USDA (1975): Typic (or Arenic) Haplohumod; sandy, siliceous, mesic, cemented.
FAO (1974): Humic Podzol; coarse textured.

Analytical data for Profile 88

Horizon	Ah	Ea	Bh1m	Bh2	Cu
Depth (cm)	18–28	53–63	88–93	98–108	138–148
sand 200 μm–2 mm %	36	34	47	43	31
60–200 μm %	62	65	50	54	68
silt 2–60 μm %	2	1	1	1	1
clay <2 μm %	<1	<1	2	2	<1
organic carbon %	2.2	0.2	1.4	0.5	0.1
pyro. ext. Fe %	0.03	0.02	0.12	0.05	0.02
pyro. ext. Al %	0.03	0.01	0.12	0.11	0.03
pyro. ext. C %	1.0	0.1	0.9	0.4	0.1
total ext. Fe %	0.08	0.07	0.17	0.10	0.07
pH (H_2O)	3.9	3.9	3.7	4.1	4.7
(0.01M $CaCl_2$)	2.9	3.2	3.1	3.7	4.3

Profile 89 (Jarvis and Findlay, 1984)

Location	:	Lyndhurst, Hampshire (SU 335070)
Climatic regime:		subhumid temperate
Altitude	:	28 m Slope: 4°
Land use and vegetation	:	rough grazing; lowland heath with c. 60 per cent heather (*Calluna vulgaris*), c. 40 per cent bell heather (*Erica cinerea*) and mosses.

Horizons (cm)

2–0   Black superficial spoil (roadside section).

0–5 H   Dark reddish brown (5YR 2/2) organic material containing uncoated sand grains; moderate medium granular; very friable; abundant fine to medium roots; abrupt smooth boundary.

5–14 Ah   Black (5YR 2/1) stoneless humose fine sand (USDA fine sand) with uncoated grains; weak granular; friable; abundant roots; abrupt smooth boundary.

14–34 Ea   Reddish grey (5YR 5/2) fine sand (USDA fine sand) containing few small flints; single grain; loose; common fine to medium roots; abrupt wavy boundary.

34–42 Bh   Black (5YR 2/1) stoneless fine sand (USDA fine sand) with coated grains; massive; firm; few fine roots; abrupt irregular boundary.

42–64 BCt   Brownish yellow (10YR 6/8) stoneless fine sand (USDA fine sand) with few strong brown (7.5YR 6/8) and dark brown (7.5YR 3/2-4) inclusions; massive; firm; few roots; abrupt irregular boundary.

64–91 BC   Yellowish brown (10YR 6/4) stoneless fine sand (USDA fine sand) with common irregular very pale brown (10YR 3/3) inclusions; massive; friable; few thin wavy humus-enriched laminae and irregular soft ferruginous concentrations and nodules; no roots; abrupt wavy boundary.

91–148 Cu   Very pale brown (10YR 7/3) stoneless fine sand (USDA fine sand) with coarse brownish yellow (10YR 6/8) laminae and few irregular dark brown (10YR 3/3) bands; massive; very friable; few soft irregular ferruginous concentrations; no roots.

Analytical data *see* below.

Classification
England and Wales (1984): humo-ferric podzol; fine sandy material passing to sand or soft sandstone (unnamed series).
USDA (1975): Typic Haplohumod; sandy, siliceous or mixed, mesic, cemented.
FAO (1974): Humic Podzol; coarse textured.

Analytical data for Profile 89							
Horizon	H	Ah	Ea	Bh	BCt	BC	Cu
Depth (cm)	0–5	5–14	14–34	34–42	42–64	64–91	91–148
sand 200 μm–2 mm %	8	5	2	2	2	2	3
60–200 μm %	73	89	96	93	95	91	95
silt 2–60 μm %	15	4	2	2	1	2	1
clay <2 μm %	4	2	<1	3	2	4	1
organic carbon %	18	8.1	0.7	4.2	0.3		
pyro. ext. Fe %	0.2	0.1	0.03	0.3	0.2	0.1	0.1
pyro. ext. Al %	0.04	0.02	0.1	0.2	0.1	0.1	0.1
pyro. ext. C %	1.8	1.3	0.2	2.4	0.3	0.1	0.1
total ext. Fe %		0.2	0.1	0.4	0.4	0.4	0.3
pH (H_2O)	3.8	3.8	4.2	4.2	4.7	4.9	5.1
(0.01M $CaCl_2$)	3.0	2.8	3.1	3.6	4.1	4.3	4.5

Profile 90 (Bown and Heslop, 1979)

Location	:	Cairnsmore of Fleet, Kirkcudbrightshire (NX 502672)
Climatic regime:		perhumid oro-arctic
Altitude	:	704 m Slope: 2°
Land use and vegetation	:	rough grazing; montane dwarf heath with cowberry (*Vaccinium vitis-idaea*), bilberry (*Vaccinium mvyrtillus*), least willow (*Salix herbacea*), stiff sedge (*Carex bigelowii*) and woolly fringe-moss (*Rhacomitrium lanuginosum*)

Horizons (cm)

0–3 H	Black (N 2/0) organic material with much mineral matter and frequent bleached granite fragments; weak granular; many roots; clear boundary.
3–10 Ah	Very dark grey (5YR 3/1) very stony humose sandy loam (USDA sandy loam) containing numerous bleached sand grains; abundant large granite fragments; weak granular; many roots; clear boundary.
10–26 Bh	Dark reddish brown (5YR 3/2) very stony humose sandy loam (USDA sandy loam); stones as above; weak subangular blocky; wet; mineral grains have black humose coatings; frequent roots; abrupt boundary.
26–28 Cu	Brown (10YR 5/3) very stony sandy loam with more silt than above; stones as above; massive; wet; no roots; gradual boundary.
28+ Cu/R	Granite rock in place.

Analytical data

Horizon	H	Ah	Bh
Depth (cm)	0–2	3–7	12–17
silt[a] 2–50 μm %		19	15
clay[a] <2 μm %		7	10
loss on ignition %	41	19	12
organic carbon %	23	10.4	6.8
nitrogen %	1.44	0.70	0.43
C/N ratio	16	15	16
pyro. ext. Fe %		0.1	0.5
pyro. ext. Al %		0.2	1.6
pyro. ext. C %		2.3	2.9
pyro. Fe+Al %/clay %		0.05	0.20
total ext. Fe %		0.2	0.8
TEB (me/100 g)	2.7	0.8	0.3
CEC (me/100 g)	29.5	10.0	27.2
% base saturation	9	8	1
pH (H$_2$O)	4.1	4.3	4.6

[a] by hydrometer method

Classification
 Scotland (1984): alpine podzol (Mulltaggart series).
 England and Wales (1984): typical humic cryptopodzol; loamy, lithoskeletal acid crystalline rock (unnamed series).
 USDA (1975): Lithic Cryohumod; loamy-skeletal, non-cemented.
 FAO (1974): Orthic Podzol; coarse textured (stony phase).

Profile 91

Location	:	Braeriach, Aberdeenshire (NN 936982)
Climatic regime:		perhumid oro-arctic
Altitude	:	1,265 m Slope: 4°
Land use and vegetation	:	rough grazing; stiff sedge–fescue grassland; stiff sedge (*Carex bigelowii*) dominant, with frequent least willow (*Salix herbacea*) and occasional moss campion (*Silene acaulis*)

Horizons (cm)

0–2 Ah	More or less humified organic material mixed with granitic sand and stones.
2–21 Ea	Grey (10YR 5/1) very or extremely stony coarse sand (USDA coarse sand); abundant granite fragments with some boulders; single grain; loose; abundant roots; mainly *Carex*; clear boundary.
21–27 Bh1	Black (5YR 2/1) stony loamy coarse sand (USDA loamy coarse sand); massive; frequent roots; mineral grains have dark organic coats; abrupt wavy boundary (horizon locally almost absent).

27–48 Bh2	Brown to dark brown (7.5YR 4/2) very stony coarse sandy loam (USDA coarse sandy loam); stones as above, some very large; massive; frequent roots; clear boundary.	
48–77 2BC(x)	Brown (7.5YR 5/4) slightly to very stone sandy silt loam (USDA sandy loam); stones as above, mainly frequent to abundant, but locally absent; massive; occasional roots; silt cappings on stones; gradual boundary.	
77–90 3Cu	Brown (7.5YR 5/4) very stony sandy loam (USDA sandy loam); stones as above, including some boulders; massive; no roots.	

Analytical data *see* below.

Classification

Scotland (1984): alpine podzol (Rinnes series).
England and Wales (1984): humo-ferric podzol; loamy lithoskeletal, acid crystalline rock (Iveshead series).
USDA (1975): Typic Cryorthod or Cryic Fragiorthod; loamy-skeletal, mixed, non-cemented.
FAO (1974): Orthic Podzol; coarse textured (stony phase).

6.2.4 Luvic Podzols

These are non-hydromorphic podzols in which Ea and Bh and/or Bs horizons appear to have formed in the clay-depleted eluvial horizon of a pre-existing luvic brown soil, as evidenced by the presence of an underlying argillic B (Bt). Bisequal profiles of this type with a brightly coloured (chromic) argillic B horizon (Chapter 3, p. 109) below the podzolic B were first identified in older river-terrace and plateau deposits in southern England (Avery, 1958, 1964; Jarvis, 1968; Mackney, 1970) and were subsequently classed as paleo-argillic podzols by Avery (1973, 1980). Others with ordinary argillic B horizons occur sporadically in coarse textured Devensian deposits and pre-Quaternary sandy formations but are relatively inextensive. A brownish or yellowish Eb horizon appears in places between the podzolic horizons and the upper boundary of the argillic B, which is generally wavy or irregular. Elsewhere this intermediate horizon is absent and a Bh or Bs is superimposed

Analytical data for Profile 91

Horizon	Ah	Ea	Bh1	Bh2	2BC(x)	3Cu
Depth (cm)	0–2	7–17	21–27	32–42	57–62	82–92
silt[a] 2–50 µm %	14	9	18	22	41	33
clay[a] <2 µm %	2	1	3	3	5	4
loss on ignition %	14.6	5.2	8.2	5.1	3.0	2.9
carbon %	7.6	2.9	3.4	1.7	0.6	0.4
nitrogen %	0.37	0.17	0.17	0.08	0.03	0.02
C/N ratio	21	17	20	21	19	19
pyro. ext. Fe %	0.02	0.04	0.3	0.2	0.04	0.04
pyro. ext. Al %	0.08	0.01	0.5	0.4	0.3	0.2
pyro. ext. C %	0.9	0.2	0.7	1.0	0.3	<0.05
pyro. Fe+Al %/clay %			0.3	0.2	0.06	
total ext. Fe %	0.1	0.1	0.7	0.7	0.8	1.0
TEB (me/100 g)	1.3	0.2	0.1	<0.05	0.1	0.1
CEC (me/100 g)	12.4	3.9	9.9	7.7	3.6	2.4
% base saturation	9	4	1	1	3	3
pH (H$_2$O)	3.7	3.6	3.9	3.8	3.9	3.9
(0.01M CaCl$_2$)	2.9	2.8	3.2	3.4	3.6	3.7

[a] by hydrometer method

on the upper part of the argillic horizon. Horizons below the podzolic B may have fragipan characteristics, and mottles attributable to gleying occur at or below 50 cm depth in intergrades to luvic gley-podzols (p. 266).

Some of these soils, especially those of the Southampton series (M.G. Jarvis *et al.*, 1984) in existing or former heathland on 'plateau gravels', have a thick Ea horizon over a more or less cemented Bh. In loamier materials the eluvial horizons are thinner and the podzolic B horizons more friable, as in profile 92 below. This woodland soil was described and sampled by S.J. Fordham in Angular Chert Drift (Dines *et al.*, 1969; Atkinson and Burrin, 1984), a quasi-residual plateau deposit which overlies Lower Cretaceous (Hythe Beds) rocks in west Kent. As in the associated chromoluvic brown soils in similar deposits, the upper horizons down to and including the Eb are developed in a lithologically distinct surface layer containing admixed loess, and the underlying Bt in pre-existing locally derived material with a considerably larger sand/silt ratio. Percentage base saturation values are extremely low throughout the depth sampled but probably increase below. The pyrophosphate-extractable iron and aluminium data show clear maxima in the Bh and Bs horizons respectively whereas the total extractable iron content is much larger in the Bt horizon in accordance with the greatly increased clay content. A thin section from this horizon was estimated to contain at least 10 per cent of preferentially oriented clay bodies, including argillans and intrapedal concentrations.

Profile 92

Location	:	Seal, Kent (TQ 564563)
Climatic regime:		subhumid temperate
Altitude	:	160 m Slope: 3°
Land use and vegetation	:	broadleaf woodland; oak standards and field layer dominated by bracken (*Pteridium aquilinum*) with some heather (*Calluna vulgaris*)

Horizons (cm)

L/F — Undecomposed and partially decomposed litter, chiefly oak leaves.

0–6 H — Black (5YR 2/1) organic material with an appreciable mineral fraction; moderate medium granular; very friable; abundant fleshy and woody roots, mainly fine and medium, some coarse; abrupt wavy boundary.

6–20 Ea — Pinkish grey (5YR 7/2) moderately stony sandy silt loam (USDA fine sandy loam) with few fine reddish brown (5YR 3/4) inclusions; many small and medium angular and tabular chert and cherty sandstone fragments; massive; friable; common roots, mainly fine; abrupt irregular boundary.

20–23 Bh — Dark reddish brown (5YR 3/2) moderately stony sandy silt loam (USDA loam); stones as above; massive; friable; few roots; abrupt irregular boundary.

23–30 Bs — Strong brown to yellowish red (7.5–5YR 5/8) moderately stony sandy silt loam (USDA loam); stones as above; weak granular; friable; few roots.

30–52 Eb(g) — Brownish yellow (10YR 6/8) moderately stony sandy silt loam (USDA loam) with few fine reddish yellow (7.5YR 6/6) and very pale brown (10YR 7/4) mottles; stones as above, some large; weak coarse angular blocky; friable to firm; few roots; clear smooth boundary.

52–62 Eb/Bt(g) — Strong brown (7.5YR 5/8) moderately stony sandy silt loam (USDA loam) with common fine yellowish red (5YR 5/8) and few very pale brown (10YR 7/4) mottles; stones as above; weak coarse angular blocky; firm; roots rare; clear irregular boundary.

62–75+ 2Bt — Yellowish red (5YR 5/6) moderately stony clay (USDA clay) with common red to yellowish red (2.5–5YR 5/8) mottles; stones as above; moderate coarse subangular blocky; very firm; clay coats on ped faces and stones; becoming very stony, with larger stones, below 75 cm.

Analytical data for Profile 92

Horizon	H	Ea	Bh	Bs	Eb(g)	Eb/Bt(g)	2Bt
Depth (cm)	0–6	6–20	20–23	23–30	30–52	52–62	62–75
sand 600 μm–2 mm %		5	5	5	3	3	4
200–600 μm %		15	13	14	13	14	11
60–200 μm %		28	25	24	25	28	25
silt 2–60 μm %		45	42	41	47	39	16
clay <2 μm %		7	15	16	12	16	44
<0.2 μm %				5	3	7	25
fine clay/total clay				0.31	0.25	0.44	0.57
organic carbon %	27	3.1	3.2	1.6	0.4		
pyro. ext. Fe %		0.3	1.4	1.3	0.2	0.1	0.1
pyro. ext. Al %		0.1	0.2	0.2	0.2	0.2	0.2
pyro. ext. C %		1.0	1.2	0.9	0.2		
pyro. Fe+Al %/clay %		0.05	0.10	0.09	0.03	0.02	<0.01
total ext. Fe %		0.7	2.2	2.3	1.3	1.1	3.3
TEB (me/100 g)		0.7		0.7	0.4	0.3	0.5
CEC (me/100 g)		11.4		14.6	7.9	7.6	15.3
% base saturation		6		5	5	4	3
pH (H$_2$O)	4.4	4.1	4.0	4.0	4.5	4.4	4.1
(0.01M CaCl$_2$)	3.5	3.4	3.4	3.6	3.2	4.0	3.8

Classification
England and Wales (1984): paleo-argillic podzol; coarse loamy over clayey, drift with siliceous stones (unnamed series).
USDA (1975): Ultic (or Alfic) Haplorthod; coarse loamy over clayey, mixed or siliceous, mesic, non-cemented.
FAO (1974): Orthic Podzol; medium textured.

6.3 Gley-podzols

6.3.1 General Characteristics, Classification and Distribution

These are podzols, mostly classed as gleyic in the FAO system and as Aquods in the US Taxonomy, that have a gleyed E, B or C horizon directly below a Bh or Bs or within 40 cm depth and no thin ironpan (placic horizon). As in the non-hydromorphic podzols which they otherwise resemble, a light coloured (albic) E horizon normally underlies the organic surface layer in uncultivated phases but is commonly absent in cultivated land.

Gley-podzols so defined include soils with Bh horizons formed in the presence of a shallow fluctuating ground-water table, and others in which periodic wetness is mainly confined to horizons below the podzolic B. The associated semi-natural vegetation is correspondingly variable, ranging from wet heath or moor with cross-leaved heath (*Erica tetralix*) and purple moor-grass (*Molinia caerulea*) as characteristic species to deciduous woodland with few distinctively hydrophilous plants. As Ea and Bh horizons formed above or below the upper limit of seasonal saturation are difficult to distinguish, however, separation of 'semi-hydromorphic' classes comparable with gleyic and stagnogleyic brown soils is considered impracticable. In any case the gley-podzols which retain their identity under cultivation have usually been drained artificially, implying that their morphology is unrelated to the current soil-water regime.

Soils of this group generally occur in close association with non-hydromorphic podzols, humic gley soils, or both, and are considerably less extensive than other kinds of podzol in the British Isles as a whole. They have not been set apart consistently in the classifications used as a basis for soil mapping in either Scotland or Ireland and the few that have been described in these countries have been given various names, including *podzols with gleying* (Birse and Robertson, 1976) and *peaty podzolized gley* (Conry and O'Shea, 1973). In England and Wales, where they are estimated to cover about 0.6 per cent of the total area (Mackney *et al.*, 1983), two main subgroups of roughly equal extent have been distinguished as typical gley-podzols (Figure 6.1) and stagnogley-podzols respectively. The former are developed in sandy aeolian or fluvial deposits with ground water at moderate depths and have a Bh horizon containing little extractable iron, apparently because any originally present has been mobilized in reduced form and removed by lateral seepage. Where the water table is or was close to the surface for long periods, the surface horizons are peaty. In the stagnogley-podzols, which are typically in higher ground on base-deficient Pleistocene or older deposits, the podzolic B is underlain by a relatively impermeable Btg (or 2Bg) horizon of finer texture that impedes downward percolation. Nearly all these soils can be considered as hydromorphic analogues of the luvic podzols (Figure 6.1). They can also be equated with the variously named intergrades between podzol and pseudogley recognized in French and German classifications (CPCS, 1967; Duchaufour, 1982; Mückenhausen *et al.*, 1977). Other less commonly occurring variants include humoferric gley-podzols (Avery, 1980) with iron-enriched Bh or Bs horizons and no immediately underlying finer textured horizon.

The inherent chemical and physical limitations to plant growth characteristic of podzols generally are accompanied in these soils by varying degrees of seasonal wetness. Lowland gley-podzols in sandy materials are easily drained, however, and are capable of producing good yields of a wide range of agricultural or horticultural crops when limed and generously manured. Intensively cultivated soils of this type are well represented in the Lancashire coastal plain (Hall and Folland, 1967; Ragg *et al.*, 1984) and the Vale of York (R.A. Jarvis *et al.*, 1984). Those with a humose Ap and a water table within root range for some part of the growing season are distinctly less droughty than cultivated non-hydromorphic podzols of similar texture, but crops can suffer severely from water shortage where rooting is restricted by a cemented Bhm horizon or by artificially induced subsurface compaction. The 'upland' gley-podzols with a slowly permeable subsurface horizon at moderate depths are less easily brought into productive agricultural use and mostly remain under semi-natural vegetation or have been planted with conifers.

Key to Subgroups

1. Gley-podzols with an argillic B horizon (p. 109) or a slowly permeable 2Bg horizon of finer texture below the podzolic B
Luvic gley-podzols
2. Other gley-podzols with a Bs horizon 5 cm or more thick
Humoferric gley-podzols
3. Other gley-podzols
Typical gley-podzols

6.3.2 Typical Gley-podzols

These are the gley-podzols in permeable coarse textured materials that have a dark coloured Bh and no underlying Bs or Bt of significant thickness. They typically occur as intermediate members of catenary sequences ranging from higher lying non-hydromorphic podzols to somewhat lower lying humic sandy gley soils (p. 297). In some gently undulating sandy plains, however, the freely draining member is missing and the gley-podzols occupy minor rises, with peat soils in the lowest positions. The underlying groundwater body may be permanent and regional in extent, as in thick sandy deposits close to sea level, or it may be 'perched' above an impervious substratum at some depth below the solum, as in parts of the Lancashire and Cheshire plains. Typical profiles have a more or less cemented Bh, often of considerable thickness, which contains little free iron and passes downwards into a greyish or mottled Cg horizon. The presence of 0.5 per cent or more extractable iron in all or part of the Bh implies either that the parent material contains significant amounts of mobilizable iron which has not been removed laterally or that iron mobilized elsewhere has precipitated within the horizon, presumably under periodically well aerated conditions. Data on two representative profiles are reproduced below.

Profile 93 was described and sampled by B.R. Hall in Shirdley Hill Sand (Wilson *et al.*, 1981), a late Devensian coversand which overlies till in the south-west Lancashire coastal plain. According to Hall and Folland (1967), peat formerly covered much of this area but substantial lowering of the water table by artificial drainage, followed by wastage and mixing of the surface layers by cultivation, has greatly reduced its extent. Wet strongly gleyed till is encountered at 137 cm depth in the profile recorded. The analytical data show that the black Ap horizon is marginally humose (4.3 per cent organic carbon) and that the thick Bh horizon owes its reddish hues to organic coats on the sand grains, as there is very little extractable iron in any of the subhorizons sampled. The bleached sand forming the E horizons of these soils has provided valuable raw material for glass making. When exploited for this purpose, the topsoil is stripped and the bleached sand layer, which ranges in thickness from 10 to 90 cm, is removed. After the laying of a tile-drainage system and replacement of the topsoil material, the land is returned to agriculture.

Profile 94 is located in a very gentle footslope in the Sherwood Sandstone area of north Nottinghamshire. It is developed in nearly stone-free sandy alluvium and has essentially similar horizons, but the topsoil contains more organic matter and the Bh is thinner but more cemented than in profile 93.

Profile 93 (Mackney, 1970)

Location	:	Ormskirk, Lancashire (SD 420070)
Climatic regime:		humid temperate
Altitude	:	40 m Slope: 2°
Land use	:	arable

Horizons (cm)

0–28 Ap	Black (10YR 2/1) stoneless humose medium sand (USDA sand); rare gravel and cinders; weak medium subangular blocky and granular; very friable; common fine roots; abrupt smooth boundary.
28–41 Ea(g)	Light brownish grey (10YR 6/2) stoneless medium sand (USDA sand) with few fine black mottles associated with root channels; single grain; loose; few fine roots; abrupt smooth boundary.
41–48 Bh1(m)	Dark reddish brown (5YR 2/2) medium sand (USDA sand) with partly coated and partly uncoated grains; massive; firm, very weakly cemented in parts; clear irregular boundary.
48–56 Bh2m	Very dusky red (10YR 2/2) stoneless medium sand (USDA sand) with coated grains; massive; firm; very weakly cemented; gradual wavy boundary.
56–81 Bh3(m)	Dark reddish brown (5YR 3/4) stoneless medium sand (USDA sand) with coated grains; massive; very weakly cemented in parts but more friable than above; gradual smooth boundary.
81–102 BC	Yellowish red (5YR 4/6) stoneless medium sand with coated grains; massive; consistence as above; wet; water table at about 1 m depth when sampled (24 June 1967); gradual smooth boundary.

Analytical data for Profile 93

Horizon	Ap	Ea(g)	Bh1(m)	Bh2	Bh3(m)	BC	2Cg
Depth (cm)	0–28	28–41	41–48	48–56	56–81	81–102	102–137
sand 500 µm–2 mm %	2	2	6	2	5	6	<1
200–500 µm %	60	48	67	60	55	61	38
100–200 µm %	29	43	24	34	35	30	55
50–100 µm %	3	5	1	2	4	2	6
silt 2–50 µm %	3	1	<1	<1	<1	<1	1
clay <2 µm %	3	1	2	2	1	1	<1
organic carbon %	4.3	0.2	0.8	0.9	0.5	0.2	0.1
pyro. ext. Fe %	0.1	0.05	0.03	0.05	0.05	0.03	0.03
pyro. ext. Al %	0.2	0.05	0.2	0.3	0.1	0.1	0.1
pyro. ext. C %	0.4	0.1	0.8	0.8	0.5	0.2	0.1
total ext. Fe %	0.4	0.1	0.1	0.1	0.1	0.1	0.1
pH (H$_2$O)	6.6	6.1	5.7	5.7	5.8	5.6	5.6
(0.01M CaCl$_2$)	6.3	5.5	4.7	4.8	4.9	4.8	5.0

102–137 2Cg — Light yellowish brown (10YR 6/4) stoneless medium sand (USDA fine sand) finer than above, with mainly uncoated grains; wet; abrupt smooth boundary.

137+ 3Cg — Wet strongly gleyed till.

Analytical data *see* above.

Classification
England and Wales (1984): typical gley-podzol; coarse or medium sandy stoneless drift (Sollom series).
USDA (1975): Typic Haplaquod; sandy, siliceous, mesic, ortstein.
FAO (1974): Gleyic Podzol; coarse textured.

Profile 94 (Robson and George, 1971)

Location:	Bilsthorpe, Nottinghamshire (SK 640618)
Climatic regime:	subhumid temperate
Altitude:	61 m Slope: 1°
Land use:	arable (sugar beet)

Horizons (cm)

0–26 Ap — Black humose loamy medium sand (USDA loamy sand) containing occasional quartzose pebbles and abundant uncoated grains; weak subangular blocky; very friable; many fine and common fleshy beet roots; earthworms active; abrupt irregular boundary.

26–37 Ea/Bh — Variegated greyish brown (7.5YR 5/2) and dark brown (7.5YR 3/2) stoneless medium sand (USDA sand) with abundant uncoated grains; single grain; very friable, locally brittle; abundant fine pores and black root traces; common roots; abrupt irregular boundary.

37–46 Bhm1 — Dark reddish brown (5YR 2/2) stoneless medium sand (USDA sand) with coated grains; massive; hard to very hard when dry; weakly to very weakly cemented; no roots; clear boundary.

46–68 Bhm2 — Brown (7.5YR 4/4) and dark brown (7.5YR 3/2) stoneless medium sand (USDA sand) with coated grains, the latter colour mainly in the upper 5 cm and in few irregular patches below; massive, firm; weakly to very weakly cemented; no roots; clear boundary.

68–90 Cg — Light brownish grey (2.5YR–10YR 6/2) to light grey stoneless medium sand (USDA sand) with few diffuse brownish mottles and abundant uncoated grains; single grain; loose; no roots.

Analytical data for Profile 94

Horizon	Ap	Ea/Bh	Bhm1	Bhm2	Cg
Depth (cm)	0–26	26–37	37–46	46–68	68–90
sand 600 μm–2 mm %	6	2	2	1	1
200–600 μm %	53	69	55	60	52
60–200 μm %	28	22	36	34	41
silt 2–60 μm %	7	5	5	2	4
clay <2 μm %	6	2	2	3	2
organic carbon %	9.9	1.4	2.1	0.7	0.3
pyro. ext. Fe %	0.20	0.02	0.03	0.01	0.01
pyro. ext. Al %	0.2	0.1	0.2	0.1	0.1
pyro. ext. C %	3.2	1.1	1.8	0.6	0.2
total ext. Fe %	0.3	0.05	0.06	0.04	0.07
pH (H$_2$O)	6.5	5.5	4.4	4.5	4.9
(0.01M CaCl$_2$)	6.0	4.7	3.6	4.0	4.3

Classification
  England and Wales (1984): typical gley-podzol; coarse or medium sandy stoneless drift (Sollom series).
  USDA (1975): Typic Haplaquod; sandy, siliceous, mesic, ortstein.
  FAO (1974): Gleyic Podzol; coarse textured.

6.3.3 Humoferric Gley-podzols

These soils differ from those of the preceding subgroup in having a Bs horizon below a Bh or a Bs only and from the freely draining humoferric podzols in having an immediately underlying gleyed horizon. Though nowhere extensive, they occur sporadically in both upland and lowland situations as members of catenary associations in acidic materials of sandy or loamy texture. Some have a relatively impermeable horizon of fragipan character below the podzolic B.

Variants without a Bh horizon have only been identified in loamy materials and are represented by the following profile, located in low lying land backed by mountains in the Dingle peninsula, County Kerry. The parent material is coarse loamy glacial drift derived from sandstone and shale and a former peaty surface horizon is incorporated in the Ap.

Although no data on pyrophosphate-extractable iron, aluminium and carbon were obtained, the dark yellowish brown horizon at 37–66 cm is reasonably designated Bs on the basis that it contains much more free iron and significantly more carbon than the E horizon, which is compact and hard when dry. The underlying horizon is also recorded as hard and possibly has fragipan characteristics.

Profile 95 (Conry and O'Shea, 1973)

Location	: Annascaul, County Kerry
Climatic regime:	perhumid temperate
Altitude	: 53 m Slope: 1–2°
Land use and vegetation	: long-term grass with Yorkshire Fog (*Holcus lanatus*) and rushes (*Juncus* spp.) dominant.

Horizons (cm)

0–23
Ap — Very dark greyish brown (10YR 3/2) to black, humose sandy loam (USDA sandy loam); good root development; wet; abrupt smooth boundary.

23–37
Eag — Greyish brown (10YR 5/2) sandy loam (USDA sandy loam) with many black streaks due to decomposed roots; massive; firm; many *Juncus* roots; gradual wavy boundary.

37–66
Bs(g) — Dark yellowish brown (nearest 10YR 4/4) sandy loam (USDA sandy loam) with black organic streaks; weak

Analytical data for Profile 95

Horizon	Ap	Eag	Bs(g)	Cg(x)
Depth (cm)	0–23	33–37	37–66	66+
sand 200 μm–2 mm %	35	35	29	29
50–200 μm %	30	28	29	33
silt 2–50 μm %	23	26	30	31
clay <2 μm %	12	11	12	7
loss on ignition %	22.0	5.4	9.2	1.6
organic carbon %	11.3	1.8	2.8	0.5
nitrogen %	0.54	0.10	0.13	
C/N ratio	21	18	22	
citrate-dithionite ext. Fe %	0.1	0.1	0.6	0.4
TEB (me/100 g)	14.3	2.2	2.3	1.2
CEC (me/100 g)	36.4	16.0	35.0	8.4
% base saturation	39	14	7	14
pH (H$_2$O)	5.4	4.7	4.6	4.9

	granular; friable; few roots; gradual wavy boundary.
66+ Cg(x)	Light brownish grey (10YR 6/2) stony sandy loam (USDA sandy loam) with yellowish brown (10YR 5/6) mottles; massive; compact *in situ*; no roots.

Analytical data *see* above

Classification
  Ireland (1980): (Peaty podzolized) gley (Derrygorman complex).
  England and Wales (1984): humo-ferric gley-podzol; coarse loamy, drift with siliceous stones (Reaseheath series).
  USDA (1975): Typic Sideraquod (or Fragiaquod); coarse-loamy or loamy-skeletal, mixed, mesic.
  FAO (1974): Gleyic Podzol; coarse to medium textured.

6.3.4 Luvic Gley-podzols

These are gley-podzols in which drainage is impeded below the podzolic B by a slowly permeable Btg or 2Bg horizon of contrastingly finer texture. They are commonest in southern England, notably on Tertiary beds and associated 'plateau gravels' in the London and Hampshire basins, on similar deposits capping the Haldon Hills in South Devon (Clayden, 1971) and on Lower Cretaceous siltstones in the High Weald of Kent and East Sussex (M.G. Jarvis *et al.*, 1984). They also occur on the North York Moors (Carroll and Bendelow, 1981) and in a few other localities in the Midlands and north of England. Most of them are in existing or former heathland but some may have formed under 'degraded' oak forest.

Typical profiles under semi-natural vegetation have well defined Ea and Bh horizons which are underlain directly by a mottled Btg horizon or separated from it by a second weakly structured eluvial horizon (Eg) containing less clay (Figure 6.1 p. 244). The Ea horizon can be absent in cultivated land, and variants with a more or less mottled Bs horizon have also been noted, for example on Dunsmore Heath near Rugby (Whitfield and Beard, 1977). As in the non-hydromorphic luvic podzols (p. 259), development of an argillic B horizon appears to have preceded podzolization, either in early Holocene times or in a previous temperate period. In some of these soils, however, the increase in clay content with depth is primarily attributable to the presence of an originally coarser textured surface layer and evidence of clay translocation is minimal. Data on two representative profiles are reproduced on pp.267–70.

Profile 96 was described and sampled by

R.G. Sturdy in former commonland on deeply weathered clayey drift, probably of Anglian age, which overlies sandy Tertiary (Bagshot) beds at 160 cm depth. There is no evidence of wetness in the upper 42 cm, which becomes dry in most summers, but holes dug in similar soils nearby contained water up to the base of the Bh horizon for at least four months (Sturdy, 1971). The Ea horizon has apparently developed in a lithologically distinct superficial gravelly layer in which stones have been concentrated, probably by some combination of periglacial frost heaving and loss of fine material resulting from lateral water movement. It has a larger silt/sand ratio than the underlying horizons, however, suggesting that it incorporates silt of loessial origin. The succeeding Bh horizon is superimposed on the upper part of a thick grey, yellow and red mottled Btg horizon which contains many argillans and intrapedal concentrations of strongly oriented clay and clearly corresponds to the base-deficient chromic argillic B horizons of unpodzolized luvic brown (Profile 77, p. 231) and luvic gley (Profile 157, p. 372) soils in similar materials.

Profile 97 (Plate III.1) is located under moist heathland, recently invaded by bracken and Scots pine, on a convex upper slope in Ashdown Forest, Sussex (Figure 6.4). It shows a similar succession of horizons formed in cryoturbated weathering products of the subjacent siltstones and fine grained sandstones (Hastings Beds), which consist largely of quartz with few weatherable minerals other than clay-size mica and kaolinite. The Btg horizon is less strongly expressed than in the preceding profile and passes downwards into a compact, very slowly permeable BC of fragipan character.

Profile 96

Location	:	Warley, Essex (TQ 598914)
Climatic regime:		subhumid temperate
Altitude	:	107 m Slope: 2°
Land use and vegetation	:	broadleaf woodland; birch/chestnut coppice with some oak; field layer dominated by bracken (*Pteridium aquilinum*)

Figure 6.4: Ashdown Forest, East Sussex. Moist heathland on luvic gley-podzols with impeded drainage in the foreground (British Geological Survey: NERC copyright).

Horizons (cm)

- L — Litter layer of bracken fronds and birch leaves, up to 6 cm thick.
- 0–3 F — Black (5YR 2/1 rubbed) weakly laminated, partially decomposed bracken litter; common fibrous and woody roots and bracken rhizomes; sharp smooth boundary.
- 3–12 H — Black (5YR 2/1) organic material containing many uncoated sand grains; weak granular; friable; many roots; abrupt smooth boundary.
- 12–42 Ea — Pinkish grey (7.5YR 6/2; 5YR 6/1 dry) very stony fine sandy loam (USDA fine sandy loam to loamy fine sand) with many coarse diffuse brown (7.5YR 4/2) mottles; abundant medium rounded flint pebbles; very weak medium blocky; friable; common roots; abrupt wavy boundary.
- 42–51 2Bh(g) — Black (5YR 2/1) slightly stony fine sandy loam (USDA fine sandy loam) with many coarse strong brown (7.5YR 5/6) and medium light brownish grey (10YR 6/2) mottles; stones as above; massive; firm; very weakly cemented in parts; few roots; abrupt smooth boundary.
- 51–59 2Bh/Btg — Light brownish grey (10YR 6/2) slightly stony sandy clay loam (USDA fine sandy loam to sandy clay loam) with many medium reddish yellow (7.5YR 6/8) mottles and inclusions of black Bh horizon material; moderate medium subangular blocky; firm; common roots; impersistent horizon with clear wavy lower boundary.
- 59–75 2Btg — Reddish yellow (7.5YR 6/8) stoneless sandy clay (USDA sandy clay to clay) with many fine red (10R 4/6) and common light grey (5Y 6/1) mottles; wet; strong very coarse angular blocky with black (10YR 2/1) organic coats on faces, breaking easily in places to medium and fine blocks; few living roots, and common dead roots; clear wavy boundary.
- 75–110 3Btg — Greenish grey (5GY 6/1) clay (USDA clay) containing occasional large rounded flint pebbles; many fine reddish yellow (7.5YR 6/8) and red (10R 4/8) mottles; strong coarse angular blocky with light grey to grey (5Y 6/1) faces, often obliquely oriented; few fine roots; dead roots on ped faces.

Analytical data for Profile 96

Horizon	H	Ea	2Bh(g)	2Bh/Btg	2Btg	3Btg
Depth (cm)	3–12	12–42	42–51	51–59	59–75	75–110
sand 200 μm–2 mm %		13	3	1	1	<1
60–200 μm %		48	68	61	42	25
silt 2–60 μm %		38	21	18	14	34
clay <2 μm %		1	8	20	43	41
<0.2 μm %			4	12	31	21
fine clay/total clay			0.50	0.60	0.72	0.51
organic carbon %	19	0.8	1.1	0.8		0.2
pyro. ext. Fe %		0.05	0.3	0.4	0.2	0.1
pyro. ext. Al %		0.02	0.1	0.3	0.2	0.2
pyro. ext. C %		0.2	0.9	0.8	0.4	0.1
total ext. Fe %		0.1	0.7	2.0	2.3	1.6
TEB (me/100 g)	3.2	0.7	0.7	1.3	1.7	1.7
CEC (me/100 g)	37	5.6	15	23	19	17
% base saturation	9	12	5	6	9	10
pH (H_2O)	3.5	3.6	3.7	3.9	3.9	4.0
(0.01M $CaCl_2$)	2.9	3.2	3.4	3.6	3.6	3.6

Classification
    England and Wales (1984): stagnogley-podzol; coarse loamy over clayey, drift with siliceous stones (Haldon series).
    USDA (1975): Ultic Sideraquod or Aquultic Haplorthod; loamy-skeletal over clayey, siliceous, mesic.
    FAO (1974): Gleyic Podzol; coarse textured (gravelly phase).

Profile 97 (M.G. Jarvis et al., 1984)

Location	:	Ashdown Forest, East Sussex (TQ 466310)
Climatic regime:		subhumid temperate
Altitude	:	193 m Slope: 5° convex
Vegetation	:	moist heath with heather (*Calluna vulgaris*), bracken (*Pteridium aquilinum*), purple moor-grass (*Molinia caerulea*) and invading Scots pine (*Pinus sylvestris*)

Horizons (cm)

L	Undecomposed leaf and twig litter, 2 cm thick.
0–3 F	Dusky red (2.5YR 3/2) partially decomposed litter; abundant fibrous roots; sharp smooth boundary.
3–4 H	Black (5YR 2/1) well decomposed litter; weakly laminated; many roots; sharp smooth boundary.
4–18 Ah	Black (7.5YR 2/1) slightly stony humose sandy silt loam (USDA silt loam) with uncoated grains; common medium subrounded and tabular sandstone fragments; massive; friable; many roots; abrupt smooth boundary.
18–32 Ea	Light grey (10YR 7/2) slightly stony sandy silt loam (USDA silt loam to very fine sandy loam) with many fine brown to dark brown (7.5YR 4/2) mottles; stones as above; massive; friable; few roots; abrupt wavy boundary.
32–38 Bh	Black (5YR 2/1) stoneless sandy silt loam (USDA silt loam); weak coarse platy; very firm; few roots; clear wavy boundary.
38–65 Eg/Btg	Grey (2.5Y 6/1) slightly stony sandy silt loam (USDA loam) with very many medium strong brown (7.5YR 5/8) mottles; common very large subangular and tabular sandstone fragments; weak medium subangular blocky; friable to firm; few roots; gradual wavy boundary. Significant inclusions, sampled separately, of

Analytical data for Profile 97

Horizon	F	Ah	Ea	Bh	Eg/Btg	Btg[a]	BCtgx
Depth (cm)	0–3	4–13	18–32	32–38	38–65	38–65	65–102
sand 200 μm–2 mm %		<1	<1	2	2	2	3
100–200 μm %		8	6	9	8	8	8
60–100 μm %		20	25	19	28	15	17
silt 2–60 μm %		68	66	62	51	47	54
clay <2 μm %		4	3	8	11	28	18
organic carbon %	50	6.9	0.4	1.3		0.2	0.2
pyro. ext. Fe %		0.1	<0.05	0.2	0.2	<0.05	0.1
pyro. ext. Al %		0.1	0.1	0.1	0.1	0.1	0.1
pyro. ext. C %		1.7	0.2	1.0	0.2	0.1	<0.05
total ext. Fe %		0.1	<0.05	0.4	0.9	1.3	1.0
TEB (me/100 g)	20.0	1.9	0.5	1.3	0.5	0.5	0.5
CEC (me/100 g)	91	18.6	2.4	11.5	4.4	7.5	4.5
% base saturation	22	10	20	11	11	6	10
pH (H$_2$O)	4.5	3.7	4.2	4.2	4.5	4.3	4.6
(0.01M CaCl$_2$)	3.4	2.8	3.4	3.5	3.8	3.6	3.9

[a] Inclusions sampled separately

	light olive-grey (5Y 6/2) firm clay loam (USDA clay loam to silty clay loam) with many coarse reddish yellow (7.5YR 6/8) mottles and strong prismatic structure with organic coats on ped faces.
65–102 BCtgx	Light grey to grey (5Y 6/1) very stony clay loam (USDA loam to silt loam) with many coarse brownish yellow (10YR 6/8) mottles; abundant large subangular and tabular sandstone fragments; weak coarse prismatic with light brownish grey (10YR 6/2) faces; very to extremely firm; roots rare.

Analytical data *see* p.269.

Classification
England and Wales (1984): stagnogley-podzol; silty material over lithoskeletal siltstone and sandstone (variant of Poundgate series).
USDA (1975): (Ultic) Fragiaquod; fine silty, siliceous, mesic.
FAO (1974): Gleyic Podzol; medium textured (Fragipan phase).

6.4 Stagnopodzols

6.4.1 General Characteristics, Classification and Distribution

The soils classed as stagnopodzols (Avery, 1973, 1980) have as essential features either a thin ironpan (Bf) directly below the eluvial horizon or below a Bh, or a gleyed eluvial horizon (Eag or Eg) overlying a Bs and no immediately underlying gleyed horizon. Most of them also have a peaty surface layer where undisturbed by cultivation and have been identified accordingly as peaty podzols in Scotland (Romans, 1970; Soil Survey of Scotland, 1984) and Ireland (Gardiner and Radford, 1980), where they are proportionately more extensive than in England and Wales. Those with a thin ironpan have been set apart as ironpan soils in Forestry Commission surveys (Pyatt, 1970, 1977, 1982).

Soils of this group are widely distributed in the humid and perhumid uplands at altitudes

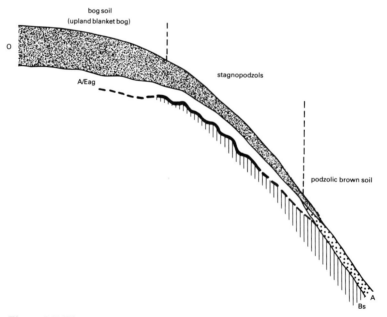

Figure 6.5: Diagrammatic representation of upland catenary sequence including podzolic brown soils, stagnopodzols and peat (bog) soils.

Figure 6.6: Howgill Fells, Cumbria. Peat (blanket bog) soils occupy the narrow summits and give place on adjacent spurs to stagnopodzols under *Nardus* grassland, shown by lighter tones. Podzolic brown soils predominate on the steep, gullied, bracken-covered slopes below and slowly permeable gley soils on dissected till in the foreground (Cambridge University Collection: copyright reserved).

ranging from the upper limit of regular cultivation to the fringe of the oro-arctic zone. They occur close to sea coasts in north and west Scotland and are also represented locally in subhumid lowland localities including the Moray Firth area and the southern English heathlands. The associated semi-natural vegetation is normally of moist heath or moorland type, with heather (*Calluna vulgaris*), purple moor-grass (*Molinia caerulea*), mat-grass (*Nardus stricta*) or deer-grass (*Trichophorum caespitosum*) as major components. As shown in Figures 6.5 and 6.6, the upland stagnopodzols commonly occur as upper-slope members of catenary sequences with podzolic brown soils or non-hydromorphic podzols on lower slopes and bog soils (blanket bog) on flatter ground at higher altitudes. They also give place to humic (peaty) gley or peat soils where the substrata are impervious and in sites receiving water from higher ground. In less humid areas and at lower altitudes, they are generally confined to gently sloping interfluves, as in parts of the North York Moors (Carroll and Bendelow, 1981).

Characteristic profiles in the main areas of occurrence have an iron-deficient eluvial horizon that is periodically saturated with water and rests sharply on a more or less continuous thin ironpan (Bf or placic horizon). Below this there is usually a friable Bs horizon, but the pan may also overlie a Bw, an indurated layer (fragipan or duric horizon) or slightly altered bedrock, or form part of a bisequal profile (p. 99) with an underlying argillic B. These *ironpan stagnopodzols* (Plate III.2) are mostly developed in stony superficial deposits derived from subjacent consolidated rocks and the upper mineral horizons are typically loamy rather than sandy. Although they have generally been regarded as podzols by British pedologists, only those in which the pan rests on a spodic horizon or a fragipan are Spodosols (Placaquods) in the US system and only those with a spodic horizon are Podzols (Placic Podzols) according to the FAO scheme. In every case, however, the upper horizons terminating in the ironpan are similar in morphology, and apparently in genesis, so placing the soils in different classes at the highest categorical level seems difficult to justify.

Studies of water and aeration regimes in German and Scottish profiles of this type by Stahr (1973) and Pyatt and Smith (1983) respectively showed that the periodically water-saturated horizons above the pan were at least partially anaerobic for much of the year, whilst those below remained unsaturated and comparatively well aerated. Stahr also concluded that waterlogging and poor aeration in the upper soil were due not so much to the ironpan, which is seldom entirely continuous, as to the adverse pore size distribution and consequently small hydraulic conductivity of the overlying horizons. This conclusion supports the hypothesis, previously advanced by Muir (1934), Crompton (1956) and Crampton (1963), that the characteristic upper mineral horizons have been superimposed on those of pre-existing well aerated soils as a result of superficial peat formation, which was promoted over much of upland Britain by the decline or clearance of forest. Once formed, the peaty surface retains water like a sponge and maintains anaerobic conditions in the weakly structured mineral soil immediately beneath it; iron is consequently mobilized in reduced form and reprecipitated to form the pan at a level determined by Eh and concentration gradients. In contrast to other iron-cemented horizons (Bgf, Cgf) attributable to reduction and segregation, however, the thin ironpans evidently contain significant amounts of translocated organic carbon (Avery *et al.*, 1977; Berrow and Goodman, 1987) and resemble the B horizons of non-hydromorphic podzols in this respect.

Although a well formed ironpan acts as a barrier to water and roots, the proposition that it is initially a result rather than a cause of poor aeration is further supported by the common occurrence, often within the same slope facet, of *ferric stagnopodzols* (Figure 6.1; Plate III.4) with similar organic and eluvial horizons, in which the pan is absent or discontinuous. These have been classed as peaty podzols (without thin ironpan) in Scotland and as intergrades between 'upland brown earths' and ironpan soils by the Forestry Commission (Pyatt, 1970, 1977, 1982). Typical profiles have a more or less massive loamy eluvial horizon (Eag or Eg) over an ochreous Bs which is usually mottled in the upper part. Duchaufour (1982) describes peaty montane soils with similar horizons as podzolic stagnogleys and considers that they represent a pre-stage in the development of ironpan stagnopodzols, which he calls humic stagnogleys with a placic horizon. In rarely occurring variants termed indurated podzols by Clayden (1971) and hardpan stagnopodzols by Avery (1973, 1980), the E horizon is strongly compacted (albic densipan). Presumably because soils resembling the ferric stagnopodzols are rare in the United States, there is no specific place for them in the US Taxonomy, those in which the Bs qualifies as spodic being grouped with gley-podzols as Aquods and others with humic stagno-orthic gley soils (p. 354) as Aquepts.

Also included in the group are *humus-ironpan stagnopodzols* (Plate III.3) in which a thin ironpan lies below a Bh horizon, either directly or separated by a more or less continuous gleyed eluvial horizon (Figure 6.1). The Bh horizons of these soils are usually iron-deficient, implying that those with an aquic moisture regime correspond to Placic Haplaquods in the US system and those without to Placohumods (Avery *et al.*, 1977). Although their precise mode of origin remains controversial, it is now generally accepted that they have evolved from non-hydromorphic (humoferric) podzols and that development of the thin ironpan involves reduction, mobilization and redeposition of iron derived from an overlying horizon. According to one explanation of their genesis, first advanced by Muir (1934) and further developed by Anderson *et al.* (1982), Righi *et al.* (1982) and Payton (1987b), pan formation is initiated at the upper surface of a Bs horizon which has become increasingly less permeable as a result of clogging of the pores by illuvial accumulation products, so promoting seasonally reducing conditions in the horizon above. However, whilst Righi *et al.* considered that the Bh horizon in a coarse-textured Belgian Placohumod formed after the pan had developed, Payton (1987b) concluded that formation of a compact Bh under heath preceded, and played a role in, the development of a hydrologic regime favourable to ironpan formation in the Northumbrian podzols that he studied.

The ericaceous or grassy vegetation which the upland stagnopodzols bear in their semi-natural state provides poor to moderate grazing over short periods and potentialities for improvement are restricted to varying degrees by climatic limitations and the low bearing strength of the topsoils when wet, coupled in many localities with steeply sloping or bouldery terrain. In marginal or moorland-fringe areas where climatic and topographic conditions are relatively favourable, however, much land with formerly similar vegetation and soil has at one time or another been reclaimed for more productive pastoral use, or in less humid areas such as eastern Scotland for tillage. Elsewhere in recent years, large areas have been afforested, chiefly with Sitka spruce. In either case the soils have been transformed to differing depths depending on the ameliorative techniques employed, which in order of increasing physical impact include the use of lime and fertilizers, paring and burning (Curtis, 1971), shallow ploughing or rotovation followed by reseeding, and deep ploughing or subsoiling to disrupt the pan and improve rooting conditions for trees (Zehetmayr, 1960; Taylor, 1970). Afforestation, especially where preceded by mechanical disruption of the upper horizons, normally leads to an appreciable decrease in surface wetness (Pyatt, 1987). 'Regraded' stagnopodzols in which an E horizon, a thin ironpan, or both, remain largely intact have usually been treated as cultivated phases for survey purposes (e.g. Staines, 1976). Others with Bs horizons at lesser depths have assumed the characteristics of podzolic brown soils, and those that have been ploughed to depths greater than 40 cm are considered here as man-made (disturbed) soils (Chapter 8, p. 394).

Key to Subgroups

1. Stagnopodzols with an albic densipan (p. 108)
 Densipan Stagnopodzols
2. Other stagnopodzols with a thin ironpan (p. 108) and an overlying Bh horizon at least 2.5 cm thick
 Humus-ironpan stagnopodzols
3. Other stagnopodzols with a thin ironpan
 Ironpan stagnopodzols
4. Other stagnopodzols
 Ferric stagnopodzols

6.4.2 Ironpan Stagnopodzols

These are the stagnopodzols, widely distributed in upland areas with perhumid or humid oroboreal climatic regimes, that have a more or less sinuous thin ironpan directly below the eluvial horizon. A peaty surface layer is normally present in undisturbed profiles and can be up to 40 cm thick. Variants with topsoils that fail to qualify as peaty occur in cultivated land and in places where the superficial organic layer has been reduced in thickness or destroyed as a result of accelerated erosion, peat cutting or moor burning. The eluvial horizon may be darkened throughout by incorporated organic matter, but an Eag or Eg horizon of higher colour value is usually discernible, especially within downward depressions of the pan, which generally has a mat of decaying roots on its upper surface. Although typically saturated with water for varying periods, this horizon is often unmottled (Eag), presumably because free iron has mostly been removed and concentrated in the pan, and it commonly has a higher clay content than underlying horizons as a result of increased weathering of rock fragments or rock-forming minerals. The horizon immediately below the pan normally shows slight or no evidence of gleying but variants transitional to humic (peaty) gley soils can be strongly gleyed at greater depths. Exposed sections and detailed surveys have revealed that the pan descends in places to more than 80 cm below the mineral surface, particularly in minor topographic depressions (flushes) where the overlying horizons are more or less permanently waterlogged. Profiles of this type resemble lithologically similar humic gley soils without a pan for practical purposes and are grouped with them in the classification used here.

Data on four representative profiles in different parent materials are reproduced on pp.276–80. Two are under semi-natural moorland vegetation, one in afforested moorland, and the fourth under grass in previously cultivated land. All have a brownish or reddish horizon below the pan but in no case does it qualify unequivocally as spodic according to the criteria used in the US (Soil Survey Staff, 1975) and FAO (1974) systems.

Profile 98 from North Wales is in loamy drift derived from Lower Palaeozoic slaty mudstones and siltstones. Similar soils (Hiraethog series) are common in the Welsh uplands (Rudeforth et al., 1984) and resemble those classed as the Dod series in southern Scotland (e.g. Muir, 1956; Ragg and Futty, 1967) and as the Black Rock Mountain series in south-east Ireland (Conry and Ryan, 1967). The profile sampled is located some 12 m above the level of a 50-year old water storage reservoir, at the edge of which a long section of the Hiraethog soil was exposed by wave erosion (Figure 6.7). Examination of the section revealed the occurrence of Mesolithic and Neolithic flint artefacts in the Eag and Bs horizons, and this finding, together with the radiocarbon dates cited in Figure 6.7, led Dr D.A. Jenkins to conclude (1) that the slightly stony slope deposit in which the E and B

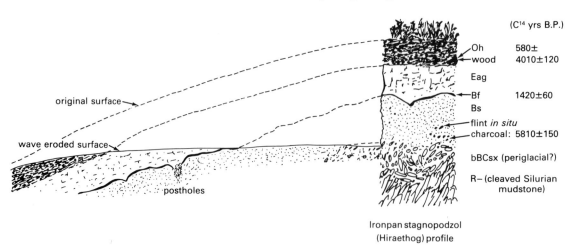

Figure 6.7: Exposures of stagnopodzols at Llyn Aled Isaf, Clwyd (from a drawing by Dr D.A. Jenkins).

horizons have formed is less than 6,000 years old; (2) that the site was formerly forested; and (3) that superficial peat formation leading to development of the Eag and Bf (ironpan) horizons took place between 2,000 and 1,000 years BP. The thick Bs horizon below the pan resembles those of podzolic brown soils in similar materials (e.g. profile 78, p. 237) in morphology and composition, suggesting strongly that the existing profile has evolved from a forest soil of that type (cf. Romans and Robertson, 1975). The chemical data show a marked concentration of both pyrophosphate and dithionite-extractable iron in the pan, unaccompanied by any appreciable increase in extractable aluminium. There is also no carbon maximum, apparently because the overlying Eag contains much irregularly distributed organic matter consisting of more or less decomposed plant remains. Proportions of free iron and organic carbon extracted by pyrophosphate are largest in the friable Bs2 and Bs3 horizons. Volumetric physical data on core samples collected from the Eag horizon show an unusual combination of low bulk density and very small air capacity, indicating that there are few coarse (<60 μm) pores but many in the size range emptied between 0.1 and 2 bars. Samples from the Bs subhorizons have similar bulk densities but the air capacity values are much larger.

Profile 99 is in rubbly head derived mainly from dolerite in the mountains south of Dublin, where O'Dubháin and Collins (1981) studied and compared a number of ironpan stagnopodzols of contrasting lithologic origin. The subsurface mineral horizons are morphologically similar to those of the preceding profile, but both the ironpan and the Bs horizon contain considerably more total free iron and amounts of extractable aluminium, here fractionated into pyrophosphate and residual citrate-dithionite extractable components, are very much larger throughout, reflecting the high proportions of weatherable ferromagnesian minerals and calcic plagioclases in the parent material. In this case, too, there is a distinct aluminium maximum coinciding with the ironpan. Conventional particle-size analyses of the Bs horizon samples gave much lower clay contents than would have been expected either from the field textures or from the high CEC values. In view of the large extractable aluminium contents, these disparities are reasonably attributable to the presence of amorphous (allophanic) clay-size material which resists dispersion and probably constitutes a dominant component of the exchange complex. As the ratio of pyrophosphate-extractable to citrate-dithionite extractable Fe + Al is much too small for the horizon to qualify as spodic (Soil Survey Staff, 1975), the soil is tentatively classed as a Placandept in the US system.

Profile 100 (Plate III.2) is in thin drift overlying weathered coarse-grained granite in the eastern Grampians. Here the ironpan rests directly on a firm stony BC horizon of loamy sand texture and there is no continuous friable horizon identifiable as Bs. Moorland soils of this type have been described in several Scottish localities north of the Forth-Clyde valley (e.g. Glentworth and Muir, 1963; Futty and Dry, 1977) but are comparatively rare further south. In some places, particularly at relatively low altitudes, the horizon below the pan is strongly indurated (duric horizon). At and above the lower altitudinal boundary of the oro-arctic (sub-alpine) zone, ironpan stagnopodzols with or without a Bs horizon become progressively less common and compacted subsoil layers, where present, are generally uncemented (Romans, 1970).

The parent material of profile 101, located in the Peak District of Derbyshire, consists of loess which has been mixed by cryoturbation with fragmentary chert derived as solution residue from the underlying Carboniferous limestone (Piggott, 1962). At this site the upper horizons terminating in the ironpan appear to have formed in the eluvial horizon of a pre-existing luvic brown soil following replacement of forest by moorland vegetation. During mediaeval times or later, the area was divided by stone walls into small enclosed fields, and the absence of an unincorporated superficial organic layer, the sharp lower boundary of the humose A horizon, and the pH values of 6.0 and above, confirm that the soil has at some time been cultivated and limed, but the characteristic Eag and Bf horizons remain intact. A yellowish Eb horizon and an underlying finer textured Bt are clearly evident below the ironpan and the uppermost few centimetres, which contain enough pyrophosphate-extractable iron and aluminium to qualify as Bs, are prominently mottled and distinctly platy. Bisequal 'ironpan soils' of this kind were first recognized and mapped by Findlay (1965) in the Mendip district of Somerset. They also occur over limestone and sandstone in the North York Moors (Bendelow and Carroll, 1979).

Profile 98 (Murphy and Bullock, 1981; Rudeforth *et al.*, 1984)

Location: beside Aled Isaf reservoir, Clwyd (SH 915597)
Climatic regime: humid oroboreal
Altitude: 375 m   Slope: 7° with moundy microrelief

Analytical data for Profile 98							
Horizon	Oh/Om	Eag	Bf	Bs1	Bs2	Bs3	Cu
Depth (cm)	0–24	24–33	33–34	34–45	45–62	62–80	80–120
sand 600 μm–2 mm %		1		5	12	13	2
200–600 μm %		2		3	9	12	6
60–200 μm %		2		3	5	9	10
silt 2–60 μm %		72		66	58	56	73
clay <2 μm %		23		23	16	10	9
organic carbon %	48	5.0	4.3	1.2	1.4	1.0	0.5
nitrogen %	2.31	0.42					
C/N ratio	21	12.5					
pyro. ext. Fe %		0.1	3.2	1.0	2.0	1.1	0.1
pyro. ext. Al %		0.3	0.3	0.3	0.3	0.4	0.2
pyro. ext. C %		1.6	1.1	0.5	0.7	0.6	0.3
pyro. Fe+Al %/clay %		0.02		0.06	0.14	0.15	0.03
total ext. Fe %		0.3	10.6	2.6	3.8	2.2	1.1
TEB (me/100 g)		0.3		0.5	0.1	0.1	0.2
CEC (me/100 g)		20.6		12.6	8.6	6.6	5.1
% base saturation		1		4	1	2	4
pH (H$_2$O)	3.5	3.9		4.3	4.5	4.8	5.3
(0.01M CaCl$_2$)	2.9	3.4		3.8	4.2	4.4	4.7
bulk density (g cm^{-3})	0.24	0.78		0.86	0.79	0.94	1.28
packing density (g cm^{-3})		0.98		1.06	0.93	1.03	1.36
retained water capacity (% vol.)	77	64		54	54	51	35
available water capacity (% vol.)	35	41		25	22	27	19
air capacity (% vol.)	9	3		13	16	13	13

[a] TEB plus exchange acidity at pH 7 by Parker's (1929) method

Land use and vegetation: rough grazing with mat-grass (*Nardus stricta*), heath rush (*Juncus squarrosus*) and wavy hair-grass (*Deschampsia flexuosa*)

Horizons (cm)

L/Om — Undecomposed and partially decomposed grass litter and moss, 5 cm thick.

0–24 Oh/Om — Black (5YR 2/1) humified peat with patches of less well decomposed dark reddish brown (5YR 3/2) semi-fibrous peat; clear irregular boundary.

24–33 Eag — Light brownish grey (10YR 6/2) very slightly stony silty clay loam (USDA silt loam) with few medium light grey (10YR 7/2) mottles and common organic inclusions and coats; few medium and small platy mudstone fragments; massive; friable; few living and many dead roots; sharp wavy to irregular boundary.

33–34 Bf — Continuous thin ironpan with black (2.5YR 2/1) cap and red (2.5YR 4/6) lower part; mat of roots on upper surface.

34–45 Bs1 — Brownish yellow (10YR 6/6) slightly stony silty clay loam (USDA silt loam) with thin and discontinuous weakly developed ironpan at base; stones as above; massive; friable; few roots; abrupt irregular boundary.

45–62 Bs2 — Strong brown (7.5YR 5/6) slightly stony sandy silt loam (USDA silt loam to loam); stones as above; weak fine subangular blocky and granular; friable; few roots; clear wavy boundary.

62–80 Bs3 — Dark yellowish brown (10YR 4/4) slightly to moderately stony sandy silt loam (USDA silt loam to loam); stones, structure and consistence as above; few roots; clear wavy boundary.

80–120 Cu — Dark greyish brown (2.5Y 4/2) very stony silt loam (USDA silt loam); abundant platy mudstone fragments; massive; friable to firm; no roots.

Analytical data *see* p.276.

Classification

England and Wales (1984): ironpan stagnopodzol; loamy material over lithoskeletal mudstone and sandstone or slate (Hiraethog series).

USDA (1975): Histic Placaquept; coarse-silty or loamy-skeletal, mixed, mesic.

FAO (1974): Humic Gleysol; medium textured.

Profile 99 (O'Dubháin and Collins, 1981)

Location : Ballinascorney Commons, Co. Dublin
Climatic regime: perhumid oroboreal
Altitude : 396 m   Slope: 1–2°
Land use and vegetation : 20-year old Sitka spruce plantation, formerly heather moor.

Horizons (cm)

L/F — Thin layer of spruce needles 1–3 cm thick.

0–10 Oh — Black (5YR 2/1) and dark reddish brown (5YR 2.5/1–2) humified peat; friable; many live roots; white fungal mycelium in cracks, sharp smooth boundary.

10–25 Ah — Very dark greyish brown (10YR 3/2) stony humose silt loam to silty clay loam (USDA silt loam); abundant bleached stones, some of quartzite; weak medium granular; friable; roots common and concentrated at base; abrupt wavy boundary (22–28 cm).

25 Bf — Hard cemented ironpan, 5–6 mm thick; black (2/0) upper part, yellowish red (5YR 3/6) below.

25–33 Bs1 — Dark reddish brown to yellowish red (5YR 3/4–4/6) very stony silt loam (USDA silt loam); weak fine granular; smeary; stones showing variable orientation; few roots; gradual boundary.

33–44 Bs2 — Brown to dark yellowish brown (7.5YR 4/4–10YR 3/6) stony silt loam to sandy silt loam (USDA silt loam to sandy loam); weak granular; smeary; clear smooth boundary.

44+ BC — Brown (10YR 5/3–4) and olive brown (2/5Y 5/3–4) very stony sandy loam (USDA sandy loam); abundant olive green dolerite fragments; very few roots.

Analytical data for Profile 99

Horizon	Oh	Ah	Bf	Bs1	Bs2	BC
Depth (cm)	0–10	10–25	25	25–33	33–34	44+
sand 50 µm–2 mm %		44		20	44	66
silt 2–50 µm %		43		71	51	31
clay <2 µm %		13		9	5	3
organic carbon %	26	11.0	8.1	5.5	3.1	1.6
pyro. ext. Fe %		0.1	6.0	0.6	0.2	0.1
pyro. ext. Al %		1.2	1.9	0.8	0.8	0.8
pyro. Fe+Al %/clay %		0.10		0.16	0.20	
total ext. Fe %		0.3	15.7	4.1	2.1	1.5
total ext. Al %		1.5	2.9	2.5	2.4	1.4
TEB (me/100 g)	5.3	2.5		0.4	0.4	0.5
CEC[a] (me/100 g)	75	50		37	22	13
% base saturation	7	5		1	2	4
pH (H_2O)	4.0	4.2		4.8	4.9	5.0
(0.01M $CaCl_2$)	3.3	3.8		4.2	4.4	4.6

[a] by leaching with N ammonium acetate at pH 7 and determining ammonium displaced by a second leaching with KCl

Classification
  Ireland (1980): peaty podzol.
  England and Wales (1984): ironpan stagnopodzol; loamy material over lithoskeletal basic crystalline rock (unnamed series).
  USDA (1975): Histic Placandept?; thixotropic?, mesic.
  FAO (1974): Humic Gleysol?; medium textured.

Profile 100 (Grant and Heslop, 1981)

Location	:	Cairn o'Mounth, Kincardineshire (NO 651831)
Climatic regime:		perhumid oroboreal
Altitude	:	305 m Slope: 8°
Land use and vegetation	:	rough grazing; bog heather moor with heather (*Calluna vulgaris*) dominant, deer-sedge (*Trichophorum caespitosum*) abundant and mat-grass (*Nardus stricta*) frequent.

Horizons (cm)

0–10 Oh — Black (5YR 2/1) humified peat containing some uncoated mineral grains, especially towards base; few small subangular granitic stones; massive; wet; abundant roots; clear wavy boundary.

10–40 Ah/Eag — Light brownish grey (10YR 6/2) slightly stony (coarse) sandy loam (USDA sandy loam) becoming dark greyish brown (10YR 4/2) towards base; discontinuous very dark greyish brown (10YR 3/2) band (Ah) at top; common subangular granitic stones, mainly very small but ranging to large; very weak blocky; friable; moist to wet; many roots; abrupt irregular boundary.

40 Bf — Continuous thin ironpan about 2 mm thick.

40–90 BC(x)1 — Reddish brown (5YR 5/4) moderately to very stony loamy (coarse) sand (USDA loamy coarse sand) with yellowish red (5YR 5/8) patches; stones as above; massive; firm; roots rare; abrupt wavy boundary.

90–120 BC(x)2 — Reddish brown (5YR 5/4) moderately to very stony loamy (coarse) sand (USDA loamy coarse sand); stones as above; massive; firm; no roots.

Analytical data for Profile 100

Horizon	Oh	Eag	BC(x)1	BC(x)2
Depth (cm)	1–9	20–30	60–70	100–110
silt[a] 2–50 μm %		20	15	13
clay[a] <2 μm %		13	5	3
organic carbon %	38	1.4	0.3	0.2
nitrogen %	1.61	0.05		
C/N ratio	23	29		
pyro. ext. Fe %		0.1	0.1	0.1
pyro. ext. Al %		0.2	0.2	0.2
pyro. ext. C %	3.4	0.5	<0.05	<0.05
total ext. Fe %		0.1	0.4	0.4
TEB (me/100 g)	4.6	0.1	0.1	0.1
CEC (me/100 g)	68	6.4	2.1	1.3
% base saturation	7	2	5	8
pH (H$_2$O)	3.8	4.5	4.6	4.7
(0.01M CaCl$_2$)	2.9	4.0	4.2	4.3

[a] by hydrometer method

Classification
  Scotland (1984): peaty podzol (Charr series).
  England and Wales (1984): ironpan stagnopodzol; loamy material over lithoskeletal acid crystalline rock (Hexworthy series).
  USDA (1975): Histic Placaquept (or Placaquod); sandy or sandy-skeletal; mixed, frigid.
  FAO (1974): Humic Gleysol; coarse textured.

Profile 101 (Johnson, 1971)

Location	:	Hucklow, Derbyshire (SK 158777)
Climatic regime:		humid oroboreal
Altitude	:	326 m Slope: 2°
Land use	:	rough grazing

Horizons (cm)

0–15 Ap — Black (N 2/–; N 3.5 dry) slightly stony humose silt loam (USDA silt loam); common angular chert fragments; moderate fine and medium granular; very friable; abundant roots; sharp smooth boundary.

15–28 Eag — Light brownish grey (10YR 6/3; 7/2 dry) moderately stony silt loam (USDA silt loam); stones as above; weak medium and fine angular blocky; friable; many roots and root mat at base; sharp wavy to irregular boundary.

28 Bf — Cemented ironpan up to 5 mm thick, very dark reddish brown in upper part and strong brown below; passes through chert fragments and ranges in depth from 25 to 38 cm.

28–31 Eg — Pale brown (10YR 6/3) and very pale brown (10YR 7/3) slightly stony silt loam (USDA silt loam) with common coarse yellowish brown (10YR 5/8) mottles and ferruginous coats associated with former root channels; stones as above; moderate medium platy; friable to firm; few roots; few fine manganiferous concentrations; clear irregular (tongued) boundary.

31–48 Eb(g) — Light yellowish brown (10YR 6/4) and yellowish brown (10YR 5/6) faintly mottled, slightly stony silt loam (USDA silt loam); stones as above; weak medium angular blocky; few roots; clear irregular boundary.

48–69 Bt1(g) — Brown (7.5YR 5/4) slightly stony silty clay loam (USDA silt loam); moderate medium subangular blocky tending to fine prismatic with smooth light yellowish brown (10YR 6/4) faces;

Analytical data for Profile 101

Horizon	Ap	Eag	Eb(g)	Bt1(g)	Bt2	2BCt
Depth (cm)	2–13	13–25	33–46	51–66	71–86	90–100
sand 200 μm–2 mm %	6	9	6	7	4	14
50–200 μm %	6	7	7	5	6	14
silt 2–50 μm %	74	71	77	69	66	38
clay <2 μm %	14	13	10	19	24	34
organic carbon %	7.7	3.6	1.0	0.4	0.6	0.7
pyro. ext. Fe %	0.2	0.1	0.3	0.1	<0.05	<0.05
pyro. ext. Al %	0.2	0.2	0.4	0.1	0.1	0.1
pyro. ext. C %	1.9	1.4	0.3	0.1		
pyro. (Fe+Al) %/clay %	0.03	0.02	0.06	0.01	<0.01	<0.01
total ext. Fe %	0.8	0.5	2.0	2.1	2.3	4.9
pH (H$_2$O)	6.2	6.0	6.4	6.9	6.7	6.0
(0.01M CaCl$_2$)	5.8	5.5	5.9	6.3	6.2	5.5

69–90 Bt2　Brown (7.5YR 5/4) and strong brown (7.5YR 5/6), moderately to very stony, silty clay loam (USDA silt loam); stones as above; moderate medium angular blocky between stones; firm; roots rare; clear wavy boundary.

common vertically oriented streaks associated with old root channels; friable to firm; few roots; clay coats on stones; gradual boundary.

90–102+ 2BCt　Brown (7.5YR 5/4) very stony clay loam (USDA clay loam); stones as above; structure ill-defined because of stoniness; firm; roots rare; many fine manganiferous concentrations.

Analytical data see above.

Classification
  England and Wales (1984): ironpan stagnopodzol; coarse silty drift with siliceous stones (variant of Priddy series).
  USDA (1975): (Typic) Placaquept; fine silty, mixed, mesic.
  FAO (1974): Humic Gleysol?, medium textured.

6.4.3 Humus-ironpan Stagnopodzols

As indicated above, these soils differ from those of the ironpan subgroup in having a dark coloured Bh horizon above the pan. They are much less extensive in the British Isles and mostly occur in less humid areas, often in close association with well drained humoferric or humus podzols. The parent materials, which are generally coarse loamy or coarser, include glaciofluvial sands and gravels of the Corby association in north-east Scotland (Glentworth and Muir, 1963; Futty and Dry, 1977; Farmer et al., 1980), weathering products of sandstone in Northumberland (Payton, 1987a,b) and the North York Moors (Carroll and Bendelow, 1981), 'plateau gravels' in the southern English heathlands (e.g. Clayden, 1971; Ragg and Clayden, 1973) and granitic drift in Cornwall (Staines, 1979). An Ea horizon is normally present under heath and is succeeded by a more or less cemented Bh. This either rests directly on the ironpan or is separated from it by a second eluvial horizon (Eag or Eg) which is most prominent above downward depressions of the pan and is periodically saturated with water wherever it is unbroken. A friable or cemented Bs horizon is usually but not invariably encountered below the pan and some of the soils, especially on older drift in southern England, have bisequal profiles with an argillic (Bt) horizon at some depth. Data on two representative heathland profiles are given below.

Profile 102 was described and sampled by the writer in a shallow drainage ditch at the edge of a Corsican pine plantation in Wareham Forest, Dorset. The parent material is a cryoturbated river-terrace deposit, probably of Anglian age, over Tertiary (Bagshot) beds, and the semi-natural vegetation is moist *Calluna* heath with purple moor-grass (*Molinia caerulea*) and cross-leaved heath (*Erica tetralix*). Although the horizons above the ironpan were dry when sampled (30 July 1976), the discontinuous Eag horizon is waterlogged after wet weather and the compact Bh also impedes drainage to some extent. An irregularly developed Bs underlies the pan and is followed by a finer textured massive to platy Bt horizon with fragipan features. Nutritional problems affecting the growth of conifer seedlings in this extremely impoverished soil were investigated over a period of more than 20 years in the adjoining forest nursery (Benzian, 1965). Bolton and Coulter (1966) showed that phosphorus supplied as superphosphate is readily leached and accumulates in the B horizon.

Profile 103, with a similar succession of horizons, is developed in thin sandy drift over Carboniferous Fell Sandstone at an altitude of 270 m in the Rothbury district of Northumberland (Plate III.3). Here the climate is distinctly cooler and more humid and the superficial organic horizons under heath are correspondingly more peaty in character. As in the preceding profile, the Bh horizon often becomes saturated after rain but the sandy Ea is never wet for long periods. The presence of an argillic B horizon, which has only been observed in the deeper soils on Fell Sandstone in this area, again suggests that the overlying podzolic horizons including the pan have formed in those of a pre-existing luvic brown soil.

Profile 102

Location : Sugar Hill, Bloxworth, Dorset (SY 877928)
Climatic regime: subhumid temperate
Altitude : 34 m   Slope: <1°
Land use and vegetation : Corsican pine plantation (c. 50 per cent cover) bordering abandoned forest nursery; field layer includes abundant purple moorgrass (*Molinia caerulea*) and heather (*Calluna vulgaris*), with cross-leaved heath (*Erica tetralix*), bristle Agrostis (*Agrostis curtisii*) and milkwort (*Polygala* spp.)

Horizons (cm)

0–13 Ap
Dark grey (5YR 3–4/1; 5YR 5–6/1 dry) loamy medium sand (USDA loamy sand) containing rare subangular flint fragments; weak granular; loose; abundant fine and many medium woody roots; abrupt wavy boundary (soil surface disturbed and covered with ditch spoil; undisturbed soil nearby has L/F layers up to 5 cm thick over black H horizon 2–4 cm thick).

13–32 Ea
Pinkish grey (7.5YR 7/2; 5YR 7/1 dry) loamy medium sand (USDA loamy sand) with common faint brown (7.5YR 5/2) mottles, particularly in upper part; rare flints as above; single grain; loose; common medium and fine woody roots; abrupt wavy boundary ranging from 27 to 40 cm.

32–37 Bh(m)
Black (5YR 2/1) to dark reddish brown (5YR 3/2) medium sandy loam (USDA sandy loam) with rare flints as above; massive, grading locally to granular; firm (very hard when dry); brittle and very weakly cemented in most parts, with many continuous organic coats on sand grains; elsewhere uncemented with organic material in fine granular aggregates; fewer roots than above; horizon 3–8 cm thick, with abrupt wavy lower boundary ranging from 30 to 47 cm.

37–41 Eag
Dark greyish brown to brown (7.5YR 4–5/2; 7.5YR 5–6/2 dry) medium sandy loam (USDA sandy loam) with faint diffuse darker and paler mottles and rare flints as above; massive; friable; common medium and fine woody roots; discontinuous horizon, up to 7 cm thick, present in 75 per

Analytical data for Profile 102

Horizon	Ap	Ea	Bh(m)	Eag	Bs	2Btx(g)
Depth (cm)	3–10	29–37	32–36	37–40	41–47	50–70
sand 600 μm–2 mm %	10	7	7	7	8	7
200–600 μm %	42	38	36	34	36	29
60–200 μm %	26	31	27	27	22	17
silt 2–60 μm %	20	23	24	25	25	34
clay <2 μm %	2	1	6	7	9	13
organic carbon %	4.1	0.5	3.9	2.9	1.7	0.7
pyro. ext. Fe %	0.1	<0.05	<0.05	<0.05	1.0	0.2
pyro. ext. Al %	0.1	<0.05	0.3	0.3	0.3	0.3
pyro. ext. C %	1.3	0.2	2.5	1.7	1.2	0.4
pyro. Fe+Al %/clay %			0.06	0.04	0.14	0.04
total ext. Fe %	0.1	0.1	0.3	0.1	1.7	1.3
pH (H$_2$O)	4.1	4.1	3.9	3.9	4.2	4.5
(0.01M CaCl$_2$)	3.1	3.2	3.1	3.4	3.6	3.9

41
Bf
cent of section; sharp wavy boundary ranging from 37 to 52 cm. Cemented ironpan 2–4 mm thick; upper 1 mm black, remainder dark reddish brown; continuous but variable in thickness and strength, strongest in downward depressions; sharp wavy boundary.

41–48
Bs
Strong brown (7.5YR 5/6), reddish brown (5YR 4/4) and yellowish red (5YR 5/6), irregularly mottled, very slightly stony medium sandy loam (USDA sandy loam); few gravel-size brown-stained subangular flint fragments; massive; mainly friable but darker parts firm and very weakly cemented; common medium woody and few fine roots; gradual wavy boundary.

48–80+
2Btx(g)
Strong brown (7.5YR 5/6) and brownish yellow (10YR 6/6), coarsely mottled, slightly stony medium sandy loam (USDA sandy loam to loam); common, medium to large, subangular and subrounded flints and few small quartz pebbles; weak platy; firm to very firm and brittle in most parts; many medium and coarse vesicular pores; few to common distinct reddish brown clay coats lining voids and bridging grains; few roots; coarse light brownish grey and ferruginous mottles appearing in places at 80 cm.

Analytical data *see* above

Classification

England and Wales (1984): humus-ironpan stagnopodzol; coarse loamy drift with siliceous stones (unnamed series).

USDA (1975): Placic Haplaquod (or Typic Placohumod); coarse-loamy, siliceous, mesic.

FAO (1974): Placic Podzol; coarse textured (fragipan phase).

Profile 103 (Payton, 1987a)

Location	:	Callaly Crags, Whittingham, Northumberland (NO 062091)

Climatic regime: humid oroboreal
Altitude : 272 m   Slope: 6° convex
Land use and vegetation : rough grazing; heather moor with *Calluna vulgaris* dominant.
Horizons (cm)

L Little decomposed heather and moss litter, 4 cm thick.

0–8
F
Very dark greyish brown (10YR 3/2) partially decomposed organic material; abundant roots; clear smooth boundary.

8–14
H/Oh
Black (10YR 2/1) humified peat; massive; many roots; sharp smooth boundary.

14–39 Ea	Pinkish grey (7.5YR 6/2) moderately stony medium sand (USDA sand); many very large subangular tabular sandstone fragments; single grain; very friable; common roots; abrupt wavy boundary.
30–38 Bh	Black (5YR 2/1) moderately stony medium sand (USDA sand) with few diffuse very dark grey (10YR 3/1) inclusions; stones as above; massive; friable; brittle when very moist or wet; few roots; many organic coats on sand grains; sharp irregular boundary.
38–40 Eag	Pale brown (10YR 6/3) moderately stony medium sandy loam (USDA fine sandy loam); stones as above; massive; wet; few roots; discontinuous horizon, best developed in downward convolutions of the succeeding ironpan.
40 Bf	Continuous convoluted thin ironpan with black (5YR 2/1) non-porous upper part 1–2 mm thick and red (10YR 4/6) base 1 mm thick; brittle and cemented; root mat on upper surface.
40–46 Bs	Dark red (2.5YR 3/6) moderately stony medium sandy loam (USDA fine sandy loam); massive grading in parts to fine granular; friable to firm; brittle and very weakly cemented in places; no roots; many sesquioxidic coats on sand grains; clear wavy boundary.
46–55 Eb	Strong brown (7.5YR 5/6) very stony loamy medium sand to sandy loam (USDA loamy fine sand); stones as above; massive; very friable; clear wavy boundary.
55–70 Bt(g)	Yellowish brown (10YR 5/4) very stony medium to fine sandy loam (USDA fine sandy loam) with common fine light yellowish brown (10YR 6/4) mottles; stones as above; massive; friable; no roots; common continuous clay coats on sand grains and lining voids; abrupt irregular boundary.
70–90 Cu	Very pale brown (10YR 7/4) extremely stony medium sand (USDA sand); extremely abundant very large angular tabular sandstone fragments; single grain; irregular boundary.
90+ R	Very pale brown (10YR 7/4) hard stratified sandstone.

Classification

England and Wales (1984): humus-ironpan stagnopodzol; sandy material over lithoskeletal sandstone (variant of Maw series).

USDA (1975): Placic Haplaquod (or Typic Placohumod); coarse-loamy or loamy-skeletal, mixed or siliceous, mesic.

FAO (1974): Placic Podzol; coarse textured.

Analytical data for Profile 103

Horizon	F	H/Oh	Ea	Bh	Eag	Bf	Bs	Eb	Bt(g)	Cu
Depth (cm)	0–8	8–14	14–30	30–38	38–40	40	40–46	46–55	55–70	70–90
sand 600 μm–2 mm %			8	3	2		2	2	1	2
200–600 μm %			56	37	26		30	27	21	34
60–200 μm %			30	51	42		42	46	45	55
silt 2–60 μm %			5	7	26		20	19	21	7
clay <2 μm %			1	2	4		6	6	12	2
organic carbon %	47	32	0.5	1.2	0.7	2.9	0.7	0.2	0.1	0.1
pyro. ext. Fe %	0.04	0.01	0.01	0.01	0.01	2.2	0.6	0.02	0.02	0.02
pyro. ext. Al %	0.08	0.02	0.01	0.1	0.1	0.2	0.2	0.1	0.2	0.04
pyro. ext. C %	3.8	2.1	<0.05	0.5	0.1	1.7	0.5	0.1	0.1	0.1
pyro. Fe+Al %/clay %							0.13	0.02		
dithionite ext. Fe %			0.01	0.03	0.04	13.4	1.8	0.8	1.0	0.3
pH (H$_2$O)	3.7	3.6	4.0	3.7	3.8		4.3	4.4	4.6	4.9
(0.01M CaCl$_2$)	2.8	2.7	3.3	3.4	3.4		3.8	3.9	4.0	4.3

6.4.4 Ferric Stagnopodzols

These are the stagnopodzols with a greyish eluvial horizon (Eag or Eg) over a more or less ochreous Bs and no intervening thin ironpan which is continuous or nearly so (Plate III.4). Periodically wet acid soils with these characteristics are widely distributed in stony loamy parent materials in humid or perhumid upland areas, usually in close association with those of the ironpan subgroup. In Wales, where they are particularly common, they are characteristically located at altitudes between about 350 and 450 m, immediately above the slopes in which podzolic brown soils predominate (Pyatt, 1977; Rudeforth et al., 1984). The semi-natural vegetation shows a corresponding altitudinal zonation, acidic grassland with bracken on the podzolic brown soils giving place on the stagnopodzols to various moorland communities in which heather (*Calluna vulgaris*), bilberry (*Vaccinium myrtillus*), western gorse (*Ulex gallii*), purple moor-grass (*Molinia caerulea*) and mat-grass (*Nardus stricta*) locally attain dominance, often with an understorey of mosses.

Undisturbed soils of this subgroup are distinguished from non-hydromorphic (humoferric) podzols in similar materials by the presence of a seasonally waterlogged peaty surface coupled with signs of gleying in the immediately underlying mineral horizons. Afforestation generally leads to a decrease in surface wetness and replacement of the moorland vegetation by improved grassland can have a similar effect, providing that grazing is carefully controlled to avoid excessive poaching. Where the organic and eluvial horizons are relatively thin, the soils quickly assume the characteristics of podzolic brown soils under cultivation, especially when transformation of the surface horizons is further promoted by the use of lime and fertilizers. Two representative profiles from Exmoor (Figure 6.8), where the historical evolution of the landscape has been well documented (Curtis, 1971), are reproduced on pp.285–7. The first is under unreclaimed heather moor and the second under a grass ley in land

Figure 6.8: Dunkery Beacon, Exmoor. Heather moors in the distance have ferric stagnopodzols on upper slopes and non-hydromorphic (humoferric) podzols on the steeper flanks. The lower lying enclosed land is mainly on well drained brown soils, with recently reclaimed podzols in the foreground (photo by D.V. Hogan).

enclosed about the middle of the nineteenth century. Both are developed in stony coarse loamy head derived from the subjacent Devonian rocks, possibly with some addition of loess.

Profile 104 has a peaty topsoil 10 cm thick and profile 105 a humose Ap horizon in which remains of the former peaty surface have been intimately mixed with the uppermost mineral soil by cultivation and earthworm activity. Although the Bs horizon of the latter contains considerably more clay, perhaps reflecting a lithologic discontinuity, the subsurface horizons are essentially similar. The chemical data show marked concentrations of pyrophosphate-extractable iron in the Bs horizons, accompanied in each case by aluminium maxima and in profile 105 by a slight increase in extractable carbon. By comparison with associated ironpan stagnopodzols, however, the eluvial horizons retain significant amounts of free iron extractable by dithionite, much of which is segregated in mottles or coatings on rock fragments. These segregations are commonest immediately above the irregular upper boundary of the Bs horizon, where they coalesce in places to form a discontinuous or incipient pan. Although the B horizons fail to qualify unequivocally as spodic (Soil Survey Staff, 1975, pp. 29–32), they are clearly enriched in sesquioxides by illuviation and differ markedly from the grey and brown mottled Bg (cambic) horizons characteristic of associated gley soils (Aquepts). Both soils are therefore tentatively identified as Spodosols (Sideraquods).

Profile 104 (Hogan and Harrod, 1982)

Location : Brendon, Devon (SS 748437)
Climatic regime: perhumid oroboreal
Altitude : 373 m Slope: 4° straight
Land use and   rough grazing; moist Atlantic
  vegetation : heather moor with heather (*Calluna vulgaris*) and purple moor-grass (*Molinia caerulea*)
Horizons (cm)
0–10   Black (5YR 2/1) stoneless humified
Oh1   peat with a thin (<3 cm) layer of dark reddish brown (5YR 2/2) peat at base; moderate fine subangular blocky; abundant roots; abrupt wavy boundary.
10–18   Black to very dark brown (7.5YR 2/1)
Oh2   stoneless humified peat; strong coarse angular blocky; firm; abundant roots; clear smooth boundary.
18–28   Pinkish grey (7.5YR 6/2) very stony
Eg   sandy silt loam (USDA loam) with common fine strong brown (7.5YR 5/6) mottles associated with soft weathered stones in lower part and medium and coarse inclusions of black organic material; abundant medium subangular tabular micaceous sandstone fragments, hard in upper part and softer below; most stones subhorizontally aligned; massive; firm; many roots; abrupt irregular boundary.
28–40   Yellowish red (5YR 5/6; 7.5YR 5/6
Bs   rubbed) moderately stony sandy silt loam (USDA loam) with common coarse dark reddish grey (5YR 4/2) and dark reddish brown (5YR 3/2) infillings; stones as above, mainly small but very small, soft and weathered near top; weak medium and fine subangular blocky; friable; few roots; clear wavy boundary.
40–66   Light brown (7.5YR 6/4) moderately
BC(x)   stony sandy silt loam (USDA loam to silt loam) with infillings as above; stones as above, mainly medium; massive; firm; few roots.
66–117   Pinkish grey to light brown (7.5YR
Cu(x)   6/3) moderately stony sandy silt loam (USDA silt loam to loam), becoming very stony below 90 cm; stones as above but mainly large; massive; firm; roots rare.

Analytical data *see* p. 286.

Classification
  England and Wales (1984): ferric stagnopodzol; loamy material over lithoskeletal mudstone and sandstone or slate (Hafren series).
  USDA (1975): (Histic) Sideraquod?; coarse loamy or loamy-skeletal, mixed, mesic.
  FAO (1974): Gleyic Podzol; medium textured.

Analytical data for Profile 104

Horizon	Eg	Bs	BC(x)	Cu(x)
Depth (cm)	18–28	28–40	40–66	66–117
sand 600 µm–2 mm %	11	18	11	11
200–600 µm %	8	8	9	8
60–200 µm %	17	15	16	16
silt 2–60 µm %	52	47	53	56
clay <2 µm %	12	12	11	9
organic carbon %	2.4	1.4		
pyro. ext. Fe %	0.4	1.7	0.4	0.2
pyro. ext. Al %	0.2	0.4	0.3	0.2
pyro. ext. C %	1.3	0.8	0.3	0.2
pyro. Fe+Al %/clay %	0.05	0.18	0.06	0.04
total ext. Fe %	1.2	3.2	1.5	1.3
pH (H$_2$O)	4.7	4.8	5.0	5.2
(0.01M CaCl$_2$)	3.8	5.1	4.3	4.3

Profile 105 (Hogan, 1981)

Location : Charles, Devon (SS 682365)
Climatic regime: humid temperate
Altitude : 262 m   Slope: 5° convex
Land use and   ley grass
  vegetation :
Horizons (cm)

0–16 Ap
Dark reddish grey (5YR 4/2) slightly stony humose sandy silt loam (USDA loam); common, very small to medium, subangular tabular sandstone and small subangular quartz fragments; moderate fine and medium subangular blocky; friable; many roots; earthworms present; clear boundary.

16–25 Eg
Pinkish grey (5YR 6/2) moderately stony sandy silt loam (USDA loam to silt loam) with common fine and medium grey (5YR 5/1) and yellowish red (5YR 5/6) mottles and dark reddish brown (5YR 2/2) inclusions in channels and pockets; stones as above, together with soft ochreous weathered sandstone fragments; weak fine and medium angular blocky; firm; common roots; earthworms present; ferruginous nodules and coats on stones forming incipient ironpan at base; negative reaction to NaF test; clear irregular boundary.

25–38 Bs
Yellowish red (5YR 5/6) slightly stony clay loam (USDA loam) with common fine and medium dark reddish grey (5YR 4/2) mottles and dark reddish brown (5YR 2/2) in channels and pockets; stones as above; weak medium and coarse angular blocky; friable; common roots; earthworms present; positive reaction to NaF test; gradual boundary.

38–85+ BC(x)
Reddish brown (2.5YR 5/4) very stony sandy silt loam (USDA silt loam to loam) with few reddish grey and dark reddish brown inclusions as above; stones as above, some very large; massive; firm; roots rare; positive reaction to NaF test.

Analytical data *see* p.287.

Classification
  England and Wales (1984): ferric stagnopodzol; reddish loamy material over lithoskeletal sandstone (Lydcott series).
  USDA (1975): Typic Sideraquod?; loamy-skeletal, mixed, mesic.
  FAO (1974): Gleyic Podzol?; medium textured.

Analytical data for Profile 105

Horizon	Ap	Eg	Bs	BC(x)
Depth (cm)	0–16	16–25	25–38	38–85
sand 600 μm–2 mm %	22	19	16	13
200–600 μm %	9	8	7	8
60–200 μm %	10	8	6	11
silt 2–60 μm %	50	54	48	58
clay <2 μm %	9	11	23	10
organic carbon %	5.6	2.8	2.2	
pyro. ext. Fe %	0.2	0.2	2.0	0.1
pyro. ext. Al %	0.1	0.1	0.5	0.1
pyro. ext. C %	0.9	1.1	1.3	0.1
pyro. Fe+Al %/clay %	0.03	0.03	0.11	0.02
total ext. Fe %	1.2	1.5	4.6	2.6
pH (H$_2$O)	5.8	6.1	5.9	5.4
(0.01M CaCl$_2$)	5.1	5.4	5.2	4.5

6.4.5 Densipan Stagnopodzols

These soils have a remarkably hard and compact grey E horizon (bleached hardpan or albic densipan) overlying a friable but more clayey Bs with organic coats on fissure faces. They have so far been identified only in moderately to steeply sloping land in and around the Yarner Wood Nature Reserve near Bovey Tracey, Devon, where they are developed in rubbly head derived from thermally altered Carboniferous mudstones, sandstones and shales with minor additions of loess (Clayden, 1971; Loveland and Clayden, 1987). Those in the Reserve bear stands of sessile oak (*Quercus petraea*) ranging from closed canopy woodland with trees 18–20 m high to often scrubby coppice, with a dense ground cover of bilberry (*Vaccinium myrtillus*). Similar soils occur nearby under moist heath and podzolic brown soils with more diversified ground vegetation in other parts of the Reserve.

Data on a representative profile under oak coppice are reproduced on pp.288–9. It has a thick well decomposed organic surface layer with granular structure (moder), which is less dense and plastic than the 'greasy' Oh horizons found in nearby heathland but exuded water when squeezed at the time of sampling (10 June 1966). The characteristic indurated Eax horizon is encountered immediately below and appears in thin section to consist of uncoated silt and sand grains closely packed in the spaces between the stones, which are often difficult to distinguish in the field from the fine earth. Its thickness can vary considerably within a profile pit and a maximum of 50 cm has been recorded. Yellow–brown mottles appear in the lower part and Clayden (1971) reports the occasional occurrence of a thin ironpan at this level. The few roots that penetrate the indurated E horizon ramify anew in the more porous Bs. This is yellower in hue than podzolic B horizons generally, but the analytical data show clearly that it is enriched in dispersed organic matter as well as in extractable aluminium and iron by comparison with horizons above and below. A much more compact BC or C horizon with fragipan features occurs in places below the Bs but is absent in the profile described.

Clayden (1971) noted that fragments of the E horizon slake readily in water, indicating that it is a 'densipan' (Smith et al., 1975) resulting from close packing of fine earth particles and not a duripan (Soil Survey Staff,

1975) cemented by silica. Using resin-coated clod samples from a nearby profile, Loveland and Clayden (1987) obtained the extremely high bulk density value of 1.97 g cm^{-3}, which is similar to those reported for densipans in Trinidad (Smith *et al.*, 1975) and New Zealand (Wells and Northey, 1985). A further sample which had been pulverized, puddled to a stiff paste with water and then allowed to dry gave a value of 1.68 g cm^{-3}, demonstrating the ability of the soil material to pack densely as a result of simple slaking, wetting and drying. This ability is attributed to the particular combination of particle sizes in the sand-silt range, the small clay content, and the low levels of free sesquioxides and organic matter, all of which are likely to facilitate close packing. Mineralogical and micromorphological data presented by the same authors suggest that the marked difference in clay content between the E and Bs horizons reflects the layered nature of the parent head deposit, coupled with some loss of clay from the E by dissolution or lateral eluviation. Although argillans were identified in the B horizon, they occupied less than 1 per cent of the thin section, indicating that downward translocation of clay-size particles has played an insignificant role in its genesis.

As this soil appears to lack both a spodic and an argillic horizon, it has to be classed as an Inceptisol in the US taxonomy as currently established. If it is assumed to have an aquic moisture regime, it keys out as a Humaquept (humic gley soil), an ascription that takes no account of the indurated E horizon. However, both Smith *et al.* (1975) and Wells and Northley 1985 proposed that additional *Densi-* classes should be introduced to accommodate such soils.

Profile 106 (Clayden, 1971)

Location	: Yarner Wood, Bovey Tracey, Devon (SX 777788)
Climatic regime:	humid temperate
Altitude	: 220 m Slope: 6°
Land use and vegetation	: national nature reserve; poor sessile oak (*Quercus petraea*) coppice (trees c. 6 m high) with some rowan (*Sorbus aucuparia*), holly (*Ilex aquifolium*) and birch (*Betula* spp.); field layer dominated by bilberry (*Vaccinium myrtillus*)

Horizons (cm)

L	Oak leaves and twigs with plentiful wood ants, c.3 cm thick.
0–1 F	Matted, partly decomposed oak leaves and twigs with wood lice, enchytraeid worms, spiders, millipedes and springtails.
1–16 H(Oh)	Reddish black (10R 2/1) stoneless organic material with small mineral fraction; strong fine granular; very friable, but water exudes when squeezed; abundant woody roots up to 1 cm diameter, abrupt smooth boundary.
16–34 Eax	Grey (N 5/1) very stony (60 per cent) sandy silt loam (USDA silt loam) with patches of dark grey (N 4/1), particularly in top 2 cm; abundant, very large to small, angular fragments of fine grained sandstone; massive; very firm; pores and fissures very rare; fine roots rare; gradual irregular boundary.
34–44 Egx	Grey (10YR 5/1, locally N 5/1) very stony sandy silt loam (USDA loam) with brown (10YR 5/3) to yellowish brown (10YR 5/6) mottles increasing in frequency and chroma with depth; stones, structure and consistence as above; rare fine pores; few fine roots; clear irregular boundary.
44–49 Bs(g)	Yellowish brown (10YR 5/4) and brown (10YR 5/3) very stony clay (USDA clay) with slight greyish mottling; stones as above; weak medium angular blocky with many black organic coats on faces; firm; common woody roots up to 1 cm diameter; clear irregular boundary.
49–59 Bs	Light olive brown (2.5Y 5/4) very stony clay loam (USDA loam) with patches of yellowish brown (10YR 5/6); stones as above; moderate gran-

59–97+ BC/Cu	ular; very friable; common roots; gradual boundary. Greyish brown (2.5Y 5/2) very stony sandy silt loam (USDA silt loam); stones as above; weak fine blocky; friable; few roots.

Classification
England and Wales (1984): hardpan stagnopodzol; loamy, lithoskeletal mudstone and sandstone or slate (Yarner series).
USDA (1975): Humaquept (or Densiaquept); loamy-skeletal, mixed, mesic.
FAO (1974): no suitable class provided.

Analytical data for Profile 106

Horizon	H(Oh)	Eax	Eg	Bs(g)	Bs	BC/Cu
Depth (cm)	1–16	16–34	34–44	44–49	49–59	59–97
sand 200 μm–2 mm %		21	22	12	16	20
50–200 μm %		19	13	8	11	13
silt 2–50 μm %		55	48	39	47	51
clay <2 μm %		5	17	41	26	16
loss on ignition %	55	2.6	4.4	11.2	10.3	5.6
organic carbon %		0.5	2.3	3.6		
pyro. ext. Fe %		<0.05	0.3	1.9	1.6	0.2
pyro. ext. Al %		0.3	0.5	0.8	0.6	0.3
pyro. ext. C %		0.1	0.5	2.1	2.2	0.3
pyro. Fe+Al %/clay %		0.06	0.05	0.07	0.08	0.03
TEB (me/100 g)		0.9			0.7	0.9
CEC (me/100 g)		3.9			20.6	8.1
% base saturation		23			3	11
pH (H$_2$O)	4.0	4.8	4.3	4.5	4.8	5.0
(0.01M CaCl$_2$)	2.9	3.9	3.4	3.5	4.3	3.9

7
Gley Soils

7.1 General Characteristics, Classification and Extent

The essential common feature of gley soils as they are defined in this account is a gleyed or hydrocalcic subsurface horizon (Chapter 3, p. x) that starts directly below the topsoil or within 40 cm depth. They are periodically or permanently saturated with water or formed under wet conditions and lack horizons characteristic of podzols. Other properties vary greatly, depending on the parent material and the extent to which it has been modified by processes other than gleying; and since the characteristic morphological features commonly persist with only minor modifications when the water table is lowered, whether naturally or by artificial drainage, the rooting zone of some of these soils is no longer saturated in any part for more than a few days at a time. Where the upper horizons are or were wet for much of the year, they are generally rich in organic matter, and intergrades to peat soils can have organic surface layers up to 40 cm thick.

Gley soils have been distinguished using similar but not identical criteria by all three soil survey organizations in the British Isles and are also placed in a single class at the highest categorical level in the French (CPCS, 1967; Duchaufour, 1982) and Canadian (Canada Soil Survey Committee, 1978) classifications. In the US and Netherlands (de Bakker and Schelling, 1966) systems, however, hydromorphism is treated as a secondary characteristic. Divisions in the highest (order) category are based on the presence or absence of diagnostic features such as the mollic epipedon and the argillic horizon, which occur in both well drained and poorly drained soils, and the latter are set apart as *aquic* or *hydro*-suborders (e.g. Aqualfs; hydrobrick soils). An intermediate approach is adopted in the FAO (1974) scheme, which groups most gley soils in recent alluvial deposits as Fluvisols, those with an argillic B and/or an albic E over a slowly permeable horizon as Luvisols, Acrisols or Planosols, and only the remainder as Gleysols.

Soils of this major group are estimated to cover about 40 per cent of England and Wales and approximately 30 per cent of Scotland and Ireland. Their distribution is governed by existing or former soil-water regimes as governed by subsoil permeability, ground–surface configuration and rainfall/evapotranspiration relationships. In the drier lowlands the main areas of occurrence (Figure 2.5, p. 78) consist of level to moderately sloping land on slowly permeable pre-Quaternary rocks, tills and glaciolacustrine deposits, on recent alluvium, and on older pervious materials affected by high ground-water levels. Apart from the coastal marshes which result from recent accretion of intertidal sediments, virtually all the ground was forested during some part of the Holocene period and small patches of semi-natural deciduous woodland remain, chiefly in the southern counties of England. Otherwise the soils are mainly in agricultural use, most of them have been artificially drained, and variants with humose or peaty surface horizons are restricted to localized depressions and flat low lying fenland receiving water from higher ground. In cooler and more humid upland areas, however, gley soils with peaty topsoils developed under moorland vegetation occur in a progressively wider range of situations as altitude and rainfall increase, and extend in places into the oro-arctic zone. Rough grazing and coniferous plantations are the main forms of land use on these soils.

In the British Isles, as in France (Duchaufour, 1982) and Germany (Mückenhausen, 1985) gley soils with or without a peaty surface horizon have been divided into surface-water (Pseudogley, Stagnogley) and ground-water (Gley *sensu stricto*) types (Figure 7.1). Those of the first type have gleyed subsurface (E and/or B) horizons with morphological features that can be attributed to restricted downward movement and consequent stagnation of water within the solum during periods when rainfall exceeds evapotranspiration. The E horizons, where present, are rusty mottled (Eg) or uniformly greyish (Eag); and the less permeable B horizons, which may or may not be significantly enriched in clay by illuviation (Bg or Btg), are blocky or prismatic, with grey or greyish ped faces and ochreous intraped mottles. In BC and C horizons below the main rooting zones, matrix colours inherited from the parent material or produced by oxidative weathering are normally dominant, with 'greying' confined to the faces of fissures or channels which become more widely spaced with increasing depth. Laboratory and field studies reported by Bloomfield (1951), Blüme (1968) and Thomasson and Bullock (1975) support the conclusion that the characteristic morphology of the slowly permeable subsurface horizons results from reduction and solution of iron by microorganisms or products of decomposing organic matter at the surfaces of voids in which water stagnates and movement of the mobilized iron into the interiors of structural aggregates, where it is precipitated as ferric hydroxide.

Typical ground-water gley soils, by contrast, show distinctive features reflecting permanent or seasonal saturation by ground water that is retained by an impervious

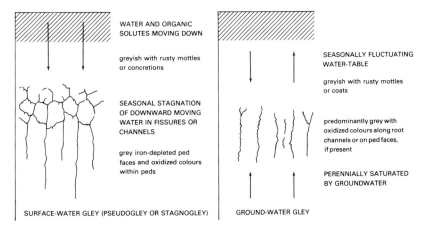

Figure 7.1: Profile patterns characteristic of surface-water and ground-water gley soils.

substratum below the solum and augmented by underground movement from higher to lower positions in the landscape. Where there is slowly moving ground water at moderate depths throughout the year and the subsoil contains appreciable amounts of clay, iron and organic matter, as in most recent alluvial deposits, a massive grey, greenish or bluish CG horizon (Chapter 3, p. 104) is generally encountered below the mean level of the water table. In other situations, particularly in permeable sandy or coarse loamy materials and in artificially drained soils, a 'reduced' CG horizon is absent and lower horizons that are or were saturated by seasonally rising ground water for substantial periods are rust-mottled (Cg) or dominantly ochreous (Cgf), or consist almost entirely of uncoated mineral grains from which any free iron originally present has been removed. In loamy or clayey alluvium accumulated under waterlogged conditions, the whole thickness of the deposit may be strongly gleyed, but an ungleyed C horizon normally appears at some depth in Pleistocene or older deposits lacking inherited organic matter.

Although the surface-water/ground-water distinction is significant ecologically (cp. Duchaufour, 1982) and in relation to land capability and drainage design (Thomasson, 1974, 1975), essentially the same processes of organically promoted reduction and segregation or removal of iron operate in each case and many medium and fine textured soils show features of both types. Given these considerations, no primary separation into surface-water and ground-water soils is made in the classification adopted here, but the distinction is retained at group or subgroup level. As in the major group of brown soils, the gley soils are first of all divided into sandy, alluvial, calcaric, orthic and luvic groups. Those of the sandy and alluvial groups are primarily of ground-water type whereas those of the remaining groups include soils of both types and intergrades between them. In the latter case variants in which gleying is at least partly attributable to the presence of a shallow fluctuating groundwater table are assigned to a *ground-water* subgroup and related soils with surface-water characteristics, which are considerably more extensive, are placed in *stagno-* and *pelo-*(clayey) subgroups. Variants with humose or peaty topsoils are also separated at this level and are identified by the term *humic* as in the England and Wales system. Those with peaty topsoils have been grouped as peaty gleys in Scotland and Ireland and the same name is used here when referring to them collectively.

The five soil groups recognized are defined as follows. Corresponding classes in the England and Wales classification are named in parentheses.

Key to Soil Groups

1. Gley soils in loamy or clayey recent alluvium (p. 113) more than 30 cm thick.
   Alluvial Gley Soils (unripened gley soils; alluvial and humic-alluvial gley soils; gleyic and humic gleyic rendzina-like alluvial soils)
2. Other gley soils with an argillic B horizon (p. 109), an abrupt textural change (p. 111), or both.
   Luvic Gley Soils (typical, paleo-argillic and some pelo-stagnogley soils; argillic, paleo-argillic and sandy stagnohumic gley soils; argillic and argillic humic gley soils)
3. Other gley soils in which at least half the upper 80 cm of mineral soil is sandy or sandy-skeletal (p. 84) and there is no loamy or clayey surface layer more than 30 cm thick.
   Sandy Gley Soils (raw sandy, sandy and humic-sandy gley soils)
4. Other gley soils with a calcareous subsurface horizon starting no more than 50 cm below the mineral soil surface.
   Calcaric Gley Soils (calcaro-cambic and calcareous humic gley soils; gleyic and humic gleyic rendzinas; some calcareous pelosols)
5. Other gley soils.
   Orthic Gley Soils (cambic and some pelo-stagnogley soils; cambic stagnohumic gley soils; typical and pelo-cambic gley soils; typical humic gley soils)

7.2 Sandy Gley Soils

7.2.1 General Characteristics, Classification and Distribution

These are ground-water gley soils in sandy or sandy–gravelly parent materials. The surface horizon may be loamy, either naturally or as a result of marling, and there is normally a finer textured substratum at some depth, but soils with an argillic B or an abrupt textural change within 80 cm are excluded. Variants with humose or peaty topsoils are set apart as humic sandy gley soils in the England and Wales system.

Soils of this group are typically developed in low lying aeolian or water-laid sands of Holocene (Flandrian) or late Pleistocene (Devensian) age. They cover rather more than one per cent of England and Wales (Mackney et al., 1983), chiefly in the Vale of York, the Cheshire Plain and the Lower Trent valley, and on the eastern side of the East Anglian Fens. They also occur sporadically in Scotland and Ireland but their total extent is small in both countries. Most of them have been drained artificially and some in the subhumid lowlands are intensively cropped, locally with vegetables. The water table is controllable in most places by ditching and often remains below 70 cm depth throughout the year (Crompton and Matthews, 1970; Robson and Thomasson, 1977), implying that the hydromorphism relates wholly or partly to a former condition.

The organic matter content and thickness of the topsoil vary in accordance with the age of the parent deposit, vegetational and land use history, and current or former water regime. The Ap horizons in cultivated land are usually dark enough and thick enough to qualify as prominent (mollic, umbric or anthropic epipedon), even when they contain no more than about 3 per cent organic matter. In some places they incorporate remnants of a former peat cover and those in northern England have commonly received additions of organic waste products derived from the industrial towns. Thinner or less prominent (ochric) A horizons, with or without a thin superficial organic layer, occur locally under semi-natural vegetation, particularly in the recently stabilized aeolian sands of dune slacks or links (e.g. Hall and Folland, 1967), and in arable land where the organic matter content has been reduced by continuous cultivation or loss of topsoil by blowing. Variants with a prominent mollic or umbric A horizon are classed as Mollisols or Inceptisols respectively in the US Taxonomy and others as Entisols (Psammaquents).

The gleyed subsurface horizons are either mottled with rusty colours or consist mainly of uncoated grains, depending on the extent to which mobilized iron has been reprecipitated or removed by lateral seepage. As peds are absent or very weakly developed, there is normally no distinct B identifiable by structure and designation of the horizons as E or C is in some cases problematical. Some profiles have a dominantly grey Cg (or CG) horizon at some depth; lower horizons of others appear to retain their original colour, presumably because reduction and mobilization of iron has been confined to the upper part of the rooting zone containing sufficient decomposable organic matter. A markedly ferruginous horizon (Cgf) can occur at variable depths, and variants intergrading to gley-podzols (p. 261) show incipient or discontinuous Bh horizons. The sandy layer generally extends to at least 80 cm depth, but a buried A or organic horizon, or a finer textured alluvial layer, may intervene between 40 and 80 cm where the parent deposit is of recent origin.

Although of low inherent fertility, these soils are well adapted to cultivation when effectively drained and are generally less drought-prone than associated sandy brown soils or podzols in similar materials, especially where ground water remains in or just below the rooting zone during all or most of the growing season. When injudiciously worked, however, those with topsoils low in organic matter are subject to slaking and plough-pan formation, which severely restricts rooting if unrectified by subsoiling. Other potential limitations are liability to wind erosion and minor element deficiencies, particularly of copper, boron and manganese, which may be promoted by overliming.

Key to Subgroups

1. Sandy gley soils with a calcareous subsurface horizon starting at no more than 50 cm below the mineral soil surface
    1.1 with a humose or peaty topsoil (p. 107)
      Humic-calcaric sandy gley soils
    1.2 without a humose or peaty topsoil
      Calcaric sandy gley soils
2. Other sandy gley soils
    2.1 with a humose or peaty topsoil (p. 107)
      Humic sandy gley soils
    2.2 without a humose or peaty topsoil
      Typical sandy gley soils

7.2.2 Typical Sandy Gley Soils

The non-calcareous sandy gley soils comprising the typical subgroup are the most extensive. Those which have not been limed are mostly moderately or strongly acid and some that are now cultivated were formerly gley-podzols (p. 261), the diagnostic Bh horizons of which have been incorporated in the Ap or destroyed as a result of increased biological activity. Data are given below on three representative profiles, the first in a pine plantation and the other two in cultivated land.

Profile 107 is sited on Holocene blown sand (Links association) in Tentsmuir Forest, Fife. The original vegetation was moist heath. Both pH and percentage base saturation are low in the upper horizons but increase markedly below 30 cm, suggesting that the freshly deposited sand was slightly calcareous.

Profile 108 from the Vale of York is in late Devensian or early Flandrian aeolian sand (Matthews, 1970) which overlies a buried A horizon in glaciolacustrine clay. There is a deep drainage ditch nearby and the water table was below 1.5 m depth at the time of sampling (18 Oct 1976).

Profile 109 is located in a small depression on Devensian glaciofluvial outwash derived mainly from reddish Permo-Triassic rocks. Although the sandy parent material is at least 2 m thick at this site, it is locally shallower over impervious till. The air capacity data indicate that the subsurface horizons transmit water readily and, as in sandy soils generally, nearly all the retained water is available to plants. The dominantly reddish brown horizon below 70 cm was saturated with water when sampled (14 Dec 1971). In both this and the preceding soil, colours attributable to gleying are most pronounced in the horizon immediately below the topsoil and there is no CG horizon as in classical ground-water gley profiles (cp. de Bakker and Schelling, 1966; Duchaufour, 1982).

Profile 107 (Laing, 1976)

Location:	Fettersloch, Tentsmuir Forest, Fife (NO 489271)
Climatic regime:	humid temperate
Altitude:	9 m Slope: <1°

Land use and vegetation: pine plantation 30–40 years old; field layer includes cross-leaved heath (*Erica tetralix*), sheeps fescue (*Festuca ovina*), heath rush (*Juncus squarrosus*) and moss (*Polytrichum commune*).

Horizons (cm)

L/F	Dark brown (7.5YR 3/2) undecomposed and partially decomposed pine litter, 5 cm thick.
0–10 Ah1	Dark greyish brown (10YR 4/2) stoneless sand (USDA sand); single grain; loose; many grass roots; clear boundary.
10–20 Ah2	Dark brown (10YR 4/3) stoneless sand (USDA sand) with many uncoated grains and few fine yellowish brown (10YR 5/6) mottles; single grain; loose; frequent grass

Analytical data for Profile 107

Horizon	Ah1	Ah2	Cg1	Cg2	Cg2	Cg3
Depth (cm)	3–10	13–20	23–30	41–53	60–65	76–91
silt[a] 2–50 μm %	9	5	3	2	1	1
clay[a] <2 μm %	3	4	<1	<1	<1	<1
loss on ignition %	7.4	4.0	0.6	0.5	0.2	0.4
organic carbon %	4.0	2.0				
nitrogen %	0.19	0.11				
C/N ratio	31	18				
TEB (me/100 g)	1.7	0.7	0.2	1.3	1.1	1.1
CEC (me/100 g)	12.9	7.3	1.7	1.3	1.2	1.1
% base saturation	13	9	13	95	96	97
pH (H$_2$O)	4.4	4.4	4.8	6.5	6.6	6.6

[a] by hydrometer method

		and few woody tree roots; abrupt smooth boundary.
20–36	Cg1	Brown (10YR 5/3) stoneless sand (USDA sand) with many coarse strong brown (7.5YR 5/8) mottles; single grain; loose; few roots; clear irregular boundary.
36–66	Cg2	Pale brown (10YR 6/3) stoneless sand (USDA sand) with common medium strong brown mottles; single grain; loose; roots rare; gradual boundary.
66–100	Cg3	Pale brown (10YR 6/3) to brown (10YR 5/3) stoneless sand (USDA sand) with few medium to fine yellowish brown (10YR 5/4) mottles; single grain; loose; few woody tree roots.

Analytical data *see* above.

Classification
  Scotland (1984): non-calcareous ground-water gley (Links association).
  England and Wales (1984): typical sandy gley soil; coarse or medium sandy stoneless drift (Formby series).
  USDA (1975): Typic Psammaquent; siliceous or mixed (non-acid), mesic, uncoated.
  FAO (1974): Dystric Gleysol; coarse textured.

Profile 108 (Furness and King, 1978)

Location : Brayton, North Yorkshire (SE 600310)
Climatic regime: subhumid temperate
Altitude : 6 m   Slope: 1°
Land use : arable (cereals)

Horizons (cm)

0–30	Ap	Very dark greyish brown (10YR 3/2) stoneless fine sand (USDA fine sand) containing rare cinder and coal fragments; weak fine granular to single grain; very friable; common fine roots; abrupt smooth boundary.
30–93	Cg	Pale brown (10YR 6/3) stoneless fine sand (USDA fine sand) with common fine and medium strong brown (7.5YR 5/8) mottles; very weak fine blocky; very friable; few fine roots; clear smooth boundary.
93–151	2Cgf	Yellowish brown (10YR 5/6) and light yellowish brown (10YR 6/4) stoneless fine sand (USDA fine sand), coarser than above; single grain; loose; abrupt smooth boundary.
151–169	3b Ahg	Very dark grey (5YR 3/1) stoneless clay with few medium yellowish red (5YR 4/8) decayed woody roots; wet.

Analytical data

Horizon		Ap	Cg	2Cgf
Depth (cm)		0–30	30–93	93–151
sand	600 μm–2 mm %	<1	1	1
	200–600 μm %	11	17	31
	60–200 μm %	79	72	62
silt	2–60 μm %	7	7	2
clay	<2 μm %	3	3	4

organic carbon %	1.6	0.3	0.2
dithionite ext Fe %	0.2	0.1	0.6
pH (H$_2$O)	5.7	5.0	5.8
(0.01M CaCl$_2$)	5.4	4.5	5.7

Classification
England and Wales (1984): typical sandy gley soil; fine sandy stoneless drift (Everingham series).
USDA (1975): (Aeric) Humaquept; sandy, siliceous or mixed (non-acid), mesic, uncoated.
FAO (1974): Humic Gleysol; coarse textured.

Profile 109 (Jones, 1975)

Location   : Seighford, Staffordshire (SJ 875264)
Climatic regime: humid temperate
Altitude   : 88 m   Slope: <1°
Land use   : arable (cereals)
Horizons (cm)

0–35 Ap — Very dark brown (7.5YR 3/2) slightly stony medium sandy loam (USDA fine sandy loam); common small and medium quartzose pebbles and few igneous rock fragments; weak fine and medium granular; very friable; common fine roots, earthworms active; clear irregular boundary.

35–70 Cg — Brown (10YR 5/3) and pale brown (10YR 6/3) slightly stony loamy medium sand (USDA loamy sand to sand) with common very pale brown (10YR 7/3), light grey (10YR 7/2) and coarse light yellowish brown (10YR 6/4) mottles; stones as above but larger; single grain; loose; few fine roots; earthworms present; gradual boundary.

70–110 Cu(g) — Reddish brown (5YR 5–4/4) and brown 7.5YR 5/4) very slightly stony loamy medium sand (USDA loamy sand to sand) with common brown (10YR 5/3) to very pale brown (10YR 7/3) mottles and yellowish brown (10YR 5/4) patches associated with weathering sandstone fragments; wet; stones otherwise as above; single grain; no roots; earthworms present (similar material by auger to 2 m).

Analytical data *see* below.

Classification
England and Wales (1984): typical sandy gley soil; sandy drift with siliceous stones (Blackwood series).
USDA (1975): Aquic Hapludoll; sandy, siliceous, mesic, coated.
FAO (1974): Gleyic Phaeozem; coarse textured.

Analytical data for Profile 109

Horizon	Ap	Cg	Cu(g)
Depth (cm)	0–35	35–70	70–110
sand 500 µm–2 mm %	10	12	7
200–500 µm %	27	45	37
100–200 µm %	16	21	29
50–100 µm %	7	8	13
silt 2–50 µm %	33	13	11
clay <2 µm %	7	1	3
organic carbon %	2.2	0.2	0.2
dithionite ext Fe %	0.4	0.1	0.1
pH (H$_2$O)	6.5	6.7	6.5
(0.01M CaCl$_2$)	6.3	6.2	6.2
bulk density (g cm^{-3})	1.31	1.68	1.60[a]
packing density (g cm^{-3})	1.37	1.69	1.62[a]
retained water capacity (% vol.)	30	20	13[a]
available water capacity (% vol.)	19	16	11[a]
air capacity (% vol.)	17	15	25[a]

[a] samples at 88–93 cm

7.2.3 Humic Sandy Gley Soils

These are similar soils with humose or peaty topsoils occurring in level or depressional sites, often marginal to peat bogs, where the water table is or was close to the surface. Where the land has been drained and cultivated, as in parts of East Anglia and the English Midlands (Hodge *et al.*, 1984; Ragg *et al.*, 1984), the topsoil usually incorporates the remains of an originally thicker and more extensive peat cover. The arable soils so formed retain more available water than those of the typical subgroup and are less liable to slaking and over-compaction, but are subject to wind erosion when the ground is bare. As the organic matter content of the Ap horizons is progressively reduced under continuous cultivation, however, they eventually fail to qualify as humose.

Immediately beneath the dark coloured topsoil there is often a loose greyish sandy horizon (Eag or Cg) with few or no ferruginous mottles. This can extend to 80 cm or more, as in Profile 110, but particularly in drained land it commonly passes at a shallower depth into a Cgf horizon in which yellowish or ochreous colours become dominant.

Variants intergrading to typical gley-podzols (p. 262) have a grey E horizon depleted of iron and an underlying brownish (e.g. 7.5YR 4/2–3) horizon which shows thin organic coats (organans) on the sand grains but is too weakly expressed to qualify as Bh. These are distinguished as *Goor earth soils* in the Netherlands (de Bakker and Schelling, 1966).

The first representative profile recorded below was described and sampled under rushy pasture near Ballylanders, County Limerick. It is formed in a coarse textured lake sediment and has a peaty surface overlying a uniformly grey subsurface horizon of medium to high base status. The water table stood at 56 cm depth when the soil was sampled.

Profile 111 is located in pump-drained 'carrland' adjoining the river Idle in north Nottinghamshire. Developed in stoneless fluvial or aeolian sand over glacio-lacustrine clay (cp. Profile 108 above), it has a humose Ap horizon with a relatively high clay content which possibly results from marling. The rust mottled Cg horizons are extremely porous (air capacity more than 20 per cent).

Profile 110 (Finch and Ryan, 1966)

Location	:	Ballylanders, County Limerick
Climatic regime:		humid temperate
Altitude	:	140 m Slope: <1°
Land use	:	long-term grass

Analytical data for Profile 110				
Horizon	Oh	Ah	Cg1	Cg2
Depth (cm)	0–21	21–38	38–56	56–81
sand 200 µm–2 mm %		30	50	57
50–200 µm %		62	43	33
silt 2–50 µm %		6	5	7
clay <2 µm %		2	2	3
organic carbon %	24.3	1.4	0.4	0.3
nitrogen %	1.86	0.14		
C/N ratio	13	10		
dithionite extractable Fe %	1.4	0.2	0.1	0.3
TEB (me/100 g)	57.0	5.4	3.2	3.5
CEC (me/100 g)	85.0	7.8	4.2	4.2
% base saturation	67	69	76	83
pH (H$_2$O)	6.1	6.0	6.1	6.6

Horizons (cm)

0–21 Very dark greyish brown (10YR 3/2)
Oh   sandy peat; moderate medium granular; friable; plentiful roots; gradual smooth boundary.

21–38 Very dark greyish brown (10YR 3/2)
Ah   fine sand (USDA fine sand); single grain; loose; few roots; gradual smooth boundary.

38–56 Grey (10YR 5/1) medium sand (USDA
Cg1 (or sand); wet; single grain; loose; no
Eag)   roots; gradual smooth boundary.

56–81 Grey to light grey (10YR 6/1) medium
Cg2   sand (USDA sand) wet (water table at 56 cm); single grain; loose; no roots.

Analytical data *see* p.297.

Classification
  Ireland (1980): peaty ground-water gley (Griston series).
  England and Wales (1984): typical humic sandy gley soil; coarse or medium sandy stoneless drift (unnamed series).
  USDA (1975): Histic Humaquept; sandy, siliceous or mixed (non-acid), mesic.
  FAO (1974): Mollic Gleysol; coarse textured.

Profile 111 (Reeve and Thomasson, 1981)

Location   : Gringley Carr, Nottinghamshire (SK 719948)

Climatic regime: subhumid temperate
Altitude   : 2 m   Slope: <1°
Land use   : ley grass

Horizons (cm)

0–35   Black (5YR 2/1, broken and rubbed;
Ap   10YR 3/2 dry) stoneless humose loamy medium sand (USDA loamy sand); moderate medium subangular blocky; friable; many fine roots; sharp wavy boundary.

35–70 Pale brown (10YR 6/3, 6/4 rubbed, 7/2
Cg1   dry) medium sand (USDA sand) with common fine strong brown (7.5YR 5/6) mottles and dark reddish brown (5YR 3/3) vertical ferruginous tubules associated with old root channels; single grain; loose; few roots; diffuse boundary.

70–145 Light yellowish brown (10YR 6/4, 6/5
Cg2   rubbed, 6/3 dry) stoneless medium sand (US sand) with many medium strong brown (7.5YR 5/6) mottles; wet; single grain; no roots.

145–160 (By auger) Brown to dark brown
2Cg   (7.5YR 4/2) clay with common ochreous mottles; wet.

Classification
  England and Wales (1984): typical humic sandy gley soil; coarse or medium sandy stoneless drift (unnamed series).

Analytical data for Profile 111

Horizon		Ap	Cg1	Cg2
Depth (cm)		0–30	40–55	100–110
sand	600 μm–2 mm %	1	1	1
	200–600 μm %	51	58	71
	60–200 μm %	28	30	24
silt	2–60 μm %	12	9	3
clay	<2 μm %	8	2	1
organic carbon %		8.5	0.4	0.1
dithionite ext. Fe %		1.4	0.4	0.3
pH (H_2O)		6.2	4.8	5.3
(0.01M $CaCl_2$)		5.8	4.3	5.0
bulk density (g cm^{-3})		0.98	1.59	1.53
packing density (g cm^{-3})		1.05	1.60	1.53
retained water capacity (% vol.)		39	19	10
available water capacity (% vol.)		23	16	8
air capacity (% vol.)		21	21	33

USDA (1975): (Aeric) Humaquept; sandy, siliceous (acid), mesic, coated.
FAO (1974): Humic (or Mollic) Gleysol; coarse textured.

7.2.4 Calcaric Sandy Gley Soils

These are sandy gley soils with non-humose topsoils and calcareous subsurface horizons. They are far less extensive than those of the typical (non-calcareous) subgroup and are mainly restricted to dune slacks and poorly drained links or machairs on wind-blown coastal sands rich in shell fragments. The largest areas recorded in the British Isles are in the Orkney Islands (Dry and Robertson, 1982) and the Outer Hebrides (Hudson et al., 1982), where they carry various semi-natural swamp and sedge communities, alternating in places with improved pasture or arable land. As on the associated well drained soils (calcareous sandy regosols and sandy brown soils), the high pH values contribute to the occurrence of manganese deficiency in crops and copper and cobalt deficiencies affecting stock (Dunn, 1980).

The following representative profile under *Carex* pasture is located east of Thurso, Caithness, in a large flat area of shelly sand which overlies till or peat at depths of about one metre.

Profile 112 (Futty and Dry, 1977)

Location: Dunnet, Caithness (ND 217688)
Climatic regime: humid boreal
Altitude: 5 m   Slope: <1°
Vegetation: species-rich pasture with rush (*Juncus balticus*), red fescue (*Festuca rubra*), carnation grass (*Carex panicea*), sedges (*Carex pulicaris* and *Carex capillaris*) and Scottish primrose (*Primula scotica*).

Horizons (cm)

0–15 Ah — Dark brown (7.5YR 3/2) stoneless calcareous sand (USDA sand) moderate granular; very friable; abundant roots; abrupt boundary.

15–36 Cg1 — Very pale brown (10YR 7/4) stoneless calcareous sand (USDA sand) containing abundant white shell fragments; few coarse strong brown (7.5YR 5/8) to yellowish red (5YR 4/6) mottles; single grain; very friable; occasional cord roots; clear boundary.

36–53 Cg2 — Light brownish grey (2.5YR 6/2) stoneless calcareous coarse sand (USDA sand or coarse sand) with common coarse yellowish brown (10YR 5/8) and yellowish red (5YR 5/8) mottles; shell fragments; structure, consistence and roots as above; clear boundary.

53–71 Cg3 — Grey (5YR 5/1) stoneless calcareous coarse sand (USDA sand or coarse sand); wet (water table at 53 cm);

Analytical data for Profile 112

Horizon	Ah	Cg1	Cg2	Cg3	2bAhg
Depth (cm)	3–13	18–28	38–48	56–66	74–84
silt[a] 2–50 µm %	4	1	1	3	13
clay[a] <2 µm %	2	1	1	3	7
organic carbon %	2.8	1.45			
nitrogen %	0.28	0.06			
C/N ratio	10	23			
pH (H$_2$O)	7.6	8.3	8.5	8.0	7.8
(0.01M CaCl$_2$)	6.8	6.9	6.9	6.8	6.7

[a] by hydrometer method

71–142 2BAhg	abundant shell fragments; occasional dead roots; abrupt boundary. Dark greyish brown (10YR 4/2 on exposure to air) stoneless calcareous loamy sand (USDA loamy sand); abundant shell fragments; wet; massive; friable to firm; common frequent dead roots and patches of organic material.

Analytical data *see* p.299.

Classification
    Scotland (1982): calcareous ground-water gley (Whitelinks series).
    England and Wales (1984): gleyic sand-pararendzina; carbonatic-sandy stoneless drift (Loggans series).
    USDA (1975): Mollic Psammaquent; carbonatic, mesic, uncoated.
    FAO (1974): Calcaric Gleysol; coarse textured.

7.2.5 Humic-calcaric Sandy Gley Soils

These are similar but less common with a humose or peaty topsoil. Examples have been noted in swampy dune slacks, in other very poorly drained sites on shelly coastal sands, and in drained peatland on coarse textured river deposits containing fragmentary limestone. The following profile, for which no analytical data are available, is of the latter type. It was recorded in the valley of the river Slea near Sleaford, Lincolnshire.

Profile 113 (George and Robson, 1978)

Location:	Quarrington, Lincolnshire (TF 051453)
Climatic regime:	subhumid temperate
Altitude:	13 m Slope: <1°
Land use:	long-term grass

Horizons (cm)
0–28 Ahg	Very dark brown (10YR 2/2) calcareous humose sandy loam with common fine dark reddish brown (5YR 3/3) mottles along root channels and abundant uncoated sand grains; strong medium granular; very friable; abundant fine roots (2 cm root mat at surface).
28–50 Cg	Pale brown (10YR 6/3), very slightly stony, calcareous medium sand with many fine and medium brownish yellow (10YR 6/6) mottles and dark greyish brown (10YR 4/2) organic staining; few small quartzite pebbles and flint fragments; abundant uncoated sand grains; very weak angular blocky; many fine roots; wet (water table at 40 cm on 7 May 1975 after wet spring); clear smooth boundary.
50–100 2Cgf	Brownish yellow (10YR 6/6) and pale brown (10YR 6/3), very stony, very calcareous medium sand; abundant very small and small subrounded limestone fragments and common small flints and quartzite pebbles; wet; single grain; common fine roots, many dead.

Classification
    England and Wales (1984): calcareous humic sandy gley soil; sandy material over calcareous gravel (Greylees series).
    USDA (1975): Aquic Hapludoll; sandy-skeletal, mixed (calcareous), mesic.
    FAO (1974): Gleyic Phaeozem; medium to coarse textured.

7.3 Alluvial Gley Soils

7.3.1 General Characteristics, Classification and Distribution

These are loamy and clayey gley soils in Holocene (Flandrian) alluvial deposits of coastal and estuarine marshes, higher lying carselands, river floodplains and former lake beds. The alluvium can range in thickness from 30 cm to several metres and generally has more inherited organic matter and a smaller bulk density than older deposits of similar texture. Many of the soils show sedimentary stratification at moderate depths and some include buried A horizons or organic layers marking pauses in mineral sedimentation. Other characteristics vary according to the composition, age and mode of origin of the alluvium and the water regime as conditioned by drainage and flood prevention measures. Soils of this group, including those separated as unripened and humic-alluvial gley soils and rendzina-like alluvial soils by Avery (1973, 1980), occupy approximately 5 per cent of England and Wales but only about 1 per cent of Scotland and Ireland. Those in Scotland have generally been grouped as alluvial soils (undifferentiated) for survey purposes.

The alluvial gley soils in coastal situations originated by progressive accumulation of muddy sediments on sheltered intertidal flats to form salt marshes, large areas of which have been embanked and brought into agricultural use from Roman times onwards, notably around the Wash, Thames, Severn and Shannon estuaries, and in Romney Marsh beside the English Channel. In early stages of salt-marsh development, when the surface is inundated daily and is only partially colonized by silt-trapping pioneer plants such as glasswort (*Salicornia* spp.) and cord-grass (*Spartina* spp.), the mud remains fluid (unripened) and more or less completely reduced (CG horizon). In later stages, and following embankment and drainage, the consistence becomes irreversibly firmer from the surface downwards (physical ripening), a distinct A horizon develops, increasing aeration causes progressive replacement of the dark grey or bluish CG by rust-mottled Cg or Cgf horizons of higher chroma, and a structural Bg horizon eventually forms as a result of seasonal drying, growth and decay of roots, and associated animal activity. As desalinization proceeds rapidly once the soil is no longer flooded by seawater, and the marine sediments are usually calcareous, exchangeable sodium and magnesium are quickly replaced by calcium and acidity produced by oxidation of sulphides is immediately neutralized to form calcium sulphate (gypsum). Where the deposits contain little or no calcium carbonate, however, as in certain estuarine environments, replacement of sodium and magnesium can be slow, and oxidation of sulphidic muds gives rise to severe acidity on drainage.

Recently accumulated marine sediments that are fully ripened no deeper than 20 cm have been set apart in England and Wales (Avery, 1980) as unripened gley soils, corresponding to *Initial Vague Soils* in the Netherlands classification (de Bakker and Schelling, 1966) and Hydraquents and Sulfaquents in the US Taxonomy. In the classification adopted here, all the soils of unembanked salt marshes are grouped as saline alluvial gley soils (Plate III.5).

Similar developmental (ripening) sequences can be observed locally in lacustrine deposits and in backswamp areas of river floodplains, but most of the riverine alluvial gley soils are fully ripened to considerable depths, either naturally or as a result of artificial drainage. Because flooding by streams is normally seasonal, the deposited material dries and cracks to some extent before the next flood and so usually has a firmer consistency and a higher bulk density than recently deposited intertidal sediment of similar composition.

The ripened alluvial gley soils that still receive fresh sediment, or did so immediately prior to embankment and drainage, mostly qualify as Fluvaquents in the US system and Fluvisols according to the FAO (1974) scheme on the basis that the epipedon (A horizon) is ochric and organic carbon contents decrease irregularly with depth or remain greater than 0.2 per cent to at least 1.25 m depth. As noted by Pons and van der Molen (1973), however, most of them diverge from the central concept of Fluvaquents in having a blocky or prismatic subsurface horizon without fine stratification, whilst variants in which a sandy or gravelly deposit, or a pre-Flandrian substratum of finer texture, appears within the requisite depth usually fail to meet the organic matter requirement and must therefore be classed as Aquepts (Haplaqepts).

Other soils included in the group have a prominent humose A horizon or a peaty surface layer. These have formed where superficial accumulation of organic matter has been promoted by the occurrence of a perennially high water table in alluvial sites receiving little or no fresh sediment. Though seldom extensive in any one locality, they are widely distributed in riverine backswamps, abandoned channels and former lake beds, and are also represented in drained land on older coastal or estuarine deposits such as the Fen Clay of East Anglia (Godwin, 1978) and the Downholland Silt of the Lancashire coastal plain (Gresswell, 1958). Cessation of sedimentation was followed in these areas by widespread accumulation of peat, most of which has now been reduced by wastage to a thin surface remnant.

The undrained humic soils in depressional sites are often wet within 40 cm depth for most of the year, typical profiles having strongly gleyed Cg or CG horizons which start more or less directly beneath the topsoil and are often incompletely ripened. The more extensive drained phases commonly have a blocky or prismatic Bg horizon, especially where the parent material is clayey. Some have developed acid-sulphate characteristics (Chapter 3, p. 111) and are mostly Sulfaquepts according to the US Taxonomy. Others, both drained and undrained, qualify as Mollisols (mostly Haplaquolls), corresponding to Mollic Gleysols in the FAO (1974) scheme, and the remainder as Humaquepts (FAO Humic Gleysols).

Though generally fertile and non-droughty, the alluvial gley soils vary greatly in land-use capability depending primarily on the current water regime and the incidence of floods (Figure 7.2). Those that are regularly flooded, or in which the water table remains less than about 60 cm below the surface throughout the year, are unsuitable for arable cropping and are mostly used for summer pasture or for hay, interspersed over small areas in some localities with plantations of moisture-tolerant trees such as poplar or willow. By contrast, those of loamy texture in subhumid areas such as the 'siltlands' around The Wash (Figure 7.3), where flooding is prevented and ground-water levels controlled, present few limitations for agricultural use and are intensively cropped. Clayey variants with slowly permeable subsurface horizons are more difficult to drain satisfactorily and can present serious cultivation and harvesting problems in wet seasons. They are therefore more often left in grass, especially in the more humid areas of occurrence.

Figure 7.2: Flooding on alluvial gley soils in the Soar valley near Sutton Bonington, Leicestershire (Copyright Aerofilms Ltd.).

Figure 7.3: Intensively cultivated 'siltland' near Terrington St Clement, Norfolk (Cambridge University Collection: copyright reserved).

Key to Subgroups

1. Alluvial gley soils with a saline (p. 96) layer at least 30 cm thick, starting within 60 cm depth
   Saline alluvial gley soils
2. Other alluvial gley soils with acid-sulphate characteristics (p. 111)
   Acid-sulphate alluvial gley soils
3. Other alluvial gley soils with a humose or peaty topsoil (p. 107)
   3.1 With a calcareous subsurface horizon starting at no more than 50 cm below the mineral soil surface
   Humic-calcaric alluvial gley soils
   3.2 With non-calcareous subsurface horizon(s) extending to more than 50 cm below the mineral soil surface
   Humic alluvial gley soils
4. Other alluvial gley soils with a calcareous subsurface horizon starting at no more than 50 cm depth
   4.1 With pelo-characteristics (p. 112)
   Pelocalcaric alluvial gley soils
   4.2 Without pelo-characteristics
   Calcaric alluvial gley soils
5. Other alluvial gley soils
   5.1 With pelo-characteristics (p. 112)
   Pelo-alluvial gley soils
   5.2 Without pelo-characteristics
   Typical (loamy) alluvial gley soils

7.3.2 Typical (Loamy) Alluvial Gley Soils

This subgroup comprises non-humic, non-calcareous, non-saline alluvial gley soils that lack acid-sulphate characteristics and are predominantly loamy in texture or have a loamy upper layer over a sandy or gravelly substratum, which can be of Pleistocene age. They are mainly in river floodplains with catchments in which most of the soils are medium or coarse textured. Variants in marine or estuarine alluvium which was originally non-calcareous or has been deeply decalcified also occur but are much less common.

Most of these soils have a grey and brown mottled, blocky or prismatic Bg horizon which is at least moderately permeable and passes downwards into a strongly gleyed C horizon (Cg and/or CG). The Bg may be weakly expressed or absent in coarse loamy or silty alluvium with little clay or where the water table is permanently within about 60 cm depth. A distinct ferruginous horizon(Cgf) appears in some profiles at or about the upper limit of the permanently saturated zone. Data on two representative profiles are given below.

Profile 114 was described and sampled under rushy pasture in the Tywi Valley, Dyfed. It is developed in fine silty river alluvium derived mainly from fine grained Lower Palaeozoic rocks and the site is regularly flooded, particularly during the winter months, but the water table falls to more than a metre below the surface in dry summers. The prismatic Bg overlies a succession of massive, grey, stony and non-stony layers with ferruginous segregations concentrated in horizontal bands (Cgf) and bordering old root channels. Packing density and air capacity values show that the soil is moderately porous in the Bg and somewhat less porous below.

Profile 115 is in Older Marine Alluvium (Mitchell *et al.*, 1973) which borders parts of the Cumbrian coast and includes raised-beach gravels and terraced 'warp deposits' lying up to 5 m above the present high-tide level as a result of recent isostatic uplift. The soil has consequently received little fresh sediment during the last few hundred years and shows more pronounced horizon development than those in younger marsh deposits of similar composition bordering the Solway Firth, which remain finely stratified to within 30 cm

of the surface (Kilgour, 1979). The physical data show that, as in the preceding profile, the gleyed C horizon is only slightly porous. Although such soils are inherently very fertile, their large silt content makes them prone to surface capping and structural damage when cultivated, and wind erosion can occur when the surface is bare and dry.

Profile 114 (Wright, 1980)

Location	: Llansadwrn, Dyfed (SN 696293)
Climatic regime:	humid temperate
Altitude	: 45 m Slope: <1°
Land use and vegetation	: long-term grass with frequent rushes (*Juncus* spp.) and occasional buttercups (*Ranunculus* spp.) and white clover (*Trifolium repens*); grasses include Yorkshire fog (*Holcus lanatus*), crested dogstail (*Cynosurus cristatus*), meadow grass (*Poa* spp.), bent (*Agrostis* spp.) and flote grass (*Glyceria plicata*)

Horizons (cm)

0–3 Ahg1	Dark greyish brown (10YR 4/2) stoneless silty clay loam (USDA silt loam) with many fine yellowish brown (10YR 5/6) mottles along root channels; moderate granular; very friable; abundant fine roots; abrupt smooth boundary.
3–17 Ahg2	Greyish brown (5YR 5/2) stoneless silty clay loam (USDA silt loam) with many fine yellowish brown (10YR 5/6) mottles along root channels; moderate fine subangular blocky; friable; many fine roots; clear smooth boundary.
17–52 Bg1	Light grey to grey (5YR 6/1); 2.5Y 6/4 rubbed) stoneless silty clay loam (USDA silty clay loam) with many very fine to coarse yellowish red and strong brown (5YR–7.5YR 5/6) mottles, mainly along root channels and on ped faces; dark brown (10YR 4/3) organic coats on some ped faces and in channels; moderate coarse to medium prismatic with fissures up to 3 mm wide; friable; many fine

52–84 Bg2	roots to 35 cm, common below; clear smooth boundary. Light grey to grey (as above) stoneless silty clay loam (USDA silt loam) with very many fine and medium light yellowish brown (10YR 6/4) and common strong brown and yellowish red mottles; moderate medium and coarse prismatic; friable; few fine rounded ferri-manganiferous nodules; few fine roots; clear smooth boundary.
84–105 BCg	Light grey to grey (as above) very slightly stony silty clay loam (USDA silt loam) with common strong brown (7.5YR 5/6) and light yellowish brown (10YR 6/4) mottles; few, small to medium, subrounded and rounded tabular siltstone and fine sandstone fragments; weak coarse prismatic; friable; common soft black and dark brown ferri-manganiferous concentrations 1–10 mm; few roots; clear smooth boundary.
105–118 2Cg	Light grey to grey (as above) moderately stony clay loam with a 3 cm zone of coarse strong brown mottles at base; stones as above, with few of red sandstone; wet; massive; slightly fluid; few ferri-manganiferous concentrations; few roots; clear smooth boundary.
118–150 3Cg	Grey (N 6/0) stoneless clay loam with many medium light yellowish brown (2.5Y 6/4) and strong brown mottles, the latter frequently bordering old root channels; massive; wet; slightly fluid.

Analytical data *see* below.

Classification
England and Wales (1984): typical alluvial gley soil; fine silty river alluvium (Conway series).
USDA (1975): Typic Fluvaquent; fine-silty, mixed or micaceous (acid), mesic.
FAO (1974): Dystric Fluvisol; medium textured.

Profile 115 (Kilgour, 1979)

Location   : Rockliffe, Cumbria (NY 340619)
Climatic regime: humid temperate
Altitude   : 7.5 m   Slope: <1°
Land use   : long-term grass
Horizons (cm)

0–22 Ap(g)	Dark brown (10YR 3/3 broken and rubbed; 10YR 5/2 dry) stoneless silt loam (USDA silt loam) with common very fine dark reddish brown (2.5YR 3/4) mottles; weak medium and coarse angular blocky; friable; abundant fine roots in upper 5 cm, many below; abrupt wavy boundary.
22–32 ABg	Dark greyish brown (10YR 4/2; 5/3 rubbed) stoneless silt loam (USDA silt loam) with many fine dark reddish

Analytical data for Profile 114

Horizon	Ahg2	Bg1	Bg2	BCg
Depth (cm)	3–17	17–52	52–84	84–105
sand 200 μm–2 mm %	1	3	5	7
60–200 μm %	2	1	2	3
silt 2–60 μm %	71	68	68	65
clay <2 μm %	26	28	25	25
organic carbon %	3.3	1.8	1.2	0.9
dithionite ext. Fe %	1.2	1.3	1.1	1.1
pH (H$_2$O)	5.1	5.0	4.8	5.1
(0.01M CaCl$_2$)	4.2	4.2	5.1	4.3
bulk density (g cm^{-3})	1.04	1.04	1.10	1.19
packing density (g cm^{-3})	1.27	1.29	1.32	1.41
retained water capacity (% vol.)	46	47	47	46
available water capacity (% vol.)	19	22	23	22
air capacity (% vol.)	14	14	12	9

	brown mottles; weak medium and coarse subangular blocky with greyish brown (10YR 5/2) faces; friable; earthworm channels present; many fine roots; sharp wavy boundary.
32–46 Bg	Brown to dark brown (10YR 4/3), dark grey (10YR 4/1) and greyish brown (10YR 5/2) stoneless silt loam (USDA silt loam) with many fine yellowish brown (10YR 5/6) mottles; weak coarse prismatic with greyish brown (10YR 5/2) faces; friable; 2 per cent coarse macropores (earthworm channels); common fine roots, mainly in channels; abrupt wavy boundary.
46–57 Cg1	Greyish brown (2.5Y 5/2) stoneless silt loam (USDA silt loam) with many yellowish brown mottles; massive; friable; macropores as above; few fine roots; clear wavy boundary.
57–125 Cg2	Greyish brown (2.5Y 5/2), yellowish brown (10YR 5/6) and brown to dark brown (7.5YR 4/4), prominently mottled, finely stratified stoneless silt loam (USDA loam) with thin (c. 1 mm) pale brown (10YR 5/3) fine sandy laminae 0.5–1 cm apart; coarse channels becoming less frequent with depth; firm; few roots.

Classification
England and Wales (1984): typical alluvial gley soil; coarse silty marine alluvium (Rockliffe series).
USDA (1975): Mollic Fluvaquent; coarse-silty, mixed (non-acid), mesic.
FAO (1974): Mollic Gleysol; medium textured.

7.3.3 Pelo-alluvial Gley Soils

These are non-humic gley soils in clayey alluvium that are non-calcareous throughout or to at least 50 cm depth but lack acid-sulphate characteristics. They are widely distributed in the floodplains of slow-moving lowland rivers and also occur in fine textured estuarine alluvium, including the early Holocene 'carse clays' mapped as the Stirling association in Scotland (Laing, 1976; Soil Survey of Scotland, 1984). Some are clayey to considerable depths; others have a coarser mineral substratum or a peat layer within 80 cm, those of the latter kind being particularly extensive in the Somerset Levels (Midelney association in Findlay et al., 1984).

Analytical data for Profile 115

Horizon	Ap(g)	ABg	Bg	Cg1	Cg2
Depth (cm)	0–22	22–32	32–46	46–57	57–125
sand 200 μm–2 mm %	4	4	<1	<1	<1
60–200 μm %	10	12	5	4	9
silt 2–60 μm %	76	74	82	82	79
clay <2 μm %	10	10	13	14	12
organic carbon %	1.9	1.1	0.4	0.4	0.3
TEB (me/100 g)	8.2	5.4	4.8	5.1	5.6
CEC (me/100 g)	14.3	9.4	6.9	7.2	7.4
% base saturation	57	57	70	71	76
pH (H$_2$O)	5.4	5.9	6.2	6.4	6.7
(0.01M CaCl$_2$)	5.0	5.1	5.4	5.4	5.7
bulk density (g cm^{-3})	1.20		1.45		1.55[a]
packing density (g cm^{-3})	1.29		1.56		1.65[a]
retained water capacity (% vol.)	42		35		35[a]
available water capacity (% vol.)	29		25		18[a]
air capacity (% vol.)	13		10		7[a]

[a] samples at 80–85 cm

Typical profiles in thick clayey alluvium have a mottled Bg horizon with angular blocky or prismatic structure and a neutral grey, bluish or greenish CG horizon is commonly encountered below the summer water-table level. In some places, especially in originally reddish material, the Bg is separated from the underlying permanently waterlogged zone by a less mottled horizon which appears to retain more of its inherited colour, suggesting that gleying in the upper horizons results mainly from impeded percolation of rainwater rather than saturation by ground water rising from below. Green and Askew (1965) showed that certain clayey alluvial soils under old pasture in Romney Marsh contained numerous interconnecting animal channels permitting rapid water movement. Elsewhere, however, as in the first representative profile recorded, air capacity data indicate that the gleyed subsurface horizons are slowly permeable in the wet condition. Seasonal shrinkage and consequent cracking, which is often pronounced in subhumid areas, varies with the amount and type of clay and the depth to which drying extends as governed by weather conditions, land use and prevailing water-table levels (Reeve et al., 1980). Variants rich in expansible clay materials display inclined pressure faces (slickensides) in the subsurface horizons and intergrade in this respect to Vertisols as defined in the US Taxonomy. Data on two representative profiles are reproduced below.

Profile 116 is located in the floodplain of the river Leadon in west Gloucestershire. The extremely clayey alluvium at this site overlies slightly weathered Triassic mudstone (Keuper Marl) at 140 cm depth and is mainly derived from similar rocks. These contain gypsum, which has apparently been mobilized and redeposited from the ground water. When examined in March 1976, following an exceptionally dry summer and winter, water rose in the profile pit to stand at 80 cm below the surface. X-ray diffraction data on clay separates from the subsurface horizons showed that they consist mainly of mica and interstratified mica-smectite, with subsidiary chlorite. Reeve et al. (1980) give coefficient of linear extensibility (COLE) values ranging from 0.20 to 0.17 for these horizons and reported the occurrence of surface cracks 5 cm wide in a soil of the same series during the dry summer of 1975. On this basis the profile is tentatively classed as a Vertic Fluvaquent in the US system.

Profile 117 was recorded under old pasture in embanked marshland with relic creeks flanking the Medway estuary in north Kent. It has a thin organic horizon at the surface but fails to qualify as humic because the succeeding dark grey horizon contains only about 3 per cent organic carbon. Probably in part as a result of salt-water flooding in 1953, sodium accounts for more than 15 per cent of the total exchangeable cations in the BCg and CG horizons and the latter is distinctly saline (Chapter 3, p. 96). Fine textured non-calcareous marsh soils of this type are called *knip* or *knick* in the Netherlands and North Germany (Pons and van der Molen, 1973; Mückenhausen, 1985). In conformity with the relatively high exchangeable sodium percentage, they show a strong tendency to structural instability marked by deflocculation and mobilization of clay, which clogs pores in the subsurface horizons. This causes blockage of infield drains, particularly if arable cropping is introduced, as has happened recently in the north Kent marshes (Hazleden et al., 1986; Loveland et al., 1987).

Profile 116 (Cope, 1986; Reeve et al., 1980)

Location : Tibberton, Gloucestershire (SO 759225)
Climatic regime: subhumid temperate
Altitude : 12 m   Slope: <1°
Land use : long-term grass
Horizons (cm)
 0–12   Very dark greyish brown (10YR 3/2)
 Ah   stoneless humose clay (USDA clay) with many very fine yellowish red (5YR 4/6) linear mottles along root

	channels; strong medium and fine subangular blocky; firm; abundant (0–2 cm) to many fine and common medium fleshy roots; abrupt smooth boundary.
12–30 Bg1	Greyish brown (10YR 5/2) stoneless clay (USDA clay) with many fine dark brown mottles; strong coarse angular blocky, breaking to medium and fine, with shiny (stress-oriented) faces; 2 per cent coarse tubular earthworm channels; firm; common fine roots; few fine rounded ferri-manganiferous nodules; abrupt smooth boundary.
30–62 Bg2	Grey (5Y 5/1; 2.5Y 5/2 rubbed) stoneless clay (USDA clay) with many fine olive-grey (5Y 5/2) and few fine brown to dark brown (7.5YR 4/4) mottles; strong coarse prismatic breaking in places to medium and fine angular blocky with many grey shiny faces and some coated with soil similar to above; earthworm channels as above; firm; few fine roots; few fine rounded ferri-manganiferous nodules and irregular soft concentrations of gypsum; abrupt wavy boundary.
62–140	Greenish grey (5GY 6/1) and olive-
Cgy/CG	grey (5Y 5/2) stoneless silty clay (USDA silty clay) with many, fine and medium, dark brown and reddish brown (7.5–5YR 4/4) mottles; wet; common fine strong brown (7.5YR 5/6) mottles associated with two lenses containing charcoal fragments at 70–80 cm; very weak (adherent) angular blocky with common shiny faces in the upper part; 1 per cent coarse macropores containing upwelling water; becoming slightly fluid (screw auger can be pushed in) below 110 cm; few roots; common irregular soft concentrations of gypsum; abrupt wavy boundary (water table at 80 cm on 10.12.1976).
140–160 2Cu(g)	(By auger) Reddish brown (5YR 4/4) and greenish grey (5G 6/1) clay containing many fine soft angular mudstone fragments; no gypsum or ferri-manganiferous concentrations.

Analytical data *see* below.

Classification

England and Wales (1984): pelo-alluvial gley soil; clayey river alluvium (Fladbury series).

USDA (1975): Vertic Fluvaquent; very-fine, mixed or montmorillonitic (non-acid), mesic.

FAO (1974): Eutric Fluvisol; fine textured.

Analytical data for Profile 116

Horizon		Ah	Bg1	Bg2	Cgy
Depth (cm)		0–12	12–30	30–65	65–100
sand 200 µm–2 mm %		1	1	<1	1
60–200 µm %		2	<1	<1	2
silt 2–60 µm %		23	15	30	46
clay <2 µm %		74	84	70	51
organic carbon %		11	2.1	1.2	0.5
dithionite ext. Fe %		1.0	1.0	0.6	0.5
pH (H_2O)		6.4	7.0	7.2	7.4
(0.01M $CaCl_2$)		6.0	6.9	7.1	7.3
clay <2 µm					
CEC (me/100 g)			49	61	58
K_2O %			3.8	3.9	2.7
bulk density (g cm^{-3})		0.64	0.88	1.04	1.03
packing density (g cm^{-3})		1.30	1.63	1.67	1.48
retained water capacity (% vol.)		62	56	60	58
available water capacity (% vol.)		24	18	19	23
air capacity (% vol.)		11	11	1	3

Profile 117 (Fordham and Green, 1976)

Location : Chetney Marsh, Iwade, Kent (TQ 895694)
Climatic regime: subhumid temperate
Altitude : 1 m   Slope: <1°
Land use : long-term grass

Horizons (cm)
0–5 H/Oh Black (10YR 2/1 rubbed) stoneless clayey organic material; strong very fine subangular blocky becoming predominantly granular near the surface; very friable; abundant fine roots; ants active; abrupt smooth boundary.

5–19 Ahg Very dark grey (10YR 3/1) stoneless clay (USDA clay) with very many fine dark reddish brown (5YR 3/4) mottles; strong very coarse prismatic parting into coarse angular blocks; firm; many roots; earthworms and ants active; clear smooth boundary.

19–49 Bg Dark grey (10YR 4/1) stoneless clay (USDA clay) with dark reddish brown mottles as above; structure as above, with some inclined shiny (stress-oriented) faces; very firm; common fine roots; few fine ferri-manganiferous concentrations near base; earthworms present; clear smooth boundary.

49–98 BCg Dark grey (5Y 4/1) stoneless clay (USDA clay) with common fine dark reddish brown (5YR 3/4) and yellowish red (5YR 4/8) mottles; compound structure of very coarse prisms parting to weak very coarse blocks; common pores partly lined by ferruginous coats and infilled with clay; very firm; few fine roots; earthworms present; gradual boundary.

98–120 CG Greenish grey (5GY 6/1) stoneless clay (USDA clay) with few medium dark reddish brown (5YR 3/4) and brown (7.5YR 5/2) mottles; wet; massive; infilled pores as above; few roots.

Analytical data for Profile 117

Horizon	H/Oh	Ahg	Bg	BCg	CG
Depth (cm)	0–5	5–19	19–49	49–98	98–120
sand 200 μm–2 mm %		<1	<1	<1	<1
60–200 μm %		1	1	<1	<1
silt 2–60 μm %		37	33	33	39
clay <2 μm %		62	65	66	60
organic carbon %	22	3.2	2.8	0.8	0.8
dithionite ext. Fe %	1.0	2.2	2.4	2.6	2.7
exchangeable cations (me/100 g)					
Ca	38.2	6.3	6.3	5.5	6.8
Mg	9.8	11.5	13.7	13.0	12.3
K	1.9	1.6	1.8	1.9	2.1
Na	1.4	2.4	3.9	6.7	8.5
exch. acidity (me/100 g)	16.4	9.8	1.7	0.3	0.0
CEC (me/100 g)	67.7	31.6	27.4	27.4	29.7
% base saturation	76	69	94	99	100
pH (H_2O)	6.6	6.5	7.7	8.0	8.3
(0.01M $CaCl_2$)	6.1	5.4	7.0	7.4	7.7
saturation extract conductivity S m^{-1}	0.16	0.14	0.22	0.35	0.47
clay <2 μm					
CEC (me/100 g)			40	43	44
K_2O %			3.5	3.2	3.1

Classification
   England and Wales (1984): pelo-alluvial gley soil; clayey marine alluvium (Wallasea series).
   USDA (1975): Mollic (or Vertic) Fluvaquent; very-fine, mixed (non-acid), mesic.
   FAO (1974): Eutric Fluvisol; fine textured (saline, sodic phase).

7.3.4 Humic Alluvial Gley Soils

These soils are also developed in non-calcareous or deeply decalcified alluvium but differ from those of the two preceding subgroups in having a humose or peaty topsoil. The mineral subsurface horizons may be loamy or clayey and lack acid-sulphate characteristics. Profiles meeting these requirements occupy small areas in riverine backswamps, minor valley bottoms and former lake beds where the water table is or was close to the surface for long periods. They also occur locally in estuarine deposits which may have been calcareous when deposited. Data on two representative profiles are reproduced below.

Profile 118 was described and sampled by V. C. Bendelow under ill-drained pasture in Wharfedale, North Yorkshire. It is sited in a backswamp which is ponded for short periods after heavy rain, and the considerable thickness of the humose Ah horizon suggests that accumulation of organic matter accompanied slow sedimentation.

Profile 119 is in drained grassland on older estuarine alluvium (post-glacial raised beach) in the lower Cree Valley, Galloway. It is located near the edge of a peat moss and may once have had a thicker cover of peat. The peaty surface horizon overlies strongly gleyed silty clay which has a weak coarse prismatic structure and is apparently calcareous below 65 cm. Similar soils of the Downholland series (Clayden and Hollis, 1984) occur in Lancashire (Hall and Folland, 1970) and in the East Anglian Fenland (Seale, 1975a).

Profile 118

Location : Buckden, North Yorkshire (SD 943763)
Climatic regime: perhumid temperate
Altitude : 210 m   Slope: 1°
Land use and vegetation : long-term grass; *Agrostis* – rye-grass pasture with buttercups (*Ranunculus* spp.) and kingcups (*Caltha palustris*)

Horizons (cm)
0–16 Ahg1 — Very dark grey (10YR 3/1 broken and rubbed; 4/1 dry) stoneless humose silty clay loam (USDA silty clay loam) with few fine ochreous mottles along root channels; weak medium subangular blocky; friable; abundant fine roots; clear smooth boundary.

16–39 Ahg2 — Dark grey (10YR 4/1; 3/1 rubbed; 5/1 dry) stoneless humose silty clay loam to silty clay (USDA silty clay loam) with common medium strong brown (7.5YR 5/6) mottles; moderate coarse subangular blocky with dark grey (10YR 4/1) faces; moderately porous (fine macropores); firm; abundant fine roots; clear smooth boundary.

39–110 Cg — Grey (N 6/0; 5Y 5/2 rubbed; 10YR 6/1 dry) stoneless clay loam (USDA clay loam) with common medium light olive-brown (2.5Y 5/4) mottles; wet (water seeping into pit); massive; few roots.

110–150 CG — Very dark grey (10YR 3/1) stoneless clay loam; wet; massive.

Analytical data

Horizon	Ahg 1	Ahg 2	Cg
Depth (cm)	0–16	16–39	39–89
sand 200 μm–2 mm %	1	6	5
60–200 μm %	9	10	21
silt 2–60 μm %	59	49	46
clay <2 μm %	31	35	28
organic carbon %	11	8.9	1.4
pH (H$_2$O)	5.5	5.6	5.7
(0.01M CaCl$_2$)	5.0	5.0	5.0

Classification
   England and Wales (1984): typical humic-alluvial

gley soil; fine loamy river alluvium (Sulham series),
USDA (1975): Fluventic Humaquept; fine-loamy, mixed, mesic.
FAO (1974): Dystric Fluvisol; medium textured.

Profile 119 (Bown and Heslop, 1979)

Location : Carslae, Wigtownshire (NX 433582)
Climatic regime: humid temperate
Altitude : 9 m   Slope: <1°
Land use : ley grass
Horizons (cm)

0–20 Op	Dark reddish brown (5YR 2/2) humified peat; moderate granular; friable; many roots; clear boundary.
20–45 Bg	Dark greyish brown (2.5Y 4/2) stoneless silty clay (USDA silty clay) with common medium yellowish red (5YR 5/6) mottles; weak coarse prismatic; firm; frequent roots; gradual boundary.
45–65 BCg	Grey (5Y 5/1) stoneless silty clay (USDA silty clay) with common fine yellowish brown (10YR 5/6) mottles; very weak prismatic; few roots; gradual boundary.
65–90+ 2Cg	Grey (10YR 5/1) stoneless silty clay loam (USDA silt loam) with common medium yellowish brown (10YR 5/4) mottles; wet; massive; few dead roots.

Analytical data

Horizon	Op	BCg	Cg	2Cg
Depth (cm)	5–15	25–35	50–60	80–90
silt^a 2–50 μm %		41	46	58
clay^a <2 μm %		47	41	25
loss on ignition %	69	5.9	5.1	6.3
organic carbon %	39	2.2		
total ext. Fe %	1.1	0.7	0.6	0.9
TEB (me/100 g)	56	17.0	17.4	
CEC (me/100 g)	96	21.3	18.3	
% base saturation	58	80	95	
pH (H$_2$O)	5.5	5.9	6.4	7.6

^{a} by hydrometer method

Classification
Scotland (1984): peaty ground-water gley (Poldar series).
England and Wales (1984): typical humic-alluvial gley soil; clayey marine alluvium (Downholland series).
USDA (1975): Histic Humaquept; fine, illitic or mixed, mesic.
FAO (1974): Mollic Gleysol; fine textured.

7.3.5 Calcaric Alluvial Gley Soils

These are non-saline alluvial gley soils that are calcareous throughout or within 40 cm of the surface and are predominantly loamy in texture. The parent materials include intertidal sediments and riverine or lacustrine deposits derived wholly or partly from calcareous rocks. Fragmented mollusc shells generally form some part of the calcium carbonate present in both cases, and some of the soils have hydrocalcic horizons (lake marl or tufa) in which carbonates have been deposited as chemical precipitates rather than as detrital particles. These resemble rendzinas, especially when drained, and have been classed as gleyic rendzina-like alluvial soils in England and Wales (Avery, 1980). Although they evidently formed under wet conditions, the extremely calcareous subsurface horizons may show no clear evidence of gleying.

Embanked and drained variants in marine alluvium, along with the associated alluvial brown soils (p. 178) into which they grade, form some of the most highly valued agricultural land in Britain, occurring extensively around the Wash and in smaller areas elsewhere. Those in recently embanked coarse silty deposits (Plate III.6) generally retain fine stratification immediately beneath the Ap; others embanked during or since the Roman occupation have a distinct structural Bg horizon and are locally non-calcareous at the surface, usually as a result of decalcification prior to or after embankment (Green, 1968; van der Sluijs, 1970). Variants in river alluvium, which are relatively inextensive, are seldom finely stratified within 40 cm depth. Of the three representative profiles recorded on pp.313–16, the first two are in intertidal deposits and the third in lake marl.

Profile 120 is located near the Lincolnshire coast in arable land reclaimed from salt marsh in 1947 and deeply drained by ditches and pumps. The permeable subsurface horizons are finely stratified throughout and below 59 cm consist dominantly of very fine sand (60–100 µm) with no more than 3 per cent clay. Those above show greyish and reddish brown mottling resembling that in unembanked salt-marsh soils nearby. The succeeding horizons designated Cgf are dominantly brown and give place abruptly at about the current water table level to a very dark grey, CG horizon which becomes brown on exposure to air and evidently contains readily oxidizable iron sulphides. The brown oxidized layer is absent in many similar marsh soils and probably reflects the rapid and considerable lowering of the water table which took place when pump drainage was installed. Coarse silty soils of this type have a very large available water capacity but are subject to compaction resulting from slaking of surface and subplough layers, especially where organic matter contents have been reduced to low levels under continuous cultivation (Hamblin and Davies, 1977; Kooistra et al., 1985).

Profile 121 was described and sampled under rushy pasture in fine silty alluvium adjoining the Shannon estuary in western Ireland. The land has been embanked and is drained by a system of ditches and sluices

Figure 7.4: Air photo showing the southern boundary of Whittlesey Mere, Cambridgeshire, drained in 1852. The light colour of the extremely calcareous lake marl deposited in the former mere contrasts strikingly with that of the bordering peat soil (Cambridge University Collection: copyright reserved).

but clearly remains much wetter than that in which the preceding profile is sited.

Profile 122 is in the largest of the extinct Fenland meres (Seale, 1975a; Godwin, 1978). These appear to have originated during or after the Romano-British marine transgression as shallow lakes ponded behind river levees, locally called *roddons* or *rodhams*, and were subsequently converted into dry land as a result of pump-assisted drainage and consequent sinking of the surrounding peat fens. Their former extent is clearly marked, particularly on air photos (Figure 7.4), by beds of shell marl which consist mainly of finely divided calcium carbonate deposited photochemically from the shallow hard water by aquatic plants, chiefly stoneworts (*Chara* spp.). In the profile described the highly porous marl rests on peat at 91 cm depth and contains 88–94 per cent $CaCO_3$, mostly as silt-size particles. Although the thick Ap horizon has nearly enough organic carbon to qualify as humose, it becomes almost white when dry. The surface of such soils readily becomes 'puffy', so seedbeds need rolling for adequate consolidation. Lime-induced manganese deficiency is also a problem and can cause serious reduction in yields of susceptible crops if uncontrolled by spraying.

Similar extremely calcareous soils, with or without a peaty substratum at 80 cm or less, have been identified and mapped in numerous other localities where clear-water streams originate from springs in calcareous rocks and calcium carbonate has been redeposited either as algal marl or as concretionary tufa, which in places forms low mounds on valley floors (e.g. Arkell, 1947; Jarvis, 1968; Hartnup, 1975; Burton, 1981; Reeve and Thomasson, 1981). Most, but not all, of these soils have mollic epipedons as defined in the US Taxonomy and qualify accordingly as Haplaquolls or Calciaquolls depending on whether one or more of the subsurface horizons is considered to meet the requirements of a calcic horizon (Soil Survey Staff, 1975, pp. 45–46).

Profile 120 (Hodge *et al.*, 1984; Robson, 1985)

Location : Friskney, Lincolnshire (TF 498543)
Climatic regime: subhumid temperate
Altitude : 4 m Slope: <1°
Land use : arable (cereal stubble)
Horizons (cm)

0–25 Ap
Dark greyish brown (10YR 4/2 broken and rubbed; 5/3 dry), stoneless, very calcareous silt loam (USDA silt loam or very fine sandy loam) with few very fine ochreous mottles; weak medium subangular blocky (fragments); friable (hard when dry); many fine roots; abrupt smooth boundary.

25–33 BCg
Greyish brown (10YR 5/2; 5/3 rubbed; 6/3 dry), stoneless, very calcareous silt loam (USDA silt loam) with many fine and common medium reddish brown (5YR 4/4) mottles; weak angular blocky superimposed on moderate medium platy; very friable (slightly hard when dry); common fine roots; abrupt smooth boundary.

33–49 2Cg1
Grey (10YR 5/1; 5/3 rubbed; 6/2 dry) stoneless calcareous very fine sandy silt loam (USDA very fine sandy loam) with mottles as above; moderate medium platy with pale brown (10YR 6/3) faces; consistence as above; common fine roots; clear smooth boundary.

49–59 2Cg2
Similar to above, but with friable consistence and few roots; abrupt smooth boundary.

59–88 3Cgf1
Brown (10YR 4/3 broken and rubbed; 6/3 dry) stoneless calcareous loamy very fine sand (USDA loamy very fine sand) with many fine dark reddish brown (5YR 2/2) mottles; strong fine platy; very friable; few roots; abrupt smooth boundary.

88–134 3Cgf2
Similar to above, but with common fine and medium strong brown (7.5YR 5/6) mottles and comminuted shell fragments; abrupt smooth boundary.

134–147 3CG
Very dark grey (N 3/0), changing to brown (10YR 5/3) and developing

Analytical data for profile 120

Horizon	Ap	BCg	2Cg1	2Cg2	3Cgf1	3Cgf2	3CG	4CG
Depth (cm)	0–25	25–33	33–49	49–59	59–88	88–134	134–147	147–159
sand 200 μm–2 mm %	<1	<1	<1	<1	<1	<1	<1	<1
100–200 μm %	1	1	1	1	3	5	10	4
60–100 μm %	17	10	37	56	71	74	71	60
silt 2–60 μm %	70	76	57	39	24	19	18	33
clay <2 μm %	12	13	5	4	2	2	1	3
organic carbon %	1.6	1.6	0.8	0.4	0.2	0.2	0.2	0.2
$CaCO_3$ equiv. %	12	11	7.4	7.0	4.5	5.0	3.1	5.4
dithionite ext. Fe %	1.1	1.3	1.0	0.8	0.7	0.4	0.6	0.7
pH (H_2O)	8.2	8.3	8.5	8.5	8.6	8.4	8.2	8.2
(0.01M $CaCl_2$)	7.6	7.8	7.9	7.9	8.0	8.1	7.9	7.9
saturation extract conductivity (S m^{-1})	0.09	0.09	0.06	0.06	0.07	0.24	0.33	0.33
bulk density (g cm^{-3})	1.35		1.35	1.45	1.40			
packing density (g cm^{-3})	1.46		1.40	1.49	1.42			
retained water capacity (% vol.)	34		39	41	36			
available water capacity (% vol.)	21		30	34	30			
air capacity (% vol.)	12		10	5	11			

common large strong brown (7.5YR 5/6) mottles within two days of exposure; stoneless, slightly calcareous, loamy very fine sand (USDA loamy very fine sand); wet; weak medium platy; no roots; smell of H_2S; abrupt smooth boundary.

147–159 As above but very fine sandy loam.
4CG

Analytical data *see* above.

Classification
  England and Wales (1984): calcareous alluvial gley soil; coarse silty marine alluvium (Wisbech series).
  USDA (1975): Typic Fluvaquent; coarse-silty, mixed (calcareous), mesic.
  FAO (1974): Calcaric Fluvisol; medium textured.

Profile 121 (Finch et al., 1971)

Location : Shannon, Co. Clare
Climatic regime: humid temperate
Altitude : 0 m   Slope: <1°
Land use and vegetation : long-term grass with crested dogstail (*Cynosurus cristatus*), meadow foxtail (*Alopecurus pratensis*), perennial ryegrass (*Lolium perenne*), white clover (*Trifolium repens*) and soft rush (*Juncus effusus*)

Horizons (cm)

0–10
Ah1
Dark greyish brown (10YR 4/2) stoneless humose calcareous silty clay loam (USDA silt loam); weak fine granular; very friable; abundant roots (root mat); clear boundary.

10–24
Ah2
Dark greyish brown (10YR 4/2) stoneless calcareous silty clay loam (USDA silt loam); weak fine subangular blocky; friable; many roots; clear smooth boundary.

24–43
Bg
Grey (10YR 5/1) stoneless calcareous silty clay loam (USDA silty clay loam) with common fine yellowish brown (10YR 5/8) mottles; weak fine subangular blocky; few roots; clear smooth boundary.

43–74
BCg
Grey (2.5Y 5/1) stoneless calcareous silty clay loam (USDA silty clay loam) with yellowish brown mottles as

Analytical data for Profile 121

Horizon	Ah1	Ah2	Bg	BCg	Cg1	Cg2
Depth (cm)	0–10	10–24	24–43	43–74	74–84	84–104
sand 200 μm–2 mm %	5	3	2	3	1	1
50–200 μm %	7	7	3	2	2	2
silt 2–50 μm %	66	70	67	66	69	75
clay <2 μm %	22	20	28	29	28	22
organic carbon %	5.5	3.1	0.8	0.7	0.7	0.5
nitrogen %	0.50	0.33				
C/N ratio	11.0	9.4				
citrate-dithionite ext. Fe %	0.9	0.9	0.7	0.9	0.5	0.2
pH (H$_2$O)	7.4	7.9	8.3	8.2	8.0	8.0

above; wet; weak coarse prismatic with some humose coatings and pin-head root holes on faces; very few roots; gradual smooth boundary.

74–84 Cg1 Grey (2.5Y 5/1) stoneless calcareous silty clay loam (USDA silty clay loam) with common fine yellowish brown and strong brown (7.5YR 5/8) mottles; wet; structure as above but weaker; no roots; gradual smooth boundary.

84–104 Cg2 Grey (2.5Y 5/1) stoneless calcareous silty clay loam (USDA silt loam) with common fine strong brown mottles; wet; massive; no roots.

Analytical data *see* above.

Classification
Ireland (1980): (alluvial) gley soil (Shannon series).
England and Wales (1984): calcareous alluvial gley soil; fine silty marine alluvium (Agney series).
USDA (1975): Typic Fluvaquent; fine-silty, mixed (calcareous), mesic.
FAO (1974): Calcaric Fluvisol; medium textured.

Profile 122 (Seale, 1975a)

Location : Redmere, Lakenheath, Norfolk (TL 678835)
Climatic regime: subhumid temperate
Altitude : 0 m   Slope: <1°
Land use : apple orchard under grass
Horizons (cm)

0–38 Ap Dark grey (10YR 4/1) stoneless, extremely calcareous silty clay loam (USDA silty clay loam) becoming slightly paler below 33 cm; weak fine and very fine subangular blocky; very friable; abundant fine roots in upper 7.5 cm; common below; few manganiferous concretions; some small freshwater shells (*Planorbis, Limnaea* spp.); abrupt smooth boundary.

38–79/84 Cgk1 Light grey to very pale brown (10YR 7/2–3), stoneless, extremely calcareous silty clay loam (USDA silt loam) with common fine light brownish grey (10YR 6/2) and few fine light yellowish brown (10YR 6/4) mottles; below 48 cm increasing proportion grey (10YR 5/1) merging to dark grey (10YR 5/1 and 3.5/1) with occasional 0.5 cm peaty patches and very few fine to very fine yellowish brown mottles; massive; very friable and porous; few fine and medium roots, some woody; none below 50 cm; shells as above; few manganiferous concretions below 50 cm; 1 cm grey (10YR 5/1.5) layer with many fine brown (10YR 5/3) mottles at base; abrupt wavy boundary.

79/84–91 Cgk2 Light grey faintly variegated with light brownish grey (2.5Y 7/2 and 6/2), stoneless, extremely calcareous silty clay loam (USDA silt loam to silty clay loam) with common fine yellowish brown (10YR 5/6) mottles and few dark red (2.5YR 3/6) mottles near bottom; massive; very friable; abrupt smooth boundary.

Analytical data for Profile 122

Horizon	Ap	Cgk1	Cgk2	Cgk2	2Oh	2Om
Depth (cm)	0–38	38–48	48–81	81–91	91–112	112–155
sand 200 µm–2 mm %	3	3	3	1		
50–200 µm %	9	11	16	13		
silt 2–50 µm %	59	61	55	59		
clay <2 µm %	29	25	26	27		
loss on ignition %	7.9	4.2	5.5	5.0	90	88
organic carbon %	4.8	2.6	2.9	3.1	44	46
nitrogen %	0.36	0.15	0.18	0.26	2.45	2.50
C/N ratio	13	17	16	12	18	18
$CaCO_3$ equiv. %	88	94	91	92		
dithionite ext. Fe %	0.3	0.1	0.2	0.4	1.9	1.5
pH (H_2O)	8.0	8.4	8.2	8.1	7.5	6.8
(0.01M $CaCal_2$)	7.2	7.4	7.5	7.6	7.0	6.4

91–112 2Oh	Black stoneless peat containing few fine fibrous plant remains, commoner below 107 cm; very friable; few dead roots; gradual boundary.
112–155 2Om	Blackish brown semi-fibrous peat with many reed (*Phragmites australis*), sedge and alder (*Alnus glutinosa*) remains.

Analytical data *see* above.

Classification
England and Wales (1984): gleyic rendzina-like alluvial soil; carbonatic-loamy tufa or lake marl (Colthrop series, formerly Willingham).
USDA (1975): Fluvaquentic (or Thapto-histic) Haplaquoll; coarse-silty, carbonatic, mesic.
FAO (1974): Mollic Gleysol; medium textured.

7.3.6 Pelocalcaric Alluvial Gley Soils

These soils have formed in calcareous clayey floodplain deposits, notably in the valleys of the Thames and its tributaries, and recent marine sediments that have been embanked and drained.

Except for the presence of carbonates within 40 cm depth, implying that the exchange complex is generally base saturated throughout, they resemble those of the pelo- subgroup (p. 306) and show similar variations in composition, horizonation and water regime. Two examples are recorded below, the first in drained and cultivated marshland in Lincolnshire and the second under old pasture in the floodplain of the river Thame east of Oxford (Figure 7.5).

The land in which profile 123 is sited was embanked during or before mediaeval times. When sampled in September 1976 after two abnormally dry years, the soil was only slightly moist throughout the upper metre. As the water level in the ditches is generally maintained at or below that depth in winter, disposal of surface water is further assisted by pipe drainage, and the subsurface horizons are estimated to be at least slightly porous, it seems certain that gley features in the Bg and Bgf horizons relate mainly to a former condition. The slightly reddish (7.5YR) hues are attributable to oxidation of pyrite present in the original waterlogged sediment, as in profile 120 from the same area. The CEC and K_2O contents of clay separates from the Bg and Bgf horizons indicate that they consist dominantly of mica, and this was confirmed by X-ray examination.

Profile 124 in very clayey riverine alluvium has a thin acid humose Ah horizon at the surface. The underlying B and C horizons are strongly gleyed and pass at about 134 cm depth into a CG horizon containing vivianite.

Figure 7.5: Summer grazing on the Thame floodplain near Shabbington, Buckinghamshire. The clayey alluvial gley soils are waterlogged for long periods in winter, there is risk of flooding, and drainage improvements are difficult to achieve (Photo by J. Hazleden).

Like profile 116 (p.307), this soil was sampled under unusually dry conditions in the spring of 1976. The subsurface horizons have low bulk densities, but the very low air capacity values indicate that they are slowly permeable when fully saturated. Judging by the clay contents and the large CEC of the clay separates, which consist mainly of interstratified mica-smectite with a high proportion of smectite layers, their shrinkage capacity is even larger than in profile 116. Wide cracks extend into the Bg in dry summers and ped faces in this horizon have prominent stress-oriented coats giving a polished appearance.

Profile 123 (Robson, 1985)

Location	:	Wainfleet All Saints, Lincolnshire (TF 462597)
Climatic regime:		subhumid temperate
Altitude	:	1 m Slope: <1°
Land use	:	arable (cereals)

Horizons (cm)

0–27 Ap	Brown to dark brown (7.5YR 4/2 broken and rubbed; 7.5YR 6/2 dry) stoneless calcareous silty clay (USDA silty clay) with few fine strong brown (7.5YR 5/6) mottles; moderate coarse subangular blocky; firm; abundant fine roots; abrupt smooth boundary.
27–41 Bg	Brown to reddish grey (7.5–5YR 5/2 broken and rubbed; 6/2 dry) stoneless calcareous silty clay (USDA silty clay) with many fine reddish brown (5YR 4/4) mottles; moderate coarse subangular blocky with grey (N 5/0) faces; very firm; many fine roots; abrupt smooth boundary.
41–76 Bgf	Yellowish red (5YR 5/6) and brown (7.5YR 5/2), prominently mottled, stoneless calcareous silty clay (USDA silty clay) with grey (N 5/0) linings to root channels; strong coarse angular blocky with brown (7.5YR 5/2) faces; very firm; many fine roots; clear smooth boundary.
76–100 2BCg	Brown (7.5YR 5/2; 5/4 rubbed; 6/2 dry) stoneless calcareous silty clay (USDA silty clay loam) with many fine brown (7.5YR 5/4) mottles; weak coarse prismatic with grey (10YR 5/1) faces; very firm; common fine roots.

Analytical data for Profile 123

Horizon	Ap	Bg	Bgf	2BCg
Depth (cm)	0–27	27–41	41–76	76–100
sand 60 μm–2 mm %	2	1	1	1
silt 2–60 μm %	49	51	51	61
clay <2 μm %	49	48	48	38
<0.2 μm %	18	18	17	14
organic carbon %	2.5	1.7	1.1	0.9
$CaCO_3$ equiv. %	5.9	6.7	7.6	6.2
dithionite ext. Fe %	2.1	2.3	3.1	1.6
saturation extract conductivity (S m^{-1})	0.12	0.10	0.09	0.11
clay <2 μm				
CEC (me/100 g)		40	42	
K_2O %		3.7	3.6	

Classification
    England and Wales (1984): pelo-calcareous alluvial gley soil; clayey marine alluvium (Newchurch series).
    USDA (1975): Typic (or Vertic) Fluvaquent; fine, illitic (calcareous), mesic.
    FAO (1974): Calcaric Fluvisol; fine textured.

Profile 124 (Hazleden, 1986)

Location	: Waterstock, Oxfordshire (SP 648064)
Climatic regime:	subhumid temperate
Altitude	: 57 m Slope: <1°
Land use and vegetation	: long-term grass with clumps of tufted hairgrass (*Deschampsia caespitosa*)

Horizons (cm)

0–11
Ahg
Very dark greyish brown (10YR 3/2) stoneless humose clay (USDA clay) with many extremely fine yellowish red (5YR 4/8) linear mottles along root channels; strong fine granular in upper 4 cm, passing to strong fine angular blocky below; very porous; 3–5 mm fissures at 5–7 cm intervals extending into horizon below; friable; abundant fine roots; abrupt smooth boundary.

11–26
Bg1
Greyish brown (10YR 5/2), stoneless, very slightly calcareous clay (USDA clay) with many fine reddish brown (5YR 4/4) mottles; strong medium prismatic with polished faces; moderately porous, with fissures as above; very firm; many fine roots concentrated in fissures; clear smooth boundary.

26–55
Bg2
Greyish brown (2.5Y 5/2) stoneless slightly calcareous clay (USDA clay) with many fine strong brown (7.5YR 5/6) mottles; moderate medium prismatic breaking to coarse angular blocky with polished faces; very firm; common roots as above; few gastropod shell fragments; few soft calcareous and ferri-manganiferous concentrations; clear smooth boundary.

55–107
BCgk/Cgk
Grey (10YR 5/1) stoneless calcareous clay (USDA clay) with many fine reddish brown (5YR 4/4) mottles; weak coarse prismatic becoming massive and wet below 85 cm; few roots and shell fragments; common soft calcareous and ferruginous or ferri-manganiferous concentrations.

136–180
CG
(by auger) Dark bluish grey (5B 4/1) and greenish grey (5BG 4/1) clay; wet; massive; slightly fluid; gives off H_2S on treatment with acid and contains white crystals of vivianite which turn blue on exposure to air.

180–200
2Cu(g)
(by auger) Greenish grey (5GY 5/1) and grey (5Y 5/1) firmer and apparently drier clay (Kimmeridge Clay *in situ*).

Analytical data for Profile 124

Horizon	Ahg	Bg1	Bg2	BCgk
Depth (cm)	0–11	11–26	26–55	55–75
sand 200 μm–2 mm %	2	1	2	3
60–200 μm %	4	2	3	11
silt 2–60 μm %	26	15	17	21
clay <2 μm %	68	82	78	65
organic carbon %	8.7	2.6	1.4	1.6
$CaCO_3$ equiv. %	0	1.0	3.6	4.6
dithionite ext. Fe %	3.0	3.2	3.5	4.2
pH (H_2O)	5.7	7.3	7.9	7.8
(0.01M $CaCl_2$)	5.2	6.7	7.3	7.2
clay <2 μm				
CEC (me/100 g)			79	77
K_2O %			2.1	1.9
bulk density (g cm^{-3})	0.68		0.97	0.97
packing density (g cm^{-3})	1.29		1.68	1.55
retained water capacity (% vol.)	52		60	59
available water (% vol.)	19		19	21
air capacity (% vol.)	21		4	5

Classification
England and Wales (1984): pelo-calcareous alluvial gley soil; clayey river alluvium (Thames series).
USDA (1975): Vertic Fluvaquent; very-fine, montmorillonitic (calcareous), mesic.
FAO (1974): Calcaric Fluvisol; fine textured.

7.3.7 Humic Calcaric Alluvial Gley Soils

These soils have formed in calcareous alluvial deposits under the same range of conditions as those of the humic subgroup and present similar variations in texture and current water regime. They have a very dark coloured humose (mollic) A horizon or a peaty surface layer and are calcareous directly below the topsoil or within 50 cm of the mineral soil surface, implying that any post-depositional decalcification has been incomplete or superficial. The parent materials include riverine and lacustrine sediments and formerly peat covered marine or estuarine deposits. Variants with extremely calcareous subsurface horizons have been classed as humic gleyic rendzina-like alluvial soils in England and Wales (Avery, 1980). These usually have a substration of lake marl and are well represented by the Drombanny series (Finch and Ryan, 1966; Finch et al., 1971) mapped in the Carboniferous limestone lowlands of west-central Ireland. There follows data on two representative profiles, the first in river alluvium overlying flint gravel and the second in a silty estuarine deposit.

Profile 125, described and sampled in March 1964 under a poplar plantation in the Pang valley, Berkshire, is located between two branches of the stream, one a slightly elevated artificial course constructed to supply a former water-meadow irrigation system (Jarvis, 1968). As with chalk streams generally, the flow of water bears little relation to short-term rainfall fluctuations and the water table remains at a fairly constant level throughout the year. Where it lies only a little below the topsoil for most of the time, a thick, very dark, biologically active, organo-mineral Ah horizon (hydro-mull) develops in the calcareous alluvium under semi-natural vegetation, as in the profile described. Such soils are difficult to consolidate when drained and cultivated, and crops grown on them are often affected by manganese deficiency.

Profile 126 is in arable land with pump-assisted drainage in Witham Fen, Lincolnshire (Robson et al., 1974). In this area, as in the Chatteris district further south (Seale, 1975b), wastage of the Upper Peat (Godwin, 1978) has revealed the underlying Fen or Buttery Clay, the surface of which is dissected by an intricate dendritic pattern of slightly elevated roddons or 'silt hills' marking the courses of distributary creeks, and it is in one of these that the profile is sited. The roddon deposits are siltier and more calcareous than the intervening clays, which have commonly developed acid-sulphate characteristics following drainage, as in profile 130 (p.326). Fine stratification remains clearly evident immediately beneath the Ap and the permeable subsurface horizons become characteristically coarser with increasing depth. Humose soils of this type have a very large available water capacity and a more water-stable structure than those of the corresponding non-humic subgroups in younger marine sediments, as represented by profile 120 (p. 313). With continuing cultivation, however, the organic-matter content of the topsoils is progressively reduced.

Profile 125 (Jarvis, 1968; Ragg and Clayden, 1973)

Location	:	Bucklebury, Berkshire (SU 537719)
Climatic regime:		subhumid temperate
Altitude	:	70 m Slope: <1°
Land use and vegetation	:	poplar plantation; field layer includes stinging nettle (*Urtica dioica*), docks (*Rumex* spp.), comfrey (*Symphytum officinale*) and reed (*Phragmites australis*)

Horizons (cm)
0–28 Ah	Very dark greyish brown (10YR 3/2) humose very calcareous sandy silt loam to clay loam (USDA loam) containing rare (<1 per cent) gravel-size flint fragments; moderate fine subangular blocky and granular; friable; abundant to many fibrous and fleshy roots; earthworms, earthworm casts and small colourless larvae present; clear smooth boundary.
28–63 BCg	Light brownish grey (2.5Y 6/2) very calcareous sandy silt loam to clay loam (USDA silt loam) with common fine and medium strong brown (7.5YR 5/6) mottles, mainly along root channels; rare stones as above; wet (water table at 43 cm); weak medium prismatic; common to few roots; abrupt smooth boundary.
63–80 2Cg	Gravel consisting mainly of small and medium subangular and angular flint fragments with a calcareous silty matrix; wet.

Analytical data

Horizon	Ah	BCg
Depth (cm)	0–28	28–63
sand 200 µm–2 mm %	18	7
50–200 µm %	24	20
silt 2–50 µm %	40	55
clay <2 µm %	18	18
organic carbon %	8.9	3.1
nitrogen %	1.01	0.39
C/N ratio	9	8
$CaCO_3$ equiv. %	11	15
pH (H_2O)	7.3	7.4
(0.01M $CaCl_2$)	6.8	7.0

Classification
  England and Wales (1984): calcareous humic-alluvial gley soil; coarse/fine loamy over calcareous gravel (variant of Gade series).
  USDA (1975): Fluvaquentic (or Typic) Haplaquoll; fine-loamy, mixed (calcareous), mesic.
  FAO (1974): Mollic Gleysol; medium textured.

Profile 126 (Hall and Heaven, 1979; Hodge et al., 1984)

Location	:	Nocton, Lincolnshire (TF 113666)
Climatic regime:		subhumid temperate
Altitude	:	2.5 m Slope: <1° (top of slight rise)

Land use : arable (winter wheat)
Horizons (cm)

0–32 Ap
Very dark brown (10YR 2/2; 5/1 dry), stoneless, humose, slightly calcareous silty clay loam (USDA silty clay loam); moderate fine subangular blocky; friable; many fine roots; abrupt smooth boundary.

32–46 2BCg
Greyish brown (10YR 5/2) and dark yellowish brown (10YR 4/4) stoneless calcareous silty clay loam (USDA silt loam) with many medium reddish brown (5YR 4/4) mottles along root channels and fine strong brown (7.5YR 5/6) mottles within peds; weak coarse prismatic with dark grey (10YR 4/1) faces superimposed on fine stratification; 2 per cent fine macropores; firm; few fine roots; gradual smooth boundary.

46–69 3Cg1
Greyish brown (10YR 5/2; 6/2 dry), stoneless, finely stratified, calcareous silt loam (USDA silt loam) with thin more clayey laminae and common medium strong brown mottles, mainly along root channels; porosity and consistence as above; few fine roots; diffuse boundary.

69–111 3Cg2
Greyish brown (10YR 5/2; 6/2 dry), stoneless, finely stratified, calcareous very fine sandy silt loam (USDA very fine sandy loam) with very dark greyish brown (10YR 3/2) laminae and common, medium and fine strong brown mottles as above; firm; fewer roots.

Classification
England and Wales (1984): calcareous humic-alluvial gley soil; silty marine alluvium (Chatteris series).
USDA (1975): Fluvaquentic Haplaquoll; coarse-silty, mixed (calcareous), mesic.
FAO (1974): Mollic Gleysol; medium textured.

7.3.8 Saline Alluvial Gley Soils

As already indicated, this subgroup includes all the unripened or partially ripened alluvial gley soils of coastal marshes (Figure 7.6) that are saline (conductivity of the saturation extract more than 4 S m^{-1} at $20°C$) in some or all horizons within 60 cm depth. In the lowest marshes the surface is flooded daily, whilst at higher levels (Plate III.5) it is reached only by occasional spring tides or salt spray, but the ground water remains saline. The vegetation commonly shows a corresponding zonation (Tansley, 1939; Adam, 1978), ranging upwards from an incomplete cover of silt-trapping pioneer plants to various closed

Analytical data for Profile 126

Horizon	Ap	2BCg	3Cg1	3Cg2
Depth (cm)	0–32	32–46	46–69	69–111
sand 100 μm–2 mm %	2	<1	1	1
60–100 μm %	2	8	16	23
silt 2–60 μm %	64	72	69	66
clay <2 μm %	32	20	14	10
organic carbon %	7.8	1.1	0.9	0.8
CaCO$_3$ equiv. %	0.9	5.1	5.6	5.8
pH (H$_2$O)	7.5	7.8	8.0	8.0
(0.01M CaCl$_2$)	7.1	7.4	7.5	7.5
bulk density (g cm^{-3})	0.99	1.32	1.29	1.27[a]
packing density (g cm^{-3})	1.28	1.50	1.41	1.36[a]
retained water capacity (% vol.)	43	40	41	44[a]
available water (% vol.)	18	18	31	32[a]
air capacity (% vol.)	18	10	11	8[a]

[a] samples at 85–90 cm.

Figure 7.6: Salt marsh with tidal creeks and pools near Pilling, Lancashire. Reclaimed land in the foreground is protected from inundation by embankments and is no longer saline (Copyright Aerofilms Ltd.).

communities dominated in some places by shrubby or herbaceous halophytes such as seablite (*Suaeda maritima*), sea aster (*Aster tripolium*) and sea purslane (*Halimione portulacoides*) and in others by salt-tolerant grasses, including 'sea poa' (*Puccinellia maritima*) and red fescue (*Festuca rubra*), which are often used for summer grazing. Elsewhere, particularly in southern England and Ireland, cordgrass (*Spartina townsendii*) forms almost pure stands on mud flats (saltings) which are flooded twice daily (e.g. Hodgson, 1967). Since it first appeared in Southampton Water about the turn of the century, this tall vigorous hybrid of *S. maritima* and *S. alterniflora* has spread widely and largely ousted the native pioneer species.

The salt-marsh soils vary considerably in composition depending on their depositional history. Textures range from loamy very fine sand to silty clay in the uppermost horizons and generally become coarser with depth, since progressively finer sediments accumulate as the age and height of the marsh increase. The typically uneven surface is usually dissected by a dendritic creek system which channels the inflowing and outflowing tide. When the creeks overflow, the sediment-laden water is partially checked by bordering vegetation, forming low banks or levees, whilst finer particles are carried further and settle in intervening 'pool' areas. Departures from these simple patterns result from erosion of earlier formed deposits by wave action and changes in creek or river courses leading to infilling of channels and closed depressions (salt pans).

The morphology of these soils varies with particle-size and the depths to which physical ripening and aeration have extended. Fine silty and clayey variants in the lowest parts of the intertidal zone and in ill drained 'pools' are dominantly greyish throughout; all horizons are unripened (fluid) and a 'reduced' CG horizon, which can be sulphidic (Chapter 3, p. 111), is usually encountered more or less immediately below the surface. In higher situations, especially near creeks, rust mottled Cg horizons extend to considerable depths. Coarse silty (very fine sandy) layers invariably show fine stratification; where textures are finer stratification is generally less evident and relatively well drained profiles have prismatic Bg or BCg horizons reflecting intermittent drying. In most places the subsurface horizons are rendered permeable by the occurrence of numerous channels produced by burrowing marine animals and by the roots of plants which colonized the mud during initial stages of soil formation. The three profiles recorded below show successive stages of physical ripening and horizon development in fine-textured calcareous marsh deposits.

Profile 127, described and sampled by the author in a sheltered inlet on the Essex coast, consists of soft slightly calcareous mud which supports an incomplete cover of glasswort (*Salicornia stricta*). It has a fluid consistence throughout and the only obvious indication of horizon development is the occurrence of numerous ferruginous mottles in the upper 15 cm. Organic matter is irregularly distributed and there is no distinct A horizon. Incipient salt-marsh soils of this kind are called *Slik Vague Soils* in the Netherlands system (de Bakker and Schelling, 1966) and Hydraquents in the US Taxonomy.

Profile 128 is located near the edge of Pagham Harbour, West Sussex, under a closed vegetative cover dominated by 'sea poa' and sea purslane. Here the soil is slightly better drained, corresponding to the *Gors Vague Soils* of de Bakker and Schelling. Although the surface remains incompletely ripened, the organic carbon data confirm that a distinct A horizon has developed and ferruginous mottles extend through the whole depth of the mud. This soil has to be classed as a Haplaquent rather than a Hydraquent in the US taxonomy because the incompletely ripened subsurface horizon rests on beach shingle at less than 50 cm depth.

Profile 129 was described and sampled by W.M. Corbett under similar vegetation in salt marsh bordering the Wash. At this site the soil is fully ripened to more than 50 cm and has A and Bg horizons with weak coarse blocky and prismatic peds.

Profile 127

Location	:	Kirby-le-Soken, Essex (TM 239229)
Climatic regime:		subhumid temperate
Altitude	:	0 m Slope: <1
Vegetation	:	salt marsh: 50 per cent cover of glasswort (*Salicornia stricta*); algal scum on surface

Horizons (cm)

0–15 Cg — Dark grey (10YR 4/1) stoneless slightly calcareous clay (USDA silty clay) with many fine yellowish red (5YR 5/6) mottles and common black patches; wet; massive; very fluid; 2–5 per cent medium macropores; abundant roots; clear boundary.

15–50 CG — Dark grey to very dark grey (5Y 4–3/1) stoneless slightly calcareous silty clay (USDA silty clay) with many coarse black (5Y 2/1) mottles; massive; very fluid; 2–5 per cent medium macropores; many roots.

Analytical data

Horizon	Cg	CG
Depth (cm)	0–15	24–40
sand 60 µm–2 mm %	3	2
silt 2–60 µm %	50	47
clay <2 µm %	47	51
organic carbon %	3.6	4.9
$CaCO_3$ equiv. %	2.0	1.4
pH (H_2O)	7.2	7.7
(0.01M $CaCl_2$)	7.2	7.7
saturation extract conductivity ($S\,m^{-1}$)	4.7	5.0

Classification
  England and Wales (1984): unripened gley soil; saline-clayey marine alluvium (unnamed series).
  USDA (1975): Typic Hydraquent; fine, mixed or illitic (calcareous), mesic.
  FAO (1974): Calcaric Fluvisol; fine textured (saline phase).

Profile 128

Location	:	Pagham Harbour, West Sussex (SZ 862971)
Climatic regime:		subhumid temperate
Altitude	:	2 m Slope: <1°
Vegetation	:	salt marsh with 'sea poa' (*Puccinellia maritima*), sea purslane (*Halimione portulacoides*), sea aster (*Aster tripolium*) and sea plantain (*Plantago maritima*)

Horizons (cm)

0–17
Ahg
Dark grey and grey (5Y 4/1 and 5/1), stoneless, slightly calcareous, humose silty clay (USDA silty clay) with many fine yellowish red (5YR 5/8) mottles, mainly around root channels; wet; massive; slightly fluid; 2–5 per cent macropores up to 10 mm; abundant fine and common medium fleshy roots; clear smooth boundary.

17–34
2Cg
Grey (N 5/0) to greenish grey (5GY 5/1), stoneless calcareous clay loam (USDA clay loam) with many fine yellowish red mottles associated with channels; wet; massive; very fluid; macropores as above; roots as above; some dead; abrupt smooth boundary.

34+
3Cg
Rounded flint gravel (beach shingle) with loamy matrix similar to above.

Analytical data

Horizon	Ahg	2Cg
Depth (cm)	0–17	17–34
sand 200 µm–2 mm %	2	3
60–200 µm %	2	21
silt 2–60 µm %	45	48
clay <2 µm %	51	28
organic carbon %	6.3	1.9
$CaCO_3$ equiv. %	4.7	6.2
pH (H_2O)	7.6	8.0
(0.01M $CaCl_2$)	7.6	7.8
saturation extract conductivity ($S\,m^{-1}$)	3.7	2.4

Classification
  England and Wales (1984): unripened gley soil; saline-clayey over non-calcareous gravel.
  USDA (1975): Typic Haplaquent; loamy-skeletal?, mixed (calcareous), mesic.
  FAO (1974): Calcaric Fluvisol; fine textured (saline phase).

Profile 129

Location	:	Wolferton, Norfolk (TF 641298)
Climatic regime:		subhumid temperate
Altitude	:	1 m Slope: <1°
Vegetation	:	salt marsh with 'sea poa' (*Puccinellia maritima*) and sea purslane (*Halimione portulacoides*)

Horizons (cm)

0–15
Ah(g)
Very dark grey (10YR 3/1), stoneless, very calcareous silty clay (USDA silty clay) with few fine yellowish red (5YR 4/6) mottles; weak coarse blocky; very porous (fine fissures and macropores); moderately firm; abundant roots; black coats on and within peds; abrupt smooth boundary.

15–80
Bg/BCg
Dark greyish brown (10YR 4/2), stoneless, very calcareous silty clay (USDA silty clay to silty clay loam) with many medium dark yellowish brown (10YR 4/4) mottles; wet (water table at *c*. 40 cm with tide out); weak coarse prismatic; many macropores; few roots; gradual smooth boundary.

Analytical data for Profile 129

Horizon	Ah(g)	Bg	BCg	2CG	3CG
Depth (cm)	0–10	15–25	28–38	90–100	120–130
sand 200 μm–2 mm %	1	<1	1	<1	<1
60–200 μm %	2	2	4	18	52
silt 2–60 μm %	56	58	58	58	38
clay <2 μm %	41	40	37	24	10
organic carbon %	3.9	2.9	2.4	1.3	0.7
$CaCO_3$ equiv. %	14	14	15	13	10
pH (H_2O)	7.6	7.8	7.8	7.9	8.0
(0.01M $CaCl_2$)	7.5	7.7	7.7	7.7	7.8
saturation extract conductivity (S m^{-1})	2.8	2.8	3.6	3.3	3.3

80–110　Dark grey (10YR 4/1), stoneless, very
2CG　　calcareous silty clay loam (USDA silty loam); wet; massive; slightly fluid, macropores as above; gradual smooth boundary.

110–140　Dark grey (10YR 4/1), stoneless, finely
3CG　　stratified, calcareous very fine sandy silt loam (USDA very fine sandy loam); wet; very fluid.

Analytical data *see* above

Classification
　England and Wales (1984): pelo-calcareous alluvial gley soil; saline-clayey alluvium (Wolferton series).
　USDA (1975): Mollic Fluvaquent; fine, illitic (calcareous), mesic.
　FAO (1974): Calcaric Fluvisol; fine textured (saline phase).

7.3.9 Acid-sulphate Alluvial Gley Soils

These are the alluvial gley soils, corresponding to Thionic Fluvisols in the FAO system and Sulfaquepts in the US Taxonomy, that have acid-sulphate characteristics (Chapter 3, p. 111) resulting from drainage and consequent oxidation of sulphides, chiefly pyrite, in the parent materials. As defined here, they are required to have a pH of 3.5 or less in 0.1.M $CaCl_2$ (or 4.0 or less in water) after drying, yellowish segregations of jarosite, or both, in one or more horizons at less than 80 cm depth. Variants with and without a humose or peaty topsoil are distinguished as sulphuric humic-alluvial and sulphuric alluvial gley soils, respectively, in the England and Wales classification, but the latter are relatively uncommon.

Most soils conforming to this subgroup originated as saline alluvial gley soils with sulphidic CG horizons formed in coastal or estuarine environments, but they can also occur in marshy river valleys where the flood deposits are non-calcareous and there is sulphur in the ground water. Those so far identified and mapped in the British Isles are in the drained Fenlands and the valleys of the river Yare and its tributaries in eastern England (Hodge *et al.*, 1984; Dent, 1985), though small patches undoubtedly occur elsewhere. In the Fens the main parent material is the Fen or Buttery Clay which lies between the Lower and the Upper Peat (Seale, 1975a; Godwin, 1978). This was deposited between about 5,000 and 4,000 years ago under brackish lagoonal conditions favouring accumulation of pyrite. Following wastage of the Upper Peat it now forms the subsoil over considerable areas. As it was partly calcareous originally, the resulting soils are variable, those with acid-sulphate characteristics occurring only where carbonates were absent or present only in small amounts. The subsurface horizons of other profiles, often within short distances, contain gypsum resulting from neutralization of the sulphuric acid formed by oxidation of pyrite but remain either calcareous (humic-calcaric alluvial gley) or slightly to moderately acid (humic alluvial gley) depending on the initial composition of the deposit.

In the Norfolk and Suffolk valleys the patterns are equally complex and the present distribution of acid-sulphate conditions is determined by the depth and length of time over which effective drainage has operated as well as by parent material variations. The potentially acid land was traditionally used as summer pasture and it is only during the last 20 years that much of it has been deeply drained in order to permit cereal growing. According to Dent (1985), this has already led to severe acidification.

As the sulphuric horizons are toxic to plant roots, the productivity of these soils and the range of crops that can be profitably grown is restricted accordingly and deep incorporation of the large amounts of lime needed to achieve effective correction of the acidity is very costly. Infield drainage, besides exacerbating the problem, is hindered by frequent blockage of tile drains by ochreous precipitates (Trafford et al., 1973) and leads to pollution of ditches and neighbouring freshwater bodies, as in parts of the Norfolk Broads (Gosling and Baker, 1980). Reclamation of the potential acid-sulphate soils for arable cropping is therefore not only difficult and expensive but undesirable on environmental grounds.

Data on two representative profiles are reproduced below, the first in deeply drained fenland near Chatteris, Cambridgeshire, and the second on pyritic alluvium in the Waveney valley close to the Norfolk–Suffolk border.

Profile 130 is developed in Fen Clay overlying a thin band of peat which rests in turn on a silty substratum. Extremely acid horizons with gypsum and jarosite occur above and below the peat. Profile 131 consists of 50 cm of humus-rich clayey alluvium overlying a succession of coarse loamy and sandy layers, the former containing numerous wood fragments. The A and Bg horizons formed in the upper layer are slightly calcareous, possibly due to addition of chalk as a liming material, but samples of the underlying layers all gave pH (H_2O) values less than 4 determined by the normal procedure. They also contain significant amounts of sulphur (sulphate-S + pyrite-S) and pyrite was identified microscopically in the sand fractions (P. Madgett, private communication). In other borings nearby, jarosite mottles were noted in the lower part of the clayey layer and black sulphidic material below the water table. Detrital pyrite in the Anglian chalky till underlying the whole area is probably the ultimate source of the sulphur at this site, which is 30 km from the sea and 15 m above Ordnance Datum. As the conductivity measurements show, however, the lower horizons of the soil are distinctly saline.

Profile 130 (Seale, 1975a)

Location	: Benson's Fen, Doddington, Cambridgeshire (TL 422895)
Climatic regime:	subhumid temperate
Altitude	: 0 m Slope: <1°
Land use	: arable (wheat stubble)
Horizons (cm)	
0–31 Ap	Black (10YR 2/1) stoneless humose silty clay (USDA silty clay) with common fine grey to olive-grey (5Y 5/1–2) and reddish brown (5YR 3–4/4) mottles; coarse subangular clods breaking in places to fine granular; friable; common fine roots; abrupt wavy boundary with discontinuous (0–8 cm) layer of semi-fibrous peat at base.
31–43 BCgy1	Dark grey (5Y 4/1) stoneless silty clay (USDA silty clay) with common medium dark reddish brown (5YR 3/3) and brown (7.5YR 4/4) mottles mainly associated with old root channels; weak fine subangular blocky; firm; common dead roots; dark reddish brown impressions of reed leaves and very small peaty inclusions; common gypsum crystals often

43–51 BCgy2	around root channels; gradual smooth boundary. Dark grey (5Y 4/1) stoneless clay (USDA clay) with mottles as above, together with brownish yellow (10YR 6/6) and pale yellow (5Y 7/3) aureoles of jarosite; weak angular blocky; fine dead roots common, but other plant remains less frequent than above; few gypsum patches; clear smooth boundary.	
51–69 2Om	Black (2.5YR 2/1) semi-fibrous peat containing dark reddish brown reed and wood remains; few roots; a little fine gypsum; abrupt boundary.	
69–76 3Cg	Grey (5Y 5/1) stoneless silty clay loam (USDA silt loam) with common light grey (2.5Y 7/2) mottles and yellow (2.5Y 8/6) vertical streaks of jarosite; very weak blocky; firm; roots rare; sedge and twig remains common; no gypsum; clear wavy boundary.	
76–91 3Cgy	Grey (N 6/0) stoneless silty clay loam (USDA silty clay loam) with many medium olive-brown to light olive-brown (2.5Y 4/4–5/5), common yellow (jarosite) and few brown to dark brown (7.5YR 4/4) mottles; massive; very firm; roots rare; few black plant remains, encrusted in places with ferruginous material; patches of fine gypsum crystals.	

Classification
England and Wales (1984): sulphuric humic-alluvial gley soil; clayey marine alluvium (Tydd series, formerly Downholland).
USDA (1975): Typic Sulfaquept; fine (over organic), illitic or mixed, mesic.
FAO (1974): Thionic Fluvisol; fine textured.

Profile 131 (Corbett, 1979)

Location   : Mendham, Suffolk (TM 264816)
Climatic regime: subhumid temperate
Altitude   : 15 m   Slope: <1°
Land use   : long-term grass

Horizons (cm)

0–10 Ahg	Very dark greyish brown (10YR 3/2) humose clay (USDA clay) with common fine brown to dark brown (7.5YR 4/4) mottles; rare very small and small rounded chalk fragments; moderate medium and coarse granular; friable; abundant fine roots; gradual boundary.
10–30 Bg	Greyish brown (10YR 5/2) stoneless clay (USDA clay) with many medium strong brown (7.5YR 5/6–8) mottles; coarse prismatic with organic coats on ped faces in upper part; firm; abundant fine roots; gradual smooth boundary.

Analytical data for Profile 130

Horizon	Ap	BCgy1	BCgy2	2Om	3Cg	3Cgy
Depth (cm)	0–30	30–43	43–51	51–69	69–76	76–914
sand 200 µm–2 mm %	1	2	2		4	6
50–200 µm %	1	3	3		7	5
silt 2–50 µm %	51	47	36		69	55
clay <2 µm %	47	48	59		20	34
loss on ignition %	24	9.5	12.4	70	6.6	6.9
organic carbon %	10.7			37.5		
nitrogen %	0.95			2.36		
C/N ratio	11			16		
pH (H$_2$O)	6.7	4.8	3.4	3.1	4.8	2.5
(0.01M CaCl$_2$)	6.5	4.6	3.3	3.0	4.7	2.4

Analytical data for Profile 131

Horizon	Ahg	Bg	2Cg	3Cg	4Cg	5CG
Depth (cm)	0–10	10–30	30–50	50–60	60–70	70–90
sand 600 μm–2 mm %	4	3	2	3	6	3
200–600 μm %	13	10	5	36	62	33
60–200 μm %	19	8	4	32	25	40
silt 2–60 μm %	23	27	45	12	4	14
clay <2 μm %	41	52	44	17	3	10
organic carbon %	5.6	3.6	5.9	9.3	1.6	0.6
$CaCO_3$ equiv. %	1.1	2.9				
dithionite ext. Fe %	3.2	5.2	1.4	0.8	0.4	0.3
total sulphur %				0.76	0.26	0.19
pH (H_2O)	6.8	7.5	6.1	3.8	3.1	3.8
(0.01M $CaCl_2$)	6.5	7.0	5.9	3.6	3.0	3.6
saturation extract conductivity (S m^{-1})	0.15	0.12	0.30	0.30	0.67	0.33

30–50 2Cg (or bAh)		Dark grey (10YR 4/1) stoneless humose silty clay (USDA silty clay) with common fine strong brown (7.5YR 5/8) mottles; wet; massive; common fine roots; common, locally abundant, wood fragments; gradual smooth boundary.
50–60 3Cg		Very dark grey (10YR 3/1) stoneless humose medium sandy loam (USDA sandy loam) containing abundant (10–50 per cent) woody fragments; wet; abrupt smooth boundary.
60–70 4Cg		Light brownish grey (10YR 6/2) stoneless medium sand (USDA sand) with few large olive-brown (2.5Y 4/4) mottles; wet; massive; gradual smooth boundary.
70–90+ 5CG		Light grey to grey (10YR 6/1) stoneless medium sandy loam (USDA fine sandy loam) containing 10–20 per cent black wood fragments; wet (below water table).

Analytical data see above.

Classification
England and Wales (1984): sulphuric alluvial gley soil; clayey over sulphuric coarse loamy or sandy, river alluvium (Shotford series).
USDA (1975): Sulfic Fluvaquent; clayey over loamy, mixed?, mesic.
FAO (1974): Thionic Fluvisol; fine textured.

7.4 Calcaric Gley Soils

7.4.1 General Characteristics, Classification and Distribution

These are loamy and clayey soils, including calcaro-cambic and calcareous humic gley soils and some rendzinas and calcareous pelosols as defined in the England and Wales classification (Avery, 1980), that have gleyed subsurface horizons formed in calcareous materials other than recent alluvial deposits. They are calcareous throughout or within 50 cm of the mineral soil surface and lack an argillic B horizon. Those having a prominent (mollic) A horizon are mostly Haplaquolls according to the US Taxonomy and Mollic Gleysols in the FAO (1974) scheme; others are mainly Haplaquepts (FAO Calcaric Gleysols). As iron is less easily reduced and mobilized in ferrous forms at high pH values than under acid conditions (Figure 1.2, p. 7), the gley features are often less clearly expressed than in texturally similar non-calcareous soils with comparable moisture regimes. Mottles higher in chroma than the matrix are commonly yellowish or olive brown rather than strong brown and may result from localized oxidation *in situ* rather than reduction and segregation of iron.

Periodically wet or artificially drained soils meeting the prescribed specification occur on calcareous argillaceous rocks and in medium to fine textured Pleistocene deposits containing fragmentary chalk or limestone. Judging from published soil survey information, they cover between 2 and 3 per cent of England and Wales and are also represented very locally in central Ireland and lowland Scotland but are far less extensive overall than their non-calcareous counterparts. Those of clayey texture normally have slowly permeable subsurface horizons and resemble the non-calcareous surface-water gley soils in this respect. Loamy variants are mainly but not exclusively of ground-water type. In either case the darkened topsoil is typically underlain within 40 cm depth by a greyish or olive-coloured Bg (cambic) horizon with brown or ochreous mottles and more or less well developed blocky or prismatic structure. This horizon, which may be partially or completely decalcified, is weakly expressed in very poorly drained sites and in extremely calcareous materials such as Chalk Marl (e.g. Burwell series in Hodge and Seale, 1966). Underlying BCg and Cg horizons commonly contain nodules, coats or soft segregations of redeposited (secondary) calcium carbonate (BCgk, Cgk). A distinct ferruginous horizon (BCgf or Cgf) related to current or former water table levels can also be identified in some profiles.

In common with lowland gley soils generally, the most important factors affecting land capability on these soils are the excess of rainfall over evapotranspiration, topsoil texture and permeability, which together determine the extent to which seasonal wetness and related workability and trafficability problems can be alleviated by drainage measures. Where the water-table levels are readily controllable, as in loamy materials with pervious substrata, the drained soils may differ little in behaviour from associated calcaric brown soils of similar composition, especially in low-rainfall areas. Clayey variants are distinctly less tractable but generally have a more stable pore structure than non-calcareous clayey gley soils and consequently respond well to mole drainage. As on certain of the calcaric alluvial gley soils, cereals, particularly oats, and other susceptible crops are often affected by manganese deficiency. Chlorosis caused by immobilization of iron in the soil or the plant is also common, especially in fruit trees grown on variants rich in clay-size calcium carbonate.

Key to Subgroups

1. Calcaric gley soils with a humose or peaty topsoil (p. 107)
   Humic calcaric gley soils
2. Other calcaric gley soils with a slowly permeable subsurface horizon (p. 110)
   2.1 With pelo- characteristics (p. 112)
   Pelocalcaric gley soils
   2.2 Without pelo- characteristics
   Stagnocalcaric gley soils
3. Other calcaric gley soils
   Ground-water calcaric gley soils

7.4.2 Ground-water Calcaric Gley Soils

These are non-humic calcaric gley soils that lack a slowly permeable subsurface horizon within 80 cm depth, implying that gleying results primarily from periodic saturation by ground water retained by an impervious substratum below the solum. The small areas they occupy are chiefly in southern and eastern England and consist for the most part of low lying land on late Pleistocene (Devensian) fluvial or head deposits which are predominantly loamy or have loamy upper horizons overlying sandy or gravelly layers. Some of these soils (Plate III.7) have prominent (mollic) A horizons that are very dark coloured but contain insufficient organic matter to qualify as humose. Data on two representative profiles in cultivated land are reproduced on pp.330–2.

Profile 132 is located in the Upper Thames valley about one metre above the level of the present floodplain. It is developed in silty 'older alluvium' overlying low-terrace gravel consisting largely of Jurassic limestone (Northmoor or Floodplain Terrace) and has a clayey topsoil which may have originated as a Holocene flood deposit. The Ap and Bg horizons are much less calcareous than the underlying BCg, the upper part of which contains numerous hard nodules of redeposited calcium carbonate. Although the water table rises nearly to the surface in winter and the silty subsurface horizons are strongly gleyed, morphological evidence of wetness in the gravelly substratum is limited to weak colour banding.

Profile 133 was described and sampled by D.W. King in very gently sloping land at the foot of the Chiltern escarpment near Ivinghoe, Buckinghamshire (Avery, 1964). The parent material is late Devensian chalky head which becomes gravelly at the base and overlies soft marly chalk (Lower Chalk). Subsurface wetness is attributable at this site to the presence of impermeable Gault clay at no great depth, coupled with seepage from higher ground. Soils of this kind were classed as gleyic rendzinas by Avery (1980) on the basis that horizons below the A are extremely calcareous (>40 per cent $CaCO_3$).

Profile 132 (Jarvis and Hazleden, 1982)

Location:	Aston, Oxfordshire (SP 339013)
Climatic regime:	subhumid temperate
Altitude:	65 m Slope: <1°
Land use:	arable (grass ley)

Horizons (cm)

0–21
Ap
Dark greyish brown (10YR 4/2) slightly calcareous silty clay (USDA silty clay) containing rare small subangular flint and rounded limestone fragments; moderate to weak fine and medium subangular blocky; friable; common fine roots; inclusions of Bg horizon material in lower part; abrupt boundary.

21–40
2Bg1
Grey (5Y 5/1), stoneless, slightly calcareous silty clay loam (USDA silty clay loam) with many fine and medium yellowish brown (10YR 5/6) mottles; moderate medium to coarse subangular blocky with some fine prismatic peds, the vertical faces having few mottles; friable to firm; common fine roots; common fine ferri-manganiferous nodules, a few up to 2 cm; gradual boundary.

40–52
2Bg2
Grey (5Y 5/1), stoneless slightly calcareous silty clay loam (USDA silt loam) with many fine and medium yellowish brown to brownish yellow (10YR 5-6/6) mottles; weak medium subangular blocky to fine prismatic; friable; common fine roots and ferri-manganiferous nodules as above; clear irregular boundary.

52–55/61
2BCgk
Light grey (10YR 7/2) and light brownish grey (10YR 6/2), stoneless, very calcareous silty clay loam with many fine yellowish brown (10YR 5/8) mottles; weak fine and medium subangular blocky; friable; few fine roots; abundant powdery calcium carbonate deposits and common hard subrounded calcareous nodules; common fine ferri-manganiferous nodules; clear irregular boundary.

55/61–75
2BCg(k)
Light brownish grey (10YR 6/2), stoneless, very calcareous sandy silt loam (USDA silt loam) with many fine and medium yellowish brown (10YR 5/8) mottles; weak medium and coarse subangular blocky; friable; few fine roots; few ferri-manganiferous nodules; a few pores partially infilled with secondary calcium carbonate; clear smooth boundary.

75–82
3Cg(k)
Grey (10YR 6/1), moderately to very stony, very calcareous sandy silt loam with many fine and medium strong brown (7.5YR 5/6) mottles; many to abundant, very small and small (<8 mm) rounded limestone and subangular flint fragments; massive; friable; few fine roots; secondary calcium carbonate infilling a few pores; clear wavy boundary.

Analytical data for Profile 132

Horizon	Ap	2Bg1	2Bg2	2BCg(k)	4Cgf/Cu
Depth (cm)	0–21	21–40	40–52	61–75	82–104
sand 600 μm–2 mm %	1	<1	1	2	34
200–600 μm %	2	3	2	4	38
60–200 μm %	4	6	5	17	13
silt 2–60 μm %	47	61	66	62	9
clay <2 μm %	46	30	26	15	6
organic carbon %	3.3	0.4	0.3	0.2	0.2
$CaCO_3$ equiv. %	1	2	3	26	63
pH (H_2O)	7.7	8.3	8.3	8.5	8.0
(0.01M $CaCl_2$)	7.4	7.7	7.7	7.8	7.5

82–104+ 4Cgf/Cu   Strong brown (7.5YR 5/6) in upper 2 cm, yellowish brown (10YR 5/4) below; gravel of rounded and subrounded limestone and subangular flint <1 cm with an extremely calcareous loamy coarse sand matrix; single grain; loose; water entered pit at 104 cm (2.8.1973).

Classification
England and Wales (1984): calcaro-cambic gley soil; silty material over calcareous gravel (Grimblethorpe series).
USDA (1975): Typic Haplaquept; fine-silty over sandy-skeletal, mixed (calcareous), mesic.
FAO (1974): Calcaric Gleysol; fine textured.

Profile 133

Location   : Pitstone, Buckinghamshire (SP 940158)
Climatic regime: subhumid temperate
Altitude   : 107 m   Slope: 1°
Land use   : arable
Horizons (cm)

0–18 Ap   Dark greyish brown (2.5Y 4/2), very slightly stony, very calcareous silty clay loam (USDA clay loam to silty clay loam); few small and medium subangular flint fragments; moderate fine subangular blocky and medium granular; friable; many fine roots; few gastropod shell fragments; abrupt smooth boundary.

18–28 Ahg   Dark greyish brown (2.5Y 4/2), very slightly stony, extremely calcareous silty clay loam (USDA clay loam to silty clay loam) with common very fine ochreous mottles; stones as above; moderate medium and fine subangular blocky; friable; common fine roots; clear smooth boundary.

28–64 Bg   Light olive-grey (5Y 6/2), very slightly stony, extremely calcareous silty clay loam (USDA silty loam) with many fine ochreous mottles; stones as above but mainly small and very small; moderate medium prismatic breaking to subangular blocky; friable; few fine roots; abrupt wavy boundary.

64–90 2Cg   Light olive-grey (5Y 6/2), very stony, extremely calcareous sandy silt loam (USDA loam) with common fine ochreous mottles; abundant small white chalk pebbles and few large subangular flint fragments; massive; wet.

Analytical data see p.332.
Classification
England and Wales (1984): gleyic rendzina; loamy material over calcareous gravel (Gubblecote series).
USDA (1975): Typic Haplaquept; fine or coarse-loamy, carbonatic, mesic.
FAO (1974): Calcaric Gleysol; medium textured.

Analytical data for Profile 133

Horizon	Ap	Ahg	Bg	2Cg
Depth (cm)	5–13	19–26	43–53	74–84
sand 200 μm–2 mm %	11	10	5	23
50–200 μm %	9	10	10	18
silt 2–50 μm %	50	50	59	44
clay <2 μm %	30	30	26	15
loss on ignition %	6.2	5.7	2.9	
$CaCO_3$ equiv. %	34	41	49	60
pH (H_2O)	7.7	7.7	7.9	8.2
(0.01M $CaCl_2$)	6.9	6.9	7.2	7.4

7.4.3 Stagnocalcaric Gley Soils

These are similar soils in which drainage is impeded within 80 cm depth by a slowly permeable substratum. They are less extensive than those of the ground-water subgroup and have not been distinguished from them at subgroup level in the England and Wales classification. The few that have been recorded occur in close association with less gleyed and/or more deeply leached soils in thin loamy drift over calcareous argillaceous rocks or chalky till. In the following example, located in level, slightly elevated land on Gault (Lower Cretaceous) clay north-east of Cambridge, the drift layer is about 56 cm thick. The junction with the clay is marked by a thin and discontinuous gravelly layer and is often irregular, with involutions presumably resulting from cryoturbation.

Profile 134 (Hodge and Seale, 1966)

Location	:	Wicken, Cambridgeshire (TL 562712)
Climatic regime:		subhumid temperate
Altitude	:	8 m Slope: <1°
Land use	:	arable (wheat stubble)

Horizons (cm)

0–23 Ap
Very dark to dark greyish brown (10YR 3.5/2) very slightly stony, slightly calcareous sandy clay loam to clay loam (USDA sandy clay loam); few small flint fragments; weak medium blocky; firm; common fine roots; earthworms common; sharp boundary.

23–56 Bg
Brown (10YR 5/3), very slightly stony, calcareous sandy clay loam (USDA sandy clay loam) with faint grey and brown mottling becoming increasingly evident with depth; stones as above and few lenticles of fine gravel at base; weak to moderate, medium subangular blocky; many pores up to 1 mm; firm; common fine roots; earthworms present; few soft manganiferous concentrations; clear smooth boundary.

56–80+ 2BCgk
Light olive-grey (5Y 6/2) extremely calcareous clay with many medium to coarse light olive-brown (2.5Y 5/4) and few light grey (5Y 7/1) mottles; few small flint fragments in upper few cm; wet (water table at 56 cm on 27 Oct. 1958); moderate medium to coarse angular blocky with grey faces; very plastic; few fine roots; few soft calcareous concentrations up to 1 cm across.

Analytical data *see* p.333.

Classification
England and Wales (1984): calcaro-cambic gley soil; fine loamy drift over clayey material passing to clay or soft mudstone (variant of St Lawrence series).
USDA (1975): Aquic Eutrochrept; fine-loamy over clayey, mixed, mesic.
FAO (1974): Gleyic Cambisol; medium textured.

Analytical data for profile 134

Horizon	Ap	Bg	Bg	2BCgk
Depth (cm)	0–23	23–38	38–56	56–80
sand 200 μm–2 mm %	32			
50–200 μm %	21			
silt 2–50 μm %	24	24[a]	16[a]	23[a]
clay <2 μm %	23	22[a]	26[a]	71[a]
organic carbon %	1.8			
$CaCO_3$ equiv. %	1.5	8.0	3.2	47
pH (H_2O)	7.8	7.9	7.9	8.1
(0.01M $CaCl_2$)	7.3	7.4	7.5	7.8

[a] by hydrometer method

7.4.4 Pelocalcaric Gley Soils

These are greyish calcareous clay soils with blocky or prismatic gleyed subsurface horizons. They have been grouped with the pelocalcaric brown soils (p. 191) as calcareous pelosols in England and Wales (Avery, 1973, 1980) and occur along with them on Jurassic and Cretaceous clays and clay shales, the East Anglian and Midland chalky boulder clays, and other fine textured calcareous deposits, including Midlandian (Devensian) tills in central Ireland (Finch and Ryan, 1966; Conry et al., 1970), but it is only on the 'solid' clay formations that they commonly form major components of the soil patterns. Water regime studies reported by Robson and Thomasson (1977) indicate that they are generally wet for longer periods at comparable depths than the associated calcaric brown soils currently placed in the same soil series, chiefly Evesham and Hanslope (Clayden and Hollis, 1984). Most of them crack strongly in dry summers and the subsurface horizons show more or less well expressed vertic features (Chapter 3, p. 112).

As in the ground-water subgroup, the A horizon is in places thick enough and dark enough to qualify as a mollic epipedon. A very fine blocky or granular tilth develops in cultivated land as the surface dries, especially after exposure to winter frosts. The underlying B horizon is typically light olive brown or greyish brown and can be unmottled in the upper part, but greyish ped faces and/or matrix colours appear at 40 cm or less, accompanied by faint to distinct mottles of higher chroma. In variants intergrading to rendzinas, which occur locally in erosional sites, the B horizon is differentiated by structure only and is at least as calcareous as those below it. Data on two representative profiles are given below. Pelocalcaric brown soils of similar origin are exemplified by profiles 52 and 53 respectively (Chapter 5, pp. 192–4).

Profile 135 was described and sampled by D. Mackney and the author in a pipe-drained apple orchard on Gault clay in the Vale of Aylesbury, about 3 km from profile 133 (p.331). As is usually the case in the clay vales of southern and eastern England, the A and Bg horizons are developed in a superficial reworked layer (head) containing occasional stones of extraneous origin, which in this soil is almost completely decalcified. There are also indications from the particle-size data that some translocation of fine clay has occurred, but a thin section of the Bg horizon showed no evidence of this. The underlying BCg horizon is much more calcareous and contains segregations of secondary calcium carbonate. Judging from the air capacity figures quoted, the subsurface horizons transmit water very slowly when the fissures are closed by swelling of the clay. Under the climatic conditions of eastern England, however, as

Nicholson (1935) showed from repeated measurements in a similar soil under grass near Cambridge, their water contents may be reduced from the maximum winter values to depths exceeding two metres in dry summers, so creating a reservoir that can absorb up to 250 mm of rainfall before the drains begin to run. The cation-exchange capacity of the silicate clay fraction exceeds 50 me/100 g in three of the four horizons sampled and X-ray examination showed that they consist dominantly of interstratified mica-smectite with a high proportion of smectite layers. These results, coupled with the structural features indicative of pressures caused by swelling, support identification of the soil as a Vertic Haplaquept in the US Taxonomy (Avery and Bullock, 1977; Reeve *et al.*, 1980), though whether it meets the prescribed cracking requirement (Chapter 3, p. 112) is uncertain.

Profile 136 is sited in a convex upper slope on Chalky Boulder Clay in Northamptonshire. It represents a variant of the widespread Hanslope series (Clayden and Hollis, 1984) in which the brownish, partially decalcified Bw horizon present in typical Hanslope soils (Plate I.8) is missing and prominently mottled calcareous clay appears directly below the Ap, probably as a result of accelerated erosion since the land was first brought into cultivation.

Profile 135 (Avery and Bullock, 1977)

Location	: Cheddington, Buckinghamshire (SP 920178)
Climatic regime:	subhumid temperate
Altitude	: 100 m Slope: 2°
Land use	: grassed apple orchard 23 years old

Horizons (cm)

L/F	Grass root mat 2–3 cm thick, sharply separated from mineral soil.
0–18 Ap	Very dark greyish brown (10YR 3/2; 4/1 rubbed), very slightly stony, slightly calcareous clay (USDA clay); few very small to medium subangular flint and rounded quartzose fragments; moderate medium angular blocky within very coarse prisms bounded by 5 mm fissures; very firm; abundant fine roots; few earthworm channels; clear smooth boundary.
18–48 Bg	Olive-grey (5Y 5/2) and greyish brown (2.5Y 5/2), very slightly stony, slightly calcareous clay (USDA clay) with common very fine and fine yellowish brown (10YR 5/4–6) mottles; stones as above; very coarse prismatic parting to moderate medium and fine angular blocky; ped faces dark greyish brown to greyish brown (2.5Y 4–5/2) with many glazed (stress-oriented) coats; common, becoming few, fine and medium roots, mainly woody; few partly infilled earthworm channels; common fine ferri-manganiferous nodules; clear wavy boundary.
48–70 2BCg(k)	Light grey (5Y 6/1–2), stoneless, extremely calcareous clay (USDA silty clay) with common to many fine yellowish brown mottles; moderate coarse prismatic with light grey and grey (5Y 5/1) glazed faces, some inclined at acute angles to the horizontal; very firm; few fine roots; few earthworm channels; few very fine ferri-manganiferous nodules and coats and nodules of secondary CaCO$_3$; gradual boundary.
70–120 2BCgk	Light grey (5Y 7/1), stoneless, extremely calcareous clay (USDA silty clay to clay), with common fine yellowish brown (10YR 5/6) mottles; structure as above, becoming weaker (adherent) in lower part, with faint laminations within prisms; firm; roots rare; common nodules, soft concentrations and coats of secondary CaCO$_3$.

Analytical data *see* p.335.

Classification
  England and Wales (1984): typical calcareous pelosol; swelling-clayey material passing to clay or soft mudstone (Evesham series).
  USDA (1975): Vertic(?) Haplaquept; fine, montmorillonitic (calcareous), mesic.
  FAO (1974): Calcaric Gleysol; fine textured.

Analytical data for Profile 135

Horizon	Ap	Bg	2BCg(k)	2BCgk
Depth (cm)	0–18	18–48	48–70	70–122
sand 200 µm–2 mm %	5	1	1	1
50–200 µm %	2	1	1	3
silt 2–50 µm %	39	37	41	40
clay <2 µm %	54	61	57	56
<0.2 µm %	34	42	22	16
fine clay/total clay	0.62	0.69	0.39	0.29
organic carbon %	3.6	1.0	0.5	0.3
$CaCO_3$ equiv.				
<2 mm %	2.0	1.2	41	46
<2 µm %			4	21
dithionite ext. Fe %	0.8	1.0	0.6	0.6
pH (H_2O)	7.8	8.0	8.5	8.5
(0.01M $CaCl_2$)	7.2	7.8	8.2	8.0
clay <2 µm				
CEC (me/100 g)	62	60		54
K_2O %	2.1	2.0		2.3
bulk density (g cm^{-3})		1.19	1.41	
packing density (g cm^{-3})		1.73	1.92	
retained water capacity (% vol.)		51	46	
available water capacity (% vol.)		16	15	
air capacity (% vol.)		4	1	

Profile 136 (Reeve, 1978)

Location: Norton, Northamptonshire (SP 607621)
Climatic regime: subhumid temperate
Altitude: 137 m Slope: 3° convex; ridge and furrow microrelief.
Land use: long-term grass

Horizons (cm)

0–16 Ah
Dark greyish brown (2.5Y 4/2), very slightly stony, slightly calcareous clay (USDA clay loam); few very small to large quartzose pebbles and subangular flints with few chalk and rare limestone and ironstone fragments; moderate fine subangular and angular blocky; firm; many fine roots; abrupt wavy boundary.

16–44 Bg
Grey (5Y 5/1) and yellowish brown (10YR 5/8) prominently mottled, very slightly stony, very calcareous clay (USDA clay loam); stones as above; moderate medium prismatic breaking in places to medium and fine angular blocky with some greyish brown (2.5Y 5/2) faces, very firm; common fine roots; gradual smooth boundary.

44–83 BCg1
Grey (2.5Y 6/0) and brown to dark brown (10YR 4/3) prominently mottled, slightly stony, very calcareous clay (USDA clay loam); stones as above; moderate coarse prismatic with grey (2.5Y 6/0) faces and common fine yellowish brown (10YR 5/6) mottles internally; extremely firm; few roots, mainly on ped faces; gradual smooth boundary.

82–120 BCg2
Grey to dark grey (2.5Y 6–5/0) and brown to dark brown (10YR 4/3) prominently mottled, very slightly stony, very calcareous clay (USDA clay loam to clay); stones as above; massive; extremely firm.

Analytical data for Profile 136

Horizon	Ah	Bg	BCg1	BCg2
Depth (cm)	0–16	23–32	60–70	105–115
sand 200 µm–2 mm %	13	13	10	9
60–200 µm %	15	14	13	13
silt 2–60 µm %	33	35	38	38
clay <2 µm %	39	38	39	40
organic carbon %	3.4	0.8		
CaCO$_3$ equiv.				
<2 mm %	3.5	19	20	20
<2 µm %		3	3	3
pH (H$_2$O)	7.7	8.2	8.3	8.2
(0.01M CaCl$_2$)	7.5	7.8	7.9	7.9

Classification
England and Wales (1984): typical calcareous pelosol; clayey chalky drift (Hanslope series).
USDA (1975): Aeric Haplaquept; fine, mixed (calcareous), mesic.
FAO (1974): Calcaric Gleysol; fine textured.

7.4.5 Humic Calcaric Gley Soils

These are calcaric gley soils with humose or peaty topsoils, occurring locally in low lying sites which are or were waterlogged to the surface for much of the year. Those in which the water table is perennially close to the surface have dominantly grey subsurface horizons and are typified by the Clowater series mapped on calcareous Midlandian till in east-central Ireland (Conry and Ryan, 1967). The main areas in England and Wales where they are extensive enough to have been identified and mapped as soil series are the 'skirtlands' surrounding the East Anglian Fens (Hodge and Seale, 1966; Seale, 1975a; Hodge et al., 1984), most of which had a peat cover more than 40 cm thick before pump assisted arterial drainage was installed. Some of these soils, exemplified by the Reach series on marly chalk, are extremely calcareous and have been classed as humic gleyic rendzinas in recent years (Avery, 1980; Clayden and Hollis, 1984). Others in less calcareous fluvial and head deposits have a mottled greyish horizon partially or completely leached of carbonates immediately beneath the topsoil, followed by a generally more calcareous horizon showing evidence of iron enrichment, either in a continuous band (Bgf or BCgf) or as a marked concentration of ochreous mottles. Data on two examples of these calcareous 'skirt soils' are reproduced on pp.337–9.

Profile 137 is in loamy chalky head over marly chalk in an area near the eastern edge of the Fens where the thickness and composition of the drift are variable over short distances. Its sand fraction is mainly derived from Devensian river deposits, remnants of which have been mixed with the underlying chalk by cryoturbation. The pale colour of the subsurface horizons and the paucity of prominent ochreous mottles are attributable to the large amounts of calcium carbonate and the small quantities of mobilizable iron they contain.

Profile 138 is on Jurassic (Oxford) clay near Chatteris, Cambridgeshire. The description and samples were derived from an undisturbed 15 cm core taken from a patch of rough grass. Adjoining ploughed ground was waterlogged at the time of sampling (13 April 1970) and water seeped into the base of the hole left by the core, but overlying horizons were unsaturated (moist), suggesting that the surface wetness nearby was induced by an impervious plough sole. Relatively large sand/silt ratios, together with occasional small flints and sandy inclusions, show that the upper 70 cm of the soil are in clayey head composed largely of locally derived material. Below the slightly calcareous subsurface horizon there is a thin, markedly more ferruginous Bgf with a sharp increase in $CaCO_3$ content at its lower boundary. This is followed by grey and light-olive brown mottled BCgk horizons, developed in the lower part of the clayey drift and in the underlying Oxford Clay, which resemble those in 'upland' soils of similar origin represented by profile 135 (p.334). The subsurface horizon is designated EBg on the basis that it may have lost iron by downward translocation. Hodge and Seale (1966) present evidence that the characteristic sequence of an apparently leached horizon followed by a ferruginous (*raseneisen*) horizon is related to the former presence of a peaty surface layer. It also seems that the ferruginous horizon generally occurs at approximately the same depth as the current winter water table, suggesting that it may have formed since the drainage was improved.

Profile 137 (Seale, 1975a)

Location	:	Hockwold, Norfolk (TL 690881)
Climatic regime:		subhumid temperate
Altitude	:	1–5 m Slope: <1°
Land use	:	arable (ploughed)

Horizons (cm)

0–30 Ap
Black (10YR 2/1) very calcareous humose clay loam (USDA sandy clay loam) containing rare small flint fragments; fine granular and very friable in upper 7.5 cm, fine subangular blocky and friable below; few fine roots; abrupt smooth boundary.

30–53 BCg
Light grey (2.5Y 7/2), very slightly stony, very calcareous sandy loam (USDA sandy loam) with common large pale yellow (2.5Y 7/3) and few fine and very fine yellow (2.5Y 7/6) mottles; few small flint fragments; pale brown sand band at 43–46 cm; massive; firm to friable; roots rare; dark grey and black tongues of A horizon material infilling channels; gradual boundary.

53–91+ 2Cg
White (2.5Y 8/1), stoneless, extremely calcareous silty clay loam (USDA silty clay loam) with common large very pale brown (10YR 7/4) mottles elongated vertically; massive; very firm; rare dead fibrous and woody roots (Chalk Marl).

Analytical data *see* below.

Classification
England and Wales (1984): calcareous humic gley soil; fine loamy drift over silty or clayey material passing to soft chalk (Blackdyke series, formerly Wilbraham).
USDA (1975): Typic Haplaquoll; fine or coarse-loamy, carbonatic, mesic.
FAO (1974): Mollic Gleysol; medium textured.

Analytical data for Profile 137

Horizon	Ap	BCg	BCg	2Cg	2Cg
Depth (cm)	0–30	30–43	43–53	53–71	71–91
sand 200 µm–2 mm %	29	27	30	10	
50–200 µm %	22	31	34	9	
silt 2–50 µm %	25	24	21	52	
clay <2 µm %	24	18	15	29	
organic carbon %	6.6	0.4			
$CaCO_3$ equiv. %	31	37	33	75	73
dithionite ext. Fe %	0.9	0.4	0.4	0.3	0.2
pH (H_2O)	7.6	8.5	8.5	8.5	8.6
(0.01M $CaCl_2$)	7.2	7.7	7.6	7.5	7.6

Profile 138 (Seale, 1975b)

Location	:	Pidley, Cambridgeshire (TL 340805)
Climatic regime:		subhumid temperate
Altitude	:	3 m Slope: 1°
Land use and vegetation	:	rough grass bordering arable land

Horizons (cm)

0–29 Ap — Black to very dark brown (10YR 2/1.5) slightly calcareous humose clay (USDA clay) with common fine to medium very dark greyish brown (10YR 3.5/2) mottles; few medium and small subangular flint fragments in top 13 cm; medium granular to very fine subangular blocky and friable to 11 cm, medium to coarse subangular blocky and firm below; many fine and few medium fleshy roots; few very fine shell fragments; clear irregular boundary with tongues extending 3 cm downwards.

29–40 ABg — Dark to very dark greyish brown (2.5Y 3.5/2–10YR 3/2) slightly calcareous clay (USDA clay) with few to common very fine brown (10YR 4/3) and few extremely fine yellowish red (5YR 4/8) mottles; rare flint fragments; medium prismatic parting to strong medium subangular blocky; firm; few fine and fleshy roots; occasional shell fragments; clear smooth boundary.

40–48 EBg — Grey to greyish brown (5Y 5/1–2.5Y 5/2), stoneless, slightly calcareous clay (USDA clay) with common fine brown and yellowish brown (10YR 5/3–4) mottles; medium prismatic parting to strong medium angular blocky; firm to very firm; some worm holes; few roots; occasional shell fragments; clear wavy boundary.

48–53 Bgf — Grey to light olive brown (5Y 5/1–2.5Y 5/3) slightly calcareous clay (USDA clay) with many medium and large dark brown (7.5YR 4/4) and few yellowish brown (10YR 5/8) mottles; rare flint fragments; moderate medium prismatic; very firm; some partly infilled worm holes; few roots, some dead; brown parts much more calcareous than grey parts; gradual boundary.

53–70 BCgk — Light olive brown (2.5Y 5/3) and grey (5Y 5/1), stoneless, very calcareous clay (USDA clay) with common brown and yellowish brown mottles and rare small (<1 cm) sandy clay loam inclusions; strong medium prismatic; very firm; few worm holes and roots as above; few soft secondary calcium carbonate concentrations; clear smooth boundary.

70–92 2BCgk — Grey (5Y 5/1), stoneless, very calcareous clay to silty clay (USDA silty clay) with common fine and medium light olive brown (2.5Y 5/3–4) mottles, some with yellowish brown centres; wet (water seeping into base of borehole); strong coarse prismatic with grey faces; few worm holes and roots as above; few calcareous concretions; occasional small fossil fragments.

Analytical data for Profile 138

Horizon	Ap	AB	EBg	Bgf	BCgk	2BCgk	2BCgk
Depth (cm)	0–27	30–40	40–48	48–53	53–70	70–82	82–92
sand 200 μm–2 mm %	9	6	7	6	7	2	4
60–200 μm %	12	11	11	6	12	5	8
silt 2–60 μm %	29	32	33	32	32	43	46
clay <2 μm %	50	51	49	56	49	50	42
organic carbon %	5.9	1.9					
$CaCO_3$ equiv. %	2.0	1.6	1.3	2.7	20	37	47
dithionite ext. Fe %	1.5	1.0	1.4	3.5	2.1	1.7	1.3
pH (H_2O)	7.4	7.8	7.9	8.1	8.1	7.9	7.8
(0.01M $CaCl_2$)	6.9	7.2	7.4	7.5	7.5	7.4	7.3

Classification
England and Wales (1984): calcareous humic gley soil; clayey material passing to clay or soft mudstone (Peacock series).
USDA (1975): Typic (or Vertic) Haplaquoll; fine, montmorillonitic (calcareous), mesic.
FAO (1974): Mollic Gleysol; fine textured.

7.5 Orthic Gley Soils

7.5.1 General Characteristics, Classification and Distribution

The gley soils grouped here as *orthic* resemble the analogous orthic brown soils in having non-calcareous loamy or clayey subsurface horizons (hydromorphic cambic horizons) that have formed in deposits other than recent alluvium and contain little or no more clay than the surface horizons. Surface-water and ground-water gley soils as previously conceived in the British Isles (Avery, 1973, 1980; Soil Survey of Scotland, 1984; Gardiner and Radford, 1980) are included, though variants of the former type (Plates III.8 and IV.2) are much the more extensive. Those of both types are mainly Aquepts (Haplaquepts, Fragiaquepts or Humaquepts) in the US Taxonomy and Gleysols (Eutric, Dystric or Humic) in the FAO (1974) system.

Judging from the published soil association maps, soils of this group cover about 10 per cent of England and Wales and around 20 per cent of Scotland and Ireland. They occur at altitudes ranging from near sea level to the oro-arctic zone and, like the orthic brown soils, are most extensive on parent materials derived from consolidated non-calcareous rocks in humid and perhumid areas. Although many of them are moderately acid in the absence of added lime, those in water-receiving sites tend to have a higher base status than associated freely draining soils in similar deposits (Glentworth and Dion, 1949). Variants derived from ultrabasic rocks have been set apart as magnesian gleys by the Soil Survey of Scotland (1984); these are characterized by high exchangeable magnesium/calcium ratios and near-neutral reaction, particularly in the lower horizons.

The morphology of these soils varies considerably, depending on their composition and permeability, and on current or former water-table levels. Where prolonged wetness is confined to the lower part of the profile, the topsoil normally contains moderate amounts of organic matter and overlies a grey and brown mottled Bg horizon with more or less well developed blocky or prismatic structure. This usually starts directly below the A horizon but is separated from it by a brownish Bw in intergrades to orthic brown soils. An Eg or Eag horizon can also intervene where a thin organic surface layer has formed under acidophile vegetation. Horizons below the solum commonly appear less modified by gleying than those within it, especially when they are slowly permeable, but some profiles of ground-water type have a dominantly grey Cg (or CG), a markedly ferruginous horizon (BCgf or Cgf) in which mobilized iron has precipitated, or both.

Where the water table is or was close to the surface for long periods, the topsoils are humose or peaty and the subsurface horizons more strongly gleyed. Unlike the humic gley soils grouped as sandy, alluvial and calcaric, those of the orthic group are widespread in upland areas where climatic wetness and low summer temperatures have led to formation of peaty surface horizons on slopes and interfluves as well as in localized depressional sites (Figure 7.7). A greyish weakly structured eluvial horizon (Eag or Eg) depleted of iron appears immediately beneath the dark coloured topsoil in many of these soils and can pass directly into a compact substratum in which the original colour of the parent material becomes evident, but is more often underlain by a strongly gleyed B horizon. The presence of a structurally differentiated Bg with grey ped faces and ochreous intraped mottles normally implies that the upper part

Figure 7.7: The upper Derwent valley from Carterway Heads, Northumberland. Medium to fine textured till derived from Carboniferous shales and sandstones mantles most of this landscape, giving rise to slowly permeable gley soils which have peaty or humose surface horizons on the higher ground in the distance (Copyright Geoffrey N. Wright).

of the rooting zone is, or has been, subject to significant seasonal changes in water content. Under wetter conditions, such horizons are weakly expressed or absent and commonly either overlie or give place to massive grey, bluish or greenish horizons in which ferruginous segregations, if present, occur chiefly as 'pipes' around old root channels (Plate IV.1). Intensely gleyed horizons of this kind may rest directly on bedrock or pass downwards into unconsolidated material with base colours or mottles of higher chroma. A dominantly ochreous horizon (BCgf or Cgf) can also occur at some depth, particularly in drained phases, where it frequently relates to the lowered water table so produced.

Although soils of this group typically show little evidence of clay translocation, some have common argillans in a lower (BCgt) horizon containing less clay than an overlying Bg. As in related orthic brown soils, this condition may reflect a lithological discontinuity, the A and Bg horizons having formed in an initially more clayey surface layer, or it may result from accumulation of clay in the upper horizons by breakdown of coarser particles, sufficient to mask any loss of eluviation. Conversely, other soils included in the group have a lithologically distinct, finer textured substratum which fails to qualify as an argillic B.

The main factors governing land capability on these soils are their texture and permeability, the climatic conditions under which they occur, and the actual water regime as influenced by the incidence and effectiveness of artificial drainage. Lowland variants of ground-water type, exemplified by profile 139 (p.342), present only minor limitations for agricultural use when adequately drained, but those with slowly permeable subsurface horizons are more difficult to drain effectively, especially where the climate is humid and textures are fine. Subsurface horizons of the undrained soils remain saturated or nearly so for more than half the year under these conditions and only limited improvement is achievable by conventional drainage measures (Figure 7.8) because water movement is mainly confined to the surface horizons and

Figure 7.8: Drainage on till-derived orthic gley soils in Wales (Photo by T.R.E. Thompson).

amounts of 'gravitational' water that can be removed from those below are proportionately very small. Land of this type is mainly used for grass production. Potential yields are high under a humid temperate climatic regime because growth is rarely restricted by drought, but full use of the crop is inhibited by the high risk of poaching or compaction by machinery, which lead in turn to deterioration in drainage, sward composition and yield.

On humic (peaty) variants at higher altitudes, grazing potential is more severely limited by wetness and low bearing strength and is further restricted by shorter growing seasons. Much of this land remains under semi-natural vegetation dominated by purple moor-grass (*Molinia caerulea*), mat grass (*Nardus stricta*) or heather (*Calluna vulgaris*), but considerable areas in both Great Britain and Ireland have been afforested during the last 50 years, chiefly with Sitka spruce, Norway spruce or lodgepole pine (Figure 7.9). In order to suppress competition by the indigenous vegetation, improve rooting conditions and help reduce premature windthrow, which is a serious hazard on such soils, the Forestry Commission recommends mouldboard ploughing up and down slope with connecting cross drains preparatory to planting (Taylor, 1970). Establishment of the young trees is also aided by application of phosphatic fertilizers.

Key to Subgroups

1. Orthic gley soils with a humose or peaty topsoil (p. 107) and
   (a) a slowly permeable subsurface horizon (p. 110) and no CG, BCg or Cg with dominant chroma of 1 or less and ferruginous deposits along root channels; or
   (b) gleyed E and/or B and less gleying in the underlying horizon; or both.
       Humic stagno-orthic gley soils
2. Other orthic gley soils with a humose or peaty topsoil (p. 107)
       Humic orthic gley soils
3. Other orthic gley soils with a slowly permeable subsurface horizon (p. 110)
   3.1 With pelo-characteristics (p. 112)
       Pelo-orthic gley soils
   3.2 Without pelo-characteristics
       Stagno-orthic gley soils
4. Other orthic gley soils
       Ground-water orthic gley soils

7.5.2 Ground-water Orthic Gley Soils

These are the non-humic orthic gley soils of ground-water type. Classed as typical cambic gley soils in England and Wales (Avery, 1973, 1980) and as non-calcareous ground-water gleys by the Soil Survey of Scotland (1984), they are widely distributed in places where medium to coarse textured deposits overlie impervious substrata below the rooting zone, but cover a much smaller total area than those of the following two subgroups. Although the A and B horizons are typically loamy, many of these soils have periodically or permanently waterlogged sandy or sandy–gravelly layers within 80 cm depth and some have clayey topsoils. As indicated above, horizons below the solum usually, but not always, show marked signs of gleying. Data on two representative profiles from England and Wales are given below.

Profile 139 is sited in a river-terrace deposit of Devensian age near Lichfield, Staffordshire. The loamy upper layer in which the A and Bg horizons are developed overlies a sandy–gravelly substratum (2Cg) which was waterlogged at the time of sampling (May 1979). Except that the subsurface horizons are gleyed from the base of the Ap downwards, the soil resembles profile 60 (p. 204) in similar deposits. It is classed as a Humaquept in the US system (FAO Humic Gleysol) on the assumption that the thick dark Ap is an umbric rather than an anthropic epipedon (Table 2.11, p. 68). As the organic carbon content is only 1.2 per cent, however, the implication that the topsoil is rich in organic matter is clearly misleading. Thin sections confirmed the field observation that argillans are present in the Bg and BCg horizons, despite their relatively coarse texture, and the air capacity and packing density data indicate that they are moderately permeable. In some gleyed soils in terrace deposits there is a more abrupt change in pore size distribution at the upper boundary of a coarse textured substratum. As this presents a barrier to downward water movement (Miller, 1973; Clothier et al., 1978), gleying in the finer textured upper layer can reflect periodic stagnation above the boundary as well as, or rather than, saturation by ground water rising from below.

Profile 140 is under old pasture on gently sloping ground below a steep ridge south of Tavistock, Devon. The parent material is slaty Head overlain by a nearly stonefree superficial layer (0–38 cm) which is possibly recent colluvium. As the structure and consistence of the subsurface horizons indicate that they are at least slightly porous (Table 3.7, p. 98), wetness is attributed at this site to downslope seepage above the impervious slate bedrock. Water levels in boreholes recorded over several periods in soils of the same series further east in Devon (Harrod et al., 1976) showed that the water table fluctuated between the surface and about 60 cm depth.

Profile 139 (Ragg et al., 1984; Hollis, 1985)

Location : Hademore, Staffordshire (SR 179084)
Climatic regime: subhumid temperate
Altitude : 57 m   Slope: <1°
Land use : ley grass
Horizons (cm)

0–31 Ap	Dark brown (10YR 3/3 broken and rubbed; 10YR 5/3 dry) very slightly stony sandy clay loam (USDA fine sandy loam); few medium quartzite pebbles; moderate medium subangular blocky; friable; inclusions of mottled material from horizon below; many fine roots; sharp smooth boundary.
31–52 Bg1	Yellowish brown (10YR 5/6) very slightly stony sandy silt loam (USDA loam) with very many medium light grey (10YR 7/2) mottles; stones as above; weak medium angular blocky with dark greyish brown (10YR 4/2) faces; friable; many fine roots; few clay coats; abrupt smooth boundary.
52–66 Bg2(t)	Yellowish brown (10YR 5/6) slightly stony medium sandy loam (USDA sandy loam to fine sandy loam) with

Analytical data for Profile 139

Horizon	Ap	Bg	B(g)	BCg	2Cg
Depth (cm)	0–31	31–52	52–66	66–84	84–107
sand 600 μm–2 mm %	4	2	7	7	11
200–600 μm %	28	16	31	27	43
60–200 μm %	20	21	25	31	30
silt 2–60 μm %	29	44	26	27	13
clay <2 μm %	19	17	11	8	3
organic carbon %	1.2	0.3	0.2		
pH (H$_2$O)	5.7	6.9	7.2	7.2	7.4
(0.01M CaCl$_2$)	5.0	6.4	6.7	6.7	6.8
bulk density (g cm^{-3})	1.36	1.50	1.40	1.62	
packing density (g cm^{-3})	1.48	1.60	1.47	1.66	
retained water capacity (% vol.)	35	31	35	30	
available water (% vol.)	16	12	17	14	
air capacity (% vol.)	12	11	10	7	

66–84 BCg(t)   Light grey to grey (10YR 6/1) moderately stony medium sandy loam (USDA sandy loam to fine sandy loam) with common medium strong brown (7.5YR 5/6) mottles; stones as above; massive to single grain; few clay coats; few irregular ferri-manganiferous nodules; abrupt smooth boundary.

84–107 2Cg   Grey (5YR 5.1) very stony loamy medium sand (USDA loamy sand to sand); abundant small quartzite pebbles; wet; single grain.

Analytical data *see* above.

Classification
   England and Wales (1984): typical cambic gley soil; coarse loamy drift with siliceous stones (Quorndon series).
   USDA (1975): Aeric Humaquept?; coarse-loamy, mixed or siliceous (non-acid), mesic.
   FAO (1974): Humic Gleysol; medium textured.

Profile 140 (Hogan, 1977)

Location   : Whitchurch, Devon (SX 476716)
Climatic regime: humid temperate
Altitude   : 117 m   Slope: 3°
Land use   : long-term grass
Horizons (cm)

0–20 Ahg   Grey to greyish brown (10YR 5/1–2) very slightly stony clay loam to silty clay loam (USDA clay loam) with common fine dark reddish brown (5YR 3/3) mottles, mainly along root channels; few small and medium subangular quartz and platy slate fragments; strong fine subangular blocky; friable; many fine roots; gradual boundary.

20–38 Bg   Grey (5Y 5/1) very slightly stony silty clay loam (USDA silty clay loam) with common medium and coarse light grey (10YR 7/2) and few fine strong brown (7.5YR 5/6) mottles and very dark grey (10YR 3/1) inclusions; stones as above; moderate medium and fine subangular blocky; friable; common fine roots; clear boundary marked by increase in stoniness.

38–63 2Bg   Light brownish grey (2.5Y 6/2) moderately to very stony clay loam (USDA clay loam) with many medium strong brown and common medium pale brown (10YR 6/3) mottles; many to abundant soft platy slate and few hard blocky slate and subangular quartz fragments; moderate medium angular blocky grading

Figure 7.9: Coniferous plantations on humic (peaty) stagno-orthic gley soils at Spithopehead, Redesdale, Northumberland (Cambridge University Collection: copyright reserved).

	in places to prismatic; common fine pores; friable; few fine roots; gradual boundary.
63–76+ 2Cg	Platy and blocky green slate and quartz fragments with interstitial loamy material similar in colour to above.

Analytical data

Horizon	Ahg	Bg	2Bg
Depth (cm)	0–20	20–38	38–63
sand 600 µm–2 mm %	8	6	8
200–600 µm %	7	5	10
60–200 µm %	6	5	8
silt 2–60 µm %	46	50	42
clay <2 µm %	33	34	32
organic carbon %	5.3	3.5	0.4
dithionite ext. Fe %	1.5	1.3	3.2
pH (H$_2$O)	5.3	5.8	6.4
(0.01M CaCl$_2$)	4.9	5.5	5.7

Classification
England and Wales (1984): typical cambic gley soil; fine loamy material over lithoskeletal mudstone and sandstone or slate (Yeolland-park series).
USDA (1975): Typic Haplaquept; loamy-skeletal, mixed (non-acid), mesic.
FAO (1974): Eutric (or Dystric) Gleysol; medium textured.

7.5.3 Stagno-orthic Gley Soils

These soils differ from those of the preceding subgroup in having a slowly permeable sub-surface horizon (Chapter 3, p. 110) within 80 cm depth. They have been classed as cambic stagnogley soils in England and Wales (Avery, 1973, 1980) and as non-calcareous surface-water gleys or brown forest soils with gleying by the Soil Survey of Scotland (1984). Soils of this subgroup are also widespread in Ireland, forming major components of associations 11, 21, 22 and 25 on the national soil association map (Gardiner and Radford, 1980). Both there and in Great Britain, they are mostly developed in more or less stony Devensian tills and compact head deposits derived from consolidated Palaeozoic rocks which yield appreciable amounts of clay on weathering. Typical profiles in such deposits have a grey and brown mottled Bg horizon, passing into a less weathered and more compact BC which often shows fragipan features (BCgx). Similar fine loamy or fine silty soils occur locally on slowly permeable pre-Quaternary rocks such as siltstone and silty shale.

Most of these soils conform both morphologically and hydrologically to the surface-water gley concept. Gleying is most pronounced in the upper part of the profile, especially in originally reddish (haematitic) materials, and seasonal wetness is attributable to impeded percolation of water which originated from precipitation *in situ* or lateral flow through the relatively permeable surface horizons from higher positions. In some intergrading situations, however, the impeding layer is of limited thickness and underlying horizons, which may or may not be strongly gleyed, are saturated by ground water retained by impervious bedrock at greater depths.

Of the three representatives profiles recorded on pp. 346–8, the first two are located 5 km apart in undulating country between Llangadog and Llandovery, Dyfed. Both are in gentle even slopes underlain by compact very stony till which in profile 141 (Plate III.8) is mainly derived from greyish Lower Palaeozoic rocks and in 142 from red Devonian siltstones and fine grained sandstones. Gleying is pronounced in the upper horizons of both soils but is barely evident in the very dense (fragic) Cx horizons, which have platy cleavage and vesicular pores believed to have originated under periglacial conditions (FitzPatrick, 1956; Van Vliet-Lanoë, 1985). In the first profile, sited in pasture reseeded after ploughing, the Ap horizon directly overlies the prismatic Bg. The second, under unploughed grassland at a higher altitude, has a thin humose Ah succeeded by a light brownish grey Eg horizon from which iron has evidently been removed. Stone contents decrease steadily from the C horizon to the surface in each case, and thin sections of similar soils clearly show progressive upward replacement of rock fragments by fine material, as in the related orthic brown soils. Water level measurements in auger holes (Robson and Thomasson, 1977) showed that, in humid to perhumid areas with average rainfall ranging from some 1,300 to 1,600 mm, undrained pasture soils typified by profile 141 (Cegin series) are waterlogged to within 40 cm of the surface for more than 200 days in most years. The duration of surface wetness can be reduced by pipe drainage, but benefits are most evident in dry years and very careful management is needed to maintain adequate permeability in the horizons above drain level.

Profile 143 from East Lothian is in fine loamy till containing materials derived from Carboniferous sedimentary rocks and lavas. It resembles the first (Cegin) profile in texture but has a thicker solum and a higher base status, and is less strongly gleyed as judged by the extent of colours of chroma 2 or less, which in this soil are mainly confined to ped faces and conducting channels. Slowly draining soils of this kind, most of which qualify as gley soils by the criteria used here and as Aquepts or Aqualfs in the US Taxonomy, have been identified and mapped as brown forest soils with gleying by the Soil Survey of Scotland (1984). In the relatively dry eastern lowlands south of the Highland line, where soils grouped under this heading occupy considerable areas, they support a wide range of arable crops and form some of the most productive land in the country, much of which was reclaimed and drained during or since the eighteenth century.

Profile 141 (Wright, 1981)

Location	:	Llangadog, Dyfed (SN 730270)
Climatic regime:		humid temperate
Altitude	:	122 m Slope: 3° straight
Land use	:	recently reseeded grass

Horizons (cm)

0–22 Apg	Greyish brown (10YR 5/2) very slightly stony silty clay loam (USDA silty clay loam) with common fine yellowish red (5YR 4/6) mottles; few small angular sandstone fragments; moderate fine subangular blocky; friable; many fine roots; clear smooth boundary.
22–47 Bg	Light brownish grey (10YR 6/2) slightly stony silty clay loam (USDA silty clay loam) with many fine strong brown (7.5YR 5/6) mottles; common small and few medium and large subangular sandstone fragments; moderate medium prismatic; firm; common fine roots; clear smooth boundary.
47–66 BCg	Light grey to grey (10YR 6/1) moderately stony silty clay loam (USDA silty clay loam) with many medium strong brown mottles; stones as above, mainly medium, together with soft shale fragments; moderate coarse prismatic; firm; few roots; clear irregular boundary.
66–100 Cx	Brown (10YR 5/3) very stony clay loam (USDA loam) with 5-8 cm tongues of above horizon extending downwards (10 per cent of upper 15 cm); stones as above with a few red sandstone fragments; massive, with platy cleavage evident in places; 0.5 per cent vesicular macropores; very firm and brittle; no roots; many very fine and medium black manganiferous and coats on stones and fracture planes.

Analytical data *see* below.

Classification

England and Wales (1984): cambic stagnogley soil; fine silty drift with siliceous stones (Cegin series).

USDA (1975): Aeric Fragiaquept; fine-silty, mixed (acid), mesic.

FAO (1974): Dystric Gleysol; medium textured (fragipan phase).

Analytical data for Profile 141

Horizon	Ahg	Bg	BCg	Cx
Depth (cm)	0–22	22–47	47–66	66–100
sand 600 µm–2 mm %	1	1	6	18
200–600 µm %	2	2	4	6
60–200 µm %	6	7	5	6
silt 2–60 µm %	62	59	53	44
clay <2 µm %	29	31	32	26
<0.2 µm %	7	9	8	7
fine clay/total clay	0.24	0.29	0.25	0.27
organic carbon %	3.9	0.8	0.2	0.2
dithionite ext. Fe %	1.6	1.9	2.0	1.6
pH (H$_2$O)	5.4	5.8	5.9	5.4
(0.01M CaCl$_2$)	4.9	5.1	4.8	4.5
bulk density (g cm^{-3})	0.87	1.45	1.65	2.00[a]
packing density (g cm^{-3})	1.13	1.72	1.93	2.23[a]
retained water capacity (% vol.)	49	40	38	24[a]
available water (% vol.)	25	9	14	8[a]
air capacity (% vol.)	17	6	<1	1[a]

[a] determined on block from 85–90 cm

Profile 142 (Wright, 1981)

Location : Llanddeusant, Dyfed (SN 748233)
Climatic regime: perhumid temperate
Altitude : 170 m   Slope: 3° straight
Land use : long-term grass
Horizons (cm)

0–7
Ah
Dark brown (7.5YR 3/2) stoneless humose silty clay loam (USDA silt loam); strong fine subangular blocky; friable; many fine roots; abrupt smooth boundary.

7–25
Eg
Light brownish grey (10YR 6/2) very slightly stony silty clay loam (USDA silt loam) with many fine light grey (10YR 7/2) and common very fine strong brown (7.5YR 5/6) mottles; few very small and small subrounded red sandstone fragments; weak coarse prismatic; firm; common fine roots; clear smooth boundary.

25–39
Bg
Pinkish grey (7.5YR 6/2) slightly stony silty clay loam to clay loam (USDA silt loam) with many fine strong brown and reddish brown (5YR 4/4) mottles; moderate coarse prismatic; firm; few fine roots; clear wavy boundary.

39–51
BC(g)
Reddish brown (2.5YR 4/4) very stony sandy silt loam (USDA silt loam) with common fine pinkish grey (5YR 7/2) and strong brown (7.5YR 5/6) mottles; abundant medium subangular red sandstone fragments; very coarse prismatic to massive; firm; few roots; clear smooth boundary.

51–80
Cx
Reddish brown (2.5YR 4/4) very stony clay loam (USDA loam); stones as above; massive with tendency to platy cleavage; 0.5 per cent vesicular macropores; extremely firm; no roots; common black manganiferous coats on stones.

Analytical data *see* below

Classification
England and Wales (1984): cambic stagnogley soil, reddish-fine silty drift with siliceous stones (Fforest series).
USDA (1975): Aeric Fragiaquept; fine-silty, mixed (acid), mesic.
FAO (1974): Dystric Gleysol; medium textured (fragipan phase).

Analytical data for Profile 142

Horizon	Ah	Eg	Bg	BCg	Cx
Depth (cm)	0–7	7–25	25–39	39–51	51–80
sand 600 µm–2 mm %	<1	<1	1	7	12
200–600 µm %	1	2	3	5	5
60–200 µm %	15	11	16	10	9
silt 2–60 µm %	59	61	58	61	51
clay <2 µm %	25	26	22	17	23
organic carbon %	12	2.4	0.7	0.3	0.1
dithionite ext. Fe %	0.5	0.3	1.8	2.2	
pH (H_2O)	4.4	4.9	5.1	5.3	5.6
(0.01M $CaCl_2$)	4.2	4.3	4.3	4.4	4.8
bulk density (g cm^{-3})	0.51	1.09	1.38		2.11[a]
packing density (g cm^{-3})	0.74	1.32	1.57		2.32[a]
retained water capacity (% vol.)	52	47	40		21[a]
available water (% vol.)	33	21	20		7[a]
air capacity (% vol.)	26	12	8		1[a]

[a] determined on block from 65–75 cm

Profile 143 (Ragg and Futty, 1967; Ragg and Clayden, 1973)

Location	:	North Berwick, East Lothian (NT 583834)
Climatic regime:		subhumid temperate
Altitude	:	40 m Slope: <1°
Land use	:	arable

Horizons (cm)

0–30 Ap — Dark brown to brown (10YR 4/3) very slightly stony clay loam (USDA sandy clay loam); medium blocky; friable; many roots; sharp boundary.

30–55 Bg — Brown (10YR 5/3) slightly stony clay loam (USDA clay loam) with many fine strong brown (7.5YR 5/8) mottles; strong coarse prismatic with dark greyish brown (10YR 4/2) faces; firm; common roots; clear boundary.

55–85 BCg(x)1 — Dark brown to brown (7.5YR 4/2) slightly stony clay loam (USDA clay loam) with many fine strong brown mottles and grey (10YR 5/1) streaks; moderate coarse prismatic with grey faces; very firm; few roots; clear boundary.

85–122 BCg(x)2 — Dark brown to brown (7.5YR 4/2) slightly stony clay loam (USDA clay loam) with faint brown (7.5YR 5/4) mottles and brown (7.5YR 5/2) streaks; stones include coal and black shale fragments; massive; very firm; no roots.

Analytical data *see* below.

Classification
  Scotland (1984): brown forest soil with gleying (Kilmarnock series).
  England and Wales (1984): cambic stagnogley soil; fine loamy drift with siliceous stones (Brickfield series).
  USDA (1975): Aeric Haplaquept; fine-loamy, mixed (non-acid), mesic.
  FAO (1974): Gleyic Cambisol; medium textured.

7.5.4 Pelo-orthic Gley Soils

These are non-calcareous non-humic gley soils derived from argillaceous rocks and fine textured glacial and glaciolacustrine deposits. They have clayey Bg horizons that do not qualify as argillic and differ in this respect from the corresponding luvic gley soils (p. 375). Slowly permeable clayey soils of both kinds are grouped as pelo-stagnogley soils in the England and Wales classification on the basis that they behave similarly for most practical purposes, and also because clay-illuvial horizons are more difficult to distinguish consistently in clay-rich materials, particularly if they are gleyed, than in those of loamy texture (Soil Survey Staff, 1975, pp. 25–26; Murphy, 1984). The division is made here in order to conform with the classification adopted for brown soils and with closely comparable separations in the US and FAO systems. Soils of the two subgroups are often closely associated in the field, however, their distribution depending on the presence or absence of thin relatively coarse textured surface layers of geological origin as well as on the balance between formation of clay by

Analytical data for Profile 143				
Horizon	Ap	Bg	BCg(x)1	BCg(x)2
Depth (cm)	10–20	35–45	65–75	112–122
silt^a 2–50 µm %	25	26	27	27
clay^a <2 µm %	27	30	30	28
organic carbon %	1.8	0.7		
TEB (me/100 g)	13.3	17.5	20.4	18.7
CEC (me/100 g)	16.6	18.2	20.4	18.7
% base saturation	80	96	sat.	sat.
pH (H$_2$O)	6.0	6.5	6.8	7.0

^{a} by hydrometer method

weathering of coarser particles and loss from the upper horizons by downward or lateral eluviation.

As in the corresponding loamy soils, the slowly permeable subsurface horizon may extend to depths well below the potential rooting zone or be underlain within it by a comparatively permeable substratum in which water moves more freely in response to prevailing suction gradients. In either case, shrinkage cracks which develop during dry summer weather close progressively in autumn as the soil moisture deficit is satisfied. Seasonal cracking is most pronounced under grass or woodland in subhumid areas, especially in variants on Mesozoic clays and clay-shales with large shrink-swell capacities as judged from linear shrinkage (COLE) or CEC measurements (Avery and Bullock, 1977; Reeve et al., 1980). Data on two representative profiles under grass are reproduced below, both with small shrinkage potential.

Profile 144 is on a footslope in mid-Devon. The parent material is compact (fragic) rubbly head derived from the subjacent Carboniferous shales and sandstones (Culm Measures). Both clay contents and fine clay/total clay ratios are relatively small in this soil and there is little evidence of pressures caused by swelling, ped faces in the Bg horizon being smooth but not shiny and predominantly vertical. Judging from physico-chemical and X-ray diffraction data on samples from soils of the same series (Avery and Bullock, 1977; Harrod, 1978, 1981) the clay fractions have low cation-exchange capacities (<40 me/100 g) and consist mainly of mica with few expansible layers. The water regimes of these and associated loamy gley soils in head over Carboniferous rocks in Devon were investigated in detail by Harrod (1981). Using water-level and hydraulic conductivity measurements in piezometers and open boreholes, he confirmed that weathered Bg and BCg horizons within the upper metre are generally very slowly permeable (K sat <5 cm/day) and are saturated with water for much of the year. Conductivities of lower horizons were very variable but in nine out of twenty sites studied, some on convex slopes, the basal head contained water under hydrostatic pressure sufficient to raise it above the ground surface in winter, and in some places throughout the year. Tile drains laid at depths below 1.2 m intercept this water but do little to reduce surface wetness, which experimental results suggest is best achieved by shallower pipe drainage with permeable fill, followed by moling or subsoiling when the soil is dry.

Profile 145 is in late Devensian glaciolacustrine clay which underlies an extensive terrace-like flat in the upper Severn valley north of Welshpool, Powys. As in the preceding profile, the A and Bg horizons contain around 40 per cent clay and show no clearly expressed vertic features. A slightly calcareous BCk horizon appears at 130 cm below the surface. All horizons below 50 cm have high bulk densities and few drainable pores.

Profile 144 (Hogan, 1978)

Location	:	North Wyke Experimental Station, North Tawton, Devon (SX 656984)
Climatic regime:		humid temperate
Altitude	:	152 m Slope: 3° straight
Land use	:	long-term grass

Horizons (cm)

0–27 Apg Greyish brown (10YR 5/2) very slightly stony clay (USDA clay loam to silty clay loam) with common very fine dark reddish brown (5YR 3/4) mottles and coarse inclusions of Bg material; few medium subrounded and tabular sandstone fragments; strong fine subangular blocky; friable to firm; many fine roots; rusty coats on ped faces; abrupt wavy boundary.

27–66 Bg Light grey (5Y 7/1) very slightly stony clay (USDA silty clay to clay) with many medium reddish yellow (7.5YR 6/8) mottles; few large subrounded tabular sandstone fragments; moderate very coarse prismatic with light

Analytical data for Profile 144

Horizon	Apg	Bg	BCg	BCg(x)1	BCg(x)2
Depth (cm)	0–27	27–66	66–84	84–107	107–131
sand 200 μm–2 mm %	4	6	13	17	13
60–200 μm %	8	8	8	9	9
silt 2–60 μm %	50	43	47	47	52
clay <2 μm %	38	43	32	27	26
<0.2 μm %	9	10	7	5	5
fine clay/total clay	0.24	0.23	0.22	0.18	0.19
organic carbon %	3.7	0.5			
pH (H_2O)	5.3	6.0	6.1	6.2	6.3
(0.01M $CaCl_2$)	4.9	5.2	5.3	5.4	5.5
bulk density (g cm^{-3})	0.99	1.30	1.55		
packing density (g cm^{-3})	1.33	1.70	1.84		
retained water capacity (% vol.)	51	46	33		
available water (% vol.)	22	17	15		
air capacity (% vol.)	12	5	8		

grey faces; very firm; common fine roots; few organic coats on ped faces above 45 cm; gradual wavy boundary.

66–84 BCg Grey (N 5/0) slightly stony clay loam (USDA clay loam) with very many medium reddish yellow (7.5YR 6/8) mottles; common stones as above and some very soft shale fragments; massive; consistence as above; common fine roots; common ferri-manganiferous coats; gradual irregular boundary tonguing down to 120 cm.

84–107 BCg(x)1 Strong brown (7.5YR 5/6) moderately stony clay loam (USDA loam to clay loam) with very many fine light grey to grey (N 6/0) mottles; many small tabular sandstone and shale fragments; massive; very firm; few roots; common ferri-manganiferous coats; abrupt smooth boundary.

107–131 BCg(x)2 Yellowish brown (10YR 5/4) moderately stony clay loam (USDA loam) with common medium light grey to grey (N 6/0) mottles, some horizontally banded; stones and soft shale fragments as above; weak medium platy; very firm; no roots; common ferri-manganiferous coats.

Analytical data *see* above.

Classification
England and Wales (1984): pelo-stagnogley soil; clayey drift with siliceous stones (Hallsworth series, formerly Tedburn).
USDA (1975): Typic Haplaquept (or Fragiaquept); fine, illitic (non-acid), mesic.
FAO (1974): Dystric or Eutric Gleysol; fine textured.

Profile 145 (Thompson, 1982)

Location   : Arddleen, Powys (SJ 252145)
Climatic regime: humid temperate
Altitude   :   61 m   Slope: 1° straight
Land use   : long-term grass
Horizons (cm)

0–12 Ah Dark greyish brown (10YR 4/2) stoneless humose silty clay (USDA silty clay); moderate fine subangular blocky; hard (dry); abundant fine roots; common ferruginous coats; abrupt smooth boundary.

12–35 Bg1 Light brownish grey (2.5Y 6/2) stoneless silty clay (USDA silty clay) with many fine yellowish brown (10YR 5/8) mottles; strong coarse prismatic with light brownish grey faces; very firm; common fine roots; few rounded ferri-manganiferous nodules up to 5 mm at 28–35 cm; slight slickensi-

35–68 Bg2 — ding on prism faces; clear smooth boundary.
Light grey to grey (5Y 6/1) stoneless silty clay (USDA silty clay) with very many medium yellowish brown (10YR 5/8) mottles; strong coarse prismatic with light grey to grey faces; very firm; common fine roots; silt with clay coats on ped faces lightly rippled; clear smooth boundary.

68–90 Bg3 Light grey to grey (5Y 6/1) stoneless silty clay (USDA silty clay loam) with very many coarse light olive brown (2.5Y 5/6) mottles; moderate very coarse prismatic with light grey to grey faces; very firm; few fine roots; no slickensiding on ped faces; clear smooth boundary.

90–100 BCg1 Light grey to grey (N 6/–) stoneless silty clay (USDA silty clay loam) with very many medium light olive brown (2.5Y 5/8) mottles; weak (adherent) very coarse prismatic; very firm; no roots; silt with clay coats on ped faces; very slight 1 mm laminations contorted and broken; clear smooth boundary.

100–130 BCg2 Light grey to grey (N 6/–) stoneless silty clay (USDA silty clay) with common medium light olive-brown (2.5Y 5/6) mottles and contorted very fine loamy sand bands up to 3 mm thick; massive; very firm; no roots; clear smooth boundary.

130–164 BCk(g) Olive-brown (2.5Y 4/4) stoneless slightly calcareous silty clay (USDA silty clay loam to silty clay); massive with weak laminations up to 1 mm thick and widely spaced grey coated fissures bordered by yellowish red (5YR 5/6); very firm; no roots; few irregular calcareous nodules; abrupt wavy boundary.

164–182 Ck(g) Olive-brown (2.5Y 4/4) stoneless slightly calcareous silty clay with grey (N 5/–) and light olive-brown (2.5Y 5/6) patches; strongly laminated; very firm; few irregular calcareous nodules.

Analytical data *see* below.

Classification
England and Wales (1984): pelo-stagnogley soil; clayey stoneless drift (Foggathorpe series).
USDA (1975): Typic Haplaquept; fine, illitic or mixed (non-acid), mesic.
FAO (1974): Eutric Gleysol; fine textured.

Analytical data for Profile 145

Horizon	Ah	Bg1	Bg2	Bg3	BCg1	BCg2	BCk(g)
Depth (cm)	0–12	12–35	35–68	68–90	90–100	100–130	130–164
sand 200 μm–2 mm %	1	3	1	<1	1	1	2
60–200 μm %	3	4	3	4	6	3	1
silt 2–60 μm %	55	51	56	58	55	53	57
clay <2 μm %	41	42	40	38	38	43	40
organic carbon %	7.5	1.6	0.6	0.4	0.3	0.3	0.2
$CaCO_3$ equiv. %	–	–	–	–	–	–	1.6
dithionite ext. Fe %	1.4	2.2	2.1	1.4			
pH (H_2O)	5.8	6.2	6.7	7.3	7.4	7.5	8.2
(0.01M $CaCl_2$)	5.5	5.7	6.3	6.9	7.0	7.0	7.7
bulk density (g cm^{-3})	0.73	1.12	1.48	1.63	1.61	1.60	1.49
packing density (g cm^{-3})	1.10	1.50	1.84	1.97	1.95	1.99	1.85
retained water capacity (% vol.)	51	44	39	38	38	41	44
available water capacity (% vol.)	18	14	13	10	12	12	17
air capacity (% vol.)	18	14	5	1	2	<1	<1

7.5.5 Humic Orthic Gley Soils

These are the orthic gley soils with humose or peaty topsoils. According to the definition adopted here, which corresponds to that of typical humic gley soils in the England and Wales system, the gleyed subsurface horizons extend to bedrock or distinctly below the rooting zone and are moderately permeable throughout, include massive grey, bluish or greenish Cg and/or CG horizons with ferruginous segregations surrounding old root channels (Plate IV.1), or both. Those without a slowly permeable subsurface horizon are clearly of ground-water type and the others, typified by profile 146 (p. 353), can be considered as intergrades between ground-water and surface-water gley soils in which intense gleying has been induced by near permanent waterlogging caused by a combination of downslope seepage and slow permeability.

Soils of this subgroup are mainly but not exclusively confined to depressions, lower slopes and localized flushes receiving water from higher ground, and are commonest in cool humid and perhumid areas, particularly where loamy drift overlies impervious bedrock. Variants described as alpine gleys by the Soil Survey of Scotland (1984) occur in rocky hollows below mountain summits where snow and meltwater accumulate. Like the more widespread oro-arctic (cryic) humus podzols (p. 254) and rankers (p. 121), these soils are characterized by the relatively loose and 'fluffy' consistence of the organic or humose surface horizons, which generally contain substantial amounts of mineral matter introduced by frost heaving or periodic colluviation. At lower altitudes, variants under semi-natural vegetation on base-deficient materials commonly have a sharply demarcated surface layer of more or less humified peat (hydromor) which can be up to 40 cm thick; others have very dark coloured topsoils in which organic and mineral fractions are intimately mixed, either naturally by biotic agencies (anmoor or hydromull) or as a result of drainage followed by cultivation and chemical amelioration. Data on two representative profiles are given on pp. 353–4.

Profile 146 is located in a gentle footslope bordering a small peat-filled depression in south-east Dyfed. It is developed in a loamy slope deposit derived from Ordovician shales and sandstones, and has a surface horizon of well humified peat containing much mineral matter. The chemical data confirm that the unmottled Eag horizon has lost iron and that, as in gley soils generally, only a small proportion of the total free iron in the succeeding horizons is extracted by pyrophosphate. Organic carbon contents exceed 0.5 per cent to 130 cm depth, suggesting that the parent deposit may be of recent colluvial origin.

The field in which this profile is sited had recently been underdrained at 40 m spacing with a perimeter ditch. When sampled in July 1976, following an abnormally dry period, the water table was apparently more than one metre below the surface. Although the grey Cg horizon with 'iron pipes' was estimated in the field to be moderately porous, laboratory measurements on core samples gave very small air capacity values. Subsequent determinations of hydraulic conductivity by the auger-hole method when the water table was at 58 cm furnished an average figure of 0.13 cm/day, showing that the horizons below that depth are very slowly permeable in the saturated condition.

Profile 147 is sited in a steeply sloping flush at an altitude of 381 m in the eastern Grampians. Here the parent material is head derived from basic igneous rocks, mainly epidiorites and diorites, fragments of which disintegrate readily under reducing conditions. The surface horizon is humose rather than peaty and the base status is much higher than in the preceding profile, especially in the greenish 2CG horizon, which contains more organic matter than the overlying horizon and is possibly a buried topsoil (bAhg). Like profile 28 (p. 154) nearby, this soil has a cryic temperature regime as defined in the US Taxonomy.

Profile 146 (Wright, 1985)

Location : Llanarthney, Dyfed (SB 487178)
Climatic regime: humid temperate
Altitude : 137 m   Slope: 1° straight
Land use and long-term grass with 25 per cent
  vegetation : rushes (*Juncus* spp.)

Horizons (cm)

0–15
Oh
Very dark brown (10YR 2/2) humified peat; weak fine subangular blocky and granular; abundant fine to medium roots in upper 4 cm, many below; clear smooth boundary.

15–22
Ah
Very dark brown (10YR 2/2) stoneless humose silty clay loam (USDA silty clay loam); weak fine subangular blocky; very friable; many fine roots; clear smooth boundary.

22–37
Eag
Light brownish grey (10YR 6/2) moderately stony sandy silt loam to clay loam (USDA silt loam) with common dark brown (10YR 3/3) organic coats on stones and lining fissures and channels; many, small to large, angular and subangular fragments of sandstone and grit, some soft and weathered; moderate medium prismatic; no visible fissures; firm; common fine roots; clear smooth boundary.

37–85
Cg1
Light grey to grey (10YR 6/1) slightly stony sandy silt loam to clay loam (USDA silt loam) with many medium strong brown (7.5YR 5/6) to yellowish red (5YR 5/8) mottles, mainly around old root channels; stones as above; massive; firm; common live roots in upper 20 cm, few below; earthworms present throughout; clear wavy boundary.

85–100
Cg2
Grey (N 5/0) slightly stony sandy silt loam (USDA loam) with common yellowish red (5YR 4/8) linear mottles lining old root channels; stones as above; wet; massive; no visible fissures; no living roots; common dead roots; clear wavy boundary.

100–130
Cg3
Dark grey to grey (N 4–5/0) moderately stony sandy silt loam (USDA loam) with common coarse olive (5Y 5/3) mottles increasing in amount

Analytical data for Profile 146

Horizon	Oh	Ah	Eag	Cg1	Cg2	Cg3
Depth (cm)	0–15	15–22	22–37	37–85	85–100	100–130
sand 200 µm–2 mm %			4	9	8	9	15
60–200 µm %			8	16	15	24	19
silt 2–60 µm %			56	57	59	52	50
clay <2 µm %			32	18	18	15	16
organic carbon %	27	12	1.2	0.6	0.5	0.6
pyro. ext. Fe %		0.3	0.1	0.1	0.05	0.1
pyro. ext. Al %		0.5	0.1	0.1	0.05	0.06
pyro. ext. C %		3.7	0.4	0.1	0.1	0.1
total ext. Fe %		0.5	0.3	1.3	0.8	1.0
TEB (me/100 g)	26		1.6		2.4	2.4
CEC (me/100 g)	83		12.2		9.8	11.1
% base saturation	31		13		24	22
pH (H$_2$O)	4.9	4.9	4.8	4.8	4.9	5.0
(0.01M CaCl$_2$)	4.5	4.2	4.1	4.0	4.0	4.1
bulk density (g cm^{-3})				1.52	1.61	1.56
packing density (g cm^{-3})				1.68	1.75	1.70
retained water capacity (% vol.)				40	37	36
available water capacity (% vol.)				20	23	14
air capacity (% vol.)				2	3	4

with depth; very small to large stones as above; wet; massive; no visible fissures; common dead roots.

Analytical data *see* p. 353.

Classification
England and Wales (1984): typical humic gley soil; fine loamy drift with siliceous stones (Freni series, formerly Ynys).
USDA (1975): Histic Humaquept; coarse-loamy, mixed (acid), mesic.
FAO (1974): Humic Gleysol; medium textured.

Profile 147 (Heslop and Bown, 1969)

Location	: Strathdon, Aberdeenshire (NJ 362134)
Climatic regime:	humid oroboreal
Altitude	: 381 m Slope: 15°
Land use and vegetation	: Scots pine/larch/Norway spruce plantation about 60 cm high; field layer includes tufted hair-grass (*Deschampsia caespitooa*) and common bent-grass (*Agrostis tenuis*)

Horizons (cm)

0–28 Ah	Very dark greyish brown to dark grey (10YR 3/1.5) humose sandy loam (USDA sandy loam); weak fine to medium subangular blocky; abundant roots; gradual boundary.
28–64 Eag	Greyish brown (10YR 5/2) stony gritty sandy loam (USDA sandy loam); small basic igneous rock fragments; weak medium to coarse subangular blocky; common roots; gradual boundary.
64–84 BCg	Grey (10YR 5/1) stony gritty sandy loam (USDA sandy loam) with yellowish brown (10YR 5/6) mottles; wet; weak subangular blocky; few fine roots; clear boundary.
84–97 2Cg	Dark greenish grey (5G 4/1) sandy silt loam (USDA fine sandy loam to loam) with common to many yellowish red (5YR 5/6) linear mottles ('iron drainpipes') around root channels; wet; massive; gradual boundary.
97–120 2CG	Greenish grey (5G 5/1) gritty sandy loam (USDA fine sandy loam); wet; massive.

Analytical data *see* below.

Classification
Scotland (1984): humic ground-water gley (Myreton series).
England and Wales (1984): typical humic gley soil; coarse loamy drift with siliceous stones (Fordham series?).
USDA (1975): Typic Cryaquoll; coarse-loamy, mixed, non-calcareous.
FAO (1974): Mollic Gleysol; medium textured.

7.5.6 Humic Stagno-orthic Gley Soils

These are similar loamy or clayey soils in which both gleying and superficial accumulation of organic matter are attributable to

Analytical data for Profile 147

Horizon	Ah	Ah	Eag	BCg	2Cg	2CG
Depth (cm)	0–10	15–25	36–46	66–76	86–96	109–119
silt[a] 2–50 μm %	30	27	21	19	33	29
clay[a] <2 μm %	11	14	11	8	16	12
loss on ignition %	17.6	10.2	3.4	3.0	7.8	5.7
organic carbon %	8.5	5.5	1.7			
nitrogen %	0.42	0.33	0.09			
C/N ratio	20	17	19			
TEB (me/100 g)	12.4	7.7	4.1	4.8	17.2	10.8
CEC (me/100 g)	22.5	15.7	7.7	8.4	19.5	13.3
% base saturation	55	49	53	57	88	81
pH (H$_2$O)	5.7	5.7	6.0	6.3	6.6	6.4

[a] by hydrometer method

some combination of high rainfall, low evaporation, acidity and slow internal drainage rather than to saturation by ground water of extraneous origin. Classed as cambic stagnohumic gley soils in the England and Wales system and as peaty or humic surface-water gley soils in Scotland and Ireland, they are widely distributed on undulating plateaux and gentle to moderate valley-side slopes in upland areas with perhumid or humid oroboreal climatic regimes, but are rare in the subhumid lowlands. Typical profiles have strongly gleyed subsurface (E and/or B) horizons passing downwards into less altered horizons that are slowly permeable, retain more of their inherited colour, or both, and feel relatively dry when exposed in the side of a pit. The substratum is usually a compact till or head deposit and often shows fragipan features where it is loamy in texture. Under cool perhumid conditions, however, it can be moderately permeable, indicating that gleying in the overlying horizons is primarily bioclimatic in origin (Clayden, 1979).

The characteristic sequence of horizons is strikingly developed on dense reddish parent materials, as in profile 148 (p. 356). In materials derived from greyish source rocks (Plate IV.2) it is often less clearly expressed; the soils are generally distinguishable from those of the preceding subgroup by the absence of an intensely gleyed horizon in which iron segregations occur chiefly as 'pipes' surrounding old root channels rather than as intraped mottles, but intergrades are frequent, particularly in perhumid areas where peaty gley soils which conform hydrologically and topographically to the surface-water type commonly have more or less permanently waterlogged subsurface horizons. Data on three representative profiles are reproduced on pp. 356–9, all under rough grazing with semi-natural acidic grassland or moorland vegetation.

Profile 148 is located in a shallow upland depression south of Llangadog, Dyfed, on reddish loamy till derived from Devonian (Old Red Sandstone) rocks. As in the related non-humic profile 142 (p. 347), the light coloured Eg horizon contains more clay but less extractable iron than the very dense (fragic) BC and C horizons in which the original matrix colour becomes dominant. Physical data on fragipans in soils of the same series are reported by Payton (1980).

Profile 149 was described and sampled by V.C. Bendelow in a mid-slope position on the Pennine moorlands west of Ripon, North Yorkshire, with boulders on the surface nearby. The parent material is greyish loamy drift derived from Carboniferous sandstone and shale. A grey Eag horizon containing little extractable iron appears directly beneath the peaty surface layer and overlies a merging succession of prominently mottled horizons (Bg, BCg and Cg) which become firmer, more coarsely structured and less porous with increasing depth. Although a few (<1 per cent) argillans were observed in thin sections of the Bg and BCg horizons, the variations in clay content indicated by the particle-size analyses are primarily attributable to the heterogeneous composition of the parent drift and the extent to which partially weathered sandstone fragments contributed material to the samples. In both this and the preceding profile, the soil is saturated to the surface for long periods and water moves laterally above the relatively impermeable lower horizons.

Profile 150, which exemplifies the clayey soils included in this subgroup, was described and sampled under unusually dry conditions (2 Sept. 1975). Located on the Cleveland Hills, North Yorkshire, it is underlain by Middle Jurassic clay shales with subordinate sandstone bands and has apparently formed in head derived from these rocks. The topsoil is humose rather than peaty and is too thin to qualify as umbric according to the criteria used in the US and FAO systems. All the subsurface horizons are very rich in clay and there is a considerable increase in clay content at the base of the Bg horizon, which evidently marks a lithological discontinuity as the relative proportions of sand and silt also change markedly at this level. The uniformly low pH values may result partly from oxidation of pyrite contained in the source rocks, but no jarosite mottles were noted and the sulphur contents recorded are too small to support recognition of a sulphuric horizon (Chapter 3, p. 111). Similar but generally less clayey upland soils are common on tills derived from Carboniferous shales or other argillaceous rocks in northern England (R.A. Jarvis et al., 1984), in the Midland Valley of Scotland (Bown et al., 1982), and in northern and western Ireland (association 2 in Gardiner and Radford, 1980).

Profile 148 (Wright, 1981)

Location	:	Llangadog, Dyfed (SN 727224)
Climatic regime:		perhumid temperate
Altitude	:	205 m Slope: 1°
Land use and vegetation	:	rough grazing dominated by purple moor-grass (*Molinia caerulea*) with rushes (*Juncus* spp.) and heather (*Calluna vulgaris*)

Horizons (cm)

0–16 Oh	Very dark brown (7.5YR 2/2 broken and rubbed) stoneless humified peat; moderate fine subangular blocky; many roots; sharp smooth boundary.
16–28 Eg	Light brownish grey (10YR 6/2) slightly stony silty clay loam (USDA silt loam) with common fine yellowish brown (10YR 5/6) mottles; common medium subangular fragments of red sandstone, some soft and weathered; moderate coarse prismatic; firm; common fine roots; clear smooth boundary.
28–46 Bg	Light grey to grey (10YR 6/1) moderately stony sandy silt loam (USDA silt loam) with many coarse reddish brown (2.5YR 5/4) and common medium yellowish red (5YR 4/6) mottles; many stones as above; moderate coarse prismatic with light grey to grey faces; firm; common fine roots; clear smooth boundary.
46–64 BCg(x)	Reddish brown (2.5YR 4/4) moderately stony sandy silt loam to clay loam (USDA silt loam to loam) with common medium light grey to grey (10YR 6/1) and fine strong brown

Analytical data for Profile 148

Horizon		Oh	Eg	Bg	BCg(x)	C(xg)
Depth (cm)		0–16	16–28	28–47	46–64	64–100
sand 200 μm–2 mm %			4	16	20	
60–200 μm %			11	9	7	
silt 2–60 μm %			62	58	55	
clay <2 μm %			23	17	18	
organic carbon %		20	2.3	1.2	0.3	
pyro. ext. Fe %			0.1	0.1	0.1	
pyro. ext. Al %			0.3	0.1	0.1	
total ext. Fe %			0.4	0.7	1.2	
pH (H$_2$O)		4.7	5.0	5.0	5.3	
(0.01M CaCl$_2$)		4.0	4.1	4.0	4.0	
bulk density (g cm^{-3})		0.44	1.18	1.83		1.89
packing density (g cm^{-3})			1.38	1.98		2.05
retained water capacity (% vol.)		63	46	27		26
available water (% vol.)		40	19	10		11
air capacity (% vol.)		15	10	1		1

| 64–100 C(xg) | (7.5YR 5/6) mottles; stones as above; moderate coarse prismatic with light grey to grey faces; very firm; few roots; gradual boundary. Reddish brown (2.5YR 4/4) moderately stony sandy silt loam to clay loam with common medium strong brown and few fine light grey to grey mottles; stones as above; massive; very firm; no roots. |

Analytical data see p. 356.

Classification
England and Wales (1984): cambic stagnohumic gley soil; reddish-loamy, drift with siliceous stones (Wenallt series).
USDA (1975): Humic Fragiaquept; fine-loamy, mixed (acid), mesic.
FAO (1974): Humic Gleysol; medium textured (fragipan phase).

Profile 149

Location	:	Fountains Earth Moor, North Yorkshire (SE 148725)
Climatic regime:		humid oroboreal
Altitude	:	384 m Slope: 3° straight
Land use and vegetation	:	Heather moor managed for grouse; heather (*Calluna vulgaris*) dominant, with frequent soft rush (*Juncus effusus*), bilberry (*Vaccinium myrtillus*) and purple moor-grass (*Molinia caerulea*)

Horizons (cm)

L/F 0–29 Oh	More or less decomposed heather and grass litter, 2 cm thick. Black (5YR 2/1 broken and rubbed; 10YR 3/1 dry) sandy humified peat; weak fine subangular blocky; common fine and few medium woody roots; abrupt wavy boundary.
29–42 Eag	Pinkish grey (7.5YR 6/2; 10YR 5/3 rubbed; 10YR 8/2 dry) very slightly stony sandy silt loam (USDA loam to sandy loam) with dark brown (7.5YR 3/2) vertical streaks following root channels; few, mainly small, partly weathered subangular sandstone fragments; weak medium subangular blocky; friable (slightly sticky and slightly plastic when wet); few fine woody roots; clear smooth boundary.
42–69 Bg	Light brownish grey (10YR 6/2; 10YR 6/3 rubbed; 10YR 7/1 dry) slightly stony sandy silt loam (USDA loam) with very many very fine brownish yellow (10YR 6/6) mottles; common small to medium subangular sandstone fragments, partly weathered; weak medium subangular blocky with greyish brown (10YR 5/2) faces; friable (slightly sticky and moderately plastic when wet); few roots, nearly all dead; gradual smooth boundary.
69–102 BCg	Light brownish grey (10YR 6/2; 10YR 6/3 rubbed; 2.5Y 8/0 dry) slightly

Analytical data for Profile 149

Horizon	Oh	Eag	Bg	BCg	2Cg(x)
Depth (cm)	0–29	29–42	42–69	69–102	102–150
sand 600 µm–2 mm %	4	4	5	4	10
200–600 µm %	27	14	17	13	23
60–200 µm %	35	29	17	26	20
silt 2–60 µm %	26	40	44	43	26
clay <2 µm %	8	13	17	14	21
organic carbon %	17	1.9	0.8	0.5	0.6
dithionite ext. Fe %		0.2	2.3	1.0	2.2
pH (H$_2$O)	3.8	4.2	4.4	4.4	4.3
(0.01M CaCl$_2$)	3.1	3.7	3.9	3.8	3.7

| 102–150 2Cg(x) | stony sandy silt loam (USDA loam) with many fine yellowish red (5YR 5/6) mottles; stones as above; moderate coarse prismatic with light grey to grey (10YR 6/1) faces; very slightly porous; firm (slightly sticky and moderately plastic when wet); gradual smooth boundary. Grey (N 5/-; 10YR 5/3 rubbed; N 6/- dry) slightly stony sandy clay loam (USDA sandy clay loam) with many medium yellowish red (5YR 5/6) mottles; common, small to medium, subangular and tabular sandstone fragments and few of shale; massive; very slightly porous; very firm; no roots. |

Analytical data *see* p. 357.

Classification
  England and Wales (1984): cambic stagnohumic gley soil; loamy drift with siliceous stones (Wilcocks series).
  USDA (1975): Histic Cryaquept; coarse-loamy, mixed or siliceous (acid).
  FAO (1974): Humic Gleysol; medium textured (stony phase).

Profile 150 (Bendelow and Carroll, 1979)

Location	:	Jenny Brewsters Moor, Osmotherley, North Yorkshire (SE 481959)
Climatic regime:	humid oroboreal
Altitude	:	271 m Slope: 2° straight

| Land use and vegetation | : | rough grazing; heather moor with very abundant *Calluna vulgaris* and occasional rushes (*Juncus* spp.) and cotton-grass (*Eriophorum vaginatum*) |

Horizons (cm)

L/F	Litter of heather roots and stems, c. 2 cm thick.
0–18 Ah	Black (5YR 2/1 moist and dry) stoneless humose clay; dry; weak medium subangular blocky; many fine roots; abrupt wavy boundary.
18–32 Eg	Greyish brown (10YR 5/2; 6/2 dry) very slightly stony clay (USDA clay) with many medium yellowish brown (10YR 5/6) mottles; few, small to very large, angular tabular sandstone fragments with bleached surfaces; dry; moderate medium angular blocky with grey (10YR 5/1) faces; firm (peds very hard when dry); many fine roots; clear smooth boundary.
32–52 Bg1	Light grey (10YR 7/2; 5Y 6/0 dry) stoneless clay (USDA clay) with many fine brownish yellow (10YR 6/6) mottles; strong medium prismatic with light brownish grey (10YR 6/2) faces; consistence as above; common fine roots; clear smooth boundary.
52–65 Bg2	Grey (10YR 6/1) stoneless clay (USDA clay) with many fine strong brown (7.5YR 5/8) mottles; moderate medium prismatic with grey (5Y 5/1) faces; firm; few fine roots; gradual

Analytical data for Profile 150

Horizon	Ah	Eg	Bg1	Bg2	2BCg1	2BCg2	2BCg3
Depth (cm)	0–18	18–32	32–52	52–65	65–84	84–106	106–120
sand 200 µm–2 mm %		1	2	3	1	1	3
60–200 µm %		14	16	29	8	7	18
silt 2–60 µm %		22	14	12	15	17	18
clay <2 µm %		63	68	56	76	75	61
organic carbon %	13	3.1	0.7		0.5		0.4
pyro. ext. Fe %		0.3	0.1	0.2	0.1	0.1	<0.05
pryo. ext. Al %		0.3	0.2	0.2	0.2	0.1	0.1
total ext. Fe %		1.6	3.5	7.1	2.3	1.3	1.1
total sulphur %			0.09	0.02	<0.01	0.02	<0.01
pH (H$_2$O)	3.7	3.7	3.9	4.0	4.0	4.0	4.0
(0.01M CaCl$_2$)	3.2	3.3	3.5	3.5	3.5	3.5	3.5

65–84 2BCg1	smooth boundary. Grey (10YR 6/1) stoneless clay (USDA clay) with common medium brown (7.5YR 5/4) mottles; weak (adherent) coarse prismatic with grey smooth faces (stress-oriented coats); firm; no roots; diffuse boundary.
84–106 2BCg2	Grey (10YR 6/1) very slightly stony clay (USDA clay) with common fine yellowish brown (10YR 5/4) mottles; few angular tabular sandstone fragments; massive; very firm; diffuse boundary.
106–120 2BCg3	Grey (10YR 6/1) very slightly stony clay (USDA clay) with common fine yellowish brown (10YR 5/6) mottles; few small platy shale fragments; massive; firm.

Analytical data *see* p. 358.

Classification
   England and Wales (1984): cambic stagnohumic gley soil; clayey material passing to clay or soft mudstone (Onecote series).
   USDA (1975): Humic Haplaquept; very-fine, mixed or illitic (acid), mesic.
   FAO (1974): Dystric Gleysol; fine textured.

7.6 Luvic Gley Soils

7.6.1 General Characteristics, Classification and Distribution

As defined here, this group comprises gley soils having an argillic (Btg) horizon, together with essentially similar soils in which there is an abrupt textural change (Chapter 3, p. 111) between an A or E horizon and an underlying slowly permeable 2 Bg horizon that shows little evidence of clay illuviation. Those in which periodic wetness and associated hydromorphic features are attributable to the relative impermeability of the B horizon or an immediately underlying layer have been grouped as surface-water (stagno- or stagnohumic) gley soils in previous British classifications. Less common variants in which deep seated mobile ground water rises into the solum during the winter half-year are set apart as argillic and argillic humic gley soils by Avery (1973, 1980). Those of both types are mostly Aqualfs (Ochraqualfs, Albaqualfs, Glossaqualfs or Fragiaqualfs) in the US Taxonomy; others are Aquolls (Argiaquolls) or Aquults. According to the FAO (1974) system, Gleyic Luvisols, Gleyic Acrisols, Gleyic Podzoluvisols and Planosols are included.

Taken together, these soils occupy about the same proportionate area (c. 10 per cent) as the closely related orthic gley soils in the British Isles as a whole, but are much more extensive in the subhumid lowlands and are estimated to cover more of England and Wales (c. 15 per cent) than any of the other soil groups distinguished in this account. The parent materials, many of which are calcareous in the unweathered state, include medium and fine textured tills, loess (brickearth) and various other Pleistocene and pre-Pleistocene deposits which contain or yield significant amounts of silicate clay. In some cases, exemplified by profile 151 (p. 362), an argillic B horizon has evidently formed in initially uniform material as a result of weathering and clay translocation; in others the increase in clay content with depth is mainly, if not entirely, attributable to the presence of an originally coarser textured surface layer of aeolian, solifluxional or fluvial origin. The upper horizons may also have lost clay by lateral eluviation or by decomposition (ferrolysis) induced by alternation of reducing and oxidizing conditions (Brinkman, 1979), though conclusive evidence of clay removal by either of these processes is difficult to obtain. As in the luvic brown soils (p. 209), the lower boundary of the eluvial horizon can be abrupt or gradual, smooth or irregular, depending in part on whether it coincides with a lithologic discontinuity and whether the discontinuity has an irregular conformation reflecting late Devensian cryoturbation. Some loamy variants have fragipan characteristics above, in, or below the B horizon, and red-mottled (chromic) argillic B horizons showing evidence of pre-Devensian pedogenesis occur locally in plateau sites on Wolstonian and older deposits.

Most lowland soils of this group have a non-humic topsoil underlain by a rust-mottled Eg horizon which is intermittently saturated with water in the absence of effective artificial drainage. Under semi-natural deciduous woodland these upper horizons are more or less strongly acid, even where there is a calcareous substratum at 60 cm or less, as in profiles 153 (p. 366) and 159 (p. 376). A brownish Eb (or Bw) horizon with little or no mottling appears immediately below the A in intergrades to luvic brown soils, and intergrades to luvic gley-podzols (p. 266) have an unincorporated superficial organic layer underlain by an unmottled Ea or Eag, an incipient Bh horizon, or both. In cultivated land the Ap horizon often rests directly on the finer textured B and can be prominent (mollic, umbric or anthropic epipedon), especially where it is coarse textured and has been heavily manured.

Variants with humose or peaty topsoils that are or were saturated with water for long periods occur in both lowland and upland situations but are less extensive than the humic gley soils conforming to the orthic group. Insofar as a seasonal soil-water deficit is a necessary prerequisite for formation of an argillic B horizon, it seems likely that those in these soils originated during Holocene or pre-Holocene periods when conditions were drier than at present, and in some cases there is clear evidence of this.

Land-use capability is restricted to varying degrees on luvic gley soils by seasonal wetness and related workability and trafficability problems, which are mitigated but seldom completely rectified by artifical drainage. For this to be effective, the soil above drain depth must be sufficiently permeable. In many of these soils, however, a slowly permeable Btg horizon is encountered within 40 cm depth. Others have loamy eluvial horizons which are weakly aggregated and therefore easily compacted when the soil is worked or trampled by stock. In either case the adverse physical conditions can be alleviated by subsoiling or moling when moisture conditions are favourable, but long-lasting substantial improvement is difficult to achieve, especially in the more humid areas of occurrence. Under subhumid conditions, crops tend to suffer more from drought than on well aerated soils with similar texture profiles because rooting is generally restricted to some extent by seasonally anaerobic conditions, adverse pore-size distribution, or both.

Key to Subgroups

1. Luvic gley soils with a humose or peaty topsoil (p. 107) and
   (a) a slowly permeable B horizon (p. 110) and no immediately underlying permeable horizon 15 cm or more thick that is gleyed (Cg or Cgf), saturated with mobile ground water at some season, or both; or
   (b) gleyed E and/or B horizon and less gleying in the underlying horizon; or both
   Humic stagnoluvic gley soils
2. Other luvic gley soils with a humose or peaty topsoil (p. 107)
   Humic luvic gley soils
3. Other luvic gley soils with a slowly permeable B horizon (p. 110) and no immediately underlying permeable horizon 15 cm or so thick that is gleyed (Cg or Cgf), saturated with mobile ground water at some season, or both
   3.1 With a chromic argillic B horizon (p. 109)
   Chromic stagnoluvic gley soils
   3.2 With pelo-characteristics (p. 112) and a non-chromic (ordinary) argillic B horizon
   Peloluvic gley soils
   3.3 Others with a non-chromic (ordinary) argillic B horizon
   Stagnoluvic gley soils
4. Other luvic gley soils
   Ground-water luvic gley soils

7.6.2 Ground-water Luvic Gley Soils

These are the non-humic luvic gley soils with relatively permeable substrata that have been grouped as argillic gley soils in England and Wales (Avery, 1973, 1980). They form significant components of the soil patterns in a number of widely scattered localities in southern and eastern England (M.G. Jarvis et al., 1984; Hodge et al., 1984) but are apparently inextensive elsewhere in the British Isles. Like the analogous *hydrobrick soils* in the Netherlands (de Bakker and Schelling, 1966), they are typically developed in late Pleistocene (Devensian) fluvial, aeolian or head deposits which are low lying and/or overlie impervious beds at depths below the main rooting zone. Without effective artificial drainage, the water table is normally near the surface in winter and falls below the base of the B horizon in summer. To the extent that water 'perches' temporarily above the argillic B, these soils exhibit both surface-water and ground-water characteristics and have been classed accordingly as Amphigleys (Ehwald et al., 1966; CPCS, 1967) or Pseudogley-Gley intergrades (Mückenhausen et al., 1977) in German and French classifications.

Typical profiles have a loamy, or occasionally sandy, A horizon over a rust-mottled Eg of similar texture, often with an irregular lower boundary. The underlying Btg normally has argillans on ped faces and in pores, especially in the lower part, and passes downwards into a coarser textured greyish or ochreous Cg or Cgf horizon which can be thin where it rests on an impervious substratum within 1.50 m depth. Redeposited iron, or iron and manganese, may be segregated as irregularly distributed nodules or concentrated at particular levels, in places forming a cemented pan. Horizons below the rooting zone can also retain their inherited colours, though field observations show that they are periodically saturated with water. Variants with humose or peaty topsoils occur locally in the East Anglian Fenland but are evidently rare elsewhere.

As these soils are mainly in level or gently sloping land in subhumid areas and have easily worked, loamy or sandy surface horizons, they are potentially suited to a wide range of uses when adequately drained. In some places, however, the ground is too low lying to provide sufficient outfall without recourse to pumping. The effectiveness of pipe drainage also depends on the hydraulic conductivity of the soil above drain depth and this can be reduced to low values under long-term cultivation as a result of slaking and compaction of the inherently weakly structured upper horizons. Data on two representative profiles are reproduced on pp. 362–4.

Profile 151 was described and sampled by A.J. Moffatt in the Lower Coastal Plain of West Sussex (Hodgson, 1967). Like profile 66 from the same area, it is developed in reworked loess (brickearth), which at this site overlies Ipswichian raised-beach sand at 114 cm depth. All horizons are non-calcareous, but pH values have been raised by past liming, and in places nearby the lower part of the loessial layer is unleached and contains numerous concretions of redeposited calcium carbonate. Air capacity and packing density data confirm that all the subsurface horizons are slightly porous (Table 3.7, p. 98). Hodgson (1967) showed from regular measurements in similar soils over several years that the water table rose to within 30 cm of the surface during the wettest periods, even in sites with tile drainage, which apparently tends to reduce the duration of wetness rather than the minimum depth at which it occurs. Although the Btg horizon impedes drainage to some extent, the general distribution of gley soils in the Lower Coastal Plain reflects the sub-drift boundary between the basal Chalk and Tertiary clay formations, indicating that seasonal wetness results primarily from a combination of low relief and impervious substratum.

Profile 152 is on the Lower Taplow Terrace (Thomas, 1961) in the Loddon valley southeast of Reading, Berkshire. It is developed in a loamy surface layer over flint gravel, which rests in turn on Tertiary London Clay and is permanently waterlogged in the lower part, the water table standing at 112 cm depth in July 1964 when the soil was sampled. The boundary between the loam and the gravel is uneven, presumably as a result of Devensian cryoturbation, and the irregularities so produced have affected the development and shape of the pedogenic horizons. The Btg horizon, in which clay coats are clearly evident, is mainly in the upper, cryoturbated part of the gravel and the Eg, which has a distinctly platy structure and hardens markedly on drying, is thickest and most clearly expressed in wedge-shaped bodies of loam occupying sharply defined depressions in its upper surface. A narrow horizontal zone of iron concentration (3Btgf) is roughly coincident with the summer water table level and discontinuous cemented iron pans occur in soils of the same series nearby.

Profile 151

Location	:	Aldingbourne, West Sussex (SU 927046)
Climatic regime:		subhumid temperate
Altitude	:	10 m Slope: <1°
Land use	:	arable (cereals)

Horizons (cm)

0–23 Ap	Brown to dark brown (10YR 4/3) silt loam (USDA silt loam) containing occasional (<1 per cent) small subangular flint fragments; weak medium subangular blocky; firm; common fine roots; abrupt smooth boundary.
23–35 Eg	Light brownish grey (2.5Y 6/2) silty clay loam (USDA silt loam) with many fine strong brown (7.5Y 5/6) mottles; rare small flints as above; weak medium subangular blocky; friable; common fine roots; few earthworm channels infilled with A horizon material; few rounded ferri-manganiferous nodules; clear smooth boundary.
35–59 Btg1	Brown (10YR 5/3) silty clay loam (USDA silty clay loam) with many fine strong brown (7.5YR 5/6) and few dark brown (7.5YR 4/4) and light brownish grey (2.5Y 6/2) mottles; rare flints as above; moderate medium to coarse subangular blocky with pale brown (10YR 6/3) ped faces; friable to firm; common roots; few rounded ferri-manganiferous nodules; clear wavy boundary.
59–103 Btg2	Brown (10YR 5/3) silty clay loam (USDA silty clay loam) with many

Analytical data for Profile 151

Horizon	Ap	Eg	Btg1	Btg2	Btg2	2BCtg	3Cgf
Depth (cm)	0–23	23–35	35–59	59–81	81–103	103–114	114–140
sand 200 μm–2 mm %	5	3	<1	2	2	6	<1
100–200 μm %	3	2	1	1	2	5	38
60–100 μm %	4	2	1	2	4	13	45
silt 2–60 μm %	72	71	68	69	70	53	9
clay <2 μm %	16	22	30	26	22	23	8
<0.2 μm %	8	9	17	14	12	13	5
fine clay/total clay	0.50	0.41	0.57	0.54	0.55	0.57	
organic carbon %	1.5	0.6	0.3	0.3			
pH (H$_2$O)	6.2	7.0	7.5	7.4	7.1	7.3	7.3
(0.01M CaCl$_2$)	6.1	6.4	6.6	6.5	6.5	6.9	6.9
bulk density (g cm^{-3})	1.39	1.50	1.39	1.40	1.48		
packing density (g cm^{-3})	1.53	1.70	1.66	1.63	1.68		
retained water capacity (% vol.)	35	36	39	37	37		
available water capacity (% vol.)	19	15	15	16	17		
air capacity (% vol.)	10	7	9	10	8		

	fine strong brown (7.5YR 4/6) and few light brownish grey mottles; rare flints as above; moderate coarse subangular blocky; firm; few fine roots; common rounded ferri-manganiferous nodules, especially in lower part; clear smooth boundary.
103–114 2BCtg	Light brownish grey (10YR 6/2) slightly stony clay loam (USDA clay loam to silty clay loam) with many fine strong brown (7.5YR 4/6) mottles; common small angular and tabular flint fragments; moist to wet; moderate fine subangular blocky with brown (10YR 5/3) faces; friable; no roots; common ferri-manganiferous nodules; abrupt wavy boundary.
114–140+ 3Cgf	Strong brown (7.5YR 4/6) stoneless loamy fine sand (USDA loamy fine sand) with light brownish grey (2.5Y 5/2) lamellae; wet; massive; no roots; water table at 128 cm (27 Nov. 1981).

Analytical data *see* p. 362.

Classification
England and Wales (1984): typical argillic gley soil; silty stoneless drift (Park Gate series).
USDA (1975): Aeric Ochraqualf; fine silty, mixed, mesic.
FAO (1974): Gleyic Luvisol; medium textured.

Profile 152 (Jarvis, 1968)

Location	: Arborfield, Berkshire (SU 748676)
Climatic regime:	subhumid temperate
Altitude	: 47 m Slope: <1°
Land use	: ley grass

Horizons (cm)

0–11 Ap1	Dark grey (10YR 4/1) slightly stony medium sandy loam (USDA fine sandy loam); common subangular and rounded flint fragments up to 3 cm; moderate medium subangular blocky, friable, many roots; added chalk; clear smooth boundary.
11–30 Ap2	Dark greyish brown and dark grey (10YR 4/1) slightly stony medium sandy loam (USDA fine sandy loam); stones as above, up to 5 cm; weak fine and medium subangular blocky; friable; common fine roots; earthworm channels and casts common; clear smooth boundary.
30–53 Eg	Greyish brown (10YR 5/2) and brown (10YR 5/3), very slightly stony, medium sandy loam (USDA fine sandy loam) with common coarse strong brown (7.5YR 5/6) mottles; stones as above; weak medium and coarse platy; friable; few fine roots; many partly infilled earthworm channels; abrupt wavy boundary. (Elsewhere in the section a clearly defined body of pale olive (5Y 6/3) sandy loam with many coarse yellowish brown mottles, dark grey infillings and numerous vesicular pores extends to 76 cm depth).
53–76 2Btg	Greyish brown (10YR 5/2) and brown (10YR 5/3) very stony sandy clay loam (USDA sandy clay loam) with common coarse strong brown (7.5YR 5/8) mottles; abundant subangular and rounded flint fragments up to 2 cm; massive; few fine roots; discontinuous reddish yellow (7.5YR 6/6) clay coats on stones; gradual irregular boundary (discontinuous horizon).
76–89 3Btg	Greyish brown to brown (10YR 5/2–3) very stony coarse sandy clay loam (USDA sandy clay loam) with many coarse strong brown mottles; stones and clay coats as above; massive; roots rare; clear irregular boundary.
89–97 3Btgf	Yellowish red (5YR 5/8) very stony coarse sandy clay loam (USDA sandy clay loam) with many medium olive-grey (5Y 6/2) mottles and discontinuous horizontal grey streaks; stones as above; moist to wet; massive; no roots; common clay and iron oxide coats; sharp wavy boundary.
97–127 3Cg/Cgf	Grey flint gravel with intercalated sandy seams and a reddish yellow (7.5YR 6/8) layer at 107–115 cm; wet (water table in trench at 112 cm).

Analytical data *see* p. 364.

Classification
England and Wales (1984): typical argillic gley soil; coarse loamy material over non-calcareous gravel (Hurst series).
USDA (1975): Aeric Ochraqualf (or Glossaqualf); loamy-skeletal, siliceous or mixed, mesic.
FAO (1974): Gleyic Luvisol (or Podzoluvisol); medium textured.

Analytical data for Profile 152

Horizon	Ap1	Ap2	Eg	2Btg	3Btg	3Btgf	3Cg/Cgf
Depth (cm)	0–11	11–30	30–53	53–76	76–89	89–97	97–115
sand 600 μm–2 mm %	7	3	5	16	27	27	16
200–600 μm %	25	24	28	21	38	34	49
60–200 μm %	24	29	32	23	6	7	21
silt 2–60 μm %	32	30	19	18	6	3	4
clay <2 μm %	12	14	16	22	23	29	10
organic carbon %	2.2	0.9	0.7	0.9	0.5	0.1	0.1
$CaCO_3$ equiv. %	0.7	0	0	0	0	0	0
dithionite ext. Fe %	0.8	0.8	1.2	1.7	4.1	7.9	1.3
TEB (me/100 g)	16.3	10.7	10.2	12.3	10.8	16.6	4.9
CEC (me/100 g)	19.0	13.0	12.5	14.7	13.1	19.5	5.9
% base saturation	86	82	82	84	82	85	83
pH (H_2O)	6.7	6.5	6.6	6.7	6.3	6.2	6.8
(0.01M $CaCl_2$)	6.3	6.1	5.9	6.2	5.6	5.5	6.3

7.6.3 Stagnoluvic Gley Soils

These are non-humic luvic gley soils with a loamy or sandy eluvial horizon and an appreciably finer textured B horizon that is slowly permeable or immediately overlies a slowly permeable substratum (Plate IV.3). Identified as typical (argillic) and sandy stagnogley soils by Avery (1973, 1980) and as Pseudogley or Staugley (Ehwald et al., 1966) in other European classifications, they are much more extensive in the British Isles than those of the ground-water subgroup. Although a major proportion of the total area covered is in England and Wales, some of the soils mapped as brown forest soils with gleying or non-calcareous gleys by the Soil Survey of Scotland and as surface-water or podzolic gleys in Ireland are also included.

The main parent materials are medium and fine textured tills and glaciolacustrine deposits, weathering products of siltstones and silty shales, and thin loamy or sandy drift over argillaceous rocks. Most of the soils have a lithologic discontinuity at some depth in the profile and some show little micromorphological evidence of clay translocation. Of the four representative profiles recorded on pp. 366–70, two are under broadleaf woodland and two in agricultural land. As the first three all have an abrupt textural change (Soil Survey Staff, 1975, p. 47) at the upper boundary of a slowly permeable argillic B horizon, they qualify as Albaqualfs in the US Taxonomy and as Planosols in the FAO system. Similar soils in which the clay increase is more gradual often occur in close proximity, however, and have been placed in the same soil series.

Profile 153 is in a flat-topped spur of the dissected plateau underlain by chalky Anglian till (Chalky Boulder Clay) which forms most of central East Anglia (Hodge et al., 1984). The Anglian deposits in this area are covered by variable thicknesses of loamy drift (cover-loam) believed to consist of Devensian cover-sand which was irregularly mixed by cryoturbation with the underlying materials (Perrin et al., 1974). The resulting composite parent materials have been modified by post-Devensian pedogenesis involving decalcification and redistribution of clay-size particles, so giving rise to a range of soils which vary considerably in horizonation and texture over short distances, but stagnoluvic gley soils (Beccles or Holderness series) predominate where the ground surface is level or gently sloping. In the woodland profile described, the prominently mottled sandy clay loam Btg horizon rests on chalky till and is overlain by strongly leached and coarser textured Eg and Ah horizons. Neighbouring soils have a clayey 2Btg horizon which has a much higher silt/sand ratio than the overlying horizons and appears to have weathered in place from the till. The undrained soils are waterlogged to within 40 cm of the surface in winter but become hard and dry to considerable depths in summer.

Profile 154 is in loamy Devensian till derived mainly from Triassic red beds in the same area as profile 109 (p. 296). The parent material is non-calcareous or deeply decalcified and the Eg horizon is thicker and more clearly differentiated by colour than in the preceding profile. Thin sections showed that as much as 10 per cent of the fabric of the Btg and BCtg horizons consists of discrete strongly oriented clay bodies, including well defined argillans and concentrations unassociated with existing voids. The marked clay increase at the base of the Eg is therefore attributable mainly to eluviation, though the sand/silt ratios show that there is also a lithologic discontinuity at about the same depth. Judging from the physical data, the Btg and BCtg horizons are both slowly permeable and dense enough to restrict rooting to a marked degree. The overlying horizons are much less dense but the Ap has a distinctly

Figure 7.10: Stagnoluvic gley soil (Macamore series) in 'Irish Sea Drift' of Devensian (Midlandian) age in County Wexford (from Gardiner and Ryan, 1964).

smaller air capacity than the Eg, probably because it has been poached by grazing stock. Stagnoluvic gley soils of similar or finer texture (Claverley, Clifton, Dunkeswick, Holderness, Pinder, Rufford and Salop series) occupy large areas on reddish and brownish Devensian and Wolstonian tills in the Midlands and north of England (Ragg *et al.*, 1984; R.A. Jarvis *et al.*, 1984) and are also represented in similar materials in southern Scotland and Ireland. In many of these soils, however, the increase in clay content with depth is due primarily to sedimentary layering. Those in Scotland have been mapped with related orthic gley soils as imperfectly or poorly drained members of a number of associations differentiated by the lithology or stratigraphic age of the source rocks (Soil Survey of Scotland, 1984). Similar till-derived soils in Ireland include some or all of those mapped as Macamore (Figure 7.10) and Rathangan series in County Wexford (Gardiner and Ryan, 1964), Gortaclareen series in Limerick and Clare (Finch and Ryan, 1966; Finch *et al.*, 1971) and Dunnstown, Greenane, Kilpatrick, Mylorstown, Street and Toberbride series in Carlow, Kildare and Westmeath (Conry and Ryan, 1967; Conry *et al.*, 1970; Finch and Gardiner, 1977).

Profile 155 is under ancient deciduous woodland in the New Forest, Hampshire. The parent material is loamy Head over greenish glauconitic Tertiary clay (Bracklesham Beds). Water seeps downslope above the lithologic discontinuity, which is marked by a sharp increase in clay content at 39 cm depth. The underlying slowly permeable clayey horizons are designated as argillic (Btg) on the basis that thin sections at 59 and 105 cm showed a few fine argillans lining pores, together with masepic plasmic fabrics and stress cutans indicative of pressures caused by swelling, but amounts of translocated clay in these horizons are probably very small.

Profile 156 is located south-west of Ludlow, Shropshire, on soft Silurian siltstone which may be calcareous in the unweathered state. Here the morphology and particle-size data are consistent with the soil having formed more or less in place from the underlying rock, and the presence of macroscopically identifiable clay coats in the B horizon further supports the conclusion that the moderate increase in clay content is mainly of pedogenic origin.

Profile 153 (Corbett and Tatler, 1970)

Location	:	Gillingham, Norfolk (TM 417937)
Climatic regime:		subhumid temperate
Altitude	:	30 m Slope: <1°
Land use and vegetation	:	broadleaf woodland; oak (*Quercus robur*) and ash (*Fraxinus excelsior*) standards with understorey of birch (*Betula* spp.) and hazel (*Corylus avellana*); field layer dominated by bramble (*Rubus fruticosus*)

Horizons (cm)

L	Litter of loose leaves and twigs, 5 cm thick.
0–1 F/H	Very dark brown (5YR 2/2), organic matter, mainly fibrous, with uncoated sand grains; sharp boundary.
1–7 Ah	Dark reddish brown (5YR 3/2) stoneless humose medium sandy loam (USDA sandy loam); weak medium and coarse granular; very friable; abundant fine and few medium and coarse woody roots; clear boundary.
7–20 Eg	Brown to dark brown (7.5YR 4/4) and greyish brown (10YR 5/2) coarsely mottled medium sandy loam (USDA fine sandy loam) containing rare small angular flint fragments; few fine strong brown (7.5YR 5/6) mottles; weak granular to massive; friable; many fine and common medium and coarse woody roots; clear boundary.
20–70 Btg	Light grey (5Y 7/2) and strong brown (7.5YR 5/6) prominently and coarsely mottled sandy clay loam (USDA sandy clay loam) with rare flint fragments as above; moderate coarse to very coarse (with increasing depth) angular blocky; firm; few to common

Analytical data for Profile 153

Horizon	Ah	Eg	Btg	2BCg[a]
Depth (cm)	1–7	7–20	20–70	70–150
sand 600 μm–2 mm %	2	2	2	1
200–600 μm %	30	28	34	7
60–200 μm %	23	24	22	9
silt 2–60 μm %	32	33	15	34
clay <2 μm %	13	13	27	49
<0.2 μm %	7	7	17	26
fine clay/total clay	0.54	0.54	0.63	0.53
organic carbon %	7.7	2.9	0.4	0.3
$CaCO_3$ equiv. %	0	0	0	45
pyro. ext. Fe %	0.2	0.3	0.1	
pyro. ext. Al %	0.1	0.1	0.1	
total ext. Fe %	0.6	0.7	2.2	
TEB (me/100 g)	4.5	1.0	6.9	
CEC (me/100 g)	24.6	15.9	13.6	
% base saturation	18	12	51	
pH (H_2O)	4.3	4.2	4.9	
(0.01M $CaCl_2$)	3.5	3.6	4.1	

[a] Particle-size distribution in decalcified soil <2 mm.

70–150 2BCg
: fine and few medium woody roots; clear boundary.
Light olive-grey (5Y 6/2), slightly to moderately stony, extremely calcareous clay loam with many medium light olive-brown (2.5Y 5/4) and yellowish brown (10YR 5/4–6) mottles becoming fewer and less prominent with depth; common to many subangular chalk fragments, mainly very small and small but ranging up to 6 cm; rare subangular flint fragments; weak coarse blocky to massive; very firm; roots rare.

Analytical data *see* above.

Classification

England and Wales (1984): typical stagnogley soil; fine loamy chalky drift (Holderness series).

USDA (1975): Aeric Albaqualf; fine loamy, mixed, mesic.

FAO (1974): Eutric Planosol; medium textured.

Profile 154 (Jones, 1975)

Location : Seighford, Staffordshire (SJ 869234)

Climatic regime: humid temperate
Altitude : 84 m   Slope: 1°
Land use : long-term grass

Horizons (cm)

0–18 Ahg
: Very dark greyish brown (10YR 3/2) very slightly stony medium sandy loam (USDA fine sandy loam) with many dark yellowish brown (10YR 3/4) mottles; few, small to large, sub-rounded quartzite fragments; weak to moderate, medium and fine granular; very friable; many fine roots; earthworms active; clear irregular boundary.

18–40 Eg
: Brown (7.5YR 4/2) slightly stony medium sandy loam (USDA sandy to fine sandy loam) with many very fine and fine brown to dark brown (7.5YR 4/4) mottles and many fine yellowish brown (10YR 5/4–6) mottles near base; stones as above; weak granular; very friable; common fine roots and many earthworm channels with very dark greyish brown infillings; abrupt irregular boundary.

40–60 Btg
: Reddish brown (5YR 4/4) slightly stony clay loam (USDA clay loam)

Analytical data for Profile 154

Horizon	Ahg	Eg	2Btg	2BCtg(x)
Depth (cm)	0–18	18–40	40–60	60–98
sand 200 μm–2 mm %	32	36	22	19
100–200 μm %	23	20	13	11
50–100 μm %	10	10	8	9
silt 2–50 μm %	20	20	25	29
clay <2 μm %	15	14	32	32
organic carbon %	4.5	0.6	0.4	0.2
CaCO₃ equiv. %	0.2	0	0	0
dithionite ext. Fe %	0.6	0.8	1.3	1.3
pH (H₂O)	6.7	6.6	6.5	7.0
(0.01M CaCl₂)	6.2	6.0	5.9	6.2
bulk density (g cm⁻³)	1.12	1.36	1.58	1.63
packing density (g cm⁻³)	1.26	1.49	1.86	1.91
retained water capacity (% vol.)	46	29	34	33
available water capacity (% vol.)	25	17	11	11
air capacity (% vol.)	9	17	5	5

with many fine and medium strong brown (7.5YR 4/8), brown (10YR 5/3) and yellowish brown, and many fine and very fine greyish brown (2.5Y 5/2) mottles; weak coarse prismatic with greyish brown and grey (5Y 5/1) faces; firm; few fine roots; many ferri-manganiferous nodules; gradual boundary.

60–98 BCtg(x)   Weak red to reddish brown (2.5YR 4/3) very slightly stony clay loam (USDA clay loam) with many medium and fine light brownish grey (2.5Y 6/2) and greyish brown (2.5Y 5/2) mottles; massive with widely spaced grey-coated cleavage faces; very firm; roots rare.

Analytical data *see* above.

Classification
England and Wales (1984): typical stagnogley soil; reddish coarse to fine loamy drift with siliceous stones (intergrade between Claverley and Clifton series).
USDA (1975): Udollic Albaqualf (or Fragiaqualf); fine loamy, mixed, mesic.
FAO (1974): Eutric Planosol; medium textured.

Profile 155 (Jarvis and Findlay, 1984)

Location : Denny Lodge Inclosure, New Forest, Hampshire (SU 344045)
Climatic regime: subhumid temperate
Altitude : 30 m   Slope: 5°
Land use and vegetation : broadleaf woodland; 80 per cent canopy of beech (*Fagus sylvatica*) and occasional holly (*Ilex aquifolium*); 50 per cent ground cover of creeping bent-grass (*Agrostis stolonifera*) with occasional mosses, sedges and rushes (*Juncus* spp.)

Horizons (cm)

0–2 L/F   Partly decomposed litter; sharp boundary.

2–3 H   Very dark greyish brown (10YR 3/2) well decomposed litter with many roots; sharp boundary.

3–4 Ahg   Very dark greyish brown (10YR 3/2) humose fine sandy loam with common fine yellowish brown (10YR 5/6) mottles; medium granular; very friable; sharp broken boundary (horizon impersistent, up to 2 cm thick).

4–16 Eag	Grey (10YR 5/1) very slightly stony sandy silt loam (USDA silt loam to loam) with few fine strong brown (7.5YR 5/8) mottles and dark brown (7.5YR 4/4) linings to root channels; few rounded and subangular flint fragments; weak coarse blocky tending to platy in upper part and breaking to medium blocky below; friable; abundant roots; sharp smooth boundary.
16–39 Eg	Light brownish grey (2.5Y 6/2) very slightly stony sandy silt loam (USDA loam) with common strong brown (7.5YR 4–5/6) mottles and yellowish red (5YR 4/8) linings to root channels; few medium subangular flint fragments; moderate coarse angular blocky with colourless sand coats and dead root traces on many ped faces; friable; abundant roots; common irregular ferri-manganiferous nodules; clear irregular boundary.
39–69 2Btg1	Greenish grey (5GY 6/1) stoneless clay (USDA silty clay) with common strong brown and few yellowish red (5YR 5/8) mottles; moderate coarse prismatic breaking to smaller prisms and angular blocks; firm; common fine and few coarse woody roots, some dead; gradual smooth boundary.
69–140 2Btg2	Greenish grey (5GY 5/1) stoneless clay (USDA clay) with many reddish yellow (7.5YR 6/8) mottles; moderate very coarse angular blocky up to 25 cm with common smooth slickensided faces inclined at c. 60° to horizontal; very firm; few fine and coarse woody roots; few soft irregular ferri-manganiferous concentrations.

Analytical data *see* below.

Classification
   England and Wales (1984): typical stagnogley soil; coarse loamy drift over clayey material passing to clay or soft mudstone (Kings Newton series).
   USDA (1975): Typic Albaqualf; fine, mixed?, mesic.
   FAO (1974): Dystric Planosol; medium textured.

Analytical data for Profile 155

Horizon		Eag	Eg	2Btg1	2Btg2
Depth (cm)		4–16	16–39	39–69	69–140
sand	200 μm–2 mm %	5	5	1	<1
	60–200 μm %	30	30	8	1
silt	2–60 μm %	53	49	43	37
clay	<2 μm %	12	16	48	62
	<0.2 μm %	6	8	28	29
	fine clay/total clay	0.50	0.50	0.58	0.47
organic carbon %		2.6	0.7		
total ext. Fe %		0.5	1.2	2.9	2.5
pH (H$_2$O)		4.4	4.8	5.0	6.6
(0.01M CaCl$_2$)		3.9	4.0	4.3	6.1
bulk density (g cm^{-3})		1.36	1.39	1.45	1.42[a]
packing density (g cm^{-3})		1.46	1.53	1.88	1.97[a]
retained water capacity (% vol.)		36	38	42	48[a]
available water capacity (% vol.)		24	20	14	16[a]
air capacity (% vol.)		10	10	3	<1[a]

[a] samples at 90–95 cm.

Profile 156 (Ragg et al., 1984)

Location	:	Aston, Hereford and Worcester (SO 462717)
Climatic regime:		humid temperate
Altitude	:	165 m Slope: 3° straight
Land use	:	long-term grass

Horizons (cm)

0–17 Apg — Dark greyish brown (2.5Y 4/2) stoneless silty clay loam (USDA silt loam) with many fine yellowish brown (10YR 5/6) mottles; strong medium subangular blocky; friable; abundant fine roots; abrupt smooth boundary.

17–37 Eg — Greyish brown (2.5Y 5/2) stoneless silty clay loam (USDA silt loam) with common very fine light olive-brown (2.5Y 5/4) mottles; moderate coarse subangular blocky with light brownish grey (2.5Y 6/2) faces; friable; many fine roots; abrupt smooth boundary.

37–65 Btg1 — Light brownish grey (2.5Y 6/2) stoneless silty clay loam (USDA silt loam) with common very fine yellowish brown (10YR 5/6) mottles; strong medium prismatic; firm; many fine roots; abrupt smooth boundary.

65–90 Btg2 — Light olive-brown (2.5Y 5/4) slightly stony clay loam (USDA silty clay loam) with many medium light olive-grey (5Y 6/2) mottles; common small angular and platy siltstone fragments; moderate medium prismatic with light olive-grey (5Y 6/2) faces; firm; common fine roots; common clay coats; common irregular soft ferri-manganiferous concentrations; gradual smooth boundary.

90–110 BCtg — Olive (5Y 5/3) slightly stony silty clay loam to silty clay (USDA silty clay loam) with many medium light grey to grey (5Y 6/1) mottles; stones as above; weak (adherent) angular blocky with some medium platy associated with weathered siltstone fragments; very firm; few fine roots; clay coats and ferri-manganiferous concentrations as above.

Classification

England and Wales (1984): typical stagnogley soil; fine silty material passing to silty shale or siltstone (Stanway series).

USDA (1975): Aeric Ochraqualf; fine-silty, mixed, mesic.

FAO (1974): Gleyic Luvisol; medium textured.

Analytical data for Profile 156

Horizon		Apg	Eg	Btg1	Btg2	BCtg
Depth (cm)		0–17	17–37	37–65	65–90	90–110
sand	200 μm–2 mm %	4	6	5	10	3
	60–200 μm %	5	4	3	7	4
silt	2–60 μm %	68	69	67	52	58
clay	<2 μm %	23	21	25	31	35
	<0.2 μm %	9	8	7	11	11
fine clay/total clay		0.39	0.38	0.28	0.35	0.31
organic carbon %		3.5	1.3			
pH (H_2O)		5.0	5.8	6.1	6.4	6.5
(0.01M $CaCl_2$)		4.4	5.0	5.3	5.4	5.7
bulk density (g cm^{-3})		1.05	1.25	1.44	1.58	1.57
packing density (g cm^{-3})		1.26	1.44	1.66	1.85	1.88
retained water capacity (% vol.)		52	39	37	36	40
available water capacity (% vol.)		30	25	15	10	11
air capacity (% vol.)		3	13	8	4	1

7.6.4 Chromic Stagnoluvic Gley Soils

Soils of this subgroup occur chiefly on older (pre-Ipswichian) drift and quasi-residual materials in relatively stable plateau sites where interglacially weathered subsurface horizons have survived with only limited disturbance or subsequent transformation. Typical profiles resemble those of the preceding subgroup in having Eg and slowly permeable Btg horizons but the latter are characterized, at least in their lower parts, by predominantly strong brown or redder colours, or a reticulate pattern of red and grey mottling. As in the related luvic brown soils (p. 228), there is generally a lithologic discontinuity at or below the level at which the clay content begins to increase, the upper horizons having formed in a more or less distinct superficial layer, often derived partly from loess, with a wavy or irregular lower boundary. In the characteristic chromic B horizon (Chapter 3, p. 109), thin sections of which normally show numerous strongly oriented clay bodies (papules or disrupted argillans) unassociated with voids, mottling becomes coarser and more prominent with increasing depth; proportions of reddish colours also generally increase and are accompanied by light grey mottles or streaks associated with clay coats lining fissures and around stones. Carbonates, if originally present, have been leached to considerable depths, and some profiles are less than 35 per cent base saturated to more than 1.25 m below the top of the B horizon, implying that they are Ultisols (Aquults) in the US system.

Surface-water gley soils with these characteristics have been identified and mapped as paleo-argillic stagnogley soils in England and Wales (Avery, 1973, 1980; Mackney *et al.*, 1983) but have not yet been recorded in Scotland or Ireland. They were first noted on Anglian and pre-Anglian deposits in the London basin (Thomasson and Avery, 1963; Sturdy, 1971; Jarvis *et al.*, 1979; Sturdy *et al.*, 1979) and have since been shown to occupy significant areas on plateau drift (Clay-with-flints-and-cherts) in East Devon (Harrod, 1971; Findlay *et al.*, 1984) and on 'Wolstonian' glacial and glaciofluvial deposits in the Midlands (Ragg *et al.*, 1984). They also occur sporadically on weathering products of Paleozoic rocks in the south-western peninsula (Harrod, 1978, 1981). Data are given on pp. 372–5 for two representative profiles, one under ancient oakwood invaded by birch and the second in a limed arable field recently converted from woodland. Both were described and sampled in Essex by R.H. Allen and R.G. Sturdy.

Profile 157 (Plate IV.4) from Epping Forest is in a slightly convex ridge top on Pebbly Clay Drift (Thomasson, 1961) over Pebble Gravel, which rests in turn on Tertiary Claygate Beds. The Pebble Gravel, which appears at about 95 cm depth, is generally interpreted as a pre-Anglian fluvial deposit (Rose, 1983). The Pebbly Clay Drift may also be pre-Anglian (Baker, 1971) though it could have formed by weathering of originally calcareous Anglian till, as in Northaw Great Wood (Catt *et al.*, 1985) 14 km to the west. There is a sharply demarcated organic layer at the surface of the soil and the strongly differentiated mineral horizons below it are all strongly acid, with percentage base saturation not exceeding 10 per cent throughout the depth of sampling. Incipient podzolization is revealed by dark brown inclusions of infiltrated organic matter in the strongly eluviated Eag horizon and confirmed by the distribution of pyrophosphate extractable iron and aluminium. Proportions of sand and gravel relative to silt are greater in the E and Btg1 horizons than in the overlying horizons, possibly because finer particles have been removed by lateral washing or coarser particles forced upwards by cryoturbation, as suggested by the occurrence of vertically oriented stones. Micromorphological data reported by Murphy and Bullock (1981) show common to many (3–6 per cent) void-argillans in each of the Btg subhorizons sampled, together with sharply defined red and yellow papules, irregular and linear intrapedal clay concentrations and common red ferruginous segregations which appear as red mottles to the unaided eye.

Profile 158 is sited in a low interfluve north-east of Chelmsford. It shows a similar succession of horizons but the eluvial horizon is finer in texture and less strongly expressed. Below it is a thick layer of prominently mottled slightly flinty clay which has Bt-horizon characteristics throughout, is intercalated with sandier inclusions at 164–192 cm, and rests sharply at 210 cm on slightly weathered Anglian till (Chalky Boulder Clay). The sand/silt ratios suggest that horizons down to and including the upper part of the Btg are derived partly from admixed loess. This was confirmed by mineralogical analyses of fine sand and coarse silt fractions from a similar profile in the same area by Sturdy et al. (1979). Their results showed that amounts of weatherable minerals, including biotite, chlorite, glauconite, apatite, collophane, pyroxenes and amphiboles, were significantly smaller in the strongly coloured 'paleo-argillic' B horizon below the discontinuity than in either the surface layer or the underlying till, indicating that fresh minerals were introduced as part of the loess deposit and that weathering since the loess addition has been less severe than that which preceded it. The data were also consistent with the lower part of the Btg horizon having developed *in situ* by decalcification of the till, presumably during the Hoxnian or Ipswichian interglacial, or both. In the profile described here, the texturally heterogeneous layer at 164–192 cm appears to have been reworked, so only the underlying horizon can be considered as a direct decalcification product. The overlying horizon (2Btg) resembles the non-calcareous residue of the till in particle-size distribution but could be a locally derived head deposit, in which case the mineral alteration may have taken place after its transportation. As very similar deeply weathered soils with red mottled B horizons occur on calcareous till dated as Wolstonian (Table 1.1, p. 14) near Moreton-in-Marsh (Ragg et al., 1984) and in the country south of Birmingham (Beard, 1984), it seems that those on both Anglian and Wolstonian tills acquired their distinctive characteristics as a result of Ipswichian pedogenesis, or that the tills were in fact all emplaced during the Anglian cold stage (Bowen et al., 1986; Rose, 1987).

Profile 157 (Murphy and Bullock, 1981)

Location	:	Deer Shelter Plain, Epping Forest, Essex (TQ 427990)
Climatic regime:		subhumid temperate
Altitude	:	107 m Slope: 1° convex
Land use and vegetation	:	broadleaf woodland; birch (*Betula* spp.) coppice with oak (*Quercus robur*) pollards; bracken (*Pteridium aquilinum*), wavy hair-grass (*Deschampsia flexuosa*) and purple moor-grass (*Molinia caerulea*) in field layer

Horizons (cm)

L	Litter of bracken fronds and stems, birch leaves and fruits and grass leaves, about 2 cm thick.
0–4 F	Dark brown (5YR 2/2) partially decomposed litter, loose and weakly laminated; abundant roots; sharp boundary.
4–8 H	Black (7/5YR 2.1) stoneless organic material containing uncoated sand grains; moderate coarse granular; very friable; common medium and fine woody roots and bracken rhizomes; sharp smooth boundary.
8–25 Eag	Light brownish grey (10YR 6/2) moderately stony sandy silt loam (USDA silt loam to loam) with common medium dark brown (7.5YR 4/2) and few ochreous mottles; many medium flint pebbles and few large subangular flint and rounded quartz fragments; some stones vertically oriented; very weak medium subangular blocky; friable; common medium woody roots; clear smooth boundary.

Analytical data to Profile 157

	H	Eag	Btg	2Btg1	2Btg2
Horizon					
Depth (cm)	4–8	8–25	25–40	40–65	65–95
sand 600 μm–2 mm %		6	5	3	1
200–600 μm %		5	6	2	<1
60–200 μm %		27	19	10	8
silt 2–60 μm %		52	41	36	40
clay <2 μm %		10	29	49	51
<0.2 μm %		5	17	32	31
fine clay/total clay		0.50	0.59	0.65	0.61
organic carbon %	25	2.8	1.0	0.6	0.2
pyro. ext. Fe %	0.3	0.4	0.5	0.2	0.1
pyro. ext. Al %	0.1	0.1	0.2	0.2	0.2
Fe +Al %/clay %		0.05	0.02	0.01	<0.01
total ext. Fe %		0.6	4.1	4.3	3.6
TEB (me/100 g)	5.3	0.4	0.6	0.8	1.7
CEC (me/100 g)	42	14.9	13.5	16.3	17.4
% base saturation	13	3	4	5	10
pH (H$_2$O)	3.8	3.7	4.0	4.0	4.1
(0.01M CaCl$_2$)	3.0	3.3	3.5	3.6	3.6
bulk density (g cm^{-3})	0.33	1.13	1.24	1.39	1.46
packing density (g cm^{-3})	0.46	1.22	1.68	1.84	1.91
retained water capacity (% vol.)	49	38	50	44	43
available water capacity (% vol.)	37	25	14	11	11
air capacity (% vol.)	38	18	3	3	2

25–40 Btg1 — Light brownish grey (2.5Y 6/2) slightly stony clay loam (USDA clay loam) with many medium strong brown (7.5YR 5/8) and few yellowish red (5YR 5/8) mottles; stones as above; moderate coarse angular blocky with light brownish grey faces; firm; common woody roots, mainly medium and some dead; clear wavy boundary.

40–65 2Btg1 — Strong brown (7.5YR 5/8) slightly stony clay (USDA clay) with many medium grey (5Y 5/1) and fine red (2.5YR 5/8) mottles; stones as above; wet; moderate coarse angular blocky with grey (10YR 5/1) faces and stone impressions; common roots as above; many dead root traces on ped faces; gradual smooth boundary.

65–95 2Btg2 — Strong brown (7–5YR 5/8) slightly stony clay (USDA clay) with many medium grey (5Y 5/1) and common but localized medium red (2.5YR 4/8) mottles; stones as above; wet; structure as above; few woody roots.

95–170 3BCt — (by auger) Reddish brown to yellowish red stony sandy clay loam.

Analytical data see above.

Classification

England and Wales (1984): paleo-argillic stagnogley soil; coarse loamy over clayey; drift with siliceous stones (Essendon series).

USDA (1975): Aeric Albaquult; fine, mixed?, mesic.

FAO (1974): Dystric Planosol; medium textured.

Profile 158 (Allen and Sturdy, 1980)

Location   : Hatfield Peverel, Essex (TL 765120)
Climatic regime: subhumid temperate
Altitude   : 43 m   Slope: 0.5°
Land use   : arable (after sugar beet), reclaimed from woodland three years previously

Horizons (cm)

Depth	Horizon	Description
0–25	Ap	Dark greyish brown (10YR 4/2) very slightly stony clay loam (USDA loam to clay loam) with very dark grey (10YR 3/1) inclusions and common brown to strong brown (7.5YR 5/5) mottles; few flint fragments and quartzose pebbles; very weak medium and coarse subangular blocky; firm (compacted); few living roots; undecayed crop residues and woody fragments; earthworms present; added lime; clear boundary.
25–42	Eg	Yellowish brown (10YR 5/4) and pale brown (10YR 6/3), intricately mottled, slightly stony clay loam (USDA loam to clay loam); also common fine reddish brown to dark brown (5–7.5YR 4/4–6) and dark greyish brown (2.5Y 4/2) mottles; stones mainly flint; weak medium and coarse subangular blocky; friable to firm; living roots absent, but few fine dead roots; few earthworm burrows; common fine black ferri-manganiferous soft concentrations; gradual boundary.
42–56	Btg	Strong brown (7.5YR 5/6–8) very slightly stony clay (USDA clay) with many coarse light brownish grey (2.5Y 6/2) and dark reddish brown (5YR 3/2) mottles; moderate coarse angular blocky parting in places to fine, with mainly light brownish grey glazed faces; firm; common tree root channels and associated organic staining; earthworms present; few black ferri-manganiferous concentrations; diffuse boundary.
56–79	2Btg1	Strong brown to yellowish brown (7.5–10YR 5/6) slightly stony clay (USDA clay) with many coarse light brownish grey (2.5Y 6/2) and few red mottles; stones, structure, consistence and root channels as above; diffuse boundary.
79–126	2Btg2	Strong brown to yellowish brown (7.5–10YR 5/6) slightly stony clay (USDA clay) with common medium red (2.5YR 4/6) and coarse greenish

Analytical data for Profile 158

Horizon	Ap	Eg	Btg	2Btg1	2Btg2	2Btg3	3BCt(g)[a]	4BCt(g)	4BC(g)[b]
Depth (cm)	0–25	25–42	42–56	56–79	79–126	126–164	164–192	192–210	210–244
sand 600 µm–2 mm %	4	5	3	4	4	4	10	6	6
200 µm–600 µm %	11	8	8	11	14	15	35	17	15
60–200 µm %	10	10	7	10	14	16	19	16	15
silt 2–60 µm %	48	50	33	24	21	21	12	23	22
clay <2 µm %	27	27	49	51	47	44	24	38	42
<0.2 µm %	14	14	31	32	25	22	14	19	21
fine clay/total clay	0.52	0.52	0.63	0.62	0.54	0.50	0.58	0.50	0.50
organic carbon %	1.4	0.8	0.6		0.3		0.1		0.2
CaCO$_3$ equiv. %	tr.	0	0	0	0	0	0	0	26
TEB (me/100 g)	16.5	7.2	15.8	18.1	16.4	14.5		17.3	
CEC (me/100 g)	19.9	14.9	26.1	27.0	26.1	22.7		18.3	
% base saturation	83	48	61	67	63	64		95	
pH (H$_2$O)	6.5	4.6	4.3	4.5	4.5	4.5	4.9	6.3	7.8
(0.01M CaCl$_2$.)	6.0	3.9	3.7	3.9	3.7	3.7	4.0	5.4	7.2

[a] Sandy clay inclusion only
[b] Particle-size distribution in decalcified soil <2 mm.

	grey (5GY 6/1) mottles, the latter occurring as streaks associated with ped faces; stones as above; weak coarse prismatic parting to coarse and medium angular blocky; firm; few very fine fissures, some containing dead flattened roots; diffuse boundary.
126–164 2Btg3	Strong brown to yellowish brown (7.5–10YR 5/6) slightly stony clay (USDA clay) streaked and mottled with greenish grey as above; common medium red mottles locally; stones as above; weak very coarse angular blocky breaking to medium; firm; few fine dead woody roots; fine black manganiferous speckling on some cleavage faces; diffuse boundary.
164–192 3BCt(g)	Strong brown (7.5YR 5/6) friable sandy clay loam (USDA sandy clay loam) occurring as inclusions in strong brown and grey mottled slightly stony clay; few large nodular flints; clear boundary.
192–210 4BCt(g)	Dark yellowish brown (10YR 4/6) to dark brown (7.5YR 4/6) slightly stony clay (USDA clay loam) with greenish grey (5GY 6/1) streaks; rounded and nodular flint fragments; abrupt boundary.
210–224 4BC(g)	Yellowish and dark yellowish brown (10YR 5–4/6) moderately stony very calcareous clay (USDA clay) with common grey (5Y 6/1) and few dark brown (7.5YR 4/6) mottles; many chalk and few flint fragments; massive; very firm.

Analytical data see p. 374.

Classification
  England and Wales (1984): paleo-argillic stagnogley soil; fine loamy over clayey; drift with siliceous stones (Oak series).
  USDA (1975): Aeric Ochraqualf; fine, mixed or montmorillonitic, mesic.
  FAO (1974): Gleyic Luvisol; medium textured.

7.6.5 Peloluvic Gley Soils

These soils are clayey throughout the profile or to within 30 cm of the surface and resemble the corresponding orthic gley soils (p. 348) in this and related properties but differ in having argillic B horizons as defined in Chapter 3 (p. 109). In England and Wales, where clayey surface-water gley soils of both types have been grouped as pelo-stagnogley soils in recent years, they are widely distributed on slowly permeable argillaceous rocks (Plate IV.5) and clayey Pleistocene deposits, including the Midland and East Anglian Chalky Boulder Clays (Ragg et al., 1984; Hodge et al., 1984) and Devensian tills and glaciolacustrine deposits in northern England (R.A. Jarvis et al., 1984). They also occur on fine textured calcareous or base-rich tills in central Ireland (associations 27 and 39 in Gardiner and Radford, 1980) and in the Scottish lowlands, for example as members of the Ashgrove association in Ayrshire (Mitchell and Jarvis, 1956; Bown et al., 1982). The argillic B horizons in these soils are normally weakly expressed; there is often no distinct E horizon, especially in cultivated land, and the increase in clay content with depth can be mainly of geological origin.

Data on two representative profiles from England and Wales are given on pp. 376–8. In each case microscopic studies of thin sections showed only a few positively identifiable illuviation argillans in or below the B horizon, which is designated Btg on the basis that it has a complex sepic plasmic fabric (Brewer, 1964; Murphy, 1984) and significantly more silicate clay, with a higher proportion of fine (<0.2 µm) clay than overlying and underlying horizons.

Profile 159, on Jurassic (Oxford) clay east of Oxford, was exposed at the edge of a young spruce plantation which replaced ancient oakwood with hazel coppice. It is non-calcareous to about 53 cm depth, below which there are less weathered BC horizons enriched successively in secondary calcium carbonate and gypsum. The unaltered parent rock below the depth of sampling is a dark grey calcareous clay-shale containing appreciable amounts of pyrite and organic carbon.

As the decalcified Btg horizon contains much more clay than the surface horizon and shows many inclined pressure faces (slickensides) in the lower part, it seems that disturbance resulting from seasonal swelling and shrinking has caused illuvial clay to be incorporated in the matrix as proposed by Nettleton *et al.* (1969), Walsh and De Coninck (1973) and Holzhey and Yeck (1974). The presence of uncoated sand and silt grains on ped faces in the overlying horizon designated EBg can also be taken as evidence of clay translocation (Soil Survey Staff, 1975, pp. 25–27), though comparison of the sand/silt ratios with those in the BCg horizons indicates that the sand-size particles are almost certainly of extraneous origin. Clay separates from the Btg and BCgk horizons analysed by Avery and Bullock (1977) consist dominantly of interstratified mica-smectite with a high proportion of smectite layers, and Reeve *et al.* (1980) obtained COLE values ranging from 0.10 to 0.19 for corresponding horizons of similar Denchworth soils. The profile is therefore tentatively identified as a Vertic Ochraqualf in the US Taxonomy.

Profile 160 on Chalky Boulder Clay was described and sampled by D. Mackney in north Buckinghamshire. By contrast with the pelo-calcaric brown soils (Hanslope series) on similar material represented by profile 53 (p. 192) from the same area, this soil has a prominently mottled decalcified B horizon containing more total and fine clay than both the Ap and the underlying BCg. The air capacity and packing density data indicate that the subsurface horizons are very slowly permeable at moisture states approximating to field capacity. Hence, as studies reviewed by Thomasson and Bullock (1975) have confirmed, water movement in the wet season is almost entirely confined to the surface horizon, those below remaining saturated or nearly so until evapotranspiration again exceeds rainfall at some time in the spring. The clay fractions consist dominantly of interstratified mica-smectite, with mica and kaolinite in smaller amounts, but shrinkage measurements on a similar soil by Reeve *et al.* (1980) showed that the profile as a whole had a distinctly smaller shrink-swell potential than those of the Denchworth soils examined.

Profile 159 (Thomasson and Bullock, 1975; Avery and Bullock, 1977)

Location	:	Worminghall, Buckinghamshire (SP 608103)
Climatic regime:		subhumid temperate
Altitude	:	69 m Slope: * <1°
Land use and vegetation	:	Sitka spruce plantation about 12 years old, replacing oakwood with hazel coppice; shrub and field layers include wild rose (*Rosa* spp.), honeysuckle (*Lonicera periclymenum*), tufted hairgrass (*Deschampsia caespitosa*), Yorkshire fog (*Holcus lanatus*) and rushes (*Juncus* spp.)

Horizons (cm)

L	Grass, moss and spruce litter, discontinuous.
0–17 Ahg	Dark greyish brown (10YR 4/2; 5/2 dry) stoneless silty clay loam to silty clay (USDA silty clay loam), darker and humose in upper 5 cm, with common fine reddish brown mottles; dry; strong medium subangular blocky, breaking in places to fine and very fine; friable when moist, hard when dry; abundant fibrous and woody roots up to 1 cm thick; clear boundary.
17–37 EBg	Grey (5Y 6/1), light brownish grey (2.5Y 6/2) and strong brown to yellowish red (7.5–5YR 4/6), prominently mottled stoneless silty clay (USDA silty clay); dry; moderate coarse angular blocky with rough grey faces, some showing uncoated silt and fine sand grains; c. 4 mm vertical fissures at about 15 cm intervals; firm when moist and very hard when dry; many roots up to 1 cm; clear boundary.
37–53 Btg1	Dark grey (5Y 4/1) and strong brown to yellowish red, prominently

53–69 Btg2	mottled stoneless clay (USDA clay); strong very coarse prismatic with some coarse angular blocky; fissures as above, decreasing in width with depth; very firm when moist and extremely hard when dry; common, woody roots, mainly medium and fine; gradual boundary. Grey (5Y 5/1) and yellowish brown (10YR 5/8), prominently mottled, stoneless slightly calcareous clay (USDA clay); moderate coarse and medium angular blocky with many grey shiny faces (slickensides) oriented at various angles to the horizontal; consistence as above; few woody roots; gradual boundary.
69–125 2BCgk	Grey (N 5.0) and yellowish brown (10YR 5/6–8), prominently mottled, stoneless very calcareous clay (USDA clay); coarse angular blocky with common inclined slickensided faces becoming more widely spaced with increasing depth; fissures closing within this horizon; consistence as above; few woody roots; common to many, irregularly distributed nodules and soft concentrations of secondary $CaCO_3$; few fossil *Gryphaea* shells; gradual boundary.
125– 135+ 2BCgy	Grey (N 5/0) stoneless calcareous clay (USDA clay) with fewer yellowish brown mottles; apparently drier than above; weak very coarse angular blocky to massive; very firm; roots rare; common irregularly distributed 'sugary' deposits of finely crystalline gypsum; few *Gryphaea* shells as above.

Analytical data *see* below.

Classification

England and Wales (1984): pelo-stagnogley soil; swelling-clayey material passing to clay or soft mudstone (Denchworth series).

USDA (1975): Vertic(?) Ochraqualf; very-fine, montmorillonitic, mesic.

FAO (1974): Gleyic (or Vertic) Luvisol; fine textured.

Analytical data for Profile 159

Horizon	Ahg	EBg	Btg1	Btg2	BCgk	BCgk	BCgy
Depth (cm)	0–17	17–37	37–53	53–69	69–79	79–125	125–135
sand[a] 200 µm–2 mm %	7	3	1	1	1	<1	1
50–200 µm %	9	7	5	4	3	1	1
silt[a] 2–50 µm %	49	49	31	32	38	37	39
clay[a] <2 µm %	35	41	63	63	58	62	59
<0.2 µm %	19	22	40	34	31		27
fine clay/total clay	0.54	0.54	0.63	0.54	0.53		0.46
organic carbon %	6.5	2.2	1.0	0.8	0.6	0.7	0.7
$CaCO_3$ equiv. %	0	0	0	3.5	23	15	5.1
dithionite ext. Fe %	1.9	2.7	3.3	2.0	2.2	1.5	
pH (H_2O)	5.3	5.3	6.7	8.0	8.5	8.3	7.5
(0.01M $CaCl_2$)	4.5	4.5	6.4	7.5	7.9	7.8	7.4
clay <2 µm							
CEC (me/100 g)			63	63	58		
K_2O %			2.7	2.7	2.9		

[a] particle-size distribution in decalcified soil <2 mm

Profile 160

Location	: Great Horwood, Buckinghamshire (SP 745321)
Climatic regime:	subhumid temperate
Altitude	: 149 m Slope: 1°
Land use	: arable (cereals), recently ploughed

Horizons (cm)

0–25
Ap
Dark greyish brown to dark brown (10YR 4/2-3) very slightly stony silty clay (USDA silty clay loam to silty clay) with irregular inclusions of prominently mottled clay (ploughed up from subjacent horizon); few weathered flint fragments and quartzite pebbles, coarse angular blocky (cloddy) with strong fine and very fine angular blocky (frost mould) in top 5 cm and massive 4 cm thick plough pan at base; firm; common roots and ploughed-in stubble; common earthworms and casts; abrupt smooth boundary.

25–67
Btg
Greyish brown (2.5Y 5/2), strong brown (7.5YR 5/6) and grey (5Y 6/1), intricately and finely mottled, very slightly stony clay (USDA silty clay to clay); moist to wet, with muddy water films on ped faces in lower part; strong medium prismatic to angular blocky with smooth, shiny and wavy, greyish brown (2.5Y 5/2) and grey (5Y 6/1) faces; very firm when moist; few fine roots; abundant fine black to dark brown ferri-manganiferous nodules; abrupt wavy boundary.

67–105
BCgk
Yellowish brown (10YR 5/4), slightly to moderately stony, very calcareous clay (USDA silty clay to clay) with many coarse grey (5Y 6/1) mottles, some as irregular sub-vertical streaks; common chalk pebbles, mainly small, and few flint and quartzite fragments as above; moist; massive; very firm; fine root channels outlined in grey (5Y 6/1); secondary carbonate as nodules and soft white streaks.

Analytical data

Horizon	Ap	Btg	BCgk
Depth (cm)	0–25	25–67	67–105
sand 200 μm–2 mm %	5	5	7
50–200 μm %	11	7	7
silt 2–50 μm %	45	40	40
clay <2 μm %	39	48	46
<0.2 μm %	20	27	17
fine clay/total clay	0.51	0.57	0.38
organic carbon %	3.4	0.8	
$CaCO_3$ equiv.			
<2 mm %	0	tr.	29
<2 μm %	0	0	6.8
dithionite ext. Fe %	2.3	3.0	2.0
pH (H_2O)	6.3	7.3	8.0
(0.01M $CaCl_2$)	6.0	7.0	7.6
clay <2 μm			
CEC (me/100 g)	51	53	41
K_2O %	1.6	1.5	1.8
bulk density (g cm^{-3})	1.04	1.41	1.68
packing density (g cm^{-3})	1.40	1.84	2.09
retained water capacity (% vol.)	47	43	34
available water capacity (% vol.)	22	14	11
air capacity (% vol.)	13	4	3

Classification
England and Wales (1984): pelo-stagnogley soil; clayey chalky drift Ragdale series).
USDA (1975): Typic (or Aeric) Ochraqualf; fine, montmorillonitic or mixed, mesic.
FAO (1974): Gleyic Luvisol; fine textured.

7.6.6 Humic Luvic Gley Soils

These are luvic gley soils with humose or peaty topsoils. Termed argillic humic gley soils by Avery (1980), they are common on Devensian fluvial or aeolian deposits in parts of the Cambridgeshire Fens (Hodge *et al.*, 1984) but have yet to be positively identified elsewhere. As in associated humic calcaric gley soils represented by profiles 137 and 138 (p. 337), the very dark coloured topsoil generally incorporates the remains of a former peat cover which has largely disappeared as a result of drainage-induced wastage, and typical luvic gley variants in which it no longer qualifies as humose occur in close proximity.

In the following representative profile, developed in river drift overlying Jurassic clay near Chatteris, the water table was about one metre below the surface at the time of sampling (19 March 1971). As in profile 152 (p. 363) in similar deposits, pedogenic features have been superimposed on an originally stratified and periglacially disturbed parent material, and their interpretation is further complicated in this soil by the marked post-depositional changes in hydrologic regime it has undergone. The relatively large clay content of the humose Ap may result either from an early Flandrian phase of alluvial sedimentation before peat formation commenced, or from artificial addition of clay-rich material. Other lithologic discontinuities revealed by the particle-size analyses occur at the base of the Eg and immediately above the calcareous sandy C horizon. The coarse loamy Eg, which hardens markedly on drying and varies considerably in thickness over short distances, contains much more silt, possibly of aeolian origin, than the horizons below. Polygonal patterns, marked by differences in crop growth and described as Weichselian (Devensian) tundra polygons by West (1977) were detected by aerial photography in similar land to the east and south (Seale, 1975a). At the margins of the polygons, where crops grow best in dry seasons, the subsurface horizons contain clay-rich inclusions interpreted as infilled frost cracks.

As in the buried loess-derived soil studied and dated by Weir et al. (1971), leaching of carbonates and downward translocation of clay-size particles giving rise to the well defined argillic horizon (Btgf) almost certainly took place during early Holocene or late Devensian times when sea level and corresponding ground-water levels were lower than at present. During the subsequent phases of marine transgression and inland peat formation (Godwin, 1978), it is probable that the water-saturated mineral horizons beneath the peat were subjected to strongly reducing conditions, in which case the marked concentrations of dithionite-extractable iron in the lower (Btg and Cgf) horizons may result mainly from redistribution and deposition of mobilized iron in response to the lowering of the water table since artificial drainage was initiated.

Profile 161 (Seale, 1975b)

Location	: Chatteris, Cambridgeshire (TL 399821)
Climatic regime:	subhumid temperate
Altitude	: 1.5 m Slope: <1°
Land use	: arable (ploughed)

Horizons (cm)

0–33
Ap
Black (7.5YR 1/2) very slightly stony humose clay loam (USDA sandy clay loam to clay loam); few small sub-angular flint fragments; moderate granular; very porous; few to common fine roots; some ploughed-in straw; sharp smooth boundary.

33–47
2Eg
Greyish brown (10YR 5/2.5) slightly to moderately stony medium sandy loam (USDA sandy loam) with common very fine and fine reddish brown (5YR 4/4) mottles; common to many, rounded and subangular flint fragments but some patches stone-free; very weak medium subangular blocky; few visible pores; hand specimen very friable when moist but compact and very hard when dry; 1–2 cm inclusions of topsoil as patches and channel fillings; few fine roots; clear wavy boundary.

47–66
3Btgf1
Brown (10YR 4.5/3.5) slightly to moderately flinty sandy clay loam (USDA sandy clay loam) with sandy loam inclusions and common fine and medium dark red (2.5YR 3/6) and reddish brown (5YR 4/4) mottles; stones as above; weak medium subangular blocky; friable to firm; roots rare; many clay coats on stones and sand grains in finer textured

Analytical data for Profile 161

Horizon	Ap	2Eg	3Btgf1	3Btgf2	4Cgf
Depth (cm)	0–33	33–49	49–66	66–80	80–100
sand 600 μm–2 mm %	10	14	37	21	53
200–600 μm %	20	27	25	36	36
60–200 μm %	11	13	10	17	2
silt 2–60 μm %	30	35	5	7	2
clay <2 μm %	29	11	23	19	7
organic carbon %	9.5	0.7	0.8		0.3
$CaCO_3$ equiv. %	0	0	0	0	11.7
dithionite ext. Fe %	2.0	1.7	5.0	5.3	6.4
pH (H_2O)	4.4	5.6	7.5	6.4	7.6
(0.01M $CaCl_2$)	4.0	5.4	6.9	6.0	7.0

66–80
3Btgf2
parts; few very fine manganiferous concentrations; gradual boundary. Brown to yellowish brown (10YR 5/3–4) moderately stony sandy clay loam (USDA sandy or coarse sandy loam) with some sandy loam inclusions and dark red to reddish brown mottles as above; weak medium and fine blocky; friable; no roots; discontinuous clay coats and blackish manganiferous coats on stones; abrupt boundary.

80–100+
4Cgf
Brown (7.5YR 4/4), very stony, very calcareous coarse sand (USDA coarse sand); abundant subrounded to subangular flint fragments with yellowish brown to brownish yellow coats; wet (water seeping in below 99 cm); single grain; very friable matrix to gravel.

Analytical data *see* above

Classification
England and Wales (1984): argillic humic gley soil; fine loamy drift with siliceous stones (fine loamy variant of Ireton series).
USDA (1975): Typic Umbraqualf; fine loamy over sandy or sandy-skeletal, mixed or siliceous, mesic.
FAO (1974): Gleyic Luvisol (or Podzoluvisol); medium textured.

7.6.7 Humic Stagnoluvic Gley Soils

These are surface-water soils with a humose or peaty topsoil and a greyish loamy, or occasionally sandy, gleyed E horizon over a finer textured B which is normally slowly permeable in some part and does not immediately overlie a pervious water-bearing substratum. Argillic, paleo-argillic and sandy stagnohumic gley soils as defined by Avery (1980) are included. Soils meeting the prescribed requirements have been identified in several widely scattered localities in England and Wales, including moist heathland at low altitudes, and also occur locally in southern Scotland and Ireland. Some have been grouped in surveys with related orthic gley soils (p. 354), which they resemble for practical purposes. The parent materials include medium to fine textured tills, weathering products of siltstones or shales, and older quasi-residual deposits such as the Clay-with-flints-and-cherts of the East Devon plateau (Harrod, 1971; Findlay et al., 1984), in which there are argillic B horizons with chromic characteristics (Chapter 3, p. 109). As in the associated non-humic stagnoluvic gley soils, the relatively fine textured B horizon appears in some places to have formed in an initially uniform material as a result of leaching and clay translocation, presumably under drier conditions than at present. In other examples there is clearly a lithologic discontinuity at or about the lower boundary of the E horizon, in which case the B may show little micromorphological evidence of clay translocation if the textural change is abrupt. Apparently indurated horizons of fragipan character occur at varying depths in some of these soils.

Profile (p. 382) below was described and sampled by P. Bullock and C.P. Murphy in rough grazing land between Penrith and Carlisle, Cumbria. It is sited in the lower sideslope of a drumlin on reddish Devensian till containing igneous erratics transported northwards from the Lake District. The particle-size data show no clear evidence of a lithologic discontinuity within the upper two metres, indicating that the well marked horizonation is primarily of pedogenic origin, but the large pit excavated at the site revealed several crudely wedge-shaped bodies of greyish material (Figure 7.11) extending to a maximum depth of 145 cm and forming a roughly polygonal pattern in horizontal section. These were interpreted by Matthews (1976) as ice-wedge casts, implying that the parent material underwent deep reaching disturbance under periglacial conditions before temperate (Holocene) pedogenesis commenced. That this involved translocation of clay as well as gleying is shown by the occurrence of common to many well oriented void argillans in thin sections taken at several depths in the B and BC horizons and at two depths in a 'grey wedge' (Murphy, 1984). Major ped faces, particularly in the upper part of the B horizon, have sandy coatings (skeletans) which may have originated by differential removal (micro-erosion) of finer particles or washing of poorly sorted material down cracks. The very dense lower horizons meet all the requirements of a fragipan. The chemical data confirm that the Ah and Eag horizons have lost iron and that nearly all the iron in the B horizon resists extraction by pyrophosphate. Other analyses, not quoted here, of material from a grey wedge penetrating the Btg2 horizon, showed that it contained about the same amount of clay as the interwedge material but only one sixth as much free iron.

Profile 163 is in cultivated land on the East Devon plateau. The field in which it is sited

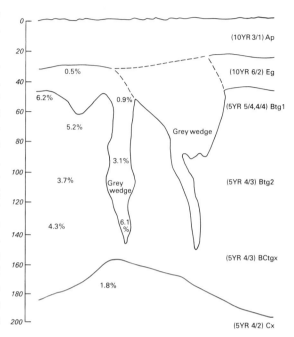

Figure 7.11: Irregular horizon boundaries and 'grey wedges' in profile 162, with illuvial clay concentrations shown (from Murphy, 1984).

was reclaimed from moorland in the recent past, and the resulting humose Ap horizon, which has evidently been limed at some time, is too thin to qualify as mollic or umbric according to the US and FAO systems. The name *Turbary* is applied to unreclaimed land nearby, so peat thick enough to cut for fuel must once have existed in the locality. As in the stagnogleyic chromoluvic brown soils exemplified by profile 76 (p. 229), the parent material apparently consists of Devensian loess mixed by cryoturbation with 'plateau drift' containing weathering products of Tertiary and Cretaceous rocks, which in this area include Upper Greensand chert beds as well as formerly superincumbent chalk with flints. The brightly coloured Btg horizon has a moderately developed fine blocky structure and is recorded as being friable or only slightly firm when moist, suggesting that it may be moderately permeable. It therefore seems possible that the strongly hydromorphic character of the overlying horizons is primarily a result of surface wetness induced by deterioration in the structure and permeability of the upper part of the mineral soil following replacement of forest by wet heath or moorland vegetation and consequent development of a peaty surface layer.

Profile 162 (Murphy, 1984)

Location : Hesket, Cumbria (NY 487407)
Climatic regime: humid temperate
Altitude : 136 m Slope: 2°
Land use and rough grazing with gorse (*Ulex*
  vegetation : *europaeus*) bushes
Horizons (cm)

0–25
Ah
Very dark grey (10YR 3/1; 3/2 rubbed) slightly stony humose medium sandy loam (USDA fine sandy loam) with abundant uncoated grains; common, very small to large, quartzose pebbles and subangular to subrounded fragments of volcanic rock, red coarse-grained sandstone and platy fine-grained sandstone; moderate medium granular; very friable; many roots; small earthworms present; abrupt irregular boundary.

25–44(69)
Eg
Light brownish grey (10YR 6/2; 6/3 rubbed) slightly stony medium sandy loam (USDA fine sandy loam) with common fine to medium strong brown (7.5YR 5/6) mottles; patches of brown (7.5YR 5/4) and few inclusions of dark humose soil in channels; stones as above; weak fine and medium subangular blocky; friable; common fine roots; clear irregular boundary wedging down to 69 cm within the described section and deeper elsewhere.

44–70
Btg1
Reddish brown (5YR 4–5/4) slightly stony sandy clay loam (USDA) sandy clay loam) with few fine and medium yellowish red (5YR 5/6) mottles; stones as above; coarse angular blocky with pinkish grey (7.5YR 6/2) faces; firm; few fine roots mainly in fissures; continuous 0.5 mm sandy coats on most ped faces; few variously sized manganiferous nodules; gradual boundary.

Analytical data for Profile 162

Horizon	Ah	Eg	Btg1	Btg2	Btg2(x)	BCtgx	Cx
Depth (cm)	5–20	28–42	48–66	79–94	100–130	140–166	174–200
sand 600 µm–2 mm %	7	5	3	5	7	4	5
200–600 µm %	24	23	17	21	18	17	23
60–200 µm %	37	36	35	31	32	33	29
silt 2–60 µm %	20	26	24	22	24	25	23
clay <2 µm %	12	10	21	21	19	21	20
<0.2 µm %	5	3	8	6	6	7	7
fine clay/total clay	0.42	0.30	0.38	0.29	0.32	0.33	0.35
organic carbon %	6.2	0.7	0.2		0.1		0.2
pyro. ext. Fe %	0.2	<0.05	<0.05	<0.05	<0.05	<0.05	<0.05
pyro. ext. Al %	0.2	0.2	0.1	<0.05	<0.05	<0.05	<0.05
total ext. Fe %	0.4	0.7	1.5	1.5	1.4	1.2	1.1
pH (H_2O)	5.1	5.4	5.6	6.1	6.3	6.4	6.7
(0.01M $CaCl_2$)	4.4	4.4	4.8	5.3	5.5	5.3	5.8
bulk density (g cm^{-3})	1.12	1.50	1.59	1.72		1.81	
packing density (g cm^{-3})	1.23	1.59	1.78	1.90		2.00	
retained water capacity (% vol.)	40	32	33	28		30	
available water capacity (% vol.)	21	18	13	11		12	
air capacity (% vol.)	17	11	7	7		1	

70–134 Btg2(x)	Reddish brown (5YR 4/3) slightly stony sandy clay loam (USDA sandy clay loam to fine sandy loam) with few to common yellowish red mottles and intricately arranged streaks and patches of lighter reddish brown (5YR 5/4) and light grey (5YR 7/1) associated with ped faces and root channels; stones as above; moderate coarse prismatic parting to coarse angular blocky; very firm; roots as above; some dead; prominent continuous clay coats overlain on major ped faces by thicker sandy coats; grey coats on upper sides of stones and reddish brown below; common manganiferous coats lining fissures and few nodules as above; few earthworm channels; gradual boundary.
134–170 BCtgx	Reddish brown (5YR 4/3) slightly stony sandy clay loam (USDA sandy clay loam) with grey (N 6/0) cleavage faces and channel linings; stones as above but generally smaller; weak coarse prismatic; very firm; few living and dead roots in fissures; thick ferruginous coats along root channels; clear irregular boundary.
170–210 Cx	Dark reddish grey (5YR 4/2) slightly stony sandy clay loam (USDA sandy clay loam to sandy loam); stones as above, mainly medium and small; moderate medium platy with rare grey coated vertical cleavage planes; very firm; no roots.

Analytical data *see* p. 382.

Classification
England and Wales (1984): argillic stagnohumic gley soil; reddish-loamy drift with siliceous stones (Lea series).
USDA (1975): Umbric Fragiaqualf; fine-loamy, mixed, mesic.
FAO (1974): Gleyic Luvisol (or Podzoluvisol); coarse textured.

Profile 163 (Harrod, 1971; Harrod *et al.*, 1973)

Location   : Dunkeswell, Devon (ST 126062)
Climatic regime: humid temperate
Altitude   : 274 m   Slope: 1°
Land use   : arable (barley stubble)

Horizons (cm)

0–17 Ap	Black (N 2/0), very slightly stony, humose silty clay loam (USDA silt loam); few, medium and small, angular and subangular chert and flint fragments; weak very fine subangular blocky; friable; many fine roots; earthworms present; abrupt irregular boundary.
17–22 Eag	Light brownish grey (10YR 6/2) slightly stony silty clay loam (USDA silt loam) with black pockets and channel fillings and few fine rusty mottles; stones as above with bleached surfaces; massive; friable to firm; few fine roots; clear irregular boundary.
22–48 Eg	Pale brown (10YR 6/3) slightly stony silty clay loam (USDA silty loam) with common fine and medium reddish yellow (7.5YR 6/6) mottles; stones as above; massive; firm; common fine soft ferruginous concentrations in upper part; few fine roots; few earthworm burrows; gradual boundary.
48–85 2Btg1	Strong brown (7.5YR 5/8) slightly stony silty clay (USDA silty clay loam) with many fine and medium light brown (7.5YR 6/4) and light grey (2.5YR 7/2) mottles; common, medium to very large, strongly patinated flint and chert fragments; moderate fine angular and subangular blocky; friable to firm; roots rare; common fine manganiferous coats on ped faces; gradual boundary.
85–110 2Btg2	Strong brown (7.5YR 5/8) slightly stony silty clay (USDA silty clay) with many mottles as above and also common red (2.5YR 4/8) mottles; stones as above; structure, roots and coats as above.

Analytical data *see* p. 384.

Classification
England and Wales (1984): paleo-argillic stagnohumic gley soil; loamy over clayey, drift with siliceous stones (Blackdown series).
USDA (1975): Aeric Paleaquult?; fine, mixed?, mesic.
FAO (1974): Gleyic Acrisol; medium textured.

Analytical data for Profile 163

Horizon	Ap	Eag	Eg	2Btg1	2Btg2
Depth (cm)	4–13	18–22	28–40	57–72	85–95
sand 200 μm–2 mm %	4	5	5	4	2
60–200 μm %	4	4	5	4	4
silt 2–60 μm %	71	68	70	56	49
clay <2 μm %	21	23	20	36	45
organic carbon %	7.6	2.5	0.8	0.5	0.4
pyro. ext. Fe %		0.1	0.1	<0.05	<0.05
pyro. ext. Al %		0.2	0.2	0.2	0.2
pyro. ext. C %		1.3	0.4	0.2	0.2
total ext. Fe %		0.7	1.3	2.2	2.5
pH (H_2O)	6.1	5.8	4.9	4.5	4.5
(0.01M $CaCl_2$)	5.7	5.2	4.2	3.9	3.9

8
Man-made Soils

8.1 General Characteristics, Classification and Extent

Although soil properties are influenced to some degree by human activity nearly everywhere in the British Isles, the term 'man-made' is conventionally restricted to predominantly mineral soils with distinctive features that can be attributed to the recent or former incidence of 'abnormal' land-use or management practices such as addition of earth-containing manures, unusually deep cultivation, or wholesale removal and re-emplacement of soil material. In terms of the differentiating criteria listed in Chapter 3, they have a thick man-made A horizon (p. 107), an artificially reworked layer (p. 111), or both. Those with a thick man-made A horizon are grouped as *cultosols*, corresponding to man-made humus soils in the England and Wales classification (Avery, 1973, 1980), and the remainder as *disturbed soils*. Only the larger areas of occurrence have been shown separately on soil maps, chiefly in England and Wales, where disturbed soils in land restored after quarrying and mining operations are the most extensive.

In other classifications which provide for separation of substantially man-modified soils, they are bracketed with other kinds of soil at the highest categorical level. The German system (Mückenhausen *et al.*, 1977), for example, divides them at the second level into *Terrestrische Anthropogene Böden (Terrestrische Kultosole)*, *Semi-Terrestrische Anthropogene Böden* and *Anthropogene Moore*, the first two classes comprising specified types of mineral soil and the third man-modified peat soils. In the Netherlands classification (de Bakker and Schelling, 1966), cultosols with an A horizon more than 50 cm thick are separated as *thick earth soils* at the suborder level but no single class embracing the disturbed soils is recognized. In the US Taxonomy, the soils classed as Plaggepts (plaggen soils) and Arents qualify as cultosols and disturbed soils respectively, but both suborders are defined in more restrictive terms, again entailing a 50 cm depth limit. The FAO system makes no provision for these soils.

Key to Soil Groups

1. Man-made soils with a thick man-made A horizon (p. 107)

    Cultosols

2. Other man-made soils

    Disturbed Soils

8.2 Cultosols

8.2.1 General Characteristics, Classification and Distribution

The essential feature of these soils, distinguishing them from those of all the other groups recognized here, is an artificially thickened A horizon that is more than 40 cm thick or overlies bedrock at a lesser depth and consists of intimately mixed mineral and organic materials. Implicit in the class concept is the inference that formation of the diagnostic surface horizon results directly from agricultural practices or from human occupation of the site. Soils with thick dark A horizons that appear to have formed naturally are therefore excluded. These are mainly restricted in the British Isles to recent alluvial or colluvial deposits, as examplified by profiles 38 (p. 173) and 59 (p. 202).

Although cultosols so defined are known to occur in many localities, the areas they occupy are generally small and have only occasionally been set apart on soil maps (e.g. Staines, 1979). In some of them the thick A horizon appears to result from the former use of earth-containing amendments, as in the plaggen soils (Plaggepts) which are extensive on sandy Pleistocene deposits in north Germany and the Low Countries (Pape, 1970; Conry, 1974; Mückenhausen et al., 1977; de Bakker, 1979). The name *Plaggen* (from the German words *Plagge*, meaning sod, and *plaggen*, to cut sods) has been applied traditionally throughout this region to thick dark surface layers produced by the long prevailing practice of bedding livestock on heather sods, turves or forest litter and spreading the manure so formed on arable land, the surface of which was thereby gradually raised. The resulting plaggen layer, or plaggen epipedon as defined in the US system, can be more than a metre thick and its colour, texture, carbon content and carbon/nitrogen ratio can be related to the nature of the bedding material and of the soil, usually a sandy podzol or gley soil, from which sods were cut.

Cultosols apparently produced by similar practices have been reported to occur in former 'infield land' in north-east Scotland and the Orkneys by Glentworth (1944, 1962), Romans (1970), Van Mensvoort and Van de Westeringh (1972) and Dry and Robertson (1982). In former heathland areas in the English lowlands, however, they are comparatively rare, probably because much less sandy land of this type was brought into cultivation during mediaeval times than in continental Europe and the use of plaggen manure was consequently never widespread or long continued.

More common in the British Isles are variants resulting from the regular use of other bulky amendments, including urban night soil, ditch dredgings, industrial refuse such as shoddy from woollen mills, and calcareous sea sand applied either alone or in conjunction with farmyard manure, seaweed or other organic materials. Those containing sea sand (Plate IV.6) are widely distributed in coastal districts of southern and western Ireland, where they have been studied in detail by Conry (1969, 1971, 1974), and also occur in Cornwall (Staines, 1979) and Devon. In the Aran Islands (Figure 8.1) and other Irish localities where population pressure increased markedly from the late seventeenth century onwards, such soils were built up laboriously on bare rock by transporting sand and seaweed (Mitchell, 1976). Following a proposal by Conry and Diamond (1971), Conry (1969, 1971, 1974) classed them with the plaggen soils (*sensu stricto*) on the basis that they have also been raised by continued application of manure containing mineral matter, and suggested that the plaggen epipedon (Soil Survey Staff, 1975) should be redefined accordingly.

Without special investigations, however, it is not always possible to distinguish soils with thick dark A horizons produced by such means from others which result from unusually deep cultivation coupled with application of organic manures only, or from incorporation of occupation residues, as in kitchen middens and other archaeological 'sites'. Those of the former type, set apart as Hortisols by Mückenhausen *et al.* (1977), occur chiefly in old gardens and market-garden

Figure 8.1: Cultosols on the Aran Islands, Co. Galway. The soil composing the cultivation ridges was made by importing sand and seaweed and spreading them on bare limestone rock (Copyright T.H. Mason).

land, and the latter in and around old settlements, including deserted mediaeval villages and sites occupied in Roman or pre-Roman times. In either case, the unusually great depth to which dark coloured intimately incorporated organic matter extends is largely attributable to earthworm activity, the role of which in mixing and burying mineral and organic occupation residues and remains of buildings is discussed at length by Limbrey (1975). The 'dark earths' (e.g. Macphail, 1983; Catt and Farrington, 1986) commonly associated with old settlements are characterized by inclusions of comminuted anthropogenic material such as brick, mortar, pottery, cinders, bone and iron slag, and by large contents of acid-soluble phosphate derived chiefly from bone fragments and other animal residues (Dekker and De Weerd, 1973). An example from the Roman city of Verulamium (St Albans) was noted by Gardner (1967).

Where the artificially thickened A horizon has been produced by manurial practices, subsurface horizons of the original profile are generally discernible beneath it. A buried A horizon corresponding to the former soil surface may also be recognizable where large amounts of mineral material have been added. In old settlement sites where the soil overlies displaced material, it is considered to be a cultosol rather than a disturbed soil if it has a dark relatively homogeneous surface layer more than 40 cm thick.

Insofar as brown soils were favoured in the past for both cultivation and settlement, many of the cultosols were originally of this group, but variants with gleyed, podzolic or lithomorphic subsurface horizons also occur. Whatever the substratum and mode of origin, both the physical and chemical properties of these soils are generally more favourable to plant growth than those of their unmodified counterparts. According to Conry (1974), however, manganese and boron deficiencies occur regularly in crops grown on the Irish 'plaggen soils' that have received additions of calcareous sand, and copper deficiency is common on the plaggen soils *sensu stricto* in continental Europe.

Following the Netherlands classification of de Bakker and Schelling (1966), soils of this group were divided by Avery (1973, 1980) into sandy and earthy (loamy or clayey) subgroups and the same arrangement is adopted here.

Key to Subgroups

1. Cultosols in which at least half the upper 80 cm of mineral soil, excluding bedrock, is sandy or sandy-gravelly.
Sandy cultosols

2. Other cultosols
Loamy and clayey cultosols

8.2.2 Sandy Cultosols

Cultosols of this subgroup, which includes most of the plaggen soils *sensu stricto* grouped as *Enk earth soils* in the Netherlands (de Bakker, 1979), are much less common in the British Isles than those of loamy texture. As indicated in Chapter 6 (p. 248), examples resulting from deep-reaching modification of sandy podzols occur very locally in the English lowlands. Others can be found in association with sandy brown soils or gley soils, and some but by no means all of the Irish 'sanded' soils also have thick man-made A horizons of sandy texture.

Variants of the latter type are exemplified by the following representative profile from the Dingle peninsula, County Kerry. Conry (1971) gives data on an 'unsanded' gley (or gley-podzol) profile with similar lower horizons in an adjacent field separated by a roadway (Figure 8.2). Both are on stony drift derived from Devonian (Old Red Sandstone) conglomerate and shale. The 'sanded' soil, the surface of which is appreciably higher than that of its 'unsanded' counterpart, has a very calcareous man-made A horizon ranging from 58 to 76 cm in thickness. Below this is a buried humose topsoil resembling that of the non-calcareous 'unsanded' soil nearby. As sea sand was apparently used extensively as stable litter in coastal areas of County Kerry until as recently as 1950, the thick man-made A horizon is likely to have been produced by repeated application of manure which originated in this way. The 'sanded' soils in the Dingle peninsula produce good yields of a wide range of crops, and the Castlegregory area in particular is noted for onion and carrot growing.

Profile 164 (Conry, 1969)

Location   : Castlegregory, Co. Kerry
Climatic regime: perhumid temperate
Altitude   : 24 m   Slope: 2–3°
Land use   : arable
Horizons (cm)

0–18 Ap1	Very dark grey (10YR 3/1), slightly stony, calcareous loamy coarse sand to coarse sand (USDA loamy coarse sand) with abundant uncoated grains; rounded beach pebbles common; moderate fine granular; very friable to loose; abundant roots.
18–58/69 Ap2	Similar to above but somewhat lighter in colour; pocket of pure sand occurs in this layer; spade marks observed; clear wavy boundary.
58/69–76 2bAp	Black (10YR 2/1) humose calcareous loamy sand (USDA loamy sand) with common uncoated grains (quartz and shell fragments); moderate fine granular; very friable; piece of broken pottery observed; plentiful roots; clear smooth boundary.
76–100 2Eag	Light grey (10YR 7/2) stony sand (USDA sand) with many blackish streaks, probably decomposed organic matter; very friable; few roots; non-calcareous; clear wavy boundary.
100–115 3Bgh	Brown (10YR 5/3 nearest) bouldery sandy loam (USDA sandy loam) with blackish streaks as above; weak fine granular to massive; compact *in situ* but friable when removed; no roots; gradual boundary.
115–136 3BCg	Very pale brown (10YR 7/4) and brownish yellow (10YR 6/6) mottled bouldery sandy loam (USDA sandy loam); weak fine granular to massive; friable; no roots; abrupt boundary.
136+	Light brown (7.5YR 6/4) and strong

3Cg   brown (7.5YR 5/6) mottled bouldery sandy loam (USDA sandy loam); massive; firm; no roots.

Analytical data *see* below.

Classification
Ireland (Conry, 1971): plaggen soil.
England and Wales (1984): sandy man-made humus soil; sandy drift with siliceous stones (unnamed series).
USDA (1975): Plaggept; sandy, mixed (calcareous), mesic.
FAO (1974): unclassified.

8.2.3 Loamy and Clayey Cultosols

These are cultosols with predominantly loamy or clayey textures, corresponding to *Tuin earth soils* in the Netherlands system (de Bakker and Schelling, 1966; de Bakker, 1979). Those that have been identified in the British Isles are mostly coarse or fine loamy and clayey variants are apparently rare. Many of the 'sanded' soils described by Conry (1969, 1971) and similar soils in the market-garden district

Analytical data for Profile 164

Horizon	Ap1	Ap2	2bAp	2Eag	3Bgh?	3BCg	3Cg
Depth (cm)	0–18	18–58/69	58/69–76	76–100	100–115	115–136	136+
sand 200 μm–2 mm %	58	56	56	52	27	32	30
50–200 μm %	29	35	27	37	36	32	35
silt 2–50 μm %	9	6	10	7	22	20	22
clay <2 μm %	4	3	7	4	15	16	13
organic carbon %	2.2	1.9	5.5	0.5	0.9	0.5	0.3
nitrogen %	0.21	0.12	0.27				
C/N ratio	11	16	20				
citrate-dithionite ext. Fe %	0.3	0.3	0.4	0.1	0.1	0.5	0.4
pH (H$_2$O)	7.5	7.5	7.4	7.4	7.1	6.8	7.0

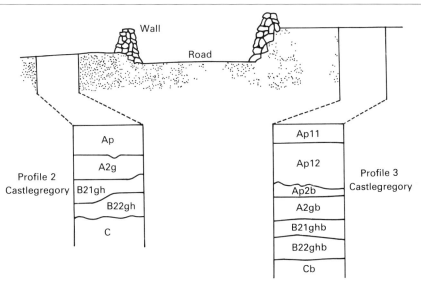

Figure 8.2: Diagrammatic representation of adjacent 'unsanded' (profile 2) and 'sanded' (profile 3) soils at Castlegregory, Co. Kerry (from Conry, 1971).

of west Cornwall mapped as Ludgvan series by Staines (1979) are coarse loamy rather than sandy and therefore conform to this subgroup. Others have thick man-made A horizons produced by differing manurial practices, with or without unusually deep cultivation, or by incorporation of occupation residues in old settlement sites. Field and analytical data on four representative profiles from widely separated localities are reproduced below.

Profile 165 is a 'sanded' soil on till derived from sandstone, shale and slate near Clonakilty, County Cork (Plate IV.6). The calcareous 'plaggen' layer is up to 84 cm thick and has a lighter brownish colour than in the preceding profile. According to Conry (1969, 1971), the difference in colour can be related to the nature of the original soil, which in profile 168 was evidently of podzolic brown type. Together with the smaller sand contents, this implies that the built-up layer incorporates pre-existing soil material throughout its depth and probably results from the practice, once common in the Clonakilty district, of composting sods, sea sand and dung.

Profile 166 was described and sampled by J.C.C. Romans in the Cromarty peninsula north of Inverness, where plaggen soils similar to those of the Netherlands were observed by Van Mensvoort and Van de Westeringh (1972). It has a dark brown man-made A horizon 69 cm thick, below which there is a buried topsoil overlying compact reddish stony till. Glentworth (1962) quotes historical evidence supporting the conclusion that cultosols of this type were produced in parts of north-east Scotland by skimming the turf off 'outbye' land, mixing it with cattle dung, and spreading the resulting compost on the regularly cropped infield. Using poll-tax returns, Walton (1950) showed a close correlation between population distribution in the Insch Valley, Aberdeenshire, in 1696 and areas of deep soil recorded by Glentworth (1954).

Profile 167, on coarse loamy raised-beach deposits in East Lothian, resembles deep, dark coloured market-garden soils described by Van Mensvoort and Van de Westeringh (1972) in the nearby Musselburgh area, where the land has been long and deeply cultivated. As the organic carbon contents decrease gradually with depth to at least 75 cm and there is no sign of a buried A horizon, it is possible that this soil may owe its distinctive characteristics to unusually deep cultivation combined with heavy organic manuring and deep reaching earthworm activity rather than to regular additions of mineral material.

Profile 168 is an 'old-settlement soil' recorded by C.A.H. Hodge and R.S. Seale at a disturbed site on a low terrace of the river Cam north of Cambridge, once occupied by Roman pottery kilns (Browne, 1978). Similar thick dark soil was found to extend over two fields at the time of sampling (23 July 1953). It contains fragments of imported Jurassic limestone as well as numerous shards, indicating that it incorporates remains of buildings.

Profile 165 (Conry, 1969)

Location : Ardfield, Clonakilty, Co. Cork
Climatic regime: humid temperate
Altitude : 46 m   Slope: <1°
Land use : arable
Horizons (cm)

0–15 Ap1	Dark greyish brown to brown (10YR 4/2-3), slightly stony, calcareous sandy loam (USDA sandy loam); rounded beach pebbles common; weak very fine granular (almost apedal); very friable to loose; abundant roots; gradual boundary.
15–30 Ap2	Similar to above horizon but colour somewhat lighter.
30–59 Ap3	Similar to Ap2, gradual wavy boundary.
59–66/84 bAp1	Dark greyish brown (10YR 4/2) calcareous sandy silt loam (USDA loam), otherwise similar to above horizon; it is a mixture of added sand and original Ap horizon; clear wavy boundary.

Analytical data for Profile 165

Horizon	Ap1	Ap2	Ap3	2bAp1	2bAp2g	2Bw	2Bs[a]	2Cu1	2Cu2
Depth (cm)	0–15	15–30	30–59	59–66	84–97	97–135	107–135	135–140	140+
sand 200 μm–2 mm %	68	67	59	40	33	40	59	28	35
50–200 μm %	8	8	11	17	14	16	9	24	18
silt 2–50 μm %	13	19	18	38	34	35	13	41	39
clay <2 μm %	11	6	13	15	19	9	19	7	8
organic carbon %	2.9	1.9	1.0	1.8	2.4	1.6	2.3	0.6	0.3
nitrogen %	0.25	0.14	0.08	0.08	0.13				
C/N ratio	12	14	13	22	17				
citrate-dithionite ext. Fe %	0.7	0.7	0.5	0.6	0.8	1.1	2.2	0.4	2.5
pH (H$_2$O)	7.6	7.9	8.2	8.3	8.1	8.2	8.1	8.1	8.0

[a] darker inclusion

66/84–97 2bAp2	Dark greyish brown (10YR 4/2) clay loam (USDA loam); moderate fine and medium granular; friable; few roots; abrupt wavy boundary.
97–135 2Bw/Bs	Yellowish brown (10YR 5/4–6) sandy loam (USDA sandy loam); weak fine granular (crumb); very friable; few roots; inclusion of dark yellowish brown (reflecting larger humus content) material between 107 and 135 cm); clear boundary.
135– 140/155 2Cu1	Olive-grey (5Y 5/2) gravelly sandy loam to sandy silt loam (USDA sandy loam to loam); massive; friable; no roots; abrupt irregular boundary.
140/155 + 2Cu2	Olive-grey (5Y 5/2) gravelly sandy loam to sandy silt loam (USDA sandy loam to loam); moist; massive; firm; no roots.

Analytical data *see* above.

Classification
   Ireland (Conry, 1971): plaggen soil.
   England and Wales (1984): earthy man-made humus soil; coarse loamy drift with siliceous stones (Ludgvan series).
   USDA (1975): Plaggept; coarse-loamy, mixed (calcareous), mesic.
   FAO (1974): unclassified.

Profile 166

Location   : Fearn, Ross and Cromarty
                 (NH 866794)
Climatic regime: humid temperate
Altitude   : 40 m   Slope: 1°
Land use   : arable

Horizons (cm)

0–25 Ap1	Dark brown (7.5YR 3/2) slightly stony sandy loam (USDA sandy loam); stones mostly schist fragments up to 2 cm; coarse subangular blocky, breaking under slight pressure to medium and fine; friable; frequent roots; earthworms present; abrupt smooth boundary.
25–69 Ap2	Dark brown (7.5YR 3/2) slightly stony sandy loam (USDA sandy loam) becoming slightly greyer in the lower 20 cm; stones mostly schist and sandstone fragments up to 10 cm; weak prismatic breaking easily to fine subangular blocky and single grain; friable; frequent roots, becoming few in lower part; earthworms present; abrupt smooth boundary.
69–74 bA1	Black (5YR 2/1) stony humose sandy loam (USDA sandy loam); stones as above; weak blocky, few roots; discontinuous horizon with wavy lower boundary.
69/74– 77 bA2	Very dark grey (10YR 3/1) slightly stony sandy loam (USDA sandy loam); stones mainly hard schist up to 7.5 cm; weak blocky; few roots; no worms seen; abrupt boundary.
77–86 bA3	Very dark greyish brown (10YR 3/2) slightly stony sandy loam (USDA

| 86–109
BCx(g) | sandy loam); weak blocky, compact in place but friable when removed; few roots; no worms seen; abrupt boundary.
Reddish brown (5YR 5/4) stony sandy loam (USDA sandy loam) with diffuse ochreous mottling and grey-coated cracks; many schist and sandstone fragments; massive; firm (indurated); no roots or worms; gradual boundary. |
| 109–130+
BCx(g)/
Cu(g) | Reddish brown (5YR 5/3–4) stony sandy loam (USDA sandy loam) with grey-coated cracks becoming more distinct with depth; stones as above, up to 10 cm; weak blocky; massive and compact, becoming less so below 120 cm; no roots or worms. |

Analytical data *see* below

Classification
  Scotland (1984): anthropogenic soil in Cromarty association.
  England and Wales (1984): earthy man-made humus soil; reddish-coarse loamy drift with siliceous stones (unnamed series).
  USDA (1975): Plaggept; coarse-loamy, mixed (non-acid), mesic.
  FAO (1974): unclassified.

Profile 167 (Ragg, 1974)

Location : Aberlady, East Lothian (NT 455795)
Climatic regime: humid temperate
Altitude : 15 m   Slope: <1°
Land use : arable
Horizons (cm)

| 0–15
Ap1 | Very dark greyish brown (10YR 3/2) stoneless sandy loam (USDA sandy loam) containing uncoated quartz grains; recently ploughed, with much decomposing barley stubble at base; common fine roots; friable; abrupt smooth boundary. |
| 15–22
Ap2 | Very dark greyish brown (10YR 3/2) stoneless sandy loam (USDA sandy loam); weak coarse and medium subangular blocky; friable; common fine roots; clear smooth boundary. |
| 22–50
Ap3 | Dark brown (7.5YR 3/2) stoneless sandy loam (USDA sandy loam) with inclusions of material similar to above; very weak medium subangular blocky; friable; common fine roots, particularly in worm channels; gradual smooth boundary. |
| 50–75
Ap4 | Dark brown (7.5YR 3/2) slightly stony sandy loam (USDA sandy loam); weak coarse blocky; friable; common roots and infilled worm channels; gradual smooth boundary. |

Analytical data for Profile 166

Horizon	Ap1	Ap2	Ap2	bAh1	bAh2	BCx(g)	BCx(g)	Cu(g)
Depth (cm)	3–10	36–43	58–66	69–74	74–77	91–99	102–109	122–130
silt[a] 2–50 μm %	27	23	31	28	23	21	21	22
clay[a] <2 μm %	11	15	10	11	11	18	17	15
organic carbon %	2.5	2.7	2.2	12.8	1.4	0.6	0.3	0.2
nitrogen %	0.18	0.16	0.15	0.50				
C/N ratio	14	17	15	26				
TEB (me/100 g)	4.3	3.4	2.5	28.3	3.9	4.2	2.1	1.9
CEC (me/100 g)	9.0	9.2	7.0	49.8	6.9	8.5	5.0	4.9
% base saturation	48	37	35	57	57	49	42	39
pH (H$_2$O)	5.6	5.9	6.1	6.1	6.1	5.8	5.5	5.0

[a] by hydrometer method

Analytical data for Profile 167

Horizon	Ap1	Ap2	Ap3	Ap4	AB?	Cg?
Depth (cm)	0–15	15–22	22–50	50–75	75–95	95–100
sand 200 μm–2 mm %	38	36	31	41	39	34
50–200 μm %	35	37	44	32	31	18
silt 2–50 μm %	12	23	11	12	15	31
clay <2 μm %	15	14	14	15	15	17
organic carbon %	2.8	2.4	1.4	1.2		
nitrogen %	0.21	0.22	0.15	0.17	0.14	0.09
C/N ratio	13	11	9	7		
TEB (me/100 g)	7.5	6.5	6.8	7.4	8.3	
CEC (me/100 g)	10.7	10.9	8.4	7.4	9.1	
% base saturation	70	60	81	sat.	91	
pH (H$_2$O)	5.9	6.0	6.5	7.0	6.9	7.3

75–95 AB? Dark brown (7.5YR 3/2) very slightly stony sandy loam (USDA sandy loam); massive; firm; few roots, mainly in worm channels; clear smooth boundary.

95–100+ Cg? Very dark grey (10YR 3/1) slightly stony sandy silt loam (USDA sandy loam) with few fine yellowish red mottles; massive; firm; few roots.

Analytical data *see* above.

Classification
  Scotland (1984): brown forest soil? (Dreghorn series; Dreghorn association).
  England and Wales (1984): earthy man-made humus soil; coarse loamy drift with siliceous stones (Ludgvan series).
  USDA (1975): Plaggept?; coarse-loamy, mixed (non-acid), mesic.
  FAO (1974): unclassified.

Profile 168

Location   : Horningsea, Cambridgeshire (TL 497636)
Climatic regime: subhumid temperate
Altitude   : 6 m   Slope: <1° overall; undulating microrelief
Land use   : long-term grass
Horizons (cm)
  0–56 Ap Very dark grey (10YR 3/1), slightly stony, slightly calcareous sandy clay loam (USDA sandy loam to sandy clay loam); stones include subangular flint, fragments of Roman pottery and Jurassic limestone fragments up to 10 cm; granular and abundant fine roots to 15 cm, subangular blocky and many roots below; friable; earthworms and millipedes present, gradual boundary.
  56–84 AC? Very dark grey (10YR 3/1), slightly to moderately stony, slightly calcareous sandy loam (USDA sandy loam) with olive (5Y 5/4–6) patches; stones as above; weak blocky to massive; friable; few roots.

Analytical data

Horizon	Ap	AC?
Depth (cm)	0–56	56–84
silt[a] 2–50 μm %	20	13
clay[a] <2 μm %	20	11
loss on ignition %	7.0	2.2
CaCO$_3$ equiv. %	1.8	1.2
pH (H$_2$O)	6.7	7.9

[a]by hydrometer method

Classification
  England and Wales (1984): earthy man-made humus soil; loamy drift with limestone (unnamed series).
  USDA (1975): Cumulic (or Vermic) Hapludoll?; coarse-loamy, mixed, mesic.
  FAO (1974): unclassified.

8.3 Disturbed Soils

8.3.1 General Characteristics and Distribution

These are man-made soils that consist for the most part of artificially displaced materials but do not have a thick dark A horizon produced by long-continued manuring or incorporation of occupation residues. They occur in a variety of situations where soil has been excavated and replaced, either in roughly the same position as elsewhere, or where the original profile has been deeply disrupted by heavy earth-moving equipment. Because of their inherently disorderly nature, no attempt has been made to divide them into subgroups on a profile basis. Some are distinguished from naturally formed soils of similar composition by the presence below normal ploughing depth of a disturbed subsurface layer in which more or less mixed and rearranged parts of pre-existing surface (O,A) or subsurface (E, B, Bc) horizons can be recognized. Disturbed soils of this kind are set apart as Rigosols in Germany (Mückenhausen et al., 1977; Mückenhausen, 1985) and as Arents in the US Taxonomy. Others in land restored after quarrying or opencast mining have a surface layer of imported soil material which overlies undisturbed rock, compacted soil showing little evidence of prior pedological reorganization, or infilled domestic or industrial waste.

The composition and morphology of such soils varies enormously, depending on the parent material, the kind and degree of disturbance, and the time that has elapsed since it took place. In freshly disturbed or restored land, both distinctive external features such as ground-surface irregularities, and subsurface indications of rearrangement, are preserved with little modification. When 'made ground' is naturally revegetated or cropped, however, superficial accumulation of organic matter, earthworm activity and other structure forming agencies, together with any weathering or translocation processes affecting particular constituents, eventually lead to differentiation of newly formed pedogenic horizons within the displaced layers and internal signs of disturbance are correspondingly obscured, initially in the upper part of the soil and subsequently at greater depths. Among examples of such soils are podzols in Bronze Age barrows (Dimbleby, 1962) and brown soils in mounds of silty material composed of discarded residues of mediaeval and Roman salt-making operations in coastal areas of Lincolnshire (Robson, 1981, 1985). Others have shallow or weakly developed profiles with organo-mineral or organic surface horizons which appear to have originated naturally in or on artificially emplaced rock waste derived more or less exclusively from pre-existing C horizons or bedrock. As indicated in Chapter 4 (p. 115), these are considered as lithomorphic (rankers, rendzinas or sandy regosols) rather than as disturbed soils, and are exemplified by profile 23 (p. 137) in chalky debris.

It is therefore only in a restricted range of man-modified sites, mainly of recent origin, that the soil qualifies unequivocally as disturbed in the sense implied above. Although ubiquitous and often numerous in long settled landscapes, many bodies of soil meeting the prescribed requirements are either of very small extent, as in cemeteries and other places where pits have been dug and refilled, or very narrow in relation to their length, as where disturbance has resulted from land drainage, embanking or road making operations. Areas of occurrence large enough to show separately on maps at scales of 1:10,000 or smaller consist mainly of land restored after mineral extraction and ground which has been deeply worked to disrupt root restricting layers such as ironpans or thin rock beds, or with the deliberate aim of bringing relatively fertile subsoil material to the surface. As a rule, however, deeply worked ground with disturbed soils has not been distinguished on published soil maps, and it is only in England

and Wales (e.g. Bridges, 1966; Hodge and Seale, 1966) that particular types of restored land have been identified and delineated. Besides disturbed soils *sensu stricto*, the generally complex soil patterns in the areas so mapped include variable and in some cases dominant proportions of soils conforming to other groups, depending on the materials used in restoration, the operations involved, and the age of the site. Areas restored after opencast coal mining are by far the most extensive, covering some 48,000 hectares in England and Wales alone (Mackney *et al.*, 1983). Others of significant total extent are in land quarried for iron ore, for phosphatic nodules ('coprolites'), and for sand, gravel or brickearth. The distribution and characteristics of these heterogeneous land types are briefly reviewed below.

8.3.2 Restored Opencast Coal Workings

Since the 1940s, opencast mining of shallow coal seams has affected many areas on Upper Carboniferous rocks in parts of northern England (R.A. Jarvis *et al.*, 1984), the Midlands (Ragg *et al.*, 1984), South Wales (Rudeforth *et al.*, 1984) and the Scottish Lowlands. With advances in mechanization, sites have become larger and some land has been re-excavated to exploit deeper seams. When farmland is worked for opencast coal, British Coal is required to restore it to a state satisfactory for continued agricultural use. In order to meet this obligation, recent practice has entailed stripping and stockpiling topsoil layers separately, and replacing them in the same order after extraction is completed to give a restored soil cover 60–120 cm thick on top of the graded overburden (Figure 8.3). These operations are normally performed with heavy earth moving machinery, however, which inevitably causes considerable compaction and consequent reduction in permeability. As soon as possible after reinstatement of the surface, grass is sown and left for five years, during which time the primary aim of management is rehabilitation of soil structure. The grass is cut for hay or silage, but any grazing is carefully regulated to avoid poaching. The land is then reseeded and a

Figure 8.3: Restoration of land after opencast coal mining. The stockpiled soil has been replaced and will be sown with grass (British Geological Survey: NERC copyright).

comprehensive drainage system is installed, or an earlier skeleton system completed, before it is returned to the farmer.

The nature of the restored land is determined by that of the soils stripped from the site and the care taken in handling and replacing them. Most opencast sites have soils derived from Carboniferous shale or sandstone, or from superincumbent till. More or less well drained brown soils, or occasionally podzols, occur where sandstone is the main source rock, but gley soils of medium to fine texture are much more common and include variants with humose or peaty topsoils in parts of the South Wales coalfield. In some early reclamations there is only a thin layer of replaced topsoil over compacted rock debris; the restored soils therefore have ranker-like profiles, often containing boulders which interfere with drainage and cultivations. Where topsoil and subsoil materials derived from medium or fine textured surface-water gley soils have been removed and replaced separately, however, the restored subsoil layers come mainly from Bg and BCg horizons of the original soils and have a similar grey and ochreous mottled appearance. Profile 169, described and sampled by M.J. Reeve one year after restoration of a Derbyshire site, is an example of a restored opencast soil of this type. It has a topsoil layer about 13 cm thick over a heterogeneous mottled subsoil which is massive and very slowly permeable, causing water to perch at or close to the surface in winter. Similar soils in earlier sites which were restored under favourable weather conditions and well managed subsequently show appreciable structural development in the subsoil.

Profile 169

Location : Mapperley, Derbyshire (SK 444428)
Climatic regime: humid temperate
Altitude : 70 m   Slope: 2° straight
Land use : ley grass

Horizons (cm)

0–13 Ap   Very dark greyish brown (10YR 3/2) very slightly stony silty clay loam (USDA clay loam to silty clay loam) with inclusions of mottled subsoil material; few small angular platy fragments of ferruginous siltstone (ironstone); weak (adherent) medium subangular blocky; firm; many fine roots; abrupt smooth boundary.

13–76 C1   Very mixed mottled grey and ochreous material, mainly silty clay loam (USDA clay loam to silty clay loam), containing large subangular sandstone fragments, small angular platy ironstone fragments and pieces of coal; mainly massive, but layers (c. 15 cm) reflecting successive additions and compressions are visible; very firm; no roots below 20 cm.

76–120 C2   Dark grey, massive silty clay loam (USDA silty clay loam) with coal and sandstone fragments.

Analytical data

Horizon	Ap	C1	C2
Depth (cm)	0–13	13–66	67–120
sand 600 μm–2 mm %	4	5	5
200–600 μm %	3	4	3
60–200 μm %	11	8	4
silt 2–60 μm %	53	51	55
clay <2 μm %	29	32	33
organic carbon %	3.2	1.1	3.4
pH (H$_2$O)	6.9	6.0	4.8
(0.01M CaCl$_2$)	6.4	5.3	4.7
bulk density (g cm^{-3})	1.21	1.60	
packing density (g cm^{-3})	1.47	1.88	
retained water capacity (% vol.)	46	38	
available water capacity (% vol.)	22	10	
air capacity (% vol.)	6	2	

Classification

England and Wales (1984): disturbed soil; fine silty drift with siliceous stones.
USDA (1975): (Aqueptic?) Arent; fine-loamy, mixed (non-acid), mesic.
FAO (1974): unclassified.

8.3.3 Restored Ironstone Workings

Restored workings in which highly ferruginous Middle and Lower Jurassic sedimentary rocks have been exploited for iron ore occur in several localities between north Oxfordshire and the Humber estuary, covering in all about 5,000 ha (Mackney et al., 1983). Variations in the nature of the overburden and in methods of restoration give a wide range of soils in these sites. Thus in former workings northwest of Banbury, where the overburden was thin and a high standard of restoration was maintained, they closely resemble surrounding orthic brown soils (Chapter 5, p. 197), though they are often more stony and the ground surface is mostly 2 m or more below its original level (M.G. Jarvis et al., 1984). Similar well drained ferruginous soils occur in reclaimed land near Northampton; in places, however, there is a clayey layer at the base, and around Corby, where several metres of chalky till were removed and replaced, the topsoils are generally thin and the subsoils slowly permeable, giving drainage problems comparable to those encountered in restored opencast coal sites (Hodge et al., 1984). Further north, in Lincolnshire and the Scunthorpe district of Humberside, ironstone has been extracted after removal of covering deposits which include Jurassic (Lincolnshire) limestone, Lower Lias Clay and blown sand, and the restored soils show corresponding variations in composition and permeability.

8.3.4 Restored 'Coprolite' Workings

These consist of land disturbed by extraction of phosphatic nodules, commonly called coprolites, from the Cambridge Greensand formation, a thin calcareous glauconitic bed of Cretaceous age which separates the Lower Chalk (Chalk Marl) from the underlying Gault

Figure 8.4: Restored coprolite workings at Stow cum Quy Fen, Cambridgeshire. The backfilled trenches from which phosphatic nodules were extracted appear as pale strips and a few of the trenches remain as elongated ponds (Cambridge University Collection: copyright reserved).

Clay in Cambridgeshire and north Hertfordshire (Figure 8.4). Quarrying of the 'coprolites' to provide raw material for superphosphate production started about 1850 and continued until 1880, followed by a short revival during the 1914–18 war, and eventually affected some 2,600 hectares, mainly in irregular strips between Burwell, Cambs. and Ashwell, Herts. (Barraud, 1951; Hodge and Seale, 1966; Hodge et al., 1984). Fields were trenched to depths up to 7 m to expose the Greensand; the phosphatic nodules were dug out, and separated from the grey or greenish loamy or clayey matrix by washing on the site. This produced a 'slurry' which was subsequently allowed to dry and used in conjunction with previously excavated topsoil material for levelling and re-soiling after the trenches had been refilled. The ground was then returned to agricultural use, the whole operation taking about three years.

The restored land contains a variety of soils on disturbed and replaced materials, which in the Fenland north-east of Cambridge included peat and gravelly superficial deposits as well as Chalk Marl. Sites of former workings are not always easy to distinguish on the ground, however, and many of the restored soils closely resemble undisturbed rendzinas or related calcaric gley soils in marly parent materials nearby. Those having a subsurface layer of 'slurry' were identified as Disturbed Lode series by Hodge and Seale (1966), who cited the following profile as an example. The restored topsoil rests abruptly on the highly calcareous slurry layer, which has developed a strong prismatic structure and overlies a less calcareous sandy loam layer with sandy and gravelly patches and ochreous segregations at around 56 cm depth.

Profile 170 (Hodge and Seale, 1966)

Location	:	Lode, Cambridgeshire (TL 521637)
Climatic regime:		subhumid temperate
Altitude	:	3 m Slope: <1° with slightly undulating microrelief
Land use	:	arable

Horizons (cm)

0–23 Ap	Very dark greyish brown (10YR 3/1–2) very calcareous clay loam (USDA loam) containing a few flint fragments and quartzite pebbles; weak medium cloddy; friable; ploughed-in stubble and few fine roots; earthworms present; abrupt wavy boundary.
23–56 2BC	Light olive-grey (5Y 6/2) extremely calcarous silty clay loam (USDA silty clay loam); strong medium prismatic; firm; few roots; abrupt irregular boundary.
56–91+ 3Cg	Pale olive (5Y 6/3) very calcareous stony sandy loam (USDA sandy loam) with yellowish brown (10YR 5/8) mottles; flint fragments, hard chalk pebbles and sandy and gravelly patches; massive; friable; ochreous segregations up to about 5 cm in diameter; no roots.

Analytical data

Horizon	Ap	2BC	3Cg
Depth (cm)	0–23	23–56	56–91
sand 200 μm–2 mm %	20	1	36
50–200 μm %	21	14	34
silt 2–50 μm %	33	55	20
clay <2 μm %	26	30	10
organic carbon %	3.1		
CaCO$_3$ equiv. %	22	43	12
pH (H$_2$O)	7.9	8.0	8.0
(0.01M CaCl$_2$)	7.4	7.4	7.4

Classification
  England and Wales (1984): disturbed soil; carbonatic-loamy.
  USDA (1975): (Aquic) Udorthent?; fine-loamy, mixed (calcareous), mesic.
  FAO (1974): unclassified.

8.3.5 Land Restored After Extraction of Sand, Gravel or Brickearth

Extraction of sand and gravel for building and other purposes is an old established industry which has vastly expanded in recent years to meet the greatly increased demand for concrete. According to Bradshaw and Chadwick (1980), it now affects about 1,000 ha annually in Britain. Smaller areas on terraces and gentle slopes, particularly in Kent (Fordham and Green, 1976) and around London, have been disturbed in the past by extraction of brickearth (Chapter 1, p. 13), which often overlies gravel.

Although some sand and gravel workings are in pre-Quaternary formations such as Lower Cretaceous and Tertiary sands, most of them are in river-terrace or glaciofluvial deposits. In either case they can be categorized as 'wet' or 'dry'. Wet pits, in which extraction takes place below the regional water table, are mainly located in the valleys of major lowland rivers such as the Thames and Trent and are usually allowed to flood when working has ceased, forming man-made lakes which provide diverse wildlife habitats and recreational facilities when appropriately managed. Particularly during the last twenty years, many dry and some wet pits have been restored to agricultural use by methods involving storage and replacement of pre-existing soil, either directly on the lowered and duly graded surface left after extraction or after infilling with waste materials such as urban refuse or power station ash. Silty or clayey washery waste is usually also returned to the pit. As heavy machinery is normally used for grading and replacement, the productivity of the restored land is influenced by the degree of compaction as well as by the composition of the replaced material. Adequate under drainage also needs to be ensured and any potentially toxic waste deeply buried. Emission of methane and carbon dioxide resulting from decomposition of organic wastes presents a major problem in some landfill sites.

9
Peat Soils

9.1 General Characteristics and Extent

These are dominantly organic soils derived for the most part from plant remains accumulated under wet conditions, either as autochthonous peat in the position of growth or as constituents of sedimentary deposits such as organic lake muds. According to the definition adopted here, which follows that of Avery (1980), they have at least 40 cm of organic material (Chapter 3, p. 82) within the upper 80 cm, or at least 30 cm if it rests directly on bedrock, and no overlying mineral layer that is more than 30 cm thick and has a non-humose B or C horizon at its base. Thus, although the organic layer usually starts at the surface, it may be overlain by 30–40 cm of mineral material. Histosols are separated in nearly the same way in the US and FAO systems, but no restriction is placed on the thickness of the organic layer if it overlies bedrock or skeletal material and in other situations the minimum required thickness is increased from 40 to 60 cm if it consists dominantly of little decomposed (fibric) moss or has a bulk density less than 0.1.

Peat soils so defined cover about 3 per cent of England and Wales (Mackney et al., 1983). They are much more extensive in Scotland and Ireland but the proportionate areas cannot be estimated on the same basis because different differentiating criteria have been used in each country. According to the 1:250,000 survey of Scotland (Soil Survey of Scotland, 1984), peat defined in more restrictive terms as organic soil containing more than 60 per cent organic matter and more than 50 cm thick occupies rather more than 10 per cent of the land area. In Ireland (Hammond, 1981) organic soil materials have been defined as in England and Wales, but the thickness limits for peat soils have been set at 45 cm in undrained land and 30 cm in drained land. Some 16 per cent of Ireland has organic soils meeting these requirements.

9.2 Mire Types

Ecologists (e.g. Bellamy and Moore, 1973; Godwin, 1975) use the general term *mire* to denote the bodies of organic material, mostly accumulated during the last 8,000 years, which give rise to these soils. The mires in the British Isles are divisible into three main types; fen, raised bog and blanket bog, each with distinctive topographic, hydrologic and phytosociological relationships. Those of the latter type are still actively growing in places, but nearly all the fens and most of the raised bogs now exist only as relics modified to varying degrees by the effects of artificial drainage and domestic or industrial exploitation of the peat for fuel or other purposes. Table 9.1 gives details of the unmodified and

Table 9.1: Classification of unmodified and modified mire types in Ireland (Hammond, 1981)

Mire types	Mire sub-types	NON-INDUSTRIAL					INDUSTRIAL	
		Area (ha)	Rainfall (mm)	Altitude (m)	Average depth (m)	Nutrient supply	Category	Area (ha)
Raised bog	True Midland	39,810[a] 1,980[b]	<1,000		7.0	Ombrotrophic	Machine peat	15,580
	Transitional	25,270 2,830	1,000 to 1,250	<152			Milled peat	24,690
	Man-modified	172,110 21,830	<1,250		2.5		Moss peat	5,050
Fen	Man-modified	92,510 9,300			1.2	Minerotrophic	Potential industrial areas	28,790
Blanket bog	Low Level Atlantic	243,610 1,130	>1,250	<152	3.0	Ombrotrophic (in small flush areas minerotrophic)	Machine peat	810
	Man-modified Atlantic	85,590 530					Milled peat	7,160
	High-level montane	321,060 100,120		>152	1.2			
	Man-modified montane	115,360 29,140						

modified mire types mapped by Hammond (1981) in Ireland.

Fen mires form in basin and valley sites under the direct influence of ground water charged with mineral plant nutrients from surrounding soils or substrata and are described accordingly as minerotrophic (eutrophic or mesotrophic). They normally consist of peat derived from reeds (*Phragmites* spp.), sedges (e.g. *Cladium mariscus*) or other semi-aquatic plants, underlain in places by organic detritus muds, which accumulated below and just above low water level. When, as a result of continuing peat growth, the ground surface emerges from the water for progressively longer periods each year, it is colonized under natural conditions by water-tolerant shrubs and trees such as sallow (*Salix cinerea*), alder (*Alnus glutinosa*) and birch (*Betula pubescens*), remains of which give rise to superincumbent layers of woody (fen-carr) peat. More or less decomposed fen or fen-carr peat occurs at or close to the surface in parts of East Anglia, the Somerset Levels and the Lancashire coastal plain, in central Ireland, and in numerous small low lying areas elsewhere (Figure 2.5, p. 78), nearly all of which have been artificially drained. The first consequence of drainage is a marked reduction in volume and a corresponding increase in bulk density due to partly irreversible dehydration of the initially water saturated peat. This is followed by progressive *wastage* resulting from oxidative decomposition of organic matter in the aerated zone above the lowered water table. A more localized effect of drainage is the development of acid-sulphate characteristics in former estuarine areas where the parent deposits contain sulphidic materials (Chapter 3, p. 111).

Raised bogs, also known as raised mosses or raised mires, result from continued upward growth of fen mires in basin sites. As peat formation carries the surface above the influence of ground water, the vegetation eventually becomes solely dependent on rainfall and is restricted accordingly to bog mosses (*Sphagnum* spp.) and other plants with small mineral requirements such as cotton-grass (*Eriophorum* spp.) and heather (*Calluna vulgaris*). Remains of these form layers of ombrotrophic (rain-nourished) peat, also described as oligotrophic or dystrophic, which is strongly acid and has a smaller ash content than fen or fen-carr peat. The characteristically domed form of the mires so formed arises mainly from the large water retaining power of fresh *Sphagnum* peat, amounting to more than ten times its dry weight. This serves to maintain water saturated conditions at or near the surface of the growing bog, whilst its lateral extension is inhibited by the collection at the margins of relatively nutrient-rich water percolating from nearby mineral soils, and eventually by better natural drainage of the sloping margins which result. In some places, however, the ombrotrophic peat has evidently spread laterally beyond the confines of the original basin. A cross-section of part of a typical raised bog in Shropshire is shown in Figure 9.1.

Raised bogs more or less modified by drainage and peat cutting are widely distributed in lowland sites with humid or subhumid climatic regimes and occur locally at higher altitudes. They are most widespread in the central plain of Ireland, where mechanized exploitation of the peat has become a major industry. The bogs which remain intact show a characteristic stratigraphic division into an upper layer of little decomposed *Sphagnum* peat, a middle layer of darker, denser and more decomposed oligotrophic peat usually containing more cotton-grass and heather remains, and a basal layer of fen or fen-carr peat (Figure 9.2). Palynological, archaeological and radiocarbon dating studies have shown that the upper boundary of the 'dark peat' commonly corresponds to the sub-boreal/sub-atlantic transition at around 2,500–3,000 years BP (Figure 1.15, p. 23), when the climate is believed to have become wetter and colder (Godwin, 1975). It is mainly this peat which is utilized as fuel, whilst the overlying

Peat soils

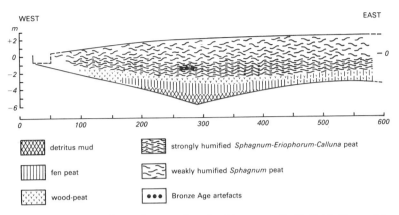

Figure 9.1: Section through part of a raised bog near Bettisfield, Shropshire (after Hardy, 1939).

Figure 9.2: Raised bog profile from Timahoe Bog, Co. Kildare, showing fen peat at the base, overlain by humified *Sphagnum* peat and the characteristic upper layer of poorly humified *Sphagnum* peat. The lighter colour towards the surface results mainly from drying (from Hammond, 1981).

Figure 9.3: The top 110 cm of a blanket bog profile at Sally Gap, Co. Wicklow, showing an upper fibrous layer dominated by recent and fossil roots of cyperaceous plants and a thick subsurface layer of semi-fibrous to amorphous peat (from Hammond, 1981).

'light peat' is better suited for horticultural use and has been increasingly exploited for this purpose in recent years.

The more widespread *blanket bogs* also consist of ombrotrophic peat but differ in having formed more or less directly on pre-existing acid mineral soils, including stagno-podzols (p. 270), stagnohumic gley soils (pp. 354, 380) and rankers (p. 120), under the influence of climatic wetness and low summer temperatures. Mires of this type occupy flat to moderately sloping land at altitudes ranging from near sea level to more than 1,000 m, but are restricted in the oro-arctic zone to occasional hollows and cols and are only extensive below 150 m in parts of northern Scotland, the northern and western Isles, and western Ireland, where climatic and topographic conditions are especially conducive to their development. In England and Wales, the south of Scotland, and eastern and central parts of Ireland, they are confined to upland plateaux with perhumid or humid oroboreal climatic regimes. The peat composing these bogs (Figure 9.3) is generally more uniform from top to bottom than in raised bogs but varies considerably in botanical composition from place to place in accordance with regional and local differences in environmental conditions. The chief peat forming plants, each of which is dominant in particular localities, include deer-grass (*Trichophorum caespitosum*), bog-rush (*Schoenus nigricans*) and purple moor-grass (*Molina caerulea*) as well as those characteristic of raised bogs. Blanket peat with deer-grass dominant is common in northern Scotland, whilst bog-rush, which also occurs in fen mires, is a major contributor to the low-level blanket bogs in western Ireland. Purple moor-grass is particularly associated with water receiving sites where flushing has caused local enrichment in nutrients. Where the drainage water is strongly acid and nutrient deficient, however, oligotrophic peat with *Sphagnum* has formed in basins as well as in unflushed areas. Similar 'valley bogs' occur on acid rocks in the subhumid lowlands but have a very restricted distribution.

Radiocarbon dating of plant remains from a number of widely distributed sites in Great

Figure 9.4: Section showing blanket bog overlying the partly collapsed wall of a cairn which contained a Court-grave at Behy, Co. Mayo (Copyright G.F. Mitchell).

Britain and Ireland has shown that whilst blanket bogs had begun to form in climatically favourable locations more than 4,000 years ago, they greatly increased in extent and thickness during the following two millenia. As tree stumps are common in the basal layers, much of the ground they now cover must previously have been forested. In certain Irish (Mitchell, 1976) and Scottish (Soulsby, 1976) localities, sites evidently occupied or cultivated by man during Neolithic or later periods have been buried beneath varying thicknesses of peat (Figure 9.4), indicating that deforestation and subsequent bog formation were conditioned by human activity as well as by climatic deterioration or natural soil impoverishment. Although actively growing blanket bog is still extensive in northern Scotland (Nature Conservancy Council, 1982), there are also large areas, particularly at low altitudes in western Ireland (Hammond, 1981), which have been cut over for fuel and subsequently reclaimed for agricultural use, and others in which the indigenous vegetation has been replaced by planted conifers.

Many of the plateau bogs, notably in the Pennines, have been affected by widespread gully erosion (Figure 9.5). This usually starts at the sloping margins and eats back into flatter areas with deeper peat, which is even-

Figure 9.5: The Forest of Bowland, Lancashire. Blanket bog covers the plateau and is severely eroded in places. The soils on the valley sides are mainly stagnopodzols (Copyright Aerofilms Ltd.).

Figure 9.6: Gullying in blanket bog on Kinder Scout, Derbyshire (Copyright Peak Planning Board).

tually dissected into steep-sided 'haggs' separated by a network of gullies in which mineral soil or bedrock is exposed (Figure 9.6). Much of the eroded material is removed by streams and the remainder forms deposits of redistributed peat. Possible causative factors include climatic change, increased instability of the water-charged peat as it grows, and modification or disruption of the vegetative cover resulting from some combination of overgrazing, heather burning, ditching to improve drainage, and air pollution. In a recent review, Tallis (1985) concluded that death of *Sphagnum* resulting from air pollution was probably the main cause of increased erosion in the southern Pennines during the last two centuries.

9.3 Classification of Peat Soils

Peat soils, as distinct from mires in their entirety, have been classified in several different ways. In the British Isles and other European countries (e.g. Kubiena, 1953; Mückenhausen, 1985), terms such as fen (*Niedermoor*), raised bog (*Hochmoor*) and blanket bog (climatic or hill peat) have traditionally been used as soil type names. Hammond's (1981) classification of Irish peatlands (Table 9.1) and the separation of basin (including fen and raised bog) and blanket peats in the 1:250,000 survey of Scotland (Soil Survey of Scotland, 1984) are examples of this approach.

In the current England and Wales classification (Avery, 1980), however, no reference is made to mire types as such and soil groups and subgroups are based instead on the degree of decomposition and other measurable properties of horizons within specified depths, using criteria derived from systems devised for the same purpose in the Netherlands (de Bakker and Schelling, 1966) and North America (Soil Survey Staff, 1975; Canada Soil Survey Committee, 1978). In accordance with the Dutch system, the soils are first divided into *raw peat soils* (or peat soils proper) and *earthy peat soils* in which drainage, with or without other ameliorative measures, has led to the formation of a fully ripened and humified earthy topsoil (Chapter 3, p. 108). The latter group includes the soils of man-modified mires distinguished by Hammond (1981). Subgroups are differenti-

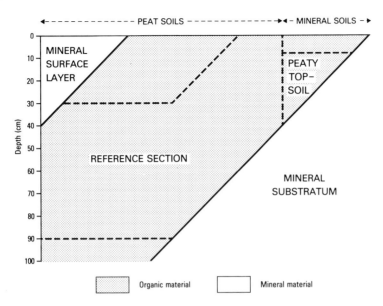

Figure 9.7: Reference section used in classifying peat soils (from Clayden and Hollis, 1984).

ated by the dominant degree of decomposition (fibrous, semi-fibrous or humified) and pH within an arbitrary reference section that extends from 30 to 90 cm below the surface where the organic layer is more than 90 cm thick and starts nearer or at the surface where it is thinner (Figure 9.7). Variants in sulphidic fen deposits which have developed acid-sulphate characteristics (Chapter 3, p. 111) as a result of drainage are also separated at this level.

In the US Taxonomy, by contrast, the division of Histosols into suborders is based primarily on the degree of decomposition of the organic deposits at depths greater than 30 cm and the surface horizons are disregarded at all categorical levels on the basis that they are readily modified by drainage and other management practices. As Dinç et al. (1976) concluded, however, this seems difficult to justify on practical grounds, as well as from a pedological standpoint. Thus, although an earthy topsoil can indeed form within a few years when initially water saturated peat is drained, it remains relatively unaffected by subsequent shrinkage and wastage because its lower boundary moves downwards at the expense of less altered horizons. It is consequently these horizons which change most in thickness and eventually disappear, as has happened over large areas in the Fenlands of eastern England.

Emphasis has been placed on degree of decomposition in recently developed systems because it can be determined with reasonable precision in the field and is well correlated with important physical properties such as bulk density, water-holding capacity, saturated hydraulic conductivity, and consistence in the wet state or after drying. A serious disadvantage of this approach, however, is that the genetically and geographically significant distinction between minerotrophic and ombrotrophic peatlands is overridden or relegated to a lower categorical level. In order to meet this objection, the classification used here features a primary division into fen (minerotrophic) and bog (ombrotrophic) groups based on identifiable profile characteristics, including pH and botanical composition. Subgroups are distinguished by differences in the degree of decomposition of surface and/or subsurface horizons and the presence or absence of acid-sulphate characteristics, using criteria similar to those employed in the England and Wales system.

In the FAO (1974) scheme, Histosols without permafrost are similarly divided into eutric (minerotrophic) and dystric (mainly ombrotrophic) units, but the pH limit of 5.5 (1:5 in water) adopted as the basis of separation is considerably higher than that (3.5 in 0.01M $CaCl_2$ or 4.0 in water) used here. As Dinç et al. noted (1976), this results in placement of many fertile soils derived from fen or fen-carr peat in the dystric class.

Key to Soil Groups

1. Peat soils that lack acid-sulphate characteristics (p. 111) and have one or both of the following:
   (a) pH ($CaCl_2$) 3.5 or less, or pH (H_2O) 4.0 or less, throughout the reference section (Figure 9.7)
   (b) macroscopically identifiable remains of *Sphagnum, Calluna, Eriophorum* or *Trichophorum* in some part of the reference section

   Bog soils

2. Other peat soils

   Fen soils

9.4 Fen Soils

9.4.1 General Characteristics, Classification and Distribution

As the name *Fen* implies, these soils are normally located in more or less man-modified minerotrophic (eutrophic or mesotrophic) mires formed under the influence of ground water enriched in mineral plant nutrients. In the undrained state the subsurface horizons are typically neutral to moderately acid in reaction but they can become strongly acid after drainage if the parent deposits contain oxidizable sulphides and little or no calcium carbonate. Conversely, drained soils with a

Figure 9.8: Somerset peat moors and Burrow Mount (Photo by D.C. Findlay).

substratum of well decomposed oligotrophic peat occasionally meet the prescribed requirements where pH values have been raised by flooding with base-rich water.

Soils of this group occur in the fenlands of eastern England and the Somerset Levels (Altcar, Adventurers and Mendham associations in Mackney *et al.*, 1983), in close association with raised bog soils in the Lancashire coastal plain (Hall and Folland, 1970) and in central Ireland (Hammond, 1981), and in numerous small low-lying areas elsewhere. Those in Scotland, where they are relatively inextensive, have been described as eutrophic and mesotrophic flushed peats in recent years (Soil Survey of Scotland, 1984). Almost all the land in the main areas of occurrence has been artificially drained; and in the East Anglian Fenland, where pump-assisted drainage was initiated from the late seventeenth century onwards and the soils are intensively cultivated, the original thickness of peat has been reduced by up to 4.5 m. As a consequence, large areas which formerly had peat soils now have humic gley ('skirt') soils of various kinds and wastage is continuing at rates estimated to range from 0.5 to 3 cm per year, depending on the thickness and organic matter content of horizons above the controlled water table (Seale, 1975a; Godwin, 1978; Burton and Hodgson, 1987). The former reclamation practice of 'paring and burning' and removal of topsoil by wind erosion have contributed to the overall loss in addition to the effects of shrinkage and wastage. Losses are much less evident in other fenland areas such as the Somerset Moors (Figure 9.8), where until recently the soils were all under grass and the water table was permanently within about 50 cm of the surface.

The upper horizons of these soils often contain mineral matter originating from intermittent alluvial sedimentation or artificial additions, and in some places they consist of mineral rather than organic material. Where the water table still remains perennially at or close to the surface, partially decomposed (fibrous or semi-fibrous) peat or soft muddy material is encountered within 20 cm depth, but the much more widespread drained soils invariably have a black earthy topsoil (Chapter 3, p. 108) or an organo-mineral A horizon containing few visually recognizable plant remains. As well as dehydration and humification, formation of an earthy topsoil from raw peat involves the complex of natural processes termed *moulding* or biological ripening by Jongerius (1962). This depends on the activity of soil animals in comminuting and ingesting plant residues, so exposing them to bacterial attack and leading to the formation of coprogenic aggregates (faecal pellets or casts), as in Ah and H horizons of mineral soils. Jongerius also distinguished mull and moder types of moulding, the former entailing the activity of earthworms and Enchytraeids. Mull formation occurs under favourable moisture conditions in topsoils having appreciable amounts of clay and a high base status, whilst moder-type horizons are formed from clay-deficient oligotrophic (bog) peats.

The thickness of the earthy layer depends on the depths to which physical, chemical and biological ripening have extended as governed by prevailing water table levels, the nature of the organic substratum, the rate of wastage and the length of time it has operated. Deep drainage and conversion to arable land hinders further biological ripening and can lead to structural deterioration, whereby the surface soil slakes and packs in wet weather to the extent that reduced infiltration causes ponding. Humified organic matter tends to be dispersed and washed downwards under these conditions, eventually giving rise to an 'organic B' or humilluvic horizon (Omh or Ohh in Hodgson, 1976) in which colloidal humus infills voids and appears as thick black coats on the walls of cracks (Van Heuveln and de Bakker, 1972). During prolonged dry periods, aggregates in the ploughed layer become hard and brittle or dusty and prone to blow, depending on the preceding state of the tilth, and are then resistant to rewetting. Intensive dehydration by roots can also reduce the water content of underlying horizons to such a degree that they lose the ability to reabsorb moisture readily. Subsurface horizons of East Anglian fen soils that exhibit this property are described as 'drummy' (Caldwell and Richardson, 1975): when brought to the surface by the plough, the drummy peat appears in the tilth as small hard angular blocks.

Where the organic layer is 90 cm or more thick, the subsurface horizons between 30 and 90 cm depth can be predominantly fibrous (Of), semi-fibrous (Om) or humified (Oh) according to their degree of decomposition, implying that the group includes Fibrists, Hemists and Saprists, respectively, in the US Taxonomy. The undrained fen soils without an earthy topsoil are not divided on this basis in the classification used here because of their limited extent in the British Isles, but drained variants qualifying as Saprists are separated at subgroup level from those with less well decomposed subsurface horizons, as in the England and Wales system (Avery, 1980).

As indicated above, nearly all the fen soils are in agricultural use. Those that remain waterlogged within 50 cm depth throughout the year are unsuited to cultivation and are mainly utilized for summer pasture or hay. Elsewhere, particularly in East Anglia and parts of the Lancashire coastal plain, the land is intensively cropped with cereals and potatoes or sugar beet, interspersed with high value vegetable crops such as brassicas, peas, carrots, onions and celery. In East Anglia the peat is often sufficiently permeable for deep ditches surrounding the fields to keep the water table below 50 cm at all times, but under-drainage is essential in the ever increasing areas where wastage has brought less permeable mineral substrata to within 90 cm of the surface. In Lancashire, where the climate is wetter, deterioration in the drainage systems resulting from shrinkage and wastage poses continuing management problems (Hall and Folland, 1970).

The suitability of these soils for intensive arable farming when adequately drained hinges on their ease of working and generally large capacity to supply nitrogen and retain water and other nutrient elements. Used in this way, however, they constitute a continually wasting resource and are subject to various physical and chemical limitations, including susceptibility to wind erosion, minor-element deficiencies, particularly of copper and manganese, and irregular development of strong acidity and 'drumminess', which affect land capability adversely wherever they occur.

Key to Subgroups

1. Fen soils with an earthy topsoil (p. 108) and acid-sulphate charcteristics (p. 111)
   Earthy acid-sulphate fen soils
2. Other fen soils with an earthy topsoil
   2.1 With humified peat (Oh) composing at least half of the reference section (Figure 9.7)
   Earthy sapric fen soils
   2.2 With semi-fibrous (Om) or fibrous (Of) peat composing more than half the reference section
   Earthy semi-fibrous fen soils
3. Other fen soils
   Raw fen soils

9.4.2 Raw Fen Soils

These are the soils of undrained fen mires without an earthy topsoil. The swampy areas where they occur commonly border open water and bear a variety of hydrophilic plant communities dominated by graminaceous, cyperaceous or woody species, the nature of which depends on local hydrologic, climatic and biotic factors, including the effects of pasturage and cutting of reeds (*Phragmites communis*) or sword sedge (*Cladium mariscus*) for thatching or other purposes. In the few profiles that have been fully described and sampled for analysis, the subsurface horizons consist predominantly of fibrous or semi-fibrous reedswamp or fen-carr peat. Variants with a subsurface layer of unripened organic lake mud (gyttja or sedimentary peat), corresponding to the raw eutro-amorphous peat soils of Avery (1980), may also occur but have not yet been positively identified in soil surveys.

All these soils are at most moderately acid in all horizons and some in which peat is interbedded with lake marl or overlies a calcareous substratum have a neutral or alkaline reaction at some depth. Others have sulphidic materials (Chapter 3, p. 111) within one metre of the surface which become extremely acid on aeration (Sulfihemists according to the US Taxonomy). No specific place was provided for such soils in the England and Wales classification (Avery, 1973, 1980) but recent surveys have shown that they occur sporadically in the river valleys of Norfolk and Suffolk where peat overlies or is interbedded with non-calcareous or slightly calcareous estuarine alluvium containing pyrites (Price, 1980; Price and Hodge, 1982; Burton and Hodgson, 1987). In the Norfolk Broads area samples were collected at kilometre intervals from depths of 30–40 cm, 60–70 cm and 100–110 cm in paired bores 50 cm apart and pH measurements made immediately after sampling and again after three months storage in porous polythene bags as recommended by Powlson (1975). Of the 75 peat profiles sampled 31 per cent gave post incubation values of 3.5 or less within 70 cm depth. The paired profiles commonly gave different results, confirming that the distribution of sulphidic material is irregular.

The following representative profile (exclusive of surface litter) was recorded under alder carr adjoining the river Bure in the same area. It shows a humified Oh horizon 8 cm thick, passing abruptly into yellowish peat which darkens on exposure and consists almost entirely of little decomposed roots, leaves and stems of reed (*Phragmites australis*). As pH values after slow oxidation were not determined, the possibility that sulphidic material is present cannot be discounted, but data on nearby sampling sites reported by Price and Hodge (Soil Survey of England and Wales, 1982) suggest that this is unlikely.

Profile 171 (Tatler and Corbett, 1977)

Location : Ranworth, Norfolk (TG 370157)
Climatic regime: subhumid temperate
Altitude : 0 m   Slope: <1°
Vegetation : reed (*Phragmites australis*) and alder (*Alnus glutinosa*) scrub

Horizons (cm)
0–8
Oh

Black to very dark brown (10YR 2/1–2) humified peat; 5–10 per cent fibrous plant remains; very moist; common fine and medium roots; abrupt smooth boundary.

8–115
Of/Om

Dark yellowish brown (10YR 4/4) fibrous (or semi-fibrous) peat; wet; 80 per cent recognizable roots, leaves and stems of reed (*Phragmites australis*); non-coherent; water table at 30 cm.

Analytical data

Horizon	Oh	Of/Om	Of/Om
Depth (cm)	0–8	8–85	85–115
loss on ignition %	83	91	72
organic carbon %	38	42	35
nitrogen %	2.6	2.2	2.6
C/N ratio	15	19	13
pH (H$_2$O)	6.7	5.1	5.4
(0.01M CaCl$_2$)	6.4	4.6	5.1

Classification
England and Wales (1984): raw eu-fibrous peat soil; grass-sedge peat (Ousby series).
USDA (1975): Typic Medifibrist (or Medihemist); euic, mesic.
FAO (1974): Eutric Histosol.

9.4.3 Earthy Semi-fibrous Fen Soils

These are drained fen soils, classed as earthy eu-fibrous peat soils by Avery (1980), that have partially decomposed (semi-fibrous or fibrous) fen or fen-carr peat directly below an earthy topsoil or within 60 cm depth and lack acid-sulphate characteristics. Variants in which the reference section (30–90 cm) consists predominantly of fibrous material (Fibrists in the US system) have been recorded in shallowly or recently drained reedswamp peat (e.g. Hall and Folland, 1967) but are uncommon. Soils of this subgroup are extensive in the Somerset Moors (Altcar association in Findlay et al., 1984) and occur widely in association with other peat soils in the East Anglian Fens (Hodge et al., 1984), in Lancashire (Ragg et al., 1984) and in central Ireland (Banagher series in Conry et al., 1970, Finch et al., 1971, and Finch and Gardiner, 1977). There are also smaller areas in northern England (R.A. Jarvis et al., 1984) and the Midlands. In some places the parent deposits were originally covered by raised bog peat which was cut and removed for fuel before the land was drained and brought into agricultural use.

Data on two representative profiles under grass are given below. The first, located near Doncaster in South Yorkshire, has an earthy topsoil 25 cm thick over dark reddish brown to black laminated fen-carr peat containing abundant woody fragments, which rests at 84 cm depth on silty mineral-organic material. There is added lime in the surface horizon and pH (CaCl$_2$) values are in the range 5.5–6.5 throughout.

Profile 173 was described and sampled by D.C. Findlay in Kings Sedgemoor, Somerset, which was enclosed and drained at the end of the eighteenth century and is no longer subject to regular floods. Here the parent material is reedswamp peat with few woody remains. It is more deeply decomposed than in the preceding profile but the black strongly humified upper horizons are similar in morphology, having a granular structure at the surface and coarser subangular blocky peds in the lower part.

Profile 172 (R.A. Jarvis, 1973)

Location	:	Hatfield, South Yorkshire (SE 644072)
Climatic regime:		subhumid temperate
Altitude	:	4 m Slope: <1°
Land use and vegetation	:	long-term grass with buttercups (*Ranunculus* spp.), some clover (*Trifolium repens*) and dandelions (*Taraxacum officinale*)

Horizons (cm)

0–7 Oh1	Black (5YR 2/1) humified peat; moderate fine granular; very friable; abundant fine roots; gradual boundary.
7–25 Oh2	Black (5YR 2/1) humified peat; moderate coarse subangular blocky with shiny ped faces, breaking in part to medium granular; friable; abundant fine roots; clear smooth boundary.

Analytical data for Profile 172

Horizon	Oh1	Oh2	Om/Of	Om	2Oc
Depth (cm)	0–7	7–25	25–36	36–84	84–93
sand 200 μm–2 mm %					2
60–200 μm %					18
silt 2–60 μm %					57
clay <2 μm %					23
loss on ignition %	74	80	92	93	51
organic carbon %	39	41	49	58	38
nitrogen %	2.7	2.8	2.5	1.9	1.7
C/N ratio	14	15	20	30	22
CaCO$_3$ equiv. %	4.3	0	0	0	0
pH (H$_2$O)	6.7	6.7	6.4	5.8	6.1
(0.01M CaCl$_2$)	6.5	6.3	6.0	5.5	6.0

25–36 Om/Of	Dark reddish brown (5YR 3/2 and 3/3) laminated semi-fibrous to fibrous peat containing abundant densely packed plant remains, including birch bark; clear boundary.
36–84 Om	Black (5YR 2/1) laminated semi-fibrous peat; coarse root remains and birch bark abundant; clear boundary.
84–93 2Oc	Olive-grey (5Y 4/2) amorphous loamy peat; wet; weak medium prismatic; slightly sticky.

Analyical data *see* p. 412.

Classification
England and Wales (1984): earthy eu-fibrous peat soil; grass-sedge peat (Altcar series, fen-carr phase).
USDA (1975): Sapric (or Limnic) Medihemist; euic, mesic.
FAO (1974): Eutric Histosol.

Profile 173

Location	: Kings Sedgemoor, Moorlinch, Somerset (ST 385352)
Climatic regime:	subhumid temperate
Altitude	: 5 m Slope: <1°
Land use	: long-term grass

Horizons (cm)

0–20 Oh1	Black (5YR 2/1) humified peat; strong medium to coarse granular; abundant fine roots; clear boundary.
20–55 Oh2	Black (5YR 2/1) humified peat with common dark reddish brown dead rootlets and vertically oriented dark brown (7.5YR 3/2) decayed reed stems; moderate coarse to medium subangular blocky; many live roots; clear boundary.
55–90 Om	Very dark brown (7.5YR 3/2) semi-fibrous peat containing many yellowish brown (straw-like) reed, stem and leaf remains; reddish brown wood fragments below 75 cm; wet; massive; few live roots.

Analytical data *see* below.

Classification
England and Wales (1984): earthy eu-fibrous peat soil; grass-sedge peat (Altcar series).
USDA (1975): Typic (or Fibric) Medihemist; euic, mesic.
FAO (1974): Dystric Histosol.

9.4.4 Earthy Sapric Fen Soils

These soils differ from those of the preceding subgroup in that not only the earthy topsoil but also the greater part of the subsurface reference section consists of strongly humified (sapric) organic material. Classed as earthy eutro-amorphous peat soils in the England and Wales system and as Saprists in the US Taxonomy, they are typically developed in mineral-rich fen deposits, including organic lake muds like those of Martin Mere, Lancashire (Crompton, 1966; Hall and Folland, 1967), which provide optimal conditions for mull-type moulding (Jongerius, 1962). Data on two representative profiles, the first under semi-natural woodland and the second in arable land, are reproduced below.

Profile 174 was described and sampled by R.H. Allen in coppiced woodland with a rich

Analytical data for Profile 173

Horizon	Oh1	Oh2	Oh2	Om	Om
Depth (cm)	0–20	20–40	40–55	55–75	75–90
loss on ignition %	73	86	90	89	89
organic carbon[a] %	34	46	45	50	46
nitrogen %	2.6	2.3	2.4	1.8	1.9
C/N ratio	13	20	19	28	24
pH (0.01M CaCl$_2$, undried)	4.9	4.6	4.5	4.8	5.2

[a] by Nömmik's (1971) method

herbaceous field layer in a valley of a small tributary of the river Stour in north-east Essex. Ash forms about two-thirds of the tree canopy and is replaced by alder on wetter ground nearby. The soil is developed in some 90 cm of eutrophic peat which accumulated under the influence of mobile calcium-saturated ground water and contains some 50 to 70 per cent mineral matter. Although the water table has been lowered, it remains within the rooting zone for most of the time. All the organic horizons are practically devoid of recognizable plant remains and the well developed structure of the earthy topsoil reflects intense earthworm activity.

Profile 175 is located in the Cambridgeshire Fens north-west of Ely, where most of the remaining peat is less than 90 cm thick. The organic material beneath the ploughed layer contains some wood fragments and is succeeded at 51 cm depth by gleyed loamy and sandy horizons formed in the same Devensian fluvial deposits as profile 161 (p. 379). Following earlier work in the Cambridge area (Hodge and Seale, 1966), Seale (1975a, b) mapped all the cultivated earthy peat soils in the Ely and Chatteris districts as depth phases of the Adventurers series but more recent studies have shown that profiles conforming to each of the three earthy subgroups recognized here are included. Acid-sulphate variants represented by profile 177 occur where the peat overlies sulphidic Fen Clay and deep phases in other situations commonly have a semi-fibrous layer within 60 cm of the surface and resemble profiles 172 and 173 in this respect.

Profile 174

Location	:	Alphamstone, Essex (TL 867359)
Climatic regime:		subhumid temperate
Altitude	:	53 m Slope: <1°
Land use and vegetation	:	broadleaf woodland; old ash-hazel coppice with field layer of dogs mercury (*Mercurialis perennis*) with frequent enchanters nightshade (*Circaea lutetiana*), ivy (*Hedera helix*) and bramble (*Rubus fruticosus*), occasional primrose (*Primula vulgaris*) and herb bennet (*Geum urbanum*), and scattered mosses.

Horizons (cm)

L	Litter of leaves and twigs averaging 1 cm thick; many earthworm casts and mole hills at surface.
0–5 Oh1	Black (2.5Y 2/1) humified peat; strong medium granular; very friable; many fine roots; many earthworms; sharp smooth boundary.
5–15 Oh2	Black (5YR 2/1) humified peat; strong fine subangular blocky and granular; friable; many fine and few medium and coarse woody roots; many earthworms; abrupt smooth boundary.
16–26 Oh3	Black (5YR 2/1) humified peat; moderate medium subangular blocky; friable; common fine and few medium to coarse woody roots; common earthworms; clear smooth boundary.
26–52 Oh4	Black (5YR 2/1) humified peat containing occasional small subangular flint fragments; wet; massive; one mole channel 6 cm in diameter; common roots; earthworms present; gradual smooth boundary.
52–93 Oh5	Black (5YR 2/1) humified peat; wet; massive; few roots; sharp smooth boundary.
93–103 2Cg	Grey (5Y 6/1) stoneless clay loam (USDA silt loam to loam) with common medium light yellowish brown (2.5Y 6/4) mottles; wet; massive; slightly fluid; no roots. (water table at 54 cm on 4 Sept. 1978)

Analytical data *see* p. 415.

Classification

England and Wales (1984): Earthy eutro-amorphous peat soil; humified peat (Adventurers series).

USDA (1975): Terric Medisaprist; loamy, mixed, euic, mesic.

FAO (1974): Eutric Histosol.

Analytical data for Profile 174

Horizon	Oh1	Oh2	Oh3	Oh4	Oh5	2Cg
Depth (cm)	0–5	5–16	16–26	26–52	52–93	93–103
sand 200 μm–2 mm %						11
60–200 μm %						14
silt 2–60 μm %						56
clay <2 μm %						19
loss on ignition %	54	52	51	57	79	
organic carbon[a] %	26	25	24	26	43	
nitrogen %	2.2	2.1	2.1	2.0	3.0	
C/N ratio	12	12	11	13	14	
pH (H$_2$O)	6.4	6.2	6.9	7.1	6.9	8.0
(0.01M CaCl$_2$, undried)	5.7	5.6	6.2	6.5	6.4	7.3

[a] by Nömmik's (1971) method

Profile 175 (Seale, 1975a)

Location : Byall Fen, Manea, Cambridgeshire (TL 462865)
Climatic regime: subhumid temperate
Altitude : 0 m   Slope: <1°
Land use : arable (sugar beet)

Horizons (cm)

0–25
Op
Black (10YR 2/1), very slightly stony, loamy humified peat containing few uncoated sand grains and small flint fragments; weak subangular blocky clods breaking to weak granular; very friable; few fine roots; sharp smooth boundary.

25–43
Oh1
Black (10YR 2/1–2) stoneless humified peat containing common (3 per cent) fine reddish brown alder-wood fragments; weak fine granular; few roots; clear boundary.

43–51
Oh2
Very dark grey to black (10YR 2–3/1) humified peat containing uncoated sand grains, otherwise as above; abrupt smooth boundary.

51–58
2Ahg
Dark grey (7.5YR 3.5/1) very slightly stony sandy silt loam (USDA loam) with common fine reddish brown (5YR 4/4) mottles and few irregular vertical black streaks formed by decayed roots; few small rounded to subangular flint fragments; weak subangular blocky; friable; few live roots; clear smooth boundary.

58–84
2Bg
Greyish brown to light brownish grey (2.5Y 5–6/2) slightly stony sandy silt loam (USDA loam to silt loam) with common, fine and medium, reddish brown to red (2.5YR 4/6) mottles; common flint fragments, becoming few below 66 cm; very weak subangular blocky; friable to firm; common decayed roots as above; few live roots; clear smooth boundary.

84–91+
3Cg
Brown to pale brown (10YR 5–6/3), slightly stony, slightly calcareous loamy sand (USDA loamy fine sand) with many yellowish brown (10YR 5/4–5) mottles; common flint fragments; wet; massive; rare decayed roots as above.

Analytical data *see* p. 416.

Classification
England and Wales (1984): earthy eutro-amorphous peat soil; humified peat (Adventurers series).
USDA (1975): Terric Medisaprist; loamy, mixed, euic, mesic.
FAO (1974): Eutric Histosol.

Analytical data for Profile 175

Horizon	Op	Oh1	Oh2	2Ahg	2Bg	2Bg	3Cg
Depth (cm)	0–25	25–43	43–51	51–58	58–64	64–84	84–91
sand 200 µm–2 mm %				29	22	16	34
50–200 µm %				22	24	15	51
silt 2–50 µm %				33	38	53	6
clay <2 µm %				16	16	16	8
loss on ignition %	40	85	55				
organic carbon %	24	43	30				
nitrogen %	1.7	3.2	2.3				
C/N ratio	14	13	13				
CaCO$_3$ equiv. %	0	0	0	0	0	0.1	4.3
pH (H$_2$O)	6.1	6.1	6.1	6.6	6.6	7.3	7.9
(0.01M CaCl$_2$)	5.8	5.8	5.8	6.4	6.4	7.1	7.6

9.4.5 Earthy Acid-sulphate Fen Soils

These are the earthy fen soils with acid-sulphate characteristics (Chapter 3, p. 111). Like the sulphidic raw fen soils from which they are derived, they are confined to areas where microbiological reduction of sulphates in the ground water led to accumulation of pyrite, either in the organic materials or in underlying or interbedded mineral layers. Those having a sulphuric horizon with yellowish segregations of jarosite and a pH less than 3.5 (1:1 in water) within 50 cm depth are classed as Sulfohemists in the US system.

The main areas of occurrence in the British Isles are in parts of the deeply drained fenland north and south of the Wash where surviving remnants of the late Flandrian Upper Peat (Godwin, 1978) overlie the estuarine Fen or Buttery Clay (Plate IV.7), much of which was non-calcareous or only slightly calcareous when deposited. Soils of this subgroup have also been identified in river valleys in Norfolk and Suffolk (Corbett, 1979; Price, 1980; Price and Hodge, 1982), in the lower Trent valley (Reeve and Thomasson, 1981) and in the south-west Lancashire coastal plain (Burton and Hodgson, 1987). As in the associated alluvial gley soils (p. 325), the presence of an extremely acid horizon at limited depths checks root development to the extent that yields of sugar beet and other sensitive crops are adversely affected, especially in dry years. Particularly in the Cambridgeshire and Lincolnshire Fens, the organic subsurface horizons are usually 'drummy', a feature attributed by Caldwell and Richardson (1975) to the presence of redeposited iron oxides, which they often contain in significant amounts. As postulated by Jongerius (1962), however, the distinctive physical properties of the drummy layers probably result primarily from shallow rooting and consequent intense dehydration above the toxic sulphuric horizon. A further adverse consequence of pyrite oxidation is the blockage of tile drains in these soils by deposits of ferruginous or aluminous material (Trafford et al., 1973; Bloomfield and Zahari, 1982).

Data on two representative profiles in eastern England are given on pp. 417–18. Profile 176, in Witham Fen, Lincolnshire, has a thick Op horizon directly over a thin extremely acid clay layer with jarosite and clearly qualifies as a Sulfohemist. The underlying peat, which extends beneath the current water table, is also extremely acid and contains more than 3 per cent oxidized or oxidizable sulphur. In profile 177, located in the Chatteris district of Cambridgeshire (Seale, 1975b), a thicker layer of peat overlies Fen Clay at 72 cm depth and was wholly above the water table at the time of sampling. Although no jarosite was observed, the subsurface horizons contain deposits of crystalline gypsum, particularly in the 'drummy' layer at 40–58 cm. This confirms that the present acidity of the peat, which originally accumulated under the influence of calcium-saturated ground water, results from pyrite oxidation induced by deep drainage.

Profile 176 (Robson *et al.*, 1974)

Location	: Potter Hanworth, Lincolnshire (TF 107690)
Climatic regime:	subhumid temperate
Altitude	: 1.5 m Slope: <1°
Land use	: arable (cereal stubble)

Horizons (cm)

0–40 Op	Black (10YR 2/1) humified peat; moderate medium and fine granular; very friable; few indistinct plant remains; abundant fine roots; abrupt smooth boundary.
40–67 2Cg	Dark grey to grey (5Y 4–5/1) stoneless clay (USDA clay) with many coarse yellowish brown (10YR 5/6) and dark reddish brown (5YR 3/3) mottles associated with old root channels; massive; common large reed stem and leaf sheath remains; many dead roots; yellow (2.5Y 8.6) deposits of jarosite in channels; abrupt smooth boundary.
67–87 3Om1	Very dark brown (10YR 2/2) laminated semi-fibrous peat containing many reed leaf and stem remains; wet; common distinct yellow veins of jarosite; abrupt smooth boundary.
87–100 3Om2	Dark brown (7.5YR 3/2) laminated semi-fibrous reed peat, more fibrous than above and rapidly becoming very dark brown (10YR 2/2) on exposure to air; wet (reduced); smell of hydrogen sulphide.

Analytical data *see* next column.

Classification
 England and Wales (1984): earthy sulphuric peat soil; humified peat with mineral layers (Prickwillow series).

USDA (1975): Typic Sulfohemist; mesic.
FAO (1974): Dystric Histosol.

Analytical data

Horizon	Op	2Cg	3Om1	3Om2
Depth (cm)	0–40	40–67	67–87	87–100
sand 200 µm–2 mm %	<1			
50–200 µm %	1			
silt 2–50 µm %	30			
clay <2µm %	69			
loss on ignition %	59	12.5	69	71
organic carbon %	28	3.1	31	
nitrogen %	1.9	0.3	1.7	
C/N ratio	15	10	18	
total sulphur %		0.7	3.6	3.5
pH (H$_2$O)	6.4	3.1	2.5	3.2
(0.01M CaCl$_2$)	6.0	2.9	2.3	3.0

Profile 177 (Seale, 1975b)

Location	: Somersham High North Fen, Cambridgeshire (TL 375808)
Climatic regime:	subhumid temperate
Altitude	: 1.2 m
Land use	: arable (potatoes)

Horizons (cm)

0–40 Op	Black (5YR 2/1.5; 10YR 3/2 dry) stoneless loamy humified peat; irregular clods breaking easily to moderate very fine and fine granular; loose in top 4 cm, very friable below, though individual granules are friable to firm; common fine roots; abrupt wavy boundary, with occasional inclusions of the underlying material between 25 and 40 cm.
40–58 Ohy	Black (5YR 2/1) humified peat becoming browner when rubbed; few wood fragments; moderate medium to coarse subangular blocky; firm (drummy); few roots; common, fine to medium, white (10YR 8/1) inclusions of fine gypsum crystals; clear smooth boundary.
58–72 Omy	Dark reddish brown (5YR 2/2) weakly laminated semi-fibrous peat with occasional leaf remains along bedding planes; very weak coarse angular blocky tending to platy; firm; roots rare; few white inclusions of crystalline gypsum; clear smooth boundary.

72–95+ Cgy	Grey (N 5.0) stoneless clay (USDA clay) with common coarse strong brown (7.5YR 5/6) linear mottles associated with plant remains; moist to wet; massive; vertical channels containing undecomposed reed remains common; no live roots; gypsum crystals as above.

Analytical data

Horizon	Op	Ohy	Omy	2Cgy
Depth (cm)	5–36	43–55	60–70	75–90
sand 200 μm–2 mm %				1
50–200 μm %				6
silt 2–50 μm %				29
clay <2 μm %				64
loss on ignition %	44	77	56	
organic carbon %	20	37	28	3.8
nitrogen %	1.5	2.0	1.9	
C/N ratio	13	19	15	
total sulphur %		3.8	1.1	3.0
pH (H$_2$O)	4.5	3.6	3.7	3.9
(0.01M CaCl$_2$)	4.3	3.4	3.6	3.8

Classification
England and Wales (1984): earthy sulphuric peat soil; humified peat (Mendham series).
USDA (1975): Terric Medisaprist; clayey, dysic, mesic.
FAO (1974): Dystric Histosol.

9.5 Bog Soils

9.5.1 General Characteristics, Classification and Distribution

These soils consist of more or less decomposed oligotrophic peat accumulated in blanket or raised bogs and in valley (flushed) bogs of limited extent. According to the definition adopted here, they have visually identifiable remains of bog-moss (*Sphagnum* spp.), cotton-grass (*Eriophorum* spp.), heather (*Calluna vulgaris*) or deer-grass (*Trichophorum caespitosum*) in some part of the reference section, pH values less than 3.5 in CaCl$_2$ or 4.0 in water throughout, or both. As noted above, the peat-forming plants listed as diagnostic are characteristic of unflushed blanket and raised bogs. Acid peat soils with subsurface horizons that are strongly decomposed (sapric), or in which only remains of other species such as purple moor-grass (*Molinia caerulea*) or black bog-rush (*Schoenus nigricans*) can be identified, are also placed in this group if the pH requirements are met and acid-sulphate characteristics are lacking.

Much of the ground underlain by these soils, including areas where the original surface has been lowered by peat cutting, carries semi-natural bog or moorland vegetation. Under these conditions the organic horizons comprising the upper 20–30 cm of the profile are for the most part perennially wet, incompletely decomposed, or both. In sites subject to seasonal drying they resemble the unincorporated superficial organic layers of acid mineral soils supporting similar vegetation, which on some lowland raised bogs has been colonized by trees, chiefly birch and Scots pine. Where the land has been systematically drained and brought into agricultural use, however, the raw upper horizons have normally been converted into a black earthy topsoil.

The subsurface horizons constituting the reference section (Figure 9.7) vary in morphology and related physical properties depending on the botanical origin and degree of decomposition of the materials composing them. Variants in which they are predominantly fibrous (Fibrists in the US Taxonomy) are comparatively uncommon in the British Isles and are mostly restricted to intact raised bogs where the characteristic upper layer of little-decomposed *Sphagnum* peat is 60 cm or more thick. Elsewhere, especially in the more widespread blanket bogs, most undisturbed profiles consist mainly of semi-fibrous material or have semi-fibrous or occasionally fibrous upper horizons overlying strongly humified (sapric) peat (Plate IV.8). The presence of a subsoil layer which is more decomposed than the overlying horizons can be attributed in some cases to formation under conditions more conducive to seasonal drying and humification than those obtaining subsequently. It may also reflect slow but continued decomposition after burial or enrichment in colloidal organic matter which has been translocated downwards in dispersed form (Duchaufour, 1982, p. 361).

In their semi-natural state, bog soils may provide extensive grazing of low quality but are otherwise of no agricultural value. On the upland blanket bogs the potential for improvement is very small because of the limitations imposed by adverse climate as well as by wetness, low bearing strength and extreme infertility. On raised bogs in the drier lowlands and to some extent on the low-level blanket bogs it is distinctly greater and substantial areas, particularly in Ireland (Hammond, 1981) and the Lancashire coastal plain (Hall and Folland, 1970), have been reclaimed for grass production or arable cropping, usually after removal of much of the peat for fuel, by drainage and addition of lime and fertilizers or other amendments such as sea sand, marl or urban night soil. The resulting earthy peat soils resemble those derived from fen deposits in their high productive capacity and ease of working, and present similar management problems under cultivation, including the effects of shrinkage and wastage, wind erosion when the ground is bare, and liability to micronutrient deficiencies.

In both Great Britain and Ireland, much bog land at altitudes below about 400 m has been afforested in recent years and a major proportion of currently projected planting is on soils of this group. As on the upland peaty gley soils, the main species used are Sitka spruce and Lodgepole pine. Intensive drainage is essential for success, as much to meet crop stability requirements as to improve growth, the usual practice being to plough deeply using wide-tracked vehicles and to plant the trees in the ridges so formed (Taylor, 1970). Once the crop is established, interception of rainfall by the canopy and extraction of water by the roots causes progressive drying of the peat and there is a consequent increase in the depth to which aeration extends (Pyatt and Craven, 1979; King et al., 1986).

In classifying peatland for afforestation, the Forestry Commission uses the existing vegetation to distinguish four main 'bog groups' (Table 2.8, p. 62) considered to represent a descending scale of fertility, namely: (1) eutrophic flushed bogs (fens); (2) mesotrophic and oligotrophic blanket and basin bogs with *Molinia* present; (3) oligotrophic (unflushed) blanket bogs on slopes with *Molinia* absent; and (4) raised bogs and unflushed blanket bogs on level sites with *Molinia* absent (Toleman, 1973; Pyatt et al., 1979). Sitka spruce grown on bogs of groups 3 and 4 commonly displays symptoms of nitrogen deficiency, as well as of P and K, without suitable fertilizer treatment. Nitrogen is seldom deficient on the 'better' bogs of group 2 but P and K deficiencies can occur in young Sitka crops.

Key to Subgroups

1. Bog soils with an earthy topsoil (p. 108)
   1.1 With humified peat (Oh) composing at least half the reference section (Figure 9.7)
   Earthy sapric bog soils
   1.2 With semi-fibrous (Om) or fibrous (Of) peat composing more than half the reference section
   Earthy semi-fibrous bog soils
2. Other bog soils with fibrous peat (Of) composing at least half the reference section (Figure 9.7)
   Raw fibrous bog soils
3. Other bog soils with humified peat (Oh) composing at least half the reference section
   Raw sapric bog soils
4. Other bog soils
   Raw semi-fibrous bog soils

9.5.2 Raw Fibrous Bog Soils

These are bog soils that consist dominantly of little decomposed peat and lack an earthy topsoil. As indicated above, they are chiefly represented in surviving areas of intact raised bog, including some retained as nature reserves, which have not been reclaimed for agriculture or exploited for fuel or other purposes. Typical profiles in such sites have a thin more or less decomposed surface horizon over a reddish Of horizon of fibrous *Sphagnum* peat with clearly recognizable cell structure. This can be more than one metre thick but usually passes at some lesser depth into denser, less permeable and more decomposed *Sphagnum – Calluna – Eriophorum* peat. It may also rest more or less directly on woody (fen-carr) or fen peat forming the basal stratum of the bog. Blanket bogs with upper layers of *Sphagnum*-rich peat give rise to similar soils in places, notably in parts of the Pennines (Longmoss association in R.A. Jarvis *et al.*, 1984) and south-east Scotland (Ragg and Futty, 1967; Ragg and Clayden, 1973). Variants with thick layers of fibrous oligotrophic peat derived dominantly from plants other than *Sphagnum* also occur but are rare in the British Isles. The following two representative profiles are both in raised bogs.

Profile 178 in Risley Moss, Cheshire (formerly Lancashire) was not sampled for analysis but is included because the description by Hall and Folland (1970) extends through the whole thickness of the bog. The fibrous *Sphagnum* layer is about 80 cm thick and overlies a thicker layer of dark reddish brown, semi-fibrous *Sphagnum – Eriophorum* peat with a few *Calluna* remains. Below this is a thin amorphous layer containing occasional wood fragments, which rests on Devensian till.

Profile 179 was described and sampled in the cut edge of a large bog in County Westmeath. Judging from the rubbed fibre determinations (Chapter 3, p. 99), all four horizons sampled qualify as fibrous but the pyrophosphate indices (Chapter 3, p. 89) suggest that the surface horizon contains appreciably more colloidal organic matter than those below.

Profile 178 (Hall and Folland, 1970)

Location	:	Risley Moss, Cheshire (SJ 670920)
Climatic regime:		humid temperate
Altitude	:	22 m Slope: 1°
Land use and vegetation	:	rough grazing dominated by purple moor-grass (*Molinia caerulea*)

Horizons (cm)

0–7.5 Oh	Black (5YR 2/1) humified peat bound by living and dead *Molinia* roots
7.5–84 Of	Yellowish red (5YR 5/8) fibrous *Sphagnum* peat; very weakly decomposed (H3); gradual boundary.
84–107 2Om1	Dark reddish brown (5YR 3/4) semi-fibrous *Eriophorum* peat; laminated; well decomposed (H6–H7); gradual boundary.
107–152 2Om2	Dark reddish brown (2.5YR 3/4) laminated semi-fibrous peat containing *Sphagnum* and *Eriophorum* with occasional woody *Calluna* remains; well decomposed (H6–H7); gradual boundary.
152–274 2Om3	Dark reddish brown (2.5YR 2/4) semi-fibrous *Sphagnum – Eriophorum* peat; laminated; well decomposed (H6–H7); abrupt boundary.
274–304 3Oh	Very dark brown (10YR 2/2) humified peat containing occasional soft wood remains.
304–320 4Cg	Grey (10YR 5/1) sandy clay loam containing occasional small subrounded stones; firm (till).

Classification

England and Wales (1984): raw oligo-fibrous peat soil; *Sphagnum* peat (Longmoss series).
USDA (1975): Typic Medihemist; dysic; mesic.
FAO (1974): Dystric Histosol.

Profile 179 (Finch and Gardiner, 1977; Hammond, 1981)

Location	:	Clonawiny, County Westmeath

Climatic regime: humid temperate
Altitude : 99 m   Slope: 1°
Vegetation : moorland with heather (*Calluna vulgaris*) and bog mosses (*Sphagnum* spp.)

Horizons (cm)

0–27 Of/Om	Dark reddish brown (5YR 3/4) fibrous to semi-fibrous *Sphagnum – Calluna* peat; living roots present; clear wavy boundary.
27–58 Of1	Dark reddish brown (5YR 3/4) fibrous *Sphagnum* peat; live roots present; clear slightly wavy boundary.
58–87 Of2	Dark reddish brown (2.5YR 2/4) fibrous *Sphagnum – Eriophorum* peat with few *Calluna* remains; live roots absent; abrupt wavy boundary.
87–118 Of3	Black (5YR 2/1) fibrous *Sphagnum – Calluna* peat with some *Eriophorum* remains; live roots absent.

Analytical data

Horizon	Of/Om	Of1	Of2	Of3
Depth (cm)	0–27	27–58	58–87	87–118
loss on ignition %	97.0	98.0	99.4	99.4
nitrogen %	1.50	1.10	0.64	0.64
rubbed fibre (% vol.)	50	41	56	64
pyrophosphate index	4	7	7	7
pH (H$_2$O)	3.4	3.4	3.4	3.5
(0.01M CaCl$_2$)	2.9	2.8	2.7	2.8
bulk density (g cm^{-3})	0.06		0.06	
saturated water content %	1548		1685	

Classification
  Ireland (Hammond, 1981): raised mire soil, Midland type (Allen series).
  England and Wales (1984): raw oligofibrous peat soil; *Sphagnum* peat (Longmoss series).
  USDA (1975): (Typic) Sphagnofibrist; dysic, mesic.
  FAO (1974): Dystric Histosol.

9.5.3 Raw Semi-fibrous Bog Soils

These soils also consist of oligotrophic peat containing many visually identifiable plant remains but have relatively well decomposed, predominantly semi-fibrous subsurface horizons which are normally denser and less permeable than fibrous materials of similar composition. They are almost certainly more extensive in the British Isles as a whole than those of any of the other peat soil subgroups distinguished here. The main areas of occurrence are in blanket bogs but they also occur in unreclaimed raised bogs where an upper layer of fibrous *Sphagnum* peat is thin or absent, either because it was never well developed or because it has been removed by cutting or transformed as a result of increased aeration. Other variants of small extent are in acid basin or flushed peats derived mainly from grasses or sedges.

Several memoirs of the Soil Survey of Scotland (Glentworth and Muir, 1963; Ragg and Futty, 1967; Bown, 1973; Futty and Dry, 1977; Bown and Heslop, 1979) contain diagrammatic representations of typical profiles conforming to this subgroup. Details of botanical composition and degree of decomposition on the Von Post scale (Table 3.5, p. 90) are given for all morphologically distinct horizons and are supplemented in most cases by chemical data, but few descriptions in standard format have been published. Data on two representative profiles from northern England and one from Scotland are given on pp. 422–4.

Profile 180 is in upland blanket bog on the Pennine moors south of Huddersfield, West Yorkshire. The extremely acid peat at this site is about 1.4 m thick and contains frequent partially decomposed remains of cotton-grass (*Eriophorum* spp.) and *Sphagnum* from 12 cm downwards. It is clearly semi-fibrous between 42 and 113 cm depth, but between 12 and 42 cm and in the basal layer it is more decomposed and qualifies as sapric in the US system if identification is based on the laboratory determinations of rubbed fibre content and pyrophosphate index.

Profile 181, also in blanket peat containing *Sphagnum*, *Eriophorum* and *Calluna* remains, was described and sampled under similar vegetation at an altitude of 641 m in the eastern Grampians. Here the surface horizon is little decomposed and the most humified material is at the base of the peat layer.

Profile 182 is in a relict raised bog north of Leeds. The site is drained by deep ditches and carried open birch woodland at the time of sampling. Below the thin well humified surface horizon there is about 70 cm of *Sphagnum – Eriophorum* peat with a very high carbon/nitrogen ratio. This is followed by successive layers of woody (fen-carr) and reedswamp (fen) peat, both with many readily recognizable plant remains. Although the diagnostic subsurface horizon contains some 40 per cent of fibres which resist disintegration by rubbing, it is classed as semi-fibrous rather than fibrous on the basis of the pyrophosphate index, which indicates a distinctly greater degree of decomposition than in the corresponding horizons of profile 179 and is outside the range specified for fibric materials in the US, Canadian and England and Wales systems. Determinations of the photometric absorbance and carbon content of the pyrophosphate extract reported by Bascomb *et al.* (1977) support this conclusion.

Profile 180 (Bascomb *et al.*, 1977; Carroll *et al.*, 1979)

Location : Gallows Moss, Dunford, South Yorkshire (SE 140005)

Climatic regime: humid oroboreal
Altitude : 418 m   Slope: 4°
Vegetation : moorland with cotton-grass (*Eriophorum vaginatum*) and some heather (*Calluna vulgaris*)

Horizons (cm)

0–12 Om1	Black (5YR 2/1) mat of living and dead cotton-grass roots with few coarser woody roots; wet; matrix soft and plastic.
12–42 Om2	Dark reddish brown (5YR 3/3) semi-fibrous *Eriophorum – Sphagnum* peat; wet; soft and plastic.
42–68 Om3	Dark reddish brown (5YR 3/2) semi-fibrous *Eriophorum – Sphagnum* peat; wet; less plastic.
68–113 Om4	Dark reddish grey to dark reddish brown (5YR 4–3/2) semi-fibrous *Eriophorum – Sphagnum* peat; wet.
113–142 Om/Oh	Dark reddish brown (5YR 3/2) semi-fibrous to humified peat; wet.
142–167 2Cu(g)	Greyish brown (10YR 5/2) very slightly stony clay loam; few small weathered sandstone fragments; massive; firm.

Analytical data *see* below.

Classification

England and Wales (1984): raw oligo-fibrous peat soil; mixed *Eriophorum* and *Sphagnum* peat (Winter Hill series).
USDA (1975): Typic Medihemist (or Borohemist); dysic, mesic.
FAO (1974): Dystric Histosol.

Analytical data for Profile 180

Horizon	Om1	Om2	Om3	Om4	Om/Oh
Depth (cm)	0–12	12–42	42–68	68–113	113–142
loss on ignition %	86	94	97	98	97
organic carbon %[a]	36	44	48	49	52
nitrogen %	1.57	1.62	1.49	1.04	1.28
C/N ratio	23	27	32	47	41
rubbed fibre (% ash-free dry matter)	14	6	14	22	6
pyrophosphate index	2	1	1	2	3
pH (0.01M CaCl$_2$, undried)	3.2	3.1	2.6	2.7	2.8
bulk density (g cm^{-3})	0.24	0.12	0.12	0.12	0.13
saturated water content %	336	232	735	747	711

[a] by Nömmik's (1971) method

Profile 181 (Heslop and Bown, 1969)

Location	:	Strathdon, Aberdeenshire (NJ 308058)
Climatic regime:		perhumid oroboreal
Altitude	:	641 m Slope: 1°
Vegetation	:	moorland; heather (*Calluna vulgaris*) and cotton-grass (*Eriophorum vaginatum*) dominant, with cross-leaved heath (*Erica tetralix*), crowberry (*Empetrum nigrum*), bog-moss (*Sphagnum* spp.), woolly fringe-moss (*Rhacomitrium lanuginosum*) and lichen (*Cladonia* spp.)

Horizons (cm)

0–13 Of	Light reddish brown (5YR 6/4) fibrous *Sphagnum* peat; fine *Calluna* roots and buried stems common; wet; clear boundary.
13–23 Om1	Very dark brown (10YR 2/2) semi-fibrous peat with *Sphagnum* and *Calluna* remains; wet; fine roots common; clear boundary.
23–163 Om2	Dark reddish brown (5YR 3/3) semi-fibrous peat with occasional *Calluna* and *Eriophorum* remains and abundant dead cord roots; wet; humification increases with depth and cord roots lose structure when squashed in lower part of horizon; sharp boundary.
163+ bAh	Partly weathered fragments of basic igneous rock with interstitial black organic loam.

Classification
  Scotland (1984): dystrophic (blanket) peat.
  England and Wales (1984): raw oligo-fibrous peat soil; mixed *Eriophorum* and *Sphagnum* peat (Winter Hill series).
  USDA (1975): Typic (or Sapric) Borohemist; dysic.
  FAO (1974): Dystric Histosol.

Profile 182 (Bascomb *et al.*, 1977)

Location	:	Alwoodley Moss, Leeds, West Yorkshire (SE 303403)
Climatic regime:		subhumid temperate
Altitude	:	139 m Slope: level (hummocky microrelief)
Land use and vegetation	:	golf course; mixed woodland dominated by birch (*Betula* spp.) with field layer of grass, sedges and bracken (*Pteridium aquilinum*)

Horizons (cm)

L/F	Grass-sedge root mat with leaves and twigs, 2 cm thick.
0–9 Oh/Om	Black (5YR 2/1) amorphous to semi-fibrous peat; abundant fibrous grass and sedge roots and common medium to coarse woody roots; clear smooth boundary.
9–16 Om1	Very dusky red (2.5YR 2/2) semi-fibrous peat; moderately well decomposed (H5); common fine and few medium to coarse woody roots; gradual smooth boundary.

Analytical data for Profile 181

Horizon	Of	Om1	Om2	Om2	Om2	Om2
Depth (cm)	0–13	13–23	41–51	91–102	132–142	152–156
loss on ignition %	92	95	81	91	91	96
organic carbon %	51	56	48	56	55	57
nitrogen %	0.69	1.52	1.05	0.91	1.25	1.39
C/N ratio	74	37	46	62	44	41
TEB (me/100 g)	14	14	11	17	12	7
CEC (me/100 g)	110	104	114	120	116	96
% base saturation	13	13	10	14	10	7
pH (H_2O)	4.0	3.7	3.7	3.6	3.8	4.3

Analytical data for Profile 182

Horizon	Oh/Om	Om1	Om2	2Om	3Om/Of
Depth (cm)	0–9	9–16	16–77	77–108	108–145
loss on ignition %	76	95	97	96	97
organic carbon[a] %	36	39	43	39	44
nitrogen %	1.43	1.00	0.63	0.91	1.27
C/N ratio	25	39	68	42	34
rubbed fibre (% ash-free dry matter)	17	25	40	20	24
pyrophosphate index	1	1	3	2	3
pH (0.01M CaCl$_2$)	3.0	2.9	2.8	2.8	2.9
bulk density (g cm^{-3})	0.33	0.11	0.10	0.15	0.11
saturated water content %	332	731	790	692	796

[a] by Nömmik's (1971) method

16–77 Om2	Dusky red (2.5YR 3/2) semi-fibrous *Sphagnum – Eriophorum* peat; slightly decomposed (H4); few roots; gradual smooth boundary.
77–108 2Om	Dark reddish brown (5YR 3/2) semi-fibrous fen-carr peat; wet; well decomposed (H6); common soft wood fragments; few medium to coarse woody roots; gradual boundary.
108–145 3Om/Of	Dark reddish brown (5YR 3/3) semi-fibrous to fibrous fen (*Phragmites – Carex*) peat; wet; slightly to very slightly decomposed (H3–4).

Analytical data *see* above.

Classification
 England and Wales (1984): raw oligofibrous peat soil; *Sphagnum – Eriophorum* peat (Winter Hill intergrading to Longmoss series).
 USDA (1975): Typic (or Fibric) Medihemist; dysic, mesic.
 FAO (1974): Dystric Histosol.

9.5.4 Raw Sapric Bog Soils

These soils have a surface horizon that is more or less fibrous, incompletely ripened, or both, over strongly decomposed oligotrophic peat containing few recognizable plant remains (Plate IV.8). They often occur in close association with those of the preceding subgroup and are particularly common in more or less flushed blanket and basin bogs with purple moor-grass (*Molina caerulea*) as a major component of the vegetation. Similar soils can also be found in 'pools' of redeposited sedimentary peat which result from erosion of plateau bogs, as described by Hall and Folland (1970).

According to the 1:250,000 soil survey of England and Wales, most of the upland peat soils in Wales (Rudeforth *et al.*, 1984) and south-west England (Findlay *et al.*, 1984) conform to this subgroup (raw oligo-amorphous peat soils in the England and Wales system). Judging from field and analytical data on representative profiles of the Glenamoy and Aughty series given by Hammond (1981), many of the soils in unmodified blanket bogs in western Ireland do also, whereas those in northern England (R.A. Jarvis *et al.*, 1984) and much of Scotland are predominantly semi-fibrous. This suggests that the degree of decomposition of the subsurface horizons is determined to some extent by climatic factors as well as by the nature of the peat-forming vegetation and the incidence of flushing.

The first representative profile recorded below is sited in the valley of a stream draining from Bodmin Moor, Cornwall. Exposed in the side of a drainage ditch, it shows 110 cm of peat separated from underlying slaty head by a thin layer of water-rounded granitic gravel. Below 35 cm the peat is predominantly amorphous, with some woody remains, and contains about 40 per cent mineral matter which is irregularly distributed in bands related to episodes of alluvial sedimentation.

Profile 184 was described and sampled by B. Clayden in an upland depression north of Llanelli, Dyfed. The diagnostic subsurface horizon (20–70 cm) again consists mainly of strongly decomposed material, much of which passes through the fingers when a handful is squeezed. Although all horizons of both this and the preceding profile qualify as sapric in the US system if identification is based on the laboratory determinations of rubbed fibre content and pyrophosphate index, neither is considered to have an earthy topsoil because incompletely ripened peat containing more than 15 per cent by volume of recognizable plant remains occurs within 20 cm depth. In profile 184 the uppermost 10 cm beneath the superficial layer of roots and moss is black and well humified and has a moderately well developed granular structure but the dark brown layer immediately below is distinctly more fibrous.

Profile 183 (Staines, 1976)

Location	: St Clether, Cornwall (SX 199828)
Climatic regime:	perhumid temperate
Altitude	: 221 m Slope: <1°
Land use and vegetation	: rough grazing dominated by purple moor-grass (*Molina caerulea*) with rushes (*Juncus* spp.), sedges (*Carex* spp.) and bog-moss (*Sphagnum* spp.)

Horizons (cm)

0–35 Om	Dark brown to brown (7.5YR 4/4) semi-fibrous peat; recognizable plant remains mainly of grass and rush; wet; abundant fine and common medium fleshy roots; clear smooth boundary.
35–110 Oh	Alternating bands of black (N 2/0) and black to very dark brown humified peat; wet; almost completely decomposed (H9) but some woody remains visible; many roots; clear smooth boundary.
110–130+ 2Cu/3Cu	5 cm layer of rounded granite gravel overlying rubbly head composed mainly of large angular tabular slate fragments.

Analytical data

Horizon	Om	Oh
Depth (cm)	3–32	40–100
loss on ignition %	84	59
organic carbon %	43	32
nitrogen %	2.6	1.1
C/N ratio	17	29
rubbed fibre (% ash-free dry matter)	7	3
pyrophosphate index	5	3
pH (H$_2$O)	4.4	3.6
(0.01M CaCl$_2$, undried)	3.4	3.3
bulk density (g cm^{-3})	0.10	0.23

Classification

England and Wales (1984): raw oligo-amorphous peat soil; humified peat (Crowdy series).
USDA (1975): Typic Medisaprist; dysic, mesic.
FAO (1974): Dystric Histosol.

Profile 184 (Bascomb et al., 1977; Rudeforth et al., 1984)

Location	: Llannon, Dyfed (SN 555084)
Climatic regime:	perhumid temperate
Altitude	: 150 m Slope: 1°
Land use and vegetation	: rough grazing; nearly total cover of rushes (*Juncus* spp.) and purple moor-grass (*Molinia caerulea*) with a surface layer of mosses, including *Polytrichum* and *Hylocomium* spp.

Horizons (cm)

L	Surface mat of roots and mosses, mainly *Polytrichum commune*, 10 cm thick.
0–10 Oh1	Black (10YR 2/1 broken and rubbed) humified peat containing a few small angular tabular stones (artefacts); wet; moderate fine granular; abundant roots; clear smooth boundary.

Analytical data for Profile 184

Horizon	Om	Oh2	Oh2	2Oh/Om
Depth (cm)	10–20	20–45	45–70	70–100
loss on ignition %	89	85	84	95
organic carbon[a] %	40	39	42	52
nitrogen %	2.5	2.3	2.5	1.7
C/N ratio	16	17	17	30
rubbed fibre (% ash-free dry matter)	9	3	3	9
pyrophosphate index	2	2	2	
pH (H$_2$O)	3.9	3.5	3.4	4.1
(0.01M CaCl$_2$, undried)	3.3	3.2	3.2	3.4

[a] by Nömmik's (1971) method

10–20 Om	Dark brown (7.5YR 3/2 broken and rubbed) semi-fibrous peat; wet; massive; 30 per cent fine fibres, mainly broken down by rubbing; well decomposed (H5–6); many roots; clear smooth boundary.
20–70 Oh2	Black (10YR 2/1 broken and rubbed) amorphous peat; wet; massive; slightly fluid; 20–25 per cent fibres almost entirely broken down by rubbing; more decomposed than above (H7–8); few roots; clear smooth boundary.
70–100+ 2Oh/Om	Very dark greyish brown (10YR 3/2) semi-fibrous to humified peat, turning black on exposure; wet; massive; very fluid; fibres as above and very common wood (*Betula*) remains up to 6 cm; few roots (mineral substratum at 130 cm).

Analytical data *see* above.

Classification
England and Wales (1984): raw oligo-amorphous peat soil; humified peat (Crowdy series).
USDA (1975): Typic (or Hemic) Medisaprist; dysic, mesic.
FAO (1974): Dystric Histosol.

9.5.5 Earthy Semi-fibrous Bog Soils

These are drained bog soils with an earthy topsoil and semi-fibrous or fibrous oligotrophic peat immediately beneath it or within 60 cm depth. They are typically represented in reclaimed raised bogs in central Ireland, the Lancashire coastal plain and elsewhere, and also occur in man-modified blanket bogs that originally had moderately or slightly decomposed subsurface horizons. Variants in which the reference section (30–90 cm) is predominantly fibrous have been noted in places where intact (uncut) raised bog with a thick upper layer of 'young *Sphagnum* peat' has been reclaimed for agriculture but are comparatively rare. Like the corresponding fen soils, deeply drained cultivated soils of this subgroup commonly show evidence that colloidal organic matter has been translocated downwards from the ploughed layer.

Data on two representative profiles, both in raised bogs, are reproduced on pp. 427–8. The first was recorded in arable land on Chat Moss, a large bog west of Manchester which is still only partially reclaimed. Apart from the sharply defined earthy topsoil, which in this case coincides with the base of the ploughed layer, it resembles profile 178 (p. 420) from the same area, but the subsurface layer of fibrous *Sphagnum* peat is too thin for it to qualify as a Fibrist according to the US Taxonomy. Where similar land has been cut over before reclamation, this horizon is absent and in some places the earthy topsoil directly overlies minerotrophic peat forming the basal layer of the bog.

Profile 186 is under pasture on cut-over bog land in County Meath. Here the subsurface horizons are more decomposed throughout than in the preceding profile but their ombrotrophic origin is confirmed by the presence of identifiable remains of *Sphagnum* and *Calluna*. The loss-on-ignition and pH data reflect the application of calcareous mineral material (marl) as an integral part of the reclamation process. According to Hammond (1981), this practice was formerly widespread in central Ireland and is well documented in the Bog Commissioner's Reports of 1810–1814.

Profile 185 (Hall and Folland, 1970)

Location	: Chat Moss, Irlam, Greater Manchester (SJ 717957)
Climatic regime:	humid temperate
Altitude	: 24 m Slope: <1°
Land use	: arable

Horizons (cm)
- 0–23 Op: Very dark brown (10YR 2/2) amorphous peat; weak medium and fine granular; very friable; very porous; common fine roots; earthworms common; few fragments of fibrous peat incorporated by ploughing; abrupt smooth boundary.
- 23–58 Of: Yellowish red (5YR 5/8) fibrous *Sphagnum* peat; very slightly decomposed (H2); gradual boundary.
- 58–109 2Om1: Dark reddish brown (2.5YR 3/4) semi-fibrous *Sphagnum – Eriophorum* peat with common woody *Calluna* remains; laminated; moderately decomposed (H5); gradual boundary.
- 109–244 2Om2: Dark brown (7.5YR 4/2) semi-fibrous *Sphagnum – Eriophorum* peat; laminated; more compact than above; well decomposed (H5–6); abrupt boundary.
- 244–269 3Of/Om: Yellowish brown, fibrous to semi-fibrous fen-carr peat with soft woody remains common; wet; very slightly to slightly decomposed (H3–4); abrupt boundary.
- 269+ 4Cg: Grey (10YR 5/1) sandy clay loam containing few small subrounded stones (till).

Classification
England and Wales (1984): earthy oligo-fibrous peat soil; mixed *Eriophorum* and *Sphagnum* peat (Turbary Moor series).
USDA (1975): Fibric Medihemist; dysic?, mesic.
FAO (1974): Dystric Histosol.

Profile 186 (Hammond, 1981)

Location	: Castletown Moor, County Meath
Climatic regime:	humid temperate
Altitude	: 97 m Slope: <1°
Land use and vegetation	: long-term grass with cocksfoot (*Dactylis glomerata*), meadowsweet (*Filipendula ulmaria*) and stinging nettle (*Urtica dioica*)

Horizons (cm)
- 0–33 Op: Black (5YR 2/1) humified peat; strong granular; many roots; no plant remains visible (much marling carried out; egg shells at 30 cm); abrupt smooth boundary.
- 33–59 Oh: Black (5YR 2/1) humified peat; strong subangular blocky; common roots; very few plant remains; abrupt smooth boundary.
- 59–80 Om: Strong brown (5YR 5/8) semi-fibrous peat rapidly turning black on exposure; wet; *Sphagnum*, *Calluna* and cyperaceous remains present; very small amount of peat exudes between fingers on squeezing.

Analytical data

Horizon	Op	Oh	Om
Depth (cm)	0–33	33–59	59–80
loss on ignition %	46	88	94
nitrogen %	0.64	0.88	0.82
rubbed fibre (% vol.)	4	10	12
pyrophosphate index	1	5	6
pH (H$_2$O)	7.3	5.8	5.3
(0.01M CaCl$_2$)	7.2	5.2	4.5
bulk density (g cm^{-3})	0.47	0.12	0.09
saturated water content %	260	835	1072

Classification
  Ireland (Hammond, 1981): man-modified raised bog soil (Gortnamona series).
  England and Wales (1984): earthy oligo-fibrous peat soil; mixed *Eriophorum* and *Sphagnum* peat (Turbary Moor series).
  USDA (1975): Sapric Medihemist; euic, mesic.
  FAO (1974): Eutric Histosol.

9.5.6 Earthy Sapric Bog Soils

These are strongly acid peat soils that have earthy topsoils and black to very dark brown, strongly decomposed subsurface horizons but lack acid sulphate characteristics. Like the analogous raw peat soils (p. 424) with which they are often closely associated, they occur chiefly in flushed oligotrophic blanket and basin bogs but are confined to sites, mainly in reclaimed land, where there has been enough seasonal drying to allow formation of an earthy topsoil.

The following representative profile was recorded under rough grazing dominated by *Molinia* in a moderately sloping lower valley side near the southern edge of Dartmoor. It is sited close to a watercourse and the upper 51 cm was only slightly moist at the time of sampling (4 June 1971). The gravelly horizon below that depth is humose rather than organic and those above all contain more than 50 per cent mineral matter, presumably resulting from over-bank deposition or slope-wash which proceeded more or less concurrently with accumulation of organic matter. Despite the low pH values, both surface and subsurface horizons are strongly humified with relatively low carbon/nitrogen ratios, and their consistence and structure indicate that they are fully ripened throughout.

Profile 187 (Harrod *et al.*, 1976)

Location	: Hartford, Devon (SX 648595)
Climatic regime:	humid temperate
Altitude	: 274 m Slope: 7°
Land use and vegetation	: rough grazing dominated by purple moor-grass (*Molinia caerulea*) with some heather (*Calluna vulgaris*)

Horizons (cm)

0–9 Oh1	Black (5Y 2/1) stoneless loamy humified peat; moderate fine subangular blocky; friable; abundant fine and few medium woody (heather) roots; gradual boundary.
9–22 Oh2	Very dark grey (5YR 3/1) stoneless loamy humified peat; moderate medium angular blocky; friable; common fine roots; gradual boundary.
22–51 Oh3	Black (10YR 2/1) loamy humified peat, slightly gritty in bands; moderate coarse angular blocky; firm; few fine roots; gradual boundary.
51–90 Ah	Black (10YR 2/1) moderately to very gravelly, humose loamy coarse sand (USDA loamy coarse sand); many to abundant, rounded to subangular granite fragments and quartz pebbles; becoming wet below 65 cm; few roots.

Analytical data

Horizon	Oh1	Oh2	Oh3	Ah
Depth (cm)	2–8	12–21	33–46	69–80
sand 600 µm–2 mm %				38
200–600 µm %				27
60–200 µm %				17
silt 2–60 µm %				13
clay <2 µm %				5
loss on ignition %	44	47	32	12
organic carbon %	21	20	18	6.6
nitrogen %	1.4	1.3	1.1	0.4
C/N ratio	15	15	17	17
pH (H$_2$O)	4.6	4.6	3.9	4.2
(0.01M CaCl$_2$)	3.6	4.0	3.6	3.8

Classification
  England and Wales (1984): earthy oligo-amorphous peat soil; humified peat (Blackland series).
  USDA (1975): Terric Medisaprist; loamy-skeletal, mixed, dysic.
  FAO (1974): Dystric Histosol.

References

Adam, P. (1978). Geographical variation in British saltmarsh vegetation. *Journal of Ecology* **66**, 339–366.

Adams, W.A. and Raza, M.A. (1978). The significance of truncation in the evolution of slope soils in mid-Wales. *Journal of Soil Science* **29**, 243–257.

Adams, W.A., Karim, M.I. and Gafoor, S.N. (1987). Composition and properties of poorly ordered minerals in Welsh soils. I. Composition. *Journal of Soil Science* **38**, 85–94.

Allen, R.H. and Sturdy, R.G. (1980). Soils in Essex III: Sheet TL 71 (Little Waltham). Soil Survey Record No. 62, Harpenden.

Allison, J.W. and Hartnup, R. (1981). Soils in North Yorkshire VI: Sheet SE 39 (Northallerton). Soil Survey Record No. 68, Harpenden.

Allison, L.E. (1965). Organic carbon. *In* C.A. Black (Ed.), *Methods of Soil Analysis*, Agronomy 9, 1367–1378. American Society of Agronomy, Madison, Wisconsin.

Anderson, M.L., Berrow, M.L., Farmer, V.C., Hepburn, A., Russell, J.D. and Walker, A.D. (1982). A reassessment of podzol formation processes. *Journal of Soil Science* **33**, 125–136.

Arkell, W.J. (1947). *The Geology of Oxford.* Clarendon Press, Oxford.

Arkley, R.J. (1976). Statistical methods in soil classification research. *Advances in Agronomy* **28**, 37–70.

Atkinson, T. and Burrin, P. (1984). Rubification, paleosols and the Wealden angular chert drift. *Quatenary Newsletter* No. 44, 21–28.

Aubert, G. and Duchaufour, Ph. (1956). Projet de classification des sols. *Transactions 6th International Congress of Soil Science* E, 597–604.

Avery, B.W. (1955). The Soils of the Glastonbury District of Somerset (Sheet 296). Memoir of the Soil Survey of Great Britain: England and Wales. HMSO, London.

Avery, B.W. (1958). A sequence of beechwood soils on the Chiltern Hills, England. *Journal of Soil Science* **9**, 210–224.

Avery, B.W. (1964). The Soils and Land Use of the District around Aylesbury and Hemel Hempstead (Sheet 238). Memoir of the Soil Survey of Great Britain: England and Wales. HMSO, London.

Avery, B.W. (1965). Soil classification in Britain. *Pédologie, Gand.* Numéro spécial **3**, 75–90.

Avery, B.W. (1968). General soil classification; hierarchical and co-ordinate systems. *Transactions 9th International Congress of Soil Science* **4**, 169–175.

Avery, B.W. (1973). Soil classification in the Soil Survey of England and Wales. *Journal of Soil Science* **24**, 324–338.

Avery, B.W. (1980). System of Soil Classification for England and Wales (Higher Categories). Soil Survey Technical Monograph No. 14, Harpenden.

Avery, B.W. (1985). Argillic horizons and their significance in England and Wales. *In* Boardman, J. (Ed.), *Soils and Quaternary Landscape Evolution*, 69–86. Wiley, Chichester.

Avery, B.W. and Bascomb, C.L. (Eds.). (1982). Soil Survey Laboratory Methods. Soil Survey Technical Monograph No. 6, Harpenden.

Avery, B.W. and Bullock, P. (1969). Morphology and classification of Broadbalk soils. *Report Rothamsted Experimental Station for 1968*, Pt 2, 63–81.

Avery, B.W. and Bullock, P. (1977). Mineralogy of Clayey Soils in relation to Soil Classification. Soil Survey Technical Monograph No. 10, Harpenden.

Avery, B.W., Bullock, P., Catt, J.A., Newman, A.C.D., Rayner, J.H. and Weir, A.H. (1972). The soil of Barnfield. *Report Rothamsted Experimental Station for 1971*, Pt 2, 5–37.

Avery, B.W., Clayden, B. and Ragg, J.M. (1977). Identification of podzolic soils (Spodosols) in upland Britain. *Soil Science* **123**, 306–318.

Avery, B.W., Findlay, D.C. and Mackney, D. (1975). Soil map of England and Wales 1:1,000,000. Ordnance Survey, Southampton.

Avery, B.W., Stephen, I., Brown, G. and Yaalon, D.H. (1959). The origin and development of brown earths on Clay-with-Flints and Coombe Deposits. *Journal of Soil Science* **10**, 177–195.

Baker, C.A. (1971). A contribution to the glacial stratigraphy of West Essex. *Essex Naturalist* **32**, 318–330.

Baldwin, M., Kellogg, C.E. and Thorp, J. (1938). Soil Classification. *In Soils and Men, Yearbook of Agriculture*, U.S. Department of Agriculture, 979–1001. U.S. Government Printing Office, Washington, DC.

Ball, D.F. and Williams, W.M. (1974). Soil development on coastal dunes at Holkham, Norfolk, England. *Transactions 10th International Congress of Soil Science* **6**(2), 380–386.

Bane, W.A. and Jones, G.H.G. (1934). Fruit growing areas on the Lower Greensand in Kent. Ministry of Agriculture and Fisheries Bulletin No. 80. HMSO, London.

Barratt, B.C. (1964). A classification of humus forms and micro-fabrics of temperate grasslands. *Journal of Soil Science* **15**, 342–356.

Barraud, E.M. (1951). Coprolites in Cambridgeshire. *Agriculture, London* **58**, 193–195.

Bartelli, L.J. and Odell, R.T. (1960). Laboratory studies and genesis of a clay-enriched horizon in the lowest part of the solum of some Brunizem and Gray-Brown Podzolic soils in Illinois. *Soil Science Society of America Proceedings* **24**, 390–395.

Bascomb, C.L. (1961). A calcimeter for routine use on soil samples. *Chemistry and Industry*, 1826–7.

Bascomb, C.L. (1968). Distribution of pyrophosphate-extractable iron and organic carbon in soils of various groups. *Journal of Soil Science* **19**, 251–268.

Bascomb, C.L. (1982). Physical and chemical analyses of <2 mm samples. *In* Avery, B.W. and Bascomb, C.L. (Eds.), Soil Survey Laboratory Methods, 14–41. Soil Survey Technical Monograph No. 6, Harpenden.

Bascomb, C.L. and Bullock, P. (1982). Sample preparation and stone content. *In* Avery, B.W. and Bascomb, C.L. (Eds.), Soil Survey Laboratory Methods, 5–13. Soil Survey Technical Monograph No. 6, Harpenden.

Bascomb, C.L., Banfield, C.F. and Burton, R.G.O. (1977). Characterization of peaty materials from organic soils (Histosols) in England and Wales. *Geoderma* **19**, 131–147.

Batey, T. (1971). Soil profile drainage. *In* Soil Field Handbook: *ADAS Advisory Papers No.* **9**, 34–37. Ministry of Agriculture, Fisheries and Food, London. © Crown Copyright 1989.

Beard, G.R. (1984). Soils in Warwickshire. V: Sheets SP 27 and 37 (Coventry South). Soil Survey Record No. 81, Harpenden.

Beckett, P.H.T. (1967). Lateral changes in soil variability. *Journal of the Australian Institute of Agricultural Science* **33**, 172–179.

Beckett, P.H.T. (1968). Method and scale of land resource surveys. *In* Stewart, G.A. (Ed.), *Land Evaluation*, 53–63. Macmillan of Australia.

Beckett, P.H.T. and Burrough, P.A. (1971). The relation between cost and utility in soil survey. IV. Comparison of the utilities of soil maps produced by different survey procedures, and to different scales. *Journal of Soil Science* **22**, 466–480.

Beckett, P.H.T. and Webster, R. (1971). Soil variability: a review. *Soils and Fertilizers* **34**, 7–15.

Bell, M. (1983). Valley sediments as evidence of prehistoric land use on The South Downs. *Proceedings of the Prehistoric Society* **49**, 119–150.

Bellamy, D.J. and Moore, P.D. (1973). *Peatlands*. Elek Science, London.

Bendelow, V.C. and Carroll, D.M. (1979). Soils in North Yorkshire V: Sheet SE 58 and parts of Sheets SE 49 and 59 (Rievaulx and Upper Ryedale). Soil Survey Record No. 58, Harpenden.

Bendelow, V.C. and Hartnup, R. (1980). Climatic Classification of England and Wales. Soil Survey Technical Monograph No. 15, Harpenden.

Benzian, B. (1965). Experiments on Nutrition Problems in Forest Nurseries. Forestry Commission Bulletin No. 37. HMSO, London.

Berrow, M.L. and Goodman, R.A. (1987). Processes of mobilization and immobilization of iron in thin ironpan formation. *Proceedings of the North of England Soils Discussion Group* **22**, 73–92.

Berry, R.A., Melville, E.M. and Louden, C. (1930). Soils and Agriculture. In *The Geology of North Ayrshire*, 347–385. Memoir of the Geological Survey of Scotland. HMSO, Edinburgh.

Bibby, J.S., Douglas, H.A., Thomasson, A.J. and Robertson, J.S. (1982). Land Capability Classification for Agriculture. Soil Survey of Scotland Monograph: Macaulay Institute for Soil Research, Aberdeen.

Bibby, J.S., Hudson, G. and Henderson, D.J. (1982). Soil and Land Capability for Agriculture: Western Scotland. Soil Survey of Scotland: Macaulay Institute for Soil Research, Aberdeen.

Bibby, J.S. and Mackney, D. (1969). Land Use Capability Classification. Soil Survey Technical Monograph No. 1. Rothamsted Experimental Station, Harpenden, and Macaulay Institute for Soil Research, Aberdeen.

Birse, E.L. (1971). Assessment of Climatic Conditions in Scotland: 3. The Bioclimatic Sub-Regions. Soil Survey of Scotland: Macaulay Institute for Soil Research, Aberdeen.

Birse, E.L. (1976). The bioclimate of Scotland in relation to a world system of classification and to land-use capability. *Transactions of The Botanical Society of Edinburgh* **42**, 463–467.

Birse, E.L. (1980). Suggested amendments to the world soil classification to accommodate Scottish mountain and aeolian soils. *Journal of Soil Science* **31**, 117–124.

Birse, E.L. (1982). Plant communities on serpentine in Scotland. *Vegetatio* **49**, 141–162.

Birse, E.L. and Dry. F.T. (1970). Assessment of climatic conditions in Scotland: 1. Based on accumulated temperature and potential water deficit. Soil Survey of Scotland: Macaulay Institute for Soil Research, Aberdeen.

Birse, E.L. and Robertson, J.S. (1976). Plant Communities and Soils of the Lowland and Southern Upland Regions of Scotland. Soil Survey of Scotland: Macaulay Institute for Soil Research, Aberdeen.

Bloomfield, C. (1951). Experiments on the mechanism of gley formation. *Journal of Soil Science* **2**, 196–211.

Bloomfield, C. (1972). The oxidation of iron sulphides in soils in relation to the formation of acid sulphate soils and of ochre deposits in field drains. *Journal of Soil Science* **23**, 1–16.

Bloomfield, C. and Coulter, J.K. (1973). Genesis and management of acid sulfate soils. *Advances in Agronomy* **25**, 265–326.

Bloomfield, C. and Zahari, A.B. (1982). Acid sulphate soils. *Outlook on Agriculture* **11**, 48–54.

Blüme, H.P. (1968). Die pedogenetische Deutung einer Catena durch die Untersuchung der Bodendynamik. *Transactions 9th International Congress of Soil Science* **4**, 441–449.

Bolton, J. and Coulter, J.K. (1966). Distribution of fertilizer residues in a forest nursery manuring experiment on a sandy podzol at Wareham, Dorset. Report on Forest Research for the year ending March 1965. HMSO, London.

Bouyoucos, G. (1927). The hydrometer as a new method for the mechanical analysis of soils. *Soil Science* **23**, 343–353.

Bouyoucos, G. (1951). A recalibration of the Bouyoucos hydrometer method for the mechanical analysis of soils. *Agronomy Journal* **43**, 434–438.

Bowen, D.Q., Rose, J., McCabe, A.M. and Sutherland, D.G. (1986). Correlation of Quaternary glaciation in England, Ireland, Scotland and Wales. *Quaternary Science Reviews* **5**, 299–340.

Bown, C.J. (1973). The Soils of Carrick and the country around Girvan (Sheets 7 and 8). Memoir of the Soil Survey of Great Britain: Scotland. HMSO, Edinburgh.

Bown, C.J. and Heslop, R.E.F. (1979). The Soils of the country around Stranraer and Wigtown (Sheets 1, 2, 3, 4 and part 7). Memoir of the Soil Survey of Great Britain: Scotland. Macaulay Institute for Soil Research, Aberdeen.

Bown, C.J., Shipley, B.M. and Bibby, J.S. (1982). Soil and Land Capability for Agriculture: Southwest Scotland. Soil Survey of Scotland: Macaulay Institute for Soil Research, Aberdeen.

Bradshaw, A.D. and Chadwick, M.J. (1980). *The Restoration of Land*. Studies in Ecology, Vol. 6. Blackwell Scientific Publications, Oxford.

Brammer, H. (1971). Coatings in seasonally flooded soils. *Geoderma* **6**, 5–16.

Brewer, R. (1964). *Fabric and Mineral Analysis of Soils*. Wiley, New York.

Brewer, R. and Sleeman, J.R. (1970). Some trends in pedology. *Earth Sciences Review* **6**, 297–335.

Bridges, E.M. (1966). The Soils and Land Use of the District north of Derby (Sheet 125). Memoir of the Soil Survey of Great Britain: England and Wales, Harpenden.

Bridges, E.M. (1978). Interaction of soil and mankind in Britain, *Journal of Soil Science* **29**, 125–139.

Bridges, E.M. and Bull, P.A. (1983). The role of silica in the formation of compact and indurated horizons in the soils of south Wales. *In* Bullock, P. and Murphy, C.P. (Eds.), *Soil Micromorphology*: Vol. 2 *Soil Genesis*, 605–614. AB Academic Publishers, Berkhamsted.

Briggs, D.J. (1976). River terraces of the Oxford area. *In* Roe, D. (Ed.), *Field Guide to the Oxford Region*. Quaternary Research Association.

Brinkman, R. (1979). *Ferrolysis, a soil-forming process in hydromorphic conditions*. Centre for Agricultural Publishing and Documentation, Wageningen, Netherlands.

British Standards Institution (1975). *BS 1377*: Methods of testing soils for engineering purposes. London.

Broadbent, F.E. (1965). Organic matter. *In* C.A. Black (Ed.), *Methods of Soil Analysis. Agronomy* 9, 1397–1400. American Society of Agronomy, Madison, Wisconsin.

Brown, J.M.B. (1953). Studies on British Beechwoods. Forestry Commission Bulletin No. 20. HMSO, London.

Browne, D.M. (1978). Roman Cambridgeshire. *In* Wilkes, J.J. and Elrington, C.R. (Eds.), *A History of Cambridgeshire and the Isle of Ely*, 7, 1–84. The University of London Institute of Historical Research, University Press, Oxford.

Bullock, P. (1971). The soils of the Malham Tarn area. *Field Studies* 3, 381–408.

Bullock, P. (1974). The use of micromorphology in the new system of soil classification for England and Wales. *In* Rutherford, G.K. (Ed.), *Soil Microscopy*, 607–631. Limestone Press, Kingston, Ontario.

Bullock, P. (1982). Micromorphology. *In* Avery, B.W. and Bascomb, C.L. (Eds.), Soil Survey Laboratory Methods, 70–81. Soil Survey Technical Monograph No. 6, Harpenden.

Bullock, P. (1985). The role of micromorphology in the study of Quaternary soil processes. *In* Boardman, J. (Ed.), *Soils and Quaternary Landscape Evolution*, 45–68. Wiley, Chichester.

Bullock, P. and Clayden, B. (1980). The morphological properties of spodosols. *In* Theng, B.K.G. (Ed.), *Soils with Variable Charge*, 45–65. New Zealand Society of Soil Science. Offset Publications, Palmerston North, New Zealand.

Bullock, P. and Loveland, P.J. (1982). Mineralogical Analyses. *In* Avery, B.W. and Bascomb, C.L. (Eds.), Soil Survey Laboratory Methods, 57–69. Soil Survey Technical Monograph No. 6, Harpenden.

Bullock, P. and Murphy, C.P. (1974). The microscopic examination of the structure of subsurface horizons of soils. *Outlook on Agriculture* 8, 348–354.

Bullock, P. and Murphy, C.P. (1979). Evolution of a paleo-argillic brown earth (Paleudalf) from Oxfordshire, England. *Geoderma* 22, 225–252.

Burgess, T.M., Webster, R. and McBratney, A.B. (1981). Optimal interpretation and isarithmic mapping of soil properties. VI. Sampling strategy. *Journal of Soil Science* 32, 643–660.

Burnham, C.P. (1983a). Soil profiles on Lullington Heath, Sussex. *Journal of the South East England Soils Discussion Group (SEESOIL)* 1, 162–171. Wye College, Ashford, Kent.

Burnham, C.P. (1983b). Soil formation on bare shingle at Dungeness, Kent. *Journal of the South East England Soils Discussion Group (SEESOIL)* 1, 42–56. Wye College, Ashford, Kent.

Burnham, C.P. and Pitman, J.I. (Eds.) (1987). Soil erosion. *Journal of the South East England Soils Discussion Group (SEESOIL)* 3. Ashford Press, Kent.

Burton, R.G.O. (1981). Soils in Cambridgeshire II: Sheet TF 00E/10W (Barnack). Soil Survey Record No. 69, Harpenden.

Burton, R.G.O. and Hodgson, J.M. (Ed.) (1987). Lowland peat in England and Wales. Soil Survey Special Survey No. 15, Harpenden.

Butler, B.E. (1980). *Soil Classification for Soil Survey*. Clarendon Press, Oxford.

Buurman, P. and Van Reeuwijk, L.P. (1984). Proto-imogolite and the process of podzol formation: a critical note. *Journal of Soil Science* 35, 447–452.

Caldwell, T.H. and Richardson, S.J. (1975). Field behaviour of lowland peats and organic soils. *In* Ministry of Agriculture, Fisheries and Food Technical Bulletin 29: Soil Physical Conditions and Crop Production, 94–111. HMSO, London.

Canada Soil Survey Committee (1978). *The Canadian System of Soil Classification*. Research Branch, Canada Department of Agriculture, Publication 1646. Supply and Services Canada, Ottawa.

Canti, M. (1983). Soils and sediments from two chalk dry valleys in Oxfordshire. *Journal of the South East England Soils Discussion Group (SEESOIL)* 1, 153–161. Wye College, Ashford, Kent.

Carroll, D.M. and Bendelow, V.C. (1981). Soils of the North York Moors. Soil Survey Special Survey No. 13, Harpenden.

Carroll, D.M., Evans, R. and Bendelow, V.C. (1977). Air-photo interpretation for soil mapping. Soil Survey Technical Monograph No. 8, Harpenden.

Carroll, D.M., Hartnup, R. and Jarvis, R.A. (1979). Soils of South and West Yorkshire. Soil Survey of England and Wales Bulletin No. 7, Harpenden.

Catt, J.A. (1977). Loess and Coversands. In Shotton, F.W. (Ed.), *British Quaternary Studies: Recent Advances*, 221–230. Clarendon Press, Oxford.

Catt, J.A. (1978). The contribution of loess to soils in lowland Britain. In Limbrey, S. and Evans, J.G. (Eds.), *The effect of man on the landscape: the lowland zone. Council for British Archaeology Research Report No. 21*, 12–20.

Catt, J.A. (1979). Soils and Quaternary geology in Britain. *Journal of Soil Science* 30, 607–642.

Catt, J.A. (1985). The nature, origin and geomorphological significance of Clay-with-Flints. In Sieveking, G. de G. and Hart, M.B. (Eds.), *The scientific study of flint and chert: papers from the Fourth International Flint Symposium*, 151–159. Cambridge University Press, Cambridge.

Catt, J.A. (1986). *Soils and Quaternary Geology. A Handbook for Field Scientists*. Clarendon Press, Oxford.

Catt, J.A. and Farrington, O. (1986). Man-made urban soils. *Report Rothamsted Experimental Station for 1985*, Pt 1, 177.

Catt, J.A., Avery, B.W., Green, C.P. and McGregor, D.F.M. (1985). The nature of the Pebbly Clay Drift at Northaw Great Wood. *Transactions Hertfordshire Natural History Society* 29, 192–203.

Catt, J.A., Weir, A.H., King, D.W., Le Riche, H.H., Pruden, G. and Norrish, R.E. (1977). The soils of Woburn Experimental Farm. II. Lansome, White Horse and School Fields. *Report Rothamsted Experimental Station for 1976*, Pt 2, 5–32.

Chartres, C.J. (1980). A Quaternary soil sequence in the Kennet valley, Central Southern England. *Geoderma* 23, 125–146.

Childs, E.C. (1943). Studies in mole draining: interim report on an experimental drainage field. *Journal of Agricultural Science, Cambridge* 33, 136–146.

Clarke, G.R. (1940). *Soil Survey of England and Wales: Field Handbook*. University Press, Oxford.

Clayden, B. (1964). The soils of the Middle Teign district of Devon. Soil Survey of Great Britain: England and Wales. Bulletin No. 1, Harpenden.

Clayden, B. (1971). Soils of the Exeter district (Sheets 325 and 339). Memoir of the Soil Survey of Great Britain: England and Wales, Harpenden.

Clayden, B. (1979). Peaty hydromorphic soils of the Welsh uplands. *Mitteilungen der Deutschen Bodenkundlichen Gesellschaft* 29, 667–682.

Clayden, B. (1982). Soil classification. In Bridges, E.M. and Davidson, D.A. (Eds.), *Principles and Applications of Soil Geography*, 58–96. Longman, London and New York.

Clayden, B. and Hollis, J.M. (1984). Criteria for differentiating soil series. Soil Survey Technical Monograph No. 17, Harpenden.

Cline, M.G. (1949). Basic principles of soil classification. *Soil Science* 67, 81–91.

Cline, M.G. (1980). Experience with Soil Taxonomy of the United States. *Advances in Agronomy* 33, 193–226.

Clothier, B.E., Pollock, J.A. and Scotter, D.R. (1978). Mottling in profiles containing a coarse textured horizon. *Soil Science Society of America Journal* 42, 761–763.

Colborne, G.J.N. and Staines, S.J. (1985). Soil erosion in south Somerset. *Journal of Agricultural Science, Cambridge* 104, 107–112.

Commission of the European Communities (1985a). *Basic statistics of the Community*. Statistical Office of the European Communities, Luxembourg.

Commission of the European Communities (1985b). *Soil Map of the European Communities 1:1000000* (9 sheets and explanatory text). Luxembourg.

Conry, M.J. (1969). Plaggen soils in Ireland. *Pédologie, Gand* 21, 152–161.

Conry, M.J. (1971). Irish Plaggen soils, their distribution, origin and properties. *Journal of Soil Science* 22, 401–416.

Conry, M.J. (1974). Plaggen soils: a review of man-made raised soils. *Soils and Fertilizers* 37, 319–326.

Conry, M.J. and Diamond, J.J. (1971). Proposed classification of Plaggen soils. *Pédologie, Gand* 21, 152–161.

Conry, M.J. and O'Shea, T. (1973). Soils of Annascaul Pilot Area. Soil Survey Bulletin 24. National Soil Survey of Ireland: An Foras Talúntais, Dublin.

Conry, M.J. and Ryan, P. (1967). Soils of County Carlow. Soil Survey Bulletin 17. National Soil Survey of Ireland: An Foras Talúntais, Dublin.

Conry, M.J., De Coninck, F., Bouma, J., Cammaerts, C. and Diamond, J.J. (1972). Some brown podzolic soils in the west and southwest of Ireland. *Proceedings of the Royal Irish Academy* **72B**, 359–402.

Conry, M.J., Hammond, R.F. and O'Shea, T. (1970). Soils of County Kildare. Soil Survey Bulletin 22. National Soil Survey of Ireland: An Foras Talúntais, Dublin.

Coope, G.R. (1977). Quaternary Coleoptera as aids in the interpretation of environmental history. *In* Shotton, F.W. (Ed.), *British Quaternary Studies: Recent Advances*, 55–68. Clarendon Press, Oxford.

Cope, D.W. (1976). Soils in Wiltshire I: Sheet SU 03 (Wilton). Soil Survey Record No. 32, Harpenden.

Cope, D.W. (1986). Soils in Gloucestershire IV: Sheet SO 72 (Newent). Soil Survey Record No. 93, Harpenden.

Corbett, W.M. (1973). Breckland Forest Soils. Soil Survey Special Survey No. 7, Harpenden.

Corbett, W.M. (1979). Soils in Norfolk IV: Sheet TM 28 (Harleston). Soil Survey Record No. 60, Harpenden.

Corbett, W.M. and Tatler, W. (1970). Soils in Norfolk I: Sheet TM 49 (Beccles North). Soil Survey Record No. 1, Harpenden.

Courtney, F.M. and Findlay, D.C. (1978). Soils in Gloucestershire II: Sheet SP 12 (Stow-on-the Wold). Soil Survey Record No. 52, Harpenden.

Courtney, F.M. and Webster, R. (1973). A taxonometric study of the Sherborne mapping unit. *Transactions of the Institute of British Geographers* **53**, 113–124.

CPCS (Commission de Pédologie et de Cartographie des Sols) (1967). *Classification des sols*. Ecole Nationale Supérieure Agronomique, Grignon.

Crampton, C.B. (1963). The development and morphology of iron pan podzols in Mid and South Wales. *Journal of Soil Science* **14**, 282–302.

Crampton, C.B. (1972). Soils of the Vale of Glamorgan (Sheets 262 and 263). Memoir of the Soil Survey of Great Britain: England and Wales, Harpenden.

Crompton, A. and Matthews, B. (1970). Soils of the Leeds district (Sheet 70). Memoir of the Soil Survey of Great Britain: England and Wales, Harpenden.

Crompton, E. (1956). The environmental and pedological relationships of peaty gleyed podzols. *Transactions 6th International Congress of Soil Science* **E**, 155–161.

Crompton, E. (1960). The significance of the weathering/leaching ratio in the differentiation of major soil groups, with particular reference to some strongly leached brown earths on the hills of Britain. *Transactions 7th International Congress of Soil Science* **4**, 406–412.

Crompton, E. (1966). The Soils of the Preston District of Lancashire (Sheet 75). Memoir of the Soil Survey of Great Britain: England and Wales, Harpenden.

Cuanalo de la C., H.E. and Webster, R. (1970). A comparative study of numerical classification and ordination of soil profiles in a locality near Oxford. I. Analysis of 85 sites. *Journal of Soil Science* **21**, 340–352.

Curtin, D. and Smillie, D. (1981). Contribution of the sand and silt fractions to the cation-exchange capacities of some Irish soils. *Journal of Earth Sciences, Royal Dublin Society* **4**, 17–20.

Curtis, L.F. (1971). Soils of Exmoor Forest. Soil Survey Special Survey No. 5, Harpenden.

Davidson, D.A. (1982). Soils and man in the past. *In* Bridges, E.M. and Davidson, D.A. (Eds.), *Principles and Applications of Soil Geography*, 1–27. Longman, New York.

Davies, D.B., Hooper, L.J., Charlesworth, R.R., Little, R.C., Evans, C. and Wilkinson, B. (1971). Copper disorders in cereals grown on chalk soils in south-eastern and central southern England. *In* Ministry of Agriculture, Fisheries and Food Technical Bulletin 21: Trace elements in soils and crops, 88–118. HMSO, London.

Deb, B.C. (1950). The estimation of free iron oxides in soils and clays and their removal. *Journal of Soil Science* **1**, 212–220.

de Bakker, H. (1979). *Major Soils and Soil Regions of the Netherlands*. Dr. W. Junk B.V. Publishers, The Hague.

de Bakker, H. and Schelling, J. (1966). *Systeem van bodem classificatie voor Nederland: de hogere niveaus*. Pudoc, Wageningen.

De Coninck, F. (1980). Major mechanisms in formation of spodic horizons. *Geoderma* **24**, 101–128.

De Coninck, F., Favrot, J.C., Tavernier, R. and Jamagne, M. (1976). Dégradation dans les sols lessivés hydromorphes sur materiaux argilo-sableux: exemple des sols de la nappe détritique bourbonnaise (France). *Pédologie, Gand.* **26**, 105–151.

de Gruijter, J.J. (1977). Numerical Classification of Soils and its Application in Survey. *Agricultural Research Report No. 885, Centre for Agricultural Publishing and Documentation*, Wageningen, Netherlands.

Dekker, L.W. and de Weerd, M.D. (1973). The value of soil survey for archaeology. *Geoderma* **10**, 169–178.

Dent, D.L. (1985). An introduction to acid sulphate soils and their occurrence in East Anglia. *Journal of the South East England Soils Discussion Group (SEESOIL)* **2**, 35–50. Ashford Press, Kent.

Dimbleby, G.W. (1962). The Development of British Heathlands and their Soils. Oxford Forestry Memoir No. 23. Clarendon Press, Oxford.

Dimbleby, G.W. (1965). Post-glacial changes in soil profiles. *Proceedings of the Royal Society of London* **B 161**, 355–362.

Dimbleby, G.W. and Gill, J.M. (1955). The occurrence of podzols under deciduous woodland in the New Forest. *Forestry* **28**, 95–106.

Dinç, V., Miedema, R., Bal, L. and Pons, L.J. (1976). Morphological and physico-chemical aspects of three soils developed in peat in the Netherlands and their classification. *Netherlands Journal of Agricultural Science* **24**, 247–265.

Dines, H.G., Buchan, S., Holmes, S.C.A. and Bristow, C.R. (1969). Geology of the country around Sevenoaks and Tonbridge. Memoir of the Geological Survey of the United Kingdom. HMSO, London.

Dines, H.G., Edwards, W., Hollingworth, S.E., Buchan, S. and Welch, F.B.A. (1940). The mapping of head deposits. *Geological Magazine* **77**, 198–226.

Dry, F.T. and Robertson, J.S. (1982). Soil and Land Capability for Agriculture: Orkney and Shetland. Soil Survey of Scotland: The Macaulay Institute for Soil Research, Aberdeen.

Duchaufour, P. (1978). *Ecological Atlas of Soils of the World*, translated from the French by G.R. Mehuys, C.R. de Kimpe and Y.A. Martel. Masson Publishing USA, New York.

Duchaufour, P. (1982). *Pedology: pedogenesis and classification*, translated from the French by T.R. Paton. George Allen and Unwin, London.

Ducloux, J. (1971). Les sols de la plaine vendéene sur substratum calcaire bathonien. *Bulletin de l'Association Française pour l'Etude du Sol* No. 3, 11–28.

Dumbleton, M.J. and West, G. (1966). Studies of the Keuper Marl: Mineralogy. Road Research Laboratory Report No. 40, Crowthorne.

Dunn, E.E. (1980). Cropping the Machair. In Ranwell, D.S. (Ed.), *Sand Dune Machair*, **3**, 6–7. Natural Environment Research Council, London.

Edelman, C.H. (1950). *Soils of the Netherlands*. North Holland Publishing Company, Amsterdam.

Ehwald, E., Lieberoth, I. and Schwanecke, W. (1966). Zur Systematik der Böden der Deutschen Demokratischen Republik, besonders in Hinblick der Bodenkartierung. *Sitzungsberichte Deutsche Akademie der Landwirtschaftwissenschaften, Berlin* **15** (18).

Ellis, J.H. (1932). A field classification of soils for use in soil survey. *Scientific Agriculture* **12**, 338–345.

Evans, J.G. (1966). Late-glacial and post-glacial subaerial deposits at Pitstone, Buckinghamshire. *Proceedings of the Geologists' Association* **77**, 347–364.

Evans, J.G. (1975). *The Environment of Early Man in the British Isles*. Paul Elek, London.

Evans, L.J. and Smillie, G.W. (1976). Extractable iron and aluminium in relation to phosphate retention in Irish soils. *Irish Journal of Agricultural Research* **15**, 65–73.

Evans, R. (1972). Air photographs for soil survey in lowland England: soil patterns. *Photogrammetric Record* **7**, 302–322.

FAO–Unesco (1974). *Soil map of the World 1:5000000, Volume 1, Legend*. Unesco, Paris.

FAO (1974–78). *Soil map of the World*. Unesco, Paris.

Farmer, V.C. (1984). Distribution of allophane and organic matter in podzol B horizons: reply to Buurman and Van Reeuwijk. *Journal of Soil Science* **35**, 453–458.

Farmer, V.C., McHardy, W.J., Robertson, L., Walker, A. and Wilson, M.J. (1985). Micromorphology and sub-microscopy of allophane and imogolite in a podzol Bs horizon: evidence for translocation and origin. *Journal of Soil*

Science **36**, 87–96.

Farmer, V.C., Russell, J.D. and Berrow, M.L. (1980). Imogolite and proto-imogolite allophane in spodic horizons: evidence for a mobile aluminium silicate complex in podzol formation. *Journal of Soil Science* **31**, 673–684.

Farmer, V.C., Russell, J.D. and Smith, B.F.L. (1983). Extraction of inorganic forms of translocated A1, Fe and Si from a podzol Bs horizon. *Journal of Soil Science* **34**, 571–576.

Fieldes, M. and Perrott, K.W. (1966). The nature of allophane in soils: Part 3. Rapid field and laboratory test for allophane. *New Zealand Journal of Science* **9**, 623–629.

Finch, T.F. and Gardiner, M.J. (1977). Soils of County Westmeath. Soil Survey Bulletin No. 33. National Soil Survey of Ireland: An Foras Talúntais, Dublin.

Finch, T.F. and Ryan, P. (1966). Soils of County Limerick. Soil Survey Bulletin No. 16. National Soil Survey of Ireland: An Foras Talúntais, Dublin.

Finch, T.F., Culleton, E. and Diamond, S. (1971). Soils of County Clare. Soil Survey Bulletin No. 23. National Soil Survey of Ireland: An Foras Talúntais, Dublin.

Findlay, D.C. (1965). The Soils of the Mendip District of Somerset: (Sheets 279 and 280). Memoir of the Soil Survey of Great Britain: England and Wales, Harpenden.

Findlay, D.C. (1976). Soils of the southern Cotswolds and surrounding country (Sheets 251 and 265). Memoir of the Soil Survey of Great Britain: England and Wales, Harpenden.

Findlay, D.C., Colborne, G.J.N., Cope, D.W., Harrod, T.R., Hogan, D.V. and Staines, S.J. (1984). Soils and their use in South West England. Soil Survey of England and Wales Bulletin No. 14, Harpenden.

FitzPatrick, E.A. (1956). An indurated soil horizon formed by permafrost. *Journal of Soil Science* **7**, 248–254.

FitzPatrick, E.A. (1963). Deeply weathered rock in Scotland, its occurrence, age and contribution to the soils. *Journal of Soil Science* **14**, 33–43.

FitzPatrick, E.A. (1986). *An Introduction to Soil Science*, 2nd edn. Longman, Harlow.

Flach, K.W., Holzhey, C.S., De Coninck, F. and Bartlett, R.J. (1980). Genesis and classification of Andepts and Spodosols. *In* B.K.G. Theng (Ed.), *Soils with Variable Charge*, 411–426. New Zealand Society of Soil Science. Offset Publications, Palmerston North, New Zealand.

Fordham, S.J. and Green, R.D. (1976). Soils in Kent III: Sheet TQ 86 (Rainham). Soil Survey Record No. 37, Harpenden.

Fordham, S.J. and Green, R.D. (1980). Soils of Kent. Soil Survey of England and Wales Bulletin No. 9, Harpenden.

Fourt, D.F. (1973). Studies on chalk soils. *In* Report on Forest Research for the year ending March 1973, 61–62. HMSO, London.

Fourt, D.F. (1975). Studies on calcareous soils in the lowlands. *In* Report on Forest Research for the year ending March 1975, 24–25. HMSO, London.

Franzmeier, D.P., Yahner, J.E., Steinhardt, G.C. and Sinclair, H.R. Jr. (1983). Colour patterns and water table levels in some Indiana soils. *Soil Science Society of America Journal* **47**, 1196–1202.

Frosterus, B. (1914). Zur Frage der Einteilung der Böden in Nord-West-Europas Moränengebieten. V. Geol. Kommis. in Finland. *Geotekn. Meddel.*, Helsinki 14.

Furneaux, B.S. (1932). The soils of the High Weald of Kent. *Journal of the South-eastern Agricultural College, Wye* **30**, 123–140.

Furness, R.R. (1985). Soils in Humberside II: Sheet TA 14 (Brandesburton). Soil Survey Record No. 82, Harpenden.

Furness, R.R. and King, S.J. (1972). Soils in Westmoreland I: Sheet SD 58 (Sedgwick). Soil Survey Record No. 10, Harpenden.

Furness, R.R. and King, S.J. (1978). Soils in North Yorkshire IV: Sheet SE 63/73 (Selby). Soil Survey Record No. 56, Harpenden.

Futty, D.W. and Dry, F.T. (1977). The Soils of the Country around Wick. Memoir of the Soil Survey of Great Britain: Scotland. Macaulay Institute for Soil Research, Aberdeen.

Gallagher, P.H. and Walsh, T. (1942). Characteristics of Irish soil types: Part 1. *Proceedings of the Royal Irish Academy* **B47**, 205–249.

Gardiner, M.J. and Culleton, E. (1978). Composition and classification of shale drift soils in Co. Wexford. *Journal of Earth Sciences, Royal Dublin Society* **1**, 143–150.

Gardiner, M.J. and Radford, T. (1980). Soil Associations of Ireland and their Land Use Potential. Soil Survey Bulletin No. 36. National Soil Survey of Ireland: An Foras Talúntais, Dublin.

Gardiner, M.J. and Ryan, P. (1962). Relic soil on limestone in Ireland. *Irish Journal of Agricultural Research* **1**, 181–188.

Gardiner, M.J. and Ryan, P. (1964). Soils of County Wexford. Soil Survey Bulletin No. 1. National Soil Survey of Ireland: An Foras Talúntais, Dublin.

Gardner, H.W. (1967). A Survey of the Agriculture of Hertfordshire. County Agricultural Surveys No. 5. Royal Agricultural Society of England, London.

Gauld, J.H. (1982). Native pinewood soils in the northern section of Abernethy Forest. *Scottish Geographical Magazine* **98**, 48–56.

George, H. (1978). Soils in Northumberland I: Sheet NZ 07 (Stamfordham). Soil Survey Record No. 53, Harpenden.

George, H. and Robson, J.D. (1978). Soils in Lincolnshire II: Sheet TF 04 (Sleaford). Soil Survey Record No. 51, Harpenden.

Gilmour, J.S.L. (1951). Taxonomy. In McLeod, A.M. and Cobley, L.S. (Eds.), *Contemporary Botanical Thought*, 27–45. Oliver and Boyd, Edinburgh.

Gilmour, J.S.L. (1962). Classification: an interdisciplinary problem. *Aslib Proceedings* **14**, 223–225.

Gimingham, C.H. (1964). Maritime and sub-maritime communities. In Burnett, J.H. (Ed.), *The Vegetation of Scotland*, 67–143. Oliver and Boyd, Edinburgh and London.

Gimingham, C.H. (1972). *Ecology of Heathlands*. Chapman and Hall, London.

Glentworth, R. (1944). Studies on the soils developed on basic igneous rocks in central Aberdeenshire. *Transactions of The Royal Society of Edinburgh* **61**, 149–170.

Glentworth, R. (1954). The Soils of the Country around Banff, Huntley and Turriff (Sheets 86 and 96). Memoir of the Soil Survey of Great Britain: Scotland. HMSO, Edinburgh.

Glentworth, R. (1962). The principal genetic soil groups of Scotland. *Transactions of Joint Meeting of International Society of Soil Science Commissions IV and V*, 480–486. Soil Bureau, Lower Hutt, New Zealand.

Glentworth, R. and Dion, H.G. (1949). The association or hydrologic sequence in certain soils of the podzolic zone of north-east Scotland. *Journal of Soil Science* **1**, 35–49.

Glentworth, R. and Muir, J.W. (1963). The Soils of the Country around Aberdeen, Inverurie and Fraserburgh (Sheets 77, 76 and 87/97). Memoir of the Soil Survey of Great Britain: Scotland. HMSO, Edinburgh.

Godwin, H. (1975). *The History of the British Flora*, 2nd edn. Cambridge University Press, Cambridge.

Godwin, H. (1978). *Fenland: its Ancient Past and Uncertain Future*. Cambridge University Press, Cambridge.

Gosling, L.M. and Baker, S.J. (1980). Acidity fluctuations at a Broadland site in Norfolk. *Journal of Applied Ecology* **17**, 479–490.

Grant, R. (1960). The Soils of the Country around Elgin (Sheet 95). Soil Survey of Great Britain: Scotland. The Macaulay Institute for Soil Research, Aberdeen.

Grant, R. and Heslop, R.E.F. (Eds.) (1981). British Society of Soil Science Autumn Meeting: Programme and Guide to Excursions. The Macaulay Institute for Soil Research, Aberdeen.

Green, F.H.W. (1964). A map of annual average potential water deficit in the British Isles. *Journal of Applied Ecology* **1**, 151–158.

Green, R.D. (1968). Soils of Romney Marsh. Soil Survey of Great Britain: England and Wales. Bulletin No. 4, Harpenden.

Green, R.D. and Askew, G.P. (1965). Observations on the biological development of macropores in the soils of Romney Marsh. *Journal of Soil Science* **16**, 342–349.

Gresswell, R.K. (1958). Hillhouse Coastal Deposits in south Lancashire. *Liverpool and Manchester Geological Journal* **2**, 60–78.

Grigal, D.F. and Arneman, H.F. (1969). Numerical classification of some forested Minnesota soils. *Soil Science Society of America Proceedings* **33**, 433–438.

Grubb, P.J., Green, H.E. and Merrifield, R.C.J. (1969). The ecology of chalk heath: its relevance to the calcicole-calcifuge and soil acidification problems. *Journal of Ecology* **57**, 175–212.

Guillet, B. (1975). Les podzols forestiers et les podzols de dégradation: relation entre l'histoire de végétation et l'evolution des podzols sur Grès Vosgien Triasique. *Revue de l'Ecologie et Biologie du Sol* **12**, 405–414.

Guillet, B., Rouiller, J. and Souchier, B. (1975). Podzolization and clay migration in Spodosols of eastern France. *Geoderma* **14**, 223–246.

Hall, A.D. and Russell, E.J. (1911). Report on the

Agriculture and Soils of Kent, Surrey and Sussex. Board of Agriculture and Fisheries. HMSO, London.

Hall, B.R. and Folland, C.J. (1967). Soils of the south-west Lancashire Coastal Plain (Sheets 74 and 83). Memoir of the Soil Survey of Great Britain: England and Wales, Harpenden.

Hall, B.R. and Folland, C.J. (1970). Soils of Lancashire. Soil Survey of Great Britain: England and Wales. Bulletin No. 5, Harpenden.

Hall, D.G.M. and Heaven, F.W. (1979). Comparison of measured and predicted soil moisture deficits. *Journal of Soil Science* **30**, 225–238.

Hall, D.G.M., Reeve, M.J., Thomasson, A.J. and Wright, V.F. (1977). Water retention, porosity and density of field soils. Soil Survey Technical Monograph No. 9, Harpenden.

Hall, I.G. (1957). The ecology of disused pit heaps in England. *Journal of Ecology* **45**, 689–720.

Hamblin, A.P. and Davies, D.B. (1977). Influence of organic matter on the physical properties of some East Anglian soils of high silt content. *Journal of Soil Science* **28**, 11–22.

Hammond, R.F. (1981). The Peatlands of Ireland. Soil Survey Bulletin No. 35, 2nd edition. National Soil Survey of Ireland: An Foras Talúntais, Dublin.

Hardy, E.M. (1939). Studies in the post-glacial history of British vegetation. V. The Shropshire and Flint Maelor mosses. *New Phytologist* **38**, 364–396.

Harrod, T.R. (1971). Soils in Devon I: Sheet ST 10 (Honiton). Soil Survey Record No. 9, Harpenden.

Harrod, T.R. (1978). Soils in Devon IV: Sheet SS 30 (Holsworthy). Soil Survey Record No. 47, Harpenden.

Harrod, T.R. (1981). Soils in Devon V: Sheet SS 61 (Chumleigh). Soil Survey Record No. 70, Harpenden.

Harrod, T.R., Catt, J.A. and Weir, A.H. (1973). Loess in Devon. *Proceedings of the Ussher Society* **2**, 554–564.

Harrod, T.R., Hogan, D.V. and Staines, S.J. (1976). Soils in Devon II: Sheet SX 65 (Ivybridge). Soil Survey Record No. 39, Harpenden.

Hartnup, R. (1975). Soils in North Yorkshire II: Sheet SE 36 (Boroughbridge). Soil Survey Record No. 30, Harpenden.

Hartnup, R. (1977). Soils in South Yorkshire I: Sheet SK 59 (Maltby). Soil Survey Record No. 42, Harpenden.

Harvey, P.N. (1963). National Agricultural Advisory Service: Experimental Husbandry Farms and Experimental Horticultural Stations, 4th Progress Report, 1–5. HMSO, London.

Hashimoto, I. and Jackson, M.L. (1960). Rapid dissolution of allophane and kaolinite-halloysite after hydration. *Clays and Clay Minerals* **7**, 102–113.

Hazleden, J. (1986). Soils in Oxfordshire II: Sheet SP 60 (Tiddington). Soil Survey Record No. 98, Harpenden.

Hazleden, J., Loveland, P.J. and Sturdy, R.G. (1986). Saline soils in North Kent. Soil Survey Special Survey No. 14, Harpenden.

Heathcote, W.R. (1951). A soil survey on warpland in Yorkshire. *Journal of Soil Science* **2**, 144–162.

Heslop, R.E.F. and Bown, C.J. (1969). The Soils of Candacraig and Glenbuchat. Soil Survey of Scotland Bulletin No. 1. Macaulay Institute for Soil Research, Aberdeen.

Hey, R.W. (1986). A re-examination of the Northern Drift of Oxfordshire. *Proceedings of the Geologists' Association* **97**, 291–302.

Hodge, C.A.H. and Seale, R.S. (1966). The Soils of the District around Cambridge (Sheet 188). Memoir of the Soil Survey of Great Britain: England and Wales, Harpenden.

Hodge, C.A.H., Burton, R.G.O., Corbett, W.M., Evans, R. and Seale, R.S. (1984). Soils and their use in Eastern England. Soil Survey of England and Wales Bulletin No. 13, Harpenden.

Hodgson, J.M. (1967). Soils of the West Sussex Coastal Plain. Soil Survey of Great Britain: England and Wales. Bulletin No. 3, Harpenden.

Hodgson, J.M. (Ed.) (1976). Soil Survey Field Handbook. Soil Survey Technical Monograph No. 5, Harpenden.

Hodgson, J.M., Catt, J.A. and Weir, A.H. (1967). The origin and development of Clay-with-Flints and associated soil horizons on the South Downs. *Journal of Soil Science* **18**, 85–102.

Hoeksema, K.J. (1953). De natuurlijke homogenisatie van het bodemprofil in Nederland. *Boor en Spade* **6**, 24–30.

Hogan, D.V. (1977). Soils in Devon III: Sheet SX 47 (Tavistock). Soil Survey Record No. 44, Harpenden.

Hogan, D.V. (Ed.) (1978). British Society of Soil Science Autumn Meeting: Programme and Guide to Excursions. University of Exeter.

Hogan, D.V. (1981). Soils in Devon VI: Sheet SS 63 (Brayford). Soil Survey Record No. 71, Harpenden.

Hogan, D.V. and Harrod, T.R. (1982). Soils in Devon VII: Sheet SS 74 (Lynton). Soil Survey Record No. 78, Harpenden.

Hollis, J.M. (1975). Soils in Staffordshire I: Sheet SK 05 (Onecote). Soil Survey Record No. 29, Harpenden.

Hollis, J.M. (1985). Soils in Staffordshire IV: Sheet SK 00/10 (Lichfield). Soil Survey Record No. 89, Harpenden.

Hollis, J.M. and Hodgson, J.M. (1974). Soils in Worcestershire I: Sheet SO 87 (Kidderminster). Soil Survey Record No. 18, Harpenden.

Hollis, J.M., Jones, R.J.A. and Palmer, R.C. (1977). The effects of organic matter and particle size on the water retention properties of some soils in the west Midlands of England. *Geoderma* **17**, 225–238.

Holmgren, G.S. (1967). A rapid citrate-dithionite extractable iron procedure. *Soil Science Society of America Proceedings* **31**, 210–211.

Holzhey, C.S. and Yeck, R.D. (1974). Micro-fabric of some argillic horizons in udic, xeric and torric environments of the United States. In G.K. Rutherford (Ed.), *Soil Microscopy*, 747–759. Limestone Press, Kingston, Ontario.

Hudson, G., Towers, W., Bibby, J.S. and Henderson, D.J. (1982). Soil and Land Capability for Agriculture: The Outer Hebrides. Soil Survey of Scotland: The Macaulay Institute for Soil Research, Aberdeen.

Jamagne, M. (1972). Caractères micromorphologiques des sols développés sur formations limoneuses. *Bulletin de l'Association Française pour l'Etude du Sol* No. 1–2, 9–32.

Jamagne, M., De Coninck, F., Robert, M. and Maucorps, J. (1984). Mineralogy of clay fractions of some soils in loess in northern France. *Geoderma* **33**, 319–342.

Jarvis, M.G. (1973). Soils of the Wantage and Abingdon District (Sheet 253). Memoir of the Soil Survey of Great Britain: England and Wales, Harpenden.

Jarvis, M.G. and Findlay, D.C. (Eds.) (1984). *Soils of the Southampton district*. British Society of Soil Science, Ashford Press.

Jarvis, M.G. and Hazleden, J. (1982). Soils in Oxfordshire I: Sheet SP 30 (Witney South). Soil Survey Record No. 77, Harpenden.

Jarvis, M.G. and Mackney, D. (Eds.). (1979). Soil Survey Applications. Soil Survey Technical Monograph No. 13, Harpenden.

Jarvis, M.G., Allen, R.H., Fordham, S.J., Hazleden, J., Moffat, A.J. and Sturdy, R.G. (1984). Soils and their use in South-East England. Soil Survey of England and Wales Bulletin No. 15, Harpenden.

Jarvis, M.G., Hazleden, J. and Mackney, D. (1979). Soils of Berkshire. Soil Survey of England and Wales Bulletin No. 8, Harpenden.

Jarvis, R.A. (1968). Soils of the Reading District (Sheet 268). Memoir of the Soil Survey of Great Britain: England Wales, Harpenden.

Jarvis, R.A. (1973). Soils in Yorkshire II: Sheet SE 60 (Armthorpe). Soil Survey Record No. 12, Harpenden.

Jarvis, R.A. (1977). Soils of the Hexham District (Sheet 19). Memoir of the Soil Survey of Great Britain: England and Wales, Harpenden.

Jarvis, R.A., Bendelow, V.C., Bradley, R.I., Carroll, D.M., Furness, R.R., Kilgour, I.N.L. and King, S.J. (1984). Soils and their use in Northern England. Soil Survey of England and Wales Bulletin No. 10, Harpenden.

Jeanroy, B. and Guillet, B. (1981). The occurrence of suspended ferruginous particles in pyrophosphate extracts of some soil horizons. *Geoderma* **26**, 95–105.

Johnson, P.A. (1971). Soils in Derbyshire I: Sheet SK 17 (Tideswell). Soil Survey Record No. 4, Harpenden.

Jones, R.J.A. (1975). Soils in Staffordshire II: Sheet SJ 82 (Eccleshall). Soil Survey Record No. 31, Harpenden.

Jones, R.J.A. (1983). Soils in Staffordshire III: Sheets SK 02/12 (Needwood Forest). Soil Survey Record No. 80, Harpenden.

Jones, R.J.A. and Thomasson, A.J. (1985). An agroclimatic data bank for England and Wales. Soil Survey Technical Monograph No. 16, Harpenden.

Jongerius, A. (1962). Soil genesis in organic soils. *Boor en Spade* **12**, 156–168.

Jongerius, A. (1970). Some morphological aspects of regrouping phenomena in Dutch Soils. *Geoderma* **4**, 311–332.

Kay, F.F. (1934). A Soil Survey of the Eastern Portion of the Vale of the White Horse. Bulletin No. 48: University of Reading Faculty of Agriculture and Horticulture.

Kay, F.F. (1939). A Soil Survey of the Strawberry District of South Hampshire. Bulletin No. 52: University of Reading Faculty of Agriculture and Horticulture.

Kerney, M.P. (1963). Late-glacial deposits on the Chalk of South-East England. *Philosophical Transactions of the Royal Society of London* **B246**, 203–254.

Kerney, M.P., Brown, E.H. and Chandler, T.J. (1964). The late-glacial and post-glacial history of the Chalk escarpment near Brook, Kent. *Philosophical Transactions of the Royal Society of London* **B248**, 135–204.

Kerney, M.P., Preece, R.C. and Turner, C. (1980). Molluscan and plant bio-stratigraphy of some Late Devensian and Flandrian deposits in Kent. *Philosophical Transactions of the Royal Society of London* **B291**, 1–43.

Kilgour, I.N.L. (1979). Soils in Cumbria II: Sheet NY 36/37 (Longtown). Soil Survey Record No. 59, Harpenden.

Kilroe, J.R. (1907). A Description of the Soil-Geology of Ireland with Notes on Climate. Memoir of the Geological Survey of Ireland. HMSO, Dublin.

King, J.A., Smith, K.A. and Pyatt, G.D. (1986). Water and oxygen regimes under conifer plantations and native vegetation on upland peaty gley soils and deep peat soils. *Journal of Soil Science* **37**, 485–497.

King, S.J. (1977). Soils in Cheshire III: Sheet SJ 45E/55W (Burwardsley). Soil Survey Record No. 43, Harpenden.

Kirkman, J.H., Mitchell, B.D. and Mackenzie, R.C. (1966). Distribution in some Scottish soils of an inorganic gel system related to 'allophane'. *Transactions of the Royal Society of Edinburgh* **66**, 393–418.

Klingebiel, A.A. and Montgomery, P.H. (1961). Land-Capability Classification. *Agriculture Handbook No. 210*: U.S. Department of Agriculture Soil Conservation Service, Washington, DC.

Knox, E.G. (1965). Soil individuals and soil classification. *Soil Science Society of America Proceedings* **29**, 79–84.

Kooistra, M.J., Bouma, J., Boersma, O.H. and Jager, A. (1985). Soil-structure differences and associated physical properties of some loamy Typic Fluvaquents in the Netherlands. *Geoderma* **36**, 215–228.

Kubiena, W.L. (1953). *The Soils of Europe*. Murby, London.

Kubiena, W.L. (1958). The classification of soils. *Journal of Soil Science* **9**, 9–19.

Kwaad, F.J.P.M. and Mücher, H.J. (1979). The formation and evolution of colluvium on arable land in northern Luxembourg. *Geoderma* **22**, 173–192.

Laing, D. (1976). The Soils of the Country round Perth, Arbroath and Dundee (Sheets 48 and 49). Memoir of the Soil Survey of Great Britain: Scotland. HMSO, Edinburgh.

Larney, F., Collins, J.F. and Walsh, M. (1981). Properties and genesis of soils developed in alluvium in Co. Monaghan. *Irish Journal of Agricultural Research* **20**, 81–100.

Lawson, T.J. (1983). A note on the significance of a red soil on dolomite in north-west Scotland. *Quaternary News Letter* No. 40, 10–11.

Lea, J.W. (1975). Soils in Powys I: Sheet SO 09 (Caersws). Soil Survey Record No. 28, Harpenden.

Lee, J. (1980). Richard Griffith's land evaluation as a basis for farm taxation. *Royal Dublin Society: Historical Studies in Irish Science and Technology* **1**, 77–101.

Limbrey, S. (1975). *Soil Science and Archaeology*. Academic Press, London.

Locket, G.H. (1946). Observations on the colonization of bare chalk. *Journal of Ecology* **33**, 205–209.

Loveday, J. (1962). Plateau deposits of the southern Chiltern Hills. *Proceedings of the Geologists' Association* **73**, 83–102.

Loveland, P.J. (1984). The characteristics of brown podzolic soils in England and Wales. *Proceedings of the North of England Soils Discussion Group* **20**, 71–82.

Loveland, P.J. and Bullock, P. (1975). Crystalline and amorphous components of the clay fractions in brown podzolic soils. *Clays and Clay Minerals* **10**, 451–469.

Loveland, P.J. and Bullock, P. (1976). Chemical and mineralogical properties of brown podzolic soils in comparison with soils of other groups. *Journal of Soil Science* **27**, 523–540.

Loveland, P.J. and Clayden, B. (1987). A hardpan podzol at Yarner Wood, Devon. *Journal of Soil Science* **38**, 357–368.

Loveland, P.J. and Findlay, D.C. (1982). Composition and development of some soils on glauconitic Cretaceous (Upper Greensand) rocks in southern England. *Journal of Soil Science* **33**, 279–294.

Loveland, P.J., Hazleden, J. and Sturdy, R.G. (1987). Chemical properties of salt-affected soils in north Kent and their relationship to soil instability. *Journal of Agricultural Science, Cambridge* **109**, 1–6.

Lozet, J.M. and Herbillon, A.J. (1971). Fragipan soils of Condroz (Belgium): mineralogical, chemical and physical aspects in relation with their genesis. *Geoderma* **5**, 325–343.

Luxmoore, C.M. (1907). The Soils of Dorset. University College Reading: Department of Agricultural Chemistry, Bulletin No. 13.

Mackney, D. (1961). A podzol development sequence in oakwoods and heath in central England. *Journal of Soil Science* **12**, 23–40.

Mackney, D. (1970). Podzols in Lowland England. *Welsh Soils Discussion Group Report* **11**, 64–87.

Mackney, D. (Ed.) (1974). Soil Type and Land Capability. Soil Survey Technical Monograph No. 4, Harpenden.

Mackney, D. and Burnham, C.P. (1964). The Soils of the West Midlands. Soil Survey of Great Britain: England and Wales. Bulletin No. 2, Harpenden.

Mackney, D. and Burnham, C.P. (1966). The Soils of the Church Stretton District of Shropshire (Sheet 166). Memoir of the Soil Survey of Great Britain: England and Wales, Harpenden.

Mackney, D., Hodgson, J.M., Hollis, J.M. and Staines, S.J. (1983). Legend for the 1:250,000 Soil Map of England and Wales. Soil Survey of England and Wales, Harpenden.

Macphail, R.I. (1983). The micromorphology of dark earth from Gloucester, London and Norwich: an analysis of urban anthropogenic deposits from the Late Roman to Early Mediaeval periods in England. *In* Bullock, P. and Murphy, C.P. (Eds.) *Soil Micromorphology* Vol. 1: *Techniques and Applications*, 245–252. AB Academic Publishers, Berkhamsted.

Mados, L. (1943). Eine Schnellmethode zu serienweisen Bestimmung der Adsorptionsungesättigheit von Böden. *Bodenkunde und Pflanzenernährung* **32**, 351–358.

MAFF (Ministry of Agriculture, Fisheries and Food) (1967). Potential Transpiration. Technical Bulletin 16. HMSO, London.

MAFF (Ministry of Agriculture, Fisheries and Food) (1970). *Modern Farming and the Soil.* HMSO, London.

MAFF (Ministry of Agriculture, Fisheries and Food) (1988). *Agricultural Statistics United Kingdom, 1986.* HMSO, London.

Matthews, B. (1970). Age and origin of aeolian sand in the Vale of York. *Nature, London* **227**, 1234–1236.

Matthews, B. (1976). Soils with discontinuous induration in the Penrith area of Cumbria. *Proceedings of the North of England Soils Discussion Group* **11**, 11–19.

Matthews, B. (1977). Soils in Cumbria I: Sheet NY 53 (Penrith). Soil Survey Record No. 46, Harpenden.

McAleese, D.M. and Mitchell, W.A. (1958). Studies on the basaltic soils of Northern Ireland V. Cation-exchange capacities and mineralogy of the silt separates. *Journal of Soil Science* **9**, 81–88.

McGowan, M. (1984). Soil water regimes of soils developed on Keuper Marl. *Journal of Soil Science* **35**, 317–322.

McKeague, J.A. (1983). Clay skins and argillic horizons. *In* Bullock, P and Murphy, C.P. (Eds.) *Soil Micromorphology* Vol. 2: *Soil Genesis*. A.B. Academic Publishers, Berkhamsted.

McKeague, J.A. and Day, J.H. (1966). Dithionite- and oxalate-extractable Fe and Al as aids in differentiating various classes of soils. *Canadian Journal of Soil Science* **46**, 13–22.

McKeague, J.A. and Kodama, H. (1981). Imogolite in cemented horizons of some British Columbia soils. *Geoderma* **25**, 189–197.

McKeague, J.A. and Sprout, P.N. (1975). Cemented subsoils (duric horizons) in some soils of British Columbia. *Canadian Journal of Soil Science* **55**, 189–203.

McKeague, J.A., Brydon, J.E. and Miles, N.M. (1971). Differentiation of forms of extractable iron and aluminium in soils. *Soil Science Society of America Proceedings* **35**, 33–38.

Mehlich, A. (1948). Determination of cation- and anion-exchange properties of soils. *Soil Science* **66**, 429–436.

Mehra, O.P. and Jackson, M.L. (1960). Iron oxide removal from soils and clays by a dithionite-citrate system buffered with sodium bicarbonate. *Clays and Clay Minerals* **5**, 317–327.

Meteorological Office (1968). *Averages of earth temperature at depths of 30 cm and 122 cm for the*

United Kingdom. HMSO, London.

Metson, A.J. (1956). Methods of chemical analysis for soil survey samples. *Soil Bureau Bulletin* **12**, DSIR, Wellington, New Zealand.

Miller, D.E. (1973). Water retention and flow in layered soil profiles. *In* Bruce, R.R. *et al.* (Eds.), *Field Soil Water Regime*, 107–117. Special Publication No. 5, Soil Science Society of America, Madison, Wisconsin.

Milne, G. (1935a). Some suggested units of classification and mapping, particularly for East African soils. *Soil Research, Berlin* **4**, 183–198.

Milne, G. (1935b). Complex units for the mapping of complex soil associations. *Transactions Third International Congress of Soil Science* **1**, 345–347.

Mitchell, B.D. and Jarvis, R.A. (1956). The Soils of the Country round Kilmarnock (Sheet 22 and part of Sheet 21). Memoir of the Soil Survey of Great Britain: Scotland. HMSO, Edinburgh.

Mitchell, B.D., Smith, B.F.L. and De Endredy, A.S. (1971). The effect of buffered sodium dithionite solution and ultrasonic agitation on soil clays. *Israel Journal of Chemistry* **9**, 45–52.

Mitchell, G.F. (1976). *The Irish Landscape*. Collins, London.

Mitchell, G.F., Penny, L.F., Shotton, F.W. and West, R.G. (1973). A Correlation of Quaternary Deposits in the British Isles. Geological Society of London, Special Report No. 4.

Mitchell, R.L. and Muir, A. (1937). Base exchange capacity and clay content of soils. *Nature, London* **139**, 552.

Mokma, D.L. and Buurman, P. (1982). Podzols and podzolization in temperate regions. ISM Monograph No. 1. International Soil Museum, Wageningen, Netherlands.

Morley Davies, W. and Owen, G. (1934). Soil Survey of North Shropshire. Pt. 1. Geography, geology, parent materials, soils. *Empire Journal of Experimental Agriculture* **2**, 178–188. Pt 2. Classification of series and types. *Ibid*, 359–379.

Morgan, R.C.P. (1977). Soil erosion in the United Kingdom: field studies in the Silsoe area, 1973–75. *National College of Agricultural Engineering, Occasional Paper* **4**.

Morgan, R.P.C. (1980). Soil erosion and conservation in Britain. *Progress in Physical Geography* **4**, 24–47.

Mückenhausen, E. (1965). The soil classification system of the Federal Republic of Germany. *Pédologie, Gand*. Numéro spécial **3**, 57–74.

Mückenhausen, E. (1985). *Die Bodenkunde und ihre geologischen, geomorphologischen, mineralogischen und petrologischen Grundlagen*, 3 ergänzte Auflage. DLG-Verlag, Frankfurt am Main.

Mückenhausen, E., unter Mitwirkung von Kohl, F., Blüme, H.P., Heinrich, F. und Müller, S. (1977). *Enstehung, Eigenschaften und Systematik der Böden der Bundesrepublik Deutschland*, 2 Auflage, DLG-Verlag, Frankfurt am Main.

Muir, A. (1934). The soils of the Teindland State Forest. *Forestry* **8**, 25–55.

Muir, A. (1961a). Soil Survey in Britain. Soil Survey of Great Britain Report 13 (1960), 1–9. Agricultural Research Council.

Muir, A. (1961b). The podzol and podzolic soils. *Advances in Agronomy* **13**, 1–56.

Muir, A. and Fraser, G.K. (1939). The soils and vegetation of the Bin and Clashindarroch Forests. *Transactions of the Royal Society of Edinburgh* **60**, 233–341.

Muir, J.W. (1956). The Soils of the Country round Jedburgh and Morebattle (Sheets 17 and 18). Memoir of the Soil Survey of Great Britain: Scotland. HMSO, Edinburgh.

Muir, J.W. (1969). A natural system of soil classification. *Journal of Soil Science* **20**, 153–166.

Mulcahy, M.J. and Humphries, A.W. (1967). Soil classification, soil surveys and land use. *Soils and Fertilizers* **30**, 1–8.

Murphy, C.P. (1984). The morphology and genesis of eight surface-water gley soils developed in till in England and Wales. *Journal of Soil Science* **35**, 251–272.

Murphy, C.P. and Bullock, P. (Eds.) (1981). International Working-Meeting on Soil Microbiology, London, England. Programme and Guide for post-meeting Excursion through England and Wales.

Nature Conservancy Council (1982). Seventh Report, covering the period 1 April 1980–31 March 1981.

Nettleton, W.D., Flach, K.W. and Brasher, B.R. (1969). Argillic horizons without clay skins. *Soil Science Society of America Proceedings* **33**, 121–125.

Newman, L.F. (1912). Soils and agriculture of Norfolk. *Transactions Norfolk and Norwich Naturalists Society* **9**, 349–393.

Nicholson, H.H. (1935). The drainage properties of heavy soils. *Transactions Third International Congress of Soil Science* **1**, 385–388.

Nömmik, H. (1971). A modified procedure for determination of organic carbon in soils by wet combustion. *Soil Science* **111**, 330-336.

Northcote, K. (1971). *A Factual Key for the Recognition of Australian Soils*, 3rd ed. Rellim Technical Publishers, Glenside, South Australia.

O'Dubháin, T. and Collins, J.F. (1981). Morphology and genesis of podzols developed in contrasting parent materials in Ireland. *Pédologie, Gand.* **31**, 81–98.

Ogg, W.G. (1920). Soil investigation work in America. *Scottish Journal of Agriculture* **3**, 287–295.

Ogg, W.G. (1935). The soils of Scotland. Pt. 1. Introduction: the Highlands and Islands. *Empire Journal of Experimental Agriculture* **3**, 174–188. Pt. 2. The north-eastern region. *Ibid*, 248–260. Pt. 3. The Central Valley and Southern Uplands. *Ibid*, 295–312.

Ollier, C.D. and Thomasson, A.J. (1957). Asymmetrical valleys of the Chiltern Hills. *Geographical Journal* **123**, 71-80.

Omond, R.T. (1910). Mean daily temperatures at the Ben Nevis and Fort William Observatories. *Transactions of the Royal Society of Edinburgh* **44**, 693–701.

Osmond, D.A., Swarbrick, T., Thompson, C.R. and Wallace, T. (1949). A Survey of the Soils and Fruit in the Vale of Evesham. Ministry of Agriculture and Fisheries Bulletin No. 116. HMSO, London.

Ovington, J.D. (1950). The afforestation of the Culbin Sands. *Journal of Ecology* **38**, 303–319.

Palmer, R.C. (1982). Soils in Hereford and Worcester I: Sheets SO 85/95 (Worcester). Soil Survey Record No. 76, Harpenden.

Pape, J.C. (1970). Plaggen soils in the Netherlands. *Geoderma* **4**, 229–256.

Parfitt, R.L. and Henmi, T. (1982). Comparison of an oxalate-extraction method and an IR spectroscopic method for determining allophane in soil clays. *Soil Science and Plant Nutrition* **28**, 183–190.

Parfitt, R.L., Russell, M. and Orbell, G.E. (1983). Weathering sequence of soils from volcanic ash involving allophane and halloysite, New Zealand. *Geoderma* **29**, 41–57.

Parker, F.W. (1929). The determination of exchangeable hydrogen in soils. *Journal of the American Society of Agronomy* **21**, 1030–1039.

Payton, R.W. (1980). Pedogenetic compaction: the character and formation of compact soil horizons. *Proceedings of the North of England Soils Discussion Group* **16**, 103–126.

Payton, R.W. (1987a). Podzolic soils of Fell Sandstone, Northumberland: their characteristics and genesis. *Proceedings of the North of England Soils Discussion Group* **22**, 1–42.

Payton, R.W. (1987b). Thin ironpan formation in Humus-Ironpan Stagnopodzols on the Fell Sandstone of Northumberland. *Proceedings of the North of England Soils Discussion Group* **22**, 43–72.

Payton, R.W. (1988). The characteristics and genesis of fragipans in British soils. Ph.D. thesis, University of Newcastle upon Tyne.

Penman, H.L. (1948). The dependence of transpiration on weather and soil conditions. *Journal of Soil Science* **1**, 74–89.

Pennington, W. (1974). *The History of British Vegetation*. 2nd edn, English Universities Press, London.

Perrin, R.M.S. (1956). Nature of chalk heath soils. *Nature, London* **178**, 31–32.

Perrin, R.M.S., Davies, H. and Fysh, M.D. (1974). Distribution of late Pleistocene aeolian deposits in eastern and southern England. *Nature, London* **248**, 320–324.

Piggott, C.D. (1962). Soil formation and development on the Carboniferous Limestone of Derbyshire 1. Parent materials. *Journal of Ecology* **50**, 145–156.

Pizer, N.H. (1931). A Survey of the Soils of Berkshire. Bulletin No. 39: University of Reading Faculty of Agriculture and Horticulture.

Pons, L.J. and Van der Molen, W.H. (1973). Soil genesis under dewatering regimes during 1000 years of polder development. *Soil Science* **116**, 228–235.

Powlson, D.S. (1975). The identification of potential acid sulphate soils. *Report Rothamsted Experimental Station for 1974*, Pt. 1, 199.

Price, J.R. (1980). Soils of Redgrave and Lopham Fens. *Transactions of the Suffolk Naturalists Society* **18**, 104–111.

Price, J.R. and Hodge, C.A.H. (1982). Acid sulphate soils in The Broads area. Soil Survey of England and Wales: Annual Report 1981, 10–12.

Pyatt, G.D. (1970). Soil Groups of Upland Forests. Forestry Commission: Forest Record No. 71. HMSO, London.

Pyatt, G.D. (1977). Guide to Site Types in Forests

of North and Mid Wales. Forestry Commission: Forest Record No. 69, 2nd edition. HMSO, London.

Pyatt, G.D. (1982). Soil classification. Research Information Note 68/82/SSN. Forestry Commission Research and Development Division.

Pyatt, G.D. (1987). The effect of afforestation on the water and oxygen regimes of stagnopodzols. *Proceedings of the North of England Soils Discussion Group* **22**, 93–108.

Pyatt, G.D. and Craven, M.M. (1979). Soil changes under even-aged plantations. In *The Ecology of Even-Aged Plantations*, 369–386. Institute of Terrestrial Ecology, Cambridge.

Pyatt, G.D., Craven, M.M. and Williams, B.L. (1979). Peatland classification for forestry in Great Britain. *Proceedings of the International Peat Society Symposium on Classification of Peat and Peatlands* Hyytiälä, Finland, 351–366.

Pyatt, G.D. and Smith, K.A. (1983). Water and oxygen regimes of four soil types at Newcastleton Forest, Scotland. *Journal of Soil Science* **34**, 465–482.

Ragg, J.M. (Ed.) (1974). British Society of Soil Science: Programme and Guide to Excursions for the Autumn Conference, Edinburgh.

Ragg, J.M. and Bibby, J.S. (1966). Frost weathering and solifluxion products in southern Scotland. *Geografiska Annaler* **48 A**, 12–23.

Ragg, J.M. and Clayden, B. (1973). The Classification of some British soils according to the Comprehensive System of the United States. Soil Survey Technical Monograph No. 3, Harpenden.

Ragg, J.M. and Futty, D.W. (1967). The Soils of the Country round Haddington and Eyemouth (Sheets 33, 34 and part 41). Memoir of the Soil Survey of Great Britain: Scotland, HMSO, Edinburgh.

Ragg, J.M., Beard, G.R., George, H., Heaven, F.W., Hollis, J.M., Jones, R.J.A., Palmer, R.C., Reeve, M.J., Robson, J.D. and Whitfield, W.A.D. (1984). Soils and their use in Midland and Western England. Soil Survey of England and Wales Bulletin No. 12, Harpenden.

Ragg, J.M., Bracewell, J.M., Logan, J. and Robertson, L. (1978). Some characteristics of the brown forest soils of Scotland. *Journal of Soil Science* **29**, 228–242.

Rayner, J.H. (1966). Classification of soil by numerical methods. *Journal of Soil Science* **17**, 79–92.

Reeve, M.J. (1978). Soils in Northamptonshire I: Sheet SP 66 (Long Buckby). Soil Survey Record No. 54, Harpenden.

Reeve, M.J. and Hall, D.G.M. (1978). Shrinkage in clayey subsoils of contrasting structure. *Journal of Soil Science* **29**, 315–323.

Reeve, M.J. and Thomasson, A.J. (1981). Soils in Nottinghamshire IV: Sheet SK 78N/79S (Gringley on the Hill). Soil Survey Record No. 72, Harpenden.

Reeve, M.J., Hall, D.G.M. and Bullock, P. (1980). The effect of soil composition and environmental factors on the shrinkage of some clayey British soils. *Journal of Soil Science* **31**, 429–442.

Renger, M. (1971). Die Ermittlung der Porengrössenverteilung aus der Körnung, dem Gehalt an organischer Substanz und der Lagerungsdichte. *Zeitschrift für Pflanzenernährung und Bodenkunde* **130**, 53–67.

Rigg, T. (1916). The soils and crops of the market-garden district of Bedfordshire. *Journal of Agriculture Science, Cambridge* **7**, 385–431.

Righi, D., Van Ranst, E., De Coninck, F. and Guillet, B. (1982). Microprobe study of a Placohumod in the Antwerp Campine (North Belgium). *Pédologie, Gand.* **32**, 117–134.

Ritchie, W. (1976). The meaning and definition of machair. *Transactions of the Botanical Society of Edinburgh* **42**, 432–440.

Roberts, R.D., Marrs, R.H., Skeffington, R.A. and Bradshaw, R.D. (1981). Ecosystem development on naturally colonized china clay wastes I. Vegetation changes and overall accumulation of organic matter and nutrients. *Journal of Ecology* **69**, 153–161.

Robinson, G.W. (1912). *A Survey of the Soils and Agriculture of Shropshire*. County of Salop: Higher Education, Shrewsbury.

Robinson, G.W. (1924). Pedology as a branch of geology. *Geological Magazine* **61**, 444–455.

Robinson, G.W. (1928). The nature of clay and its significance in the weathering cycle. *Nature, London* **121**, 903–904.

Robinson, G.W. (1930). The development of the soil profile in North Wales as illustrated by the character of the clay fraction. *Journal of Agricultural Science, Cambridge* **20**, 618–639.

Robinson, G.W. (1934). The dispersion of soils in mechanical analysis. *Comptes Rendus de la Conférence de la Première Commission (Physique du Sol) de l'Association Internationale de la Science du*

Sol, 13–18. Imprimerie Nationale, Paris.
Robinson, G.W. (1949). *Soils, their Origin, Constitution and Classification: an Introduction to Pedology*, 3rd edn. Murby, London.
Robinson, G.W., Hughes, D.O. and Jones, B. (1930). Soil Survey of Wales: Progress Report 1927–29. *Welsh Journal of Agriculture* **6**, 249–265.
Robinson, G.W., Hughes, D.O. and Roberts, E. (1949). Podzolic soils of Wales. *Journal of Soil Science* **1**, 50–62.
Robinson, K.L. (1948). The soils of Dorset. In Good, R. *A Geographical Handbook of the Dorset Flora*, 19–28. Dorset Natural History and Field Club, Dorchester.
Robson, J.D. (1981). Salterns: collaborative work with archaeologists in Lincolnshire. Soil Survey of England and Wales: Annual Report 1980, 15–18.
Robson, J.D. (1985). Soils in Lincolnshire IV: Sheet TF 45 (Friskney). Soil Survey Record No. 88, Harpenden.
Robson, J.D. and George, H. (1971). Soils in Nottinghamshire I: Sheet SK 66 (Ollerton). Soil Survey Record No. 8, Harpenden.
Robson, J.D. and Thomasson, A.J. (1977). Soil Water Regimes. Soil Survey Technical Monograph No. 11, Harpenden.
Robson, J.D., George, H. and Heaven, F.W. (1974). Soils in Lincolnshire I: Sheet TF 16 (Woodhall Spa). Soil Survey Record No. 22, Harpenden.
Romans, J.C.C. (1970). Podzolization in a zonal and altitudinal context in Scotland. *Welsh Soils Discussion Group Report* **11**, 88–101.
Romans, J.C.C. (1974). Indurated layers. *Proceedings of the North of England Soils Discussion Group* **11**, 20–30.
Romans, J.C.C. and Robertson, L. (1975). Some genetic characteristics of the freely drained soils of the Ettrick Association in east Scotland. *Geoderma* **14**, 297–317.
Romans, J.C.C., Stevens, J.H. and Robertson, L. (1966). Alpine soils of north-east Scotland. *Journal of Soil Science* **17**, 184–199.
Rose, J. (Ed.) (1983). Quaternary Research Association. Field Guide: Annual Field Meeting 1983. The Diversion of the Thames.
Rose, J. (1987). Status of the Wolstonian glaciation in the British Quaternary. *Quaternary Newsletter* No. 53, 1–9.
Rose, J. and Allen, P. (1977). Middle Pleistocene stratigraphy in south-east Suffolk. *Journal of the Geological Society* **133**, 83–102.
Rozov, N.N. and Ivanova, E. (1967). Classification of the soils of the U.S.S.R. *Soviet Soil Science* **2**, 147–156.
Rudeforth. C.C. (1970). Soils of North Cardiganshire (Sheets 163 and 178). Memoir of the Soil Survey of Great Britain: England and Wales, Harpenden.
Rudeforth, C.C. (1982). Handling soil survey data. In Bridges, E.M. and Davidson, D.A. (Eds.), *Principles and Applications of Soil Geography*, 97–131. Longman, London and New York.
Rudeforth, C.C., Hartnup, R., Lea, J.W., Thompson, T.R.E. and Wright, P.S. (1984). Soils and their use in Wales. Soil Survey of England and Wales Bulletin No. 11, Harpenden.
Russell, E.W. (1973). *Soil Conditions and Plant Growth*, 10th edn. Longman, London.
Rutter, A.J. and Fourt, D.F. (1965). Studies in the water relations of *Pinus sylvestris* in plantation conditions III. A comparison of soil water changes and estimates of total evaporation on four afforested sites and one grass-covered site. *Journal of Applied Ecology* **2**, 197–209.
Salisbury, E.J. (1922). The soils of Blakeney Point: a study of soil reaction and succession in relation to the plant covering. *Annals of Botany* **36**, 392–431.
Salisbury, E.J. (1925). Note on the edaphic succession in some dune soils with special reference to the time factor. *Journal of Ecology* **13**, 322–328.
Salisbury, E.J. (1952). *Downs and Dunes*. Bell, London.
Scheys, G., Dudal, R. and Baeyens, L. (1954). Une interprétation de la morphologie de podzols humo-ferriques. *Transactions 5th International Congress of Soil Science* **4**, 274–281.
Schweikle, V. (1973). Die Stellung der Stagnogleye in der Bodengesellschaft der Schwarzwaldhochfläche. In Pseudogley and Gley: *Transactions of International Society of Soil Science Commissions V and VI*, Stuttgart, 181–186.
Schwertmann, U. (1964). Differenzierung der Eisenoxide des Bodens durch Extraktion mit Ammoniumoxalat-Lösung. *Zeitschrift für Pflanzenernährung und Bodenkunde* **105**, 194–202.
Schwertmann, U. (1973). Use of oxalate for Fe extraction from soils. *Canadian Journal of Soil Science* **53**, 244–246.

Schwertmann, U. and Taylor, R.M. (1977). Iron oxides. *In* Dixon, J.B., Weed, S.B., Kittrick, J.A., Milford, M.H. and White, J.L. (Eds.), *Minerals in the Soil Environment*, 145–180. Soil Science Society of America, Madison, Wisconsin.

Schwertmann, U., Murad, E. and Schulze, D.G. (1982). Is there Holocene reddening (hematite formation) in soils of axeric temperate areas? *Geoderma* **27**, 209–223.

Scott, G.A.M. (1965). The shingle succession at Dungeness. *Journal of Ecology* **53**, 21–31.

Seale, R.S. (1975a). Soils of the Ely District (Sheet 173). Memoir of the Soil Survey of Great Britain: England and Wales, Harpenden.

Seale, R.S. (1975b). Soils of the Chatteris district of Cambridgeshire (Sheet TL 38). Soil Survey Special Survey No. 9, Harpenden.

Shotton, F.W., Goudie, A.S., Briggs, D. and Osmaston, H.A. (1980). Cromerian interglacial deposits at Sugworth, near Oxford, England, and their relation to the Plateau Drift of the Cotswolds and the terrace sequence of the Upper and Middle Thames. *Philosophical Transactions of the Royal Society of London* **B289**, 55–86.

Sibertsiev, N.M. (1901). Russian soil investigations. *In* Finkl, C.W.Jnr. (Ed.), *Benchmark Papers in Soil Science* 1. *Soil Classification*, 15–35. Hutchinson Ross Publishing Company, Stroudsburg, Pennsylvania.

Simonson, R.W. (1964). The Soil Series as used in the U.S.A. *Transactions 8th International Congress of Soil Science* **5**, 17–24.

Smith, C.J. (1980). *Ecology of the English Chalk*. Academic Press, London.

Smith, G.D., Arya, L.M. and Stark, J. (1975). The densipan: a diagnostic horizon for Densiaquults for Soil Taxonomy. *Soil Science Society of America Proceedings* **39**, 369–370.

Soil Survey of Scotland (1984). Organization and Methods of the 1:250,000 Soil Survey of Scotland. Macaulay Institute for Soil Research, Aberdeen.

Soil Survey Staff (1951). Soil Survey Manual. US Department of Agriculture, Agricultural Handbook No. 18, Washington, DC.

Soil Survey of Great Britain (1960). Field Handbook.

Soil Survey Staff (1960). Soil Classification: a Comprehensive System: 7th Approximation. US Department of Agriculture, Washington, DC.

Soil Survey Staff (1975). Soil Taxonomy: a Basic System of Soil Classification for Making and Interpreting Soil Surveys. US Department of Agriculture, Handbook No. 436, Washington, DC.

Soil Survey Staff (1983). National Soils Handbook. Soil Conservation Service. U.S. Department of Agriculture, Washington, D.C.

Soil Survey Staff (1987). Keys to Soil Taxonomy (third printing). SMSS Technical Monograph No. 6. Ithaca, New York.

Sokal, R.S. and Sneath, P.H.A. (1963). *Principles of Numerical Taxonomy*. Freeman, San Francisco.

Soulsby, J.A. (1976). Paleoenvironmental interpretation of a buried soil at Achnaree, Argyll. *Transactions of the Institute of British Geographers*, new series **I**, 279–283.

Stace, H.C.T., Hubble, G.D., Brewer, R., Northcote, K.H., Sleeman, J.R., Mulcahy, M.J. and Hallsworth, E.G. (1968). *A Handbook of Australian Soils*. Rellim Technical Publishers, Glenside, South Australia.

Stahr, K. (1973). Der Einfluss eines Fe-Bandchen-Mikrohorizont (thin ironpan) auf den Wasser- und Lufthaushalt von Mittelgebirgsboden. *In* Pseudogley and Gley. *Transactions of the International Society of Soil Science Commissions V and VI*, Stuttgart, 521–528.

Staines, S.J. (1976). Soils in Cornwall I: Sheet SX 18 (Camelford). Soil Survey Record No. 34, Harpenden.

Staines, S.J. (1979). Soils in Cornwall II: Sheet SW 53 (Hayle). Soil Survey Record No. 57, Harpenden.

Staines, S.J. (1984). Soils in Cornwall III: Sheets SW 61, 71 and parts of SW 62, 72, 81 and 82 (The Lizard). Soil Survey Record No. 79, Harpenden.

Steur, G.C.L. (1961). Methods of surveying in use at the Netherlands Soil Survey Institute. *Boor en Spade* **11**, 59–77.

Steven, H.M. and Carlisle, A. (1959). *The Native Pinewoods of Scotland*. Oliver and Boyd, Edinburgh.

Storrier, R.R. and Muir, A. (1962). The characteristics and genesis of a ferritic brown earth. *Journal of Soil Science* **13**, 259–270.

Sturdy, R.G. (1971). Soils in Essex I: Sheet TQ 59 (Harold Hill). Soil Survey Record No. 7, Harpenden.

Sturdy, R.G. and Allen, R.H. (1981). Soils in Essex IV: Sheet TM12 (Weeley). Soil Survey Record No. 67, Harpenden.

Sturdy, R.G., Allen, R.H., Bullock, P., Catt, J.A. and Greenfield, S. (1979). Paleosols developed on Chalky Boulder Clay in Essex. *Journal of Soil Science* **30**, 117–137.

Tait, J.M., Yoshinaga, N. and Mitchell, B.D. (1978). The occurrence of imogolite in Scottish soils. *Soil Science and Plant Nutrition* **24**, 145–151.

Tallis, J.H. (1985). Erosion of blanket peat in the southern Pennines: new light on an old problem. *In* Johnson, R.H. (Ed.), *The Geomorphology of North-west England*, 313–336. Manchester University Press, Manchester.

Tansley, A.G. (1939). *The British Islands and their Vegetation*. Cambridge University Press, Cambridge.

Tatler, W. and Corbett, W.M. (1977). Soils in Norfolk III: Sheet TG 31 (Horning). Soil Survey Record No. 41, Harpenden.

Tavernier, R. and Maréchal, R. (1962). Soil survey and soil classification in Belgium. *Transactions of Joint Meeting of International Society of Soil Science Commissions IV and V*, 298–301. Soil Bureau, Lower Hutt, New Zealand.

Taylor, G.G.M. (1970). Ploughing Practice in the Forestry Commission. Forest Record No. 73. HMSO, London.

Temple, M.S. (1929). *A Survey of the Soils of Buckinghamshire*. Bulletin No. 38: University of Reading Faculty of Agriculture and Horticulture.

Thomas, M.F. (1961). River terraces and drainage development in the Reading area. *Proceedings of the Geologists' Association* **72**, 415–436.

Thomasson, A.J. (1961). Some aspects of the drift deposits and geomorphology of south-east Hertfordshire. *Proceedings of the Geologists' Association* **72**, 287–302.

Thomasson, A.J. (1971). Soils of the Melton Mowbray District (Sheet 142). Memoir of the Soil Survey of Great Britain: England and Wales, Harpenden.

Thomasson, A.J. (1974). Soil type, water conditions and drainage. *In* Mackney, D. (Ed.), Soil Type and Land Capability, 43–52. Soil Survey Technical Monograph No. 4, Harpenden.

Thomasson, A.J. (Ed.). (1975). Soils and Field Drainage. Soil Survey Technical Monograph No. 7, Harpenden.

Thomasson, A.J. (1979). Assessment of soil droughtiness. *In* Jarvis, M.G. and Mackney D. (Eds.), Soil Survey Applications, 43–50. Soil Survey Technical Monograph No. 13, Harpenden.

Thomasson, A.J. and Avery, B.W. (1963). The Soils of Hertfordshire. *Transactions of the Hertfordshire Natural History Society and Field Club* **25**, 247–263.

Thomasson, A.J. and Bullock, P. (1975). Pedology and hydrology of some surface-water gley soils. *Soil Science* **119**, 339–348.

Thomasson, A.J. and Robson, J.D. (1971). The moisture regimes of soils developed on Keuper Marl. *Journal of Soil Science* **18**, 329–340.

Thomasson, A.J. and Youngs, E.G. (1975). Water movement in soil. *In* Soil Physical Conditions and Crop Production: Ministry of Agriculture, Fisheries and Food Technical Bulletin **29**, 228–239. HMSO, London.

Thompson, T.R.E. (1982). Soils in Powys II: Sheet SJ 21 (Arddleen). Soil Survey Record No. 75, Harpenden.

Thompson, T.R.E. (1983). Translocation of fine earth in some soils from an area in mid-Wales. *In* Bullock, P. and Murphy, C.P. (Eds.), *Soil Micromorphology*: Vol. 2, *Soil Genesis*, 531–540. AB Academic Publishers, Berkhamsted.

Thorp, J. and Smith, G.D. (1949). Higher categories of soil classification: order, suborder and great soil group. *Soil Science* **67**, 117–126.

Tinsley, J. (1950). The determination of organic carbon in soils by dichromate mixtures. *Transactions 4th International Congress of Soil Science* **1**, 161–164.

Tiurin, I.V. (1965). The system of soil classification in the USSR. Main stages in the development of the soil classification problem in the USSR. *Pédologie, Gand*. Numéro spécial **3**, 7–24.

Toleman, R.D.L. (1973). A peat classification for forest use in Great Britain. *International Peat Society Symposium, Glasgow*, 84–97.

Toleman, R.D.L. and Pyatt, G.D. (1974). Site classification as an aid to silviculture in the Forestry Commission of Great Britain. *Proceedings of the 10th Commonwealth Forestry Conference*, 1–12. Forestry Commission, London.

Trafford, B.D., Bloomfield, C., Kelso, W.I. and Pruden, G. (1973). Ochre formation in field drains in pyritic soils. *Journal of Soil Science* **24**, 453–460.

Turner, J.S. and Watt, A.S. (1939). The oakwoods

(*Quercetum Sessiliflorae*) of Killarney, Ireland. *Journal of Ecology* **27**, 202–233.

Valentine, K.W.G. and Dalrymple, J.B. (1976). The identification of a buried paleosol developed in place at Pitstone, Buckinghamshire. *Journal of Soil Science* **27**, 541–553.

Vancouver, C. (1794). *General View of the Agriculture of the County of Cambridge*. London.

Vancouver, C. (1808). *General View of the Agriculture of the County of Devon*. London.

van der Sluijs, P. (1970). Decalcification of marine clay soils connected with decalcification during silting. *Geoderma* **4**, 209–228.

Van Heuveln, B. and de Bakker, H. (1972). Soil-forming processes in Dutch peat soils with special reference to humus-illuviation. *Proceedings 4th International Peat Congress*, I–IV, 289–297, Helsinki.

Van Mensvoort, M.E.F. and Van de Westeringh, W. (1972). Report on a study tour in Ireland and Scotland. Afdelig Geologie und Bodemkunde, Landbouwhogescool, Wageningen, Netherlands.

van Vliet-Lanoë, B. (1985). Frost effects in soils. *In* Boardman, J. (Ed.), *Soils and Quaternary Landscape Evolution*, 117–158. Wiley, Chichester.

Vepraskas, M.J. and Wilding, L.P. (1983). Albic neoskeletans in argillic horizons as indices of seasonal saturation and iron reduction. *Soil Science Society of America Journal* **47**, 1202–1208.

von Post, L. (1924). Das genetische System der organogenen Bildungen Schwedens. *Memoires sur la Nomenclature et la Classification des Sols. International Committee of Science, Helsingfors*, 287–304.

Walker, P.H., Hall, G.F. and Protz, R. (1968). Soil trends and variability across selected landscapes in Iowa. *Soil Science Society of America Proceedings* **32**, 97–101.

Walkley, A. and Black, I.A. (1934). An examination of the Degtjareff method for determining soil organic matter, and a proposed modification of the chromic acid titration method. *Soil Science* **37**, 29–38.

Wallace, T., Spinks, G.J. and Ward, E. (1931). Fruit-growing Areas on the Old Red Sandstone of the West Midlands. Ministry of Agriculture and Fisheries Bulletin No. 15. HMSO, London.

Walsh, M. and De Coninck, F. (1973). Study of soils in the drumlin belt of north-central Ireland. *In* Pseudogley and Gley: *Transactions of the International Society of Soil Science Commissions V and VI*, Stuttgart, 237–246.

Walton, K. (1950). The distribution of population in Aberdeenshire in 1696. *Scottish Geographical Magazine* **66** (1), 17.

Watt, A.S., Perrin, R.M.S. and West, R.G. (1966). Patterned Ground in Breckland: structure and composition. *Journal of Ecology* **54**, 239–258.

Webster, R. (1968). Fundamental objections to the 7th Approximation. *Journal of Soil Science* **19**, 354–366.

Webster, R. (1975). Sampling, classification and quality control. *In* Bie, S.W. (Ed.), Soil Information Systems: *Proceedings of the Meeting of the International Society of Soil Science Working Group on Soil Information Systems*, 65–72. Centre for Agricultural Publishing and Documentation, Wageningen, Netherlands.

Webster, R. (1977). *Quantitative and numerical methods in soil classification and survey*. Clarendon Press, Oxford.

Webster, R. and Beckett, P.H.T. (1972). Matric suctions to which soils in south central England drain. *Journal of Agricultural Science, Cambridge* **78**, 379–387.

Webster, R. and Cuanalo de la C., H.E. (1975). Soil transect correlograms of north Oxfordshire and their interpretation. *Journal of Soil Science* **26**, 176–194.

Weir, A.H., Catt, J.A. and Ormerod, E.C. (1969). The mineralogy of Broadbalk soils. *Report Rothamsted Experimental Station for 1968*, Pt 2, 81–89.

Weir, A.H., Catt, J.A. and Madgett, P. (1971). Postglacial soil formation in the loess of Pegwell Bay, Kent (England). *Geoderma* **5**, 131–149.

Wellings, S.R. (1984). Recharge of the Upper Chalk aquifer at a site in Hampshire, England I. Water balance and unsaturated flow. *Journal of Hydrology* **48**, 119–136.

Wells, M. and Northey, R.D. (1985). Strengths of a densipan, humus-pan and clay-pan in a Spodosol developed under kauri (*Agathis australis*) and the implications for soil classification. *Geoderma* **35**, 1–14.

West, R.G. (1977). *Pleistocene Geology and Biology, with special reference to the British Isles*, 2nd edn. Longman, London and New York.

White, R.E. (1987). *Introduction to the Principles and Practice of Soil Science*, 2nd edn. Blackwell Scientific Publications, Oxford.

Whitfield, W.A.D. (1974). Soils in Warwickshire I: Sheet SP 36 (Leamington Spa). Soil Survey Record No. 19, Harpenden.

Whitfield, W.A.D. and Beard, G.R. (1977). Soils in Warwickshire III: Sheet SP 47/48 (Rugby West/Wolvey). Soil Survey Record No. 45, Harpenden.

Wilding, L.P. and Drees, L.R. (1983). Spatial variability and pedology. *In* Wilding, L.P., Smeck, N.E. and Hall, G.F. (Eds.), *Pedogenesis and Soil Taxonomy*, Vol. 1. *Concepts and Interactions*, 82–116. Elsevier, Amsterdam, Oxford and New York.

Wilson, K. (1960). The time factor in the development of dune soils at South Haven Peninsula, Dorset. *Journal of Ecology* **48**, 341–359.

Wilson, P., Bateman, R.M. and Catt, J.A. (1981). Petrography, origin and environment of deposition of the Shirdley Hill Sand of southwest Lancashire, England. *Proceedings of the Geologists' Association* **92**, 211–229.

Wischmeier, W.H. and Mannering, J.V. (1969). Relation of soil properties to erodibility. *Soil Science Society of America Proceedings* **33**, 131–137.

Wood, R.F. and Nimmo, M. (1962). Chalk Downland Afforestation. Forestry Commission Bulletin No. 34. HMSO, London.

Wright, C. and Ward, J.F. (1929). A Survey of the Soils and Fruit of the Wisbech Area. Ministry of Agriculture and Fisheries Research Monograph 6. HMSO, London.

Wright, P.S. (1980). Soils in Dyfed IV: Sheet SN 62 (Llandeilo). Soil Survey Record No. 61, Harpenden.

Wright, P.S. (1981). Soils in Dyfed VI: Sheet SN 72 (Llangadog). Soil Survey Record No. 74, Harpenden.

Wright, P.S. (1985). Soils in Dyfed VII: Sheet SN 50 (Llanelli North). Soil Survey Record No. 85, Harpenden.

Wright, T.W. (1955). Profile development in the sand dunes of the Culbin Forest, Morayshire I. Physical Properties. *Journal of Soil Science* **6**, 270–283.

Yarilov, A.A. (1927). Brief review of the progress of applied soil science in the USSR. *Russian Pedological Investigations* No. 11, 1–22.

Young, A. (1804). *General View of the Agriculture of Hertfordshire*. London.

Young, A. (1979). Mapping the world's soils. *New Scientist*, 252–254.

Zehetmayr, J.W.L. (1960). Afforestation of Upland Heaths. Forestry Commission Bulletin No. 32. HMSO, London.

Index

Numbers in bold indicate main entry

Aberdeenshire 40, 42, 56, 147, 163, 252, 258, 354, 390, 423
Aberford series 183
abrupt textural change **111**, 292, 359, 364
acid oxalate extract **94**, 197, 234–6
acid sulphate soils (characteristics) 95, **111**, 302–3, 306, 320, 325–8, 407–8, 410, 416, 418, 428
Acrisols 71–2, 76, 156, 209, 290
    Orthic Acrisols 231–2
    Gleyic Acrisols 359, 383
Adventurers association 409; series 414
aeolian deposits (see also blown sand, coversand, loess) 13, 145, 159, 293–4, 359, 361, 378
Agney series 315
agric horizon 69
agricutans 212
air capacity **98**, 111, 176, 178, 191, 203, 218, 221, 275, 294, 307, 317, 333, 352, 361, 365, 376
air photos 26, 45, 59, 313
air pollution 406
albic densipan 105–6, **108**, 245, 272, 287–8
albic (E) horizon 69, **108**, 146–7, 237, 243, 245–6, 261
alder *Alnus glutinosa* 22, 402, 411, 414
Alfisols **67**, 68, 70–1, 156, 209, 211, 228–9
Allen series 421
allophane 8, 94, 234, 236, 248
alluvial soils 37, 60–1, 301 *see also* 152–5, 169–81, 300–28
    mineral alluvial soils (Scotland) 61, 154, 169
    saline alluvial soils 61, 301
alluvium (Recent or Holocene alluvial deposits) 14–15, 59, 109–10, **113**, 116, 129, 156, 158, 169, 176, 181, 203, 290, 328, 386
    see also marine alluvium
alpine lichen heath 255
alpine soils (Scotland) 61, 255

alpine brown soils 233
alpine gleys 352
alpine podzols 246, 254, 258–9
Altcar association 409; series 413
aluminium 8, 101–2, 161, 197, 234–7, 241, 248–9, 254–5, 260, 265, 275–6, 285, 287, 371
Alun series 173
amphigleys 361
Andepts 67, 233–6
    Placandept 275
andesite 235
Andosols 72, 233–5
Andover series 139, 141
Anglezarke association 246
Anglian stage (deposits) **14**, 228–9, 281, 364, 371–2
Angular Chert Drift 260
Angus 202
anmoor 352
anthropic epipedon 68, 71, 107, 178, 293, 342, 360
Aqualfs 67, 76, 290, 345, 359
    Albaqualfs 359, 364, 367–9
    Fragiaqualfs 359, 368, 383
    Glossaqualfs 369, 363
    Ochraqualfs 359, 363, 370, 375–6, 377–8
    Umbraqualf 380
Aquents 67, 115
    Fluvaquents 76, 181, 206, 301, 305–6, 308–10, 314–15, 318–19, 325, 328
    Haplaquents 76, 324
    Hydraquents 76, 301, 323–4
    Psammaquents 76, 293, 295, 300
    Sulfaquents 301
Aquepts 67, 76, 272, 285, 301, 339, 345
    Cryaquepts 358
    Fragiaquepts 339, 346–7, 357
    Haplaquepts 76, 301, 328, 331, 334–7, 339, 344, 348, 350–1, 359

Humaquepts 76, 288–9, 296–9, 302, 311, 339, 342–3, 354
Placaquepts 76, 243, 277–80
Sulfaquepts 302, 325, 327
aquic moisture regime 69, 272, 288
Aquods 67, 261, 272
Fragiaquods 266, 270
Haplaquods 76, 264–5, 282–3
Placaquods 76, 272
Sideraquods 76, 266, 269, 285–6
Aquolls 67, 359
Argiaquolls 76, 359
Calciaquolls 313
Cryaquolls 354
Haplaquolls 76, 302, 313, 316, 320–1, 328
Aquults 67, 76, 359, 371
Albaquult 373
Paleaquult 383
Aran Islands 386
archaeology, in relation to soil 22–3, 137, 245, 274, 386, 402
Ardington series 212
Arenosols 72, 76, 143, 156, 159
Albic Arenosols 143, 149
Cambic Arenosols 148, 165, 167–8
Luvic Arenosols 143, 149
Arents 67, 385, 394, 396
Argillans (clay coats) 88, **91–2** *et seq.*
Argillic (B) horizon 30, 68–9, 80, **109** *et seq.*
Arrow series 205
Armagh (Co.) 236
Arnold series 190
artificial drainage *see* drainage
artificially reworked layers **111**, 385
Ash *Fraxinus excelsior* 213, 221, 366, 414
Ashdown Forest 267, 269
Ashgrove association 375
Ashley series 223
Ashwell, Herts. 398
association *see* soil association
Athy complex 217
Auchenblae association 157
Aughty series 424
Australia 30
Austrian pine *see* pine
available water capacity **98**, 116, 130, 139, 159, 161, 170–2, 178, 211, 223, 312, 320
Avon (Co.) 129
Ayrshire 375
azonal soils 29–30, 65, 116

Baggotstown series 185
Bagshot Beds 267, 281
Ballincurra series 196
Banagher series 412
Banbury series 54, 197, 200

Banffshire 56
Bangor series 126
Barton Beds 255
Barton series 196, 198
Baschurch series 54
base status (percentage base-saturation) 69–71, **95** *et seq.*
basic igneous (crystalline) rocks 53, 123, 194–6, 234, 249, 352
Batcombe series 228, 231
Beccles series 54, 364, 367
Beckfoot series 148–9
Bedfordshire 39, 203, 205
beech *Fagus sylvatica* 123, 132, 167, 199, 212–3, 226, 228, 230, 368
bell-heather *Erica cinerea* 252, 257
Benbecula 149
bent-grass *Agrostis* spp. 124, 185, 231, 237, 354, 368
Berkshire 40, 147, 149, 183, 319–20, 361, 363
Berwickshire (Berwick-on-Tweed) 196, 198
Beta horizon 161
bilberry *Vaccinium myrtillus* 126, 252, 258, 284, 287–8, 357
biotite 5, 372
birch *Betula* spp. 22, 256, 267, 288, 366, 372, 402, 418, 422–3
bisequal profiles **99**, 234, 259, 271, 276, 280
Blackdyke series 337
Blackland series 428
Black Rock Mountain Series 274
Blacktoft series 179
Blackwood series 296
Blakeney, Norfolk 144
blanket bog 62, 65–6, 271, 400, **404–6**, 421, 424, 426, 428
Blewbury series 183
blown sand 11, 59, 143, 147, 165, 254, 294, 397
Bodmin Moor 236, 424
bog moss *see Sphagnum*
bog-rush *Schoenus nigricans* 404, 418
boron 248, 294, 387
Borovinas 152, 155
Bovey Tracey, Devon 287
Bowden series 234, 239, 249
box *Buxus sempervirens*, Box Hill 135
bracken *Pteridium aquilinum* 125, 129, 163, 197, 231, 233, 235, 237–8, 244, 250, 256, 260, 267–9, 284, 372, 423
Bracklesham Beds 366
bramble *Rubus fruticosus* agg. 144, 197, 199, 206, 213, 221, 230, 366, 414
Braunerde 42, 194
Breckland 146, 161, 165
brickearth 13, 211–12, 217, 359, 361, 395, 398
brittle consistence **87**, 106, 206
Bromsgrove series 196

Bronze Age 22, 394
broom *Cytisus* spp. 146, 153
brown alluvial soils 49, 158, 173, 175, 177
brown calcareous alluvial soils 48, 158, 178–81
brown calcareous earths 48, 156, 158, 181, 183–5, 187, 189, 191
brown calcareous sands 48, 158–9, 168
brown calcareous soils (Scotland) 61, 159, 167–8, 181
brown earths 41–2, 46, 48, 59, 61–2, 65–6, 156, 158, 169, 181, 194, 196, 198–9, 202, 205–9, 233, 272
   argillic brown earths 49, 62, 66, 80, 158, 209, 211, 214–18, 219–20, 222–3
   ferritic brown earths 195–7, 200
   paleo-argillic brown earths 49, 80, 158, 209–10, 227–8, 231–2
brown forest soils (Scotland) 61, 156, 159, 194, 196, 199, 202, 210, 233–4, 238, 393
brown forest soils with gleying (Scotland) 61, 206, 208, 344–5, 348, 364
brown magnesian soils (Scotland) 61, 156, 194–6
brown podzolic soils 49, 60, 65, 73, 156–8, 233, 236, 238–42, 245, 249, 252
brown sands 48, 158–9, 161–6
Buckinghamshire 133, 135, 186, 191–2, 213, 229–30, 330, 334, 376
bulk density **97–9**, 106, 159, 176, 191, 275, 288, 349, 400, 402, 407
buried soil horizons 2, **99**, 114, 146, 172, 186, 201, 352, 390
Burren district, Co. Clare 119–21
Burren series 120
Burwell, Cambs. 398
Burwell series 3229

Cairngorms 247, 255
Cairn O'Mounth, Kincardineshire 278
Cairnsmore of Fleet 255
Caithness 299
Calcareous soils, class of 42, 60–62, 156
Calcic horizon 69, 313
calcium carbonate (equivalent) content 40, 94
calcium carbonate, redeposited (secondary) 105, 139, 182, 188, 329–30, 333, 361, 375
cambic horizon 69, 80, 109–10, 114, 138, 169, 172 186, 285, 339
Cambisols 72, 76, 139, 156, 169, 194, 233
   Calcic Cambisols 76, 141, 181, 183–4, 189
   Chromic Cambisols 199, 200
   Dystric Cambisols 173, 198, 238–9
   Eutric Cambisols 173, 192, 199, 208
   Gleyic Cambisols 175, 179, 181, 189, 192–4, 205, 207, 209, 332, 348
   Humic Cambisols 76, 159, 163, 169, 173, 194, 202, 241

Cambridge Greensand
Cambridgeshire 180, 326, 332, 336–8, 378–9, 390, 398, 414, 416
Cannamore series 191
carbon *see* organic matter
   pyrophosphate-extractable 8, 101, 108, 161, 249, 255, 275, 285
Carboniferous limestone 118, 120, 124, 128–9, 133, 140, 177, 181, 194, 212, 226, 276
carbon/nitrogen ratio 94, 233, 235, 248, 386, 422, 428
Carex bigelowii; *see* stiff sedge
Carlow (Co.) 65, 212, 216, 366
Carpour association 159
Carstens series 227
Castle Gregory, Co. Kerry 388
Castleton series 40
catena (catenary sequence or association) 20–1, 26, 56, 203, 262, 265, 271
cation-exchange capacity (CEC) 40, 66, **95**, 108, 113, 195, 197, 234, 275, 334, 349
Cegin series 345–6
cementation, of soil **87**, 101–2, 105–6, 246, 248–9, 263, 288, 361–2
chalk 5, 104, 129–30, 132, 134–5, 138–9, 181–2, 185, 211–12, 225, 329, 381
chalk downland 131–2, 139–40
Chalk Marl 329, 397–8
Chalky Boulder Clay 175, 180, 188, 190–1, 221, 223, 333–4, 365, 372, 375–6
Chara. see stoneworts
charcoal 114, 186, 200–1
Charr series 279
Chat Moss 426
Chatteris, Cambs. 336, 379, 414, 416
Chatteris series 321
Chelmsford, Essex 188
Chernozem 71–2
Cheshire 164, 263, 293, 420
chestnut *Castanea sativa* 206, 267
Cheviot Hills 235
Chilterns 132, 135, 186, 212, 228, 330
china clay 146
chlorite 5, 95, 208, 223, 226, 229, 372
chlorosis 131, 329
chromic argillic B horizon **109**, 210–11, 225, 228, 259, 267, 359–60, 380
chromium 195
Cladium mariscus; *see* sword sedge
Clare, (Co). 119–20, 140–1, 183, 185, 314, 366
Claverley series 366, 368
clay mineralogy 60, 66, 96, 195, 208, 223, 236, 307, 317, 334, 349, 376
clay translocation 6–8, 157, 210, 225–6, 228, 266, 288, 340, 359, 364, 376, 380–1
Clay-with-flints 15, 21, 210, 212, 225, 228

Clay-with-flints-and-cherts 371, 380
Cleveland Hills 355
Clifton series 366, 368
climatic phases **80**
climatic regimes **18**, 80–1 *et seq.*
Clonakilty, Co. Cork 390
Clonroche series 196
Clowater series 336
Clwyd 276
Clwyd series 175
cobalt 6, 66, 150, 195, 248, 299
coefficient of linear extensibility (COLE) 113, 176, 307, 349, 376
Collieston association 159
colluvium (Recent or Holocene colluvial deposits) 11, 14, 42, 109, **114**, 131, 137, 156, 181–2, 186–8, 194–6, 200–1, 203, 254, 342, 386
colour/structure B horizon **109**, 114, 156, 158, 169, 181–2
Colthrop series 316
concretions **89**, 92, 313, 315
conifers (planted) 22, 132, 144, 157, 159, 244, 246, 249, 281, 405
Conway series 40, 305
Coombe series 183
copper 6, 66, 132, 150, 248, 294, 299, 410
coprolites 395, 397–8
Corby, Northants 397
Corby association 280
cord-grass *Spartina* spp. 301, 322
Cork (Co.) 390
Cornwall 144, 146, 149–51, 236, 239, 280, 386, 390, 424–5
Corsican pine *see* pine
Cotswolds 132, 139
cotton-grass *Eriophorum* spp. 358, 402, 418, 421–3
Countesswells association 236
coversands 13–14, 161, 190, 211, 263, 364
cowberry *Vaccinium vitis-idaea* 252–3, 258
cracking in clay soils 112–3, 176, 180, 191, 224, 301, 307, 317, 333–4, 349
Cranwell series 184
Craven district 119, 121
Crediton series 196
Crewe series 54
Cromarty association 392
Cromarty peninsula 390
Cromerian stage 229
Cross Fell 255
cross-leaved heath *Erica tetralix* 125, 261, 281, 294, 423
crowberry *Empetrum nigrum* 147, 252, 423
Crowdy series 425–6
Crwbin series 129
cryoturbation 12–13, 16, 121, 139, 161, 181, 183, 190, 211, 225, 275, 332, 336, 352, 359, 371, 381

Cuckney series 163
Culbin Sands 144
Culm Measures 349
Cumbria 118, 240, 249–50, 305, 381–2
cumulic A horizon **107**, 115, 156, 158, 169, 172, 177
Curtisden series 207
cutans (*see also* argillans, matrans, etc.) 87, 91, 174, 246

dark earths (anthropogenic) 387
Darleith/Kirktonmoor association 234
Darnaway association 159
Dartmoor 236, 428
Darvel association 159
deep cultivation 385, 390
deer-grass *Trichophorum caespitosum* 271, 278, 404, 418
Denbigh series 54, 196
Denchworth series 376–7
densipan *see* albic densipan
Derbyshire 226, 275, 279, 396
Derrygorman complex 266
Devensian stage (deposits) 11, 13, **14**, 15 *et seq.*
Devon 123, 128, 146, 196, 199, 225, 266, 285, 287–8, 342–3, 349, 371, 380–1, 383, 386, 428
Devonian rocks *see also* Old Red Sandstone 52, 128, 196, 201, 212, 220, 285, 345, 355, 388
diagnostic horizons 30, 51, 68–9, 71, **106–11**
Dingle peninsula 236, 265, 388
Disturbed Lode series 398
disturbed soils 50, 75, 273, 385, **394–9**
Dod series 274
Dokuchaiev 3, 29, 38
dolerite 123, 196, 275
Domesday Book 22
Dorset 281
Down (Co.) 236
Downholland series 310
Downholland Silt 302
drainage (artificial) 22, 24, 80, 171, 178, 218, 261–3, 290, 292–3, 326, 336–7, 340, 345, 349, 352, 360–1, 379, 394, 402, 406–7, 409, 419
  drainage measures 191, 329
  mole drainage 191, 329, 349, 360
drainage class (Scotland) 56–8
Dreghorn association 159
Drombanny series 319
droughtiness 98, 121, 131, 152, 159, 161, 164, 182, 211, 262
drumlins 65, 150, 381
dry valleys 114, 137, 185–6, 200
Dublin (Co.) 275, 277
Dullingham series 187
dunes 63, 143–5, 149–50
dune slacks 143, 293, 299–300

Dunkeswick series 366
Dunnstown series 366
Dunsmore Heath, Warwicks. 266
duric horizon 106, 234, 246, 249, 275
duripan 69, 287
Durness Limestone 128
Dyfed 124, 150–1, 172, 241, 304, 345–7, 352–3, 356, 425

Eardiston series 196, 199
earthworms 3–4, 6, 132, 156, 178, 182, 212, 235, 387, 390, 394, 409, 414
earthy topsoil, in peat soils **108**, 406–11, 413–14, 418–20, 425–8
East Anglia 146, 159, 190, 297, 302, 310, 333, 336, 361, 364, 375, 402, 410
East Lothian 165–6, 168, 345, 348, 390, 392
Eckford association 60, 159
Eh 6, 272
electrical conductivity **96**, 321
Elibank Forest, Peebles-shire 237
Elmton series 139
Ely, Cambs. 414
Elymus arenaria; *see* sand couch grass
Empetrum nigrum; *see* crowberry
Enk earth soils 388
Entisols 67, 115, 156, 169, 186, 293
Epping Forest, Essex 371–2
Erica cinerea; *see* bell-heather
Erica tetralix; *see* cross-leaved heath
Eriophorum spp.; *see* cotton-grass
Erisey series 199
erosion (*see also* wind erosion) 15, 22, 25, 116, 121, 134–5, 144–5, 150, 152, 159, 181–2, 186, 200, 210, 234, 254, 294, 297, 304, 334, 405–6
Escrick series 216
Essendon series 373
Essex 188–9, 219, 226–8, 267, 323, 371–4, 414
Ettrick association (series) 60
Everingham series 296
Evesham series 191, 333–4
exchangeable cations 40, 60, 66, **95**, 195, 301, 307, 339
exchange acidity **95**
Exmoor 249, 284

faecal pellets 3, 92, 409
FAO *see* soil classification
Fell Sandstone 281
Fen Clay 302, 320, 325–6, 414, 416
fens (fenlands) 40, 293, 313, 325, 336, 361, 378, 398, 400–2, 406, 409–10, 412, 414, 416
ferri-argillans *see* argillans
ferrihydrite 6, 248
ferrolysis 359
fescue-woody fringe moss heath 255

Fforest series 347
Fibrists 68, 410, 418, 426
    Medifibrist 411
    Sphagnofibrist 421
field capacity 86, 98
Fife 294
Fladbury series 308
Flandrian stage (deposits) *see* Holocene
flint(s) 183, 186, 212, 225, 228, 337, 381
fluid soil consistence **87**, 113, 301, 323
flushes 352
Fluvents 67, 137, 152, 169
    Udifluvents 76, 138, 186–7
Fluvisols 72, 76, 115, 137, 152, 169, 290, 301
    Calcaric Fluvisols 138, 187, 314–15, 318–19, 324–5
    Dystric Fluvisols 154, 305, 311
    Eutric Fluvisols 154, 202, 308–10
    Thionic Fluvisols 325, 327–8
Foggathrope series 351
Folists 68, 125, 153, 155
Folkestone, Kent 188
Fordoles series 190
Forestry Commission 23, 43, 60–4, 341, 419
Foudland association 58
fragipan 69, 103–5, **106**, 111, 127, 174, 194–6, 206, 234, 236, 241, 260, 267, 272, 281, 287, 345, 355, 381
France *see* soil classification
Fraserburgh association 159; series 167
Freni series 354
Frog Moor series 40

Gault (clay) 330, 332–3, 397
Gade series 320
Galloway 153, 255, 310
gelifluction deposits *see also* head deposits 12
genetic soil groups 41–2, 57–8
Germany *see* soil classification
glacial drift (*see also* till) 11–12, 181, 194, 196, 209, 212, 223, 371
glaciofluvial deposits 159, 164, 181, 183, 209, 211–12, 235, 240, 246, 249, 280, 371, 399
glaciolacustrine deposits 13–14, 164, 223, 290, 348–9, 364, 375
Glassonby series 241
glass-wort *Salicornia* spp. 301, 323
glauconite (glauconitic materials) **56**, 212, 366, 372, 397
Gleadthorpe Experimental Husbandry Farm, Notts. 161–2
Glenamoy series 424
Gleneagles association 159
gleyed subsurface horizon **109–10**, 112, 115–16, 158, 169, 245, 290–1, 307, 352
gleyic features **112**, 160, 182, 195, 203, 211, 234

gleying (process of) **6**, 7–9 *et seq.*
gley-podzols 46, 49, 75, 77, 80, 243, 245, 249, **261–70**, 272, 293–4, 297, 360
gley soils (gleys) 41, 59–61, 65, 75, 77, 80, 114–16, 156, 182, 210, **290–384**
  alluvial gley soils 50, 77, 80, 152, 169–70, 178, 203, 292, **300–28**, 416
  argillic gley soils 50, 292, 359, 361–3
  calcaro-cambic gley soils 50, 328, 331–2
  cambic gley soils 50, 292, 343–4
  ground-water gley soils 41, 50, 61–2, 65, 73, **291**, 293, 295, 298, 300, 329, 339, 342, 352, 354, 360
  humic-alluvial gley soils 50, 80, 292, 301, 311, 320, 327–8
  humic-sandy gley soils 50, 292, 298–300
  humic gley soils 46, 50, 61, 274, 288, 292–3, 328, 339, 352, 354, 359–60, 380, 409
  raw gley soils 47, 73
  peaty gley soils 41, 61, 63, 66, 271, 274, 291, 311, 355
  surface-water gley soils 41, 49, 61–2, 65, 73, **291**, 329, 339, 344, 352, 355, 359, 364, 371, 375, 380, 396
Gleysols 72, 77, 290, 328, 339
  Calcaric Gleysols 300, 328, 331, 334–6
  Dystric Gleysols 295, 344, 346–7, 350, 359
  Eutric Gleysols 344, 350–1
  Humic Gleysols 277–80, 296, 299, 302, 339, 342–3, 354, 357–8
  Mollic Gleysols 206, 298–9, 302, 306, 311, 316, 320–1, 328, 337, 339, 354
Gloucestershire 140, 142, 201, 208, 307
gneiss 225
goethite 6, 94, 139, 197
Gore series 138
gorse *Ulex* spp. 123, 129, 146, 153, 284, 382
Gortaclareen series 366
Grampians 127, 249, 254, 275, 352, 422
granite 146, 153, 172, 183, 213, 275
Greater Manchester 427
Great Ouse (river) 180
great soil groups 29–30, 65–6, 69
Greenane series 366
grey-brown podzolic soils 65–6, 156, 209, 212, 217
Greylees series 300
greywacke 148, 153, 235
Grimblethorpe series 331
Grove series 187
Gubblecote series 331
Gurteen series 240
gypsum 88, 96, 106, 111, 301, 307, 325–6, 375, 419
gyttja 90, 411

haematite 6, 94, 111, 161
Hafren series 285
Haldon Hills 266

Haldon series 269
Halimione portulacoides; *see* sea purslane
Hallsworth series 360
Hamble series 215
Hampshire 50, 255, 257, 366, 368
Hanslope series 191, 194, 333–4, 336, 376
Harwell series 54, 212
Haselor series 191–2
Hastings Beds. 267
hawthorn *Crataegus monogyna* 129, 231
hazel *Corylus avellana* 22, 197, 221, 226, 366, 376, 414
head (deposits) **12**, 14–15 *et seq.*
heather *Calluna vulgaris* 118, 125–6, 132, 144, 147–8, 238, 245, 252–3, 257, 260, 269, 278, 281–2, 285, 341, 356–8, 402, 418, 421–3, 428
heather moor 125, 247, 249, 277–9, 282, 284–5, 357–8
heathland 147–8, 244, 246, 254, 260–1, 266, 287, 380, 386
heath rush *Juncus squarrosus* 125, 277, 294
Hebrides 143–4, 149–50, 165, 299
Hemists 68, 410
  Borohemists 422–3
  Cryohemists 126
  Medihemists 126, 411, 413, 420, 422, 424, 428
  Sulfihemists 411
  Sulfohemists 416–17
Hereford and Worcester 40, 148, 204, 212, 215, 370
Hertfordshire 45, 136, 398
Hexworthy series 279
Highlands, Scottish 16, 59, 116, 121, 152, 254
High Weald 206, 266
Highweek series 52
Hiraethog series 274, 277
histic epipedon 68, 107, 110, 115
Histosols 68, 71–2, 77, 115, 120, 125, 153, 400, 407
  Dystric Histosols 126, 413, 417–8, 420–8
  Eutric Histosols 411, 413–5, 428
Hobkirk series 199
hoggin 226
Holderness series 365, 367
Holidays Hill series 54
Holkham, Norfolk 144
Holocene (Flandrian) deposits 11–12, 14–15, 113–14, 152, 293–4, 300
Hornbeam series 232
Hortisols 386
Howe of the Mearns 206
Hoxnian interglacial stage **14**, 229, 372
Huddersfield 421
Humber (estuary) 169–70, 178
Humberside 161–2, 190, 397
humilluvic horizon 409
Humods 67, 76, 246, 248, 254–5
  Cryohumod 258

Haplohumods 256–7
Placohumods 273
humose topsoil 49–50, 73, **107**, 233, 292–4, 297, 300, 310, 325, 329
Hurley, Berks. 183
Hurst series 54, 363
hydraulic conductivity 110, 349, 352, 361, 407
hydrobrick soils 361
hydrocalcic subsurface horizon **110**, 114–16, 169, 290, 311
hydrologic sequences 18–20
hydromor 4, 233, 352
hydromull 4, 352
Hythe Beds. 260

ice-wedge casts 12, 210–11, 381
Icknield series 132–5
igneous rocks 194, 247, 249
imogolite 8, 94, 106, 234, 238
Inceptisols 66–7, 115, 139, 156, 169, 186, 233, 243, 248, 293
indurated layers (Scotland) 106, 122, 247, 249, 271, 275
Insch valley, Aberdeenshire 390
Insch association 234
intrazonal soils 29–30, 129
Inverness-shire 253, 390
involutions 12, 332
Ipswichian interglacial stage **14**, 228–9, 361, 372
Ireton series 380
iron, dithionite-extractable **94–5**, 101–2, 104–5, 108, 195, 236, 275, 285, 379
  pyrophosphate-extractable 8, **94–5**, 102, 108, 161, 197, 234–5, 237, 241, 249, 254–5, 260, 275–6, 285, 352, 371
iron ore 395, 397
iron oxides (oxyhydroxides) 6, 66, 88, 94, 139, 157–8, 169, 183, 191, 194, 197, 234, 416
ironpan (*see also* thin ironpan) 9, 272, 280–1, 362
ironpan soils (Forestry Commission) **62**, 270, 272
ironstone 175, 195, 397
Iveshead series 259
Iwade, Kent 309

jarosite 111, 325–6, 356, 416–17
Jurassic clays (clay-shales) 175, 180, 208, 333, 336, 355, 375, 379
Jurassic limestones 129, 139, 183, 185–6, 330, 390, 397

kandic horizon 69
kaolin (kaolinite) 96, 146, 197, 208, 229, 267, 376
Kent 38, 40, 137–8, 178, 206, 260, 266, 307, 309, 399
Kerry (Co.) 236, 240, 265, 388
Keuper Marl (Mercia Mudstone) 223, 307
Kidderminster terrace 203

Kilcolgan series 139
Kildare (Co.) 172–3, 366
Kilmarnock series 348
Kilpatrick series 366
Kincardineshire 206–7, 278
Kings Newton series 369
Kings Sedgemoor 412
Kirkcolm association 159
Kirkcudbrightshire 153, 258
kitchen middens 386

Lake District 116, 125, 240, 255
lake marl 14, 105, 110, 311–13, 319, 411
Lancashire 45, 127, 144, 155, 262–3, 302, 310, 402, 409–10, 412–13, 416, 419, 426
land capability 45, 59, 65, 121, 302, 329, 340, 360
land evaluation, in Ireland 37
larch *Larix decidua* 250, 354
lateral eluviation 359
Laurencekirk association 206–7
Lawford series 54
leaching (of carbonates) 5, 7, 26, 157, 169, 178, 181, 183, 197, 379
least willow *Salix herbacea* 258
Leeds, W. Yorks. 423
Leicestershire 199
lessivage, *see* clay translocation
levees (river) 152, 170, 172
Leziate series 256
lichens 117, 144, 147–8, 151, 423
Lichfield, Staffs. 342
Liffey (river) 172; series 173
Limerick (Co.) 65, 366
limestone (*see also* Carboniferous, Jurassic and Permian limestones) 10, 118, 120, 124, 128–9, 132, 134, 139–40, 183, 211, 225, 329
limestone pavement 119, 124
liming 22, 71, 158, 161, 206, 248–9, 276, 361
Lincolnshire 178–9, 183–4, 186–7, 190, 300, 312–3, 316–7, 320, 394, 397, 416–7
Linhope series 235
links (Scotland) 143, 146, 293
Links association 159, 294–5
lithological discontinuity **99**, 102, 172, 190, 229, 285, 355, 359, 364–5, 371, 379–80
Lithosols 61, 65, 72, 74, 76, 80, **115–20**, 120–1, 124, 255
litter layers 3, **100**, 194
Littleshalloch series 60
Lizard peninsula 195
Llandeilo, Dyfed 172, 241
Llanelli, Dyfed 425
Llangadog, Dyfed 345–6, 356
Llangendeirne series 208
Loddon (river) 361
lodgepole pine *see* pine

loess 11, 13–14, 129, 138–9, 142–3, 183, 188, 210–12, 218, 225–6, 228–9, 260, 285, 287, 359, 361, 371–2, 379, 381
Loggans series 300
London Clay 188, 218, 361
Longmoss association 420; series 421, 424
Longtown series 174
loss on ignition 40, 60, 94, 427
Lower Chalk 182, 188, 330, 397
Lower Greensand 159, 203
Lower Lias 191, 397
Lower Taplow Terrace 152
Ludgvan series 390, 393
Ludlow, Shropshire 366
Luffness series 168
Luvisols 71–2, 76, 156, 209, 290
   Chromic Luvisols 214, 216, 222, 225, 227
   Gleyic Luvisols 359, 363, 370, 375, 377–8, 380, 383
   Orthic Luvisols 215, 217, 220, 223

Macamore series 366
Macaulay Institute (for Soil Research) 39–40, 43, 56, 60
machair(s) 143, 149, 299
made ground 394; see also disturbed soils
magnesian gleys (Scotland) 61, 339
magnesian soils (Scotland) 61
magnesium 131, 195, 248
major soil groups 47–51, 61, 73–5
Maltby series 190
Malvern series 234
manganese 6, 88, 150, 248, 294, 299, 313, 329, 361, 387, 410
man-made soils 50–1, 62, 75, 77, 273, **385–399**
   man-made humus soils 50–1, 389, 391–3
man-modified mires 406
Manod series 235
Marcham series 139
marling 161, 248–9, 293, 297
Marlow, Bucks. 186
marine (estuarine) alluvium 169, 174–5, 178, 301, 304, 306, 311, 318, 321–7
marram grass *Ammophila arenaria* 143–4, 146
Martin Mere, Lancs. 413
mat-grass *Nardus stricta* 125, 271, 277–8, 284, 341
matrans 91, 174
Maw series 283
Meath (Co.) 427
Mendham association (series) 409
Mendip (Hills) 120, 129, 276
Mercia Mudstone see Keuper Marl
meres (fenland) 313
methane, emission of 399
Methwold series 168

micas (including illite and interstratified minerals) 6, 146, 208, 225, 229, 267, 307, 316, 349, 376
micromorphology see soil micromorphology
micro-podzols 155
Middle Lias 'marlstone' 197
Midlandian deposits (Ireland) **14**, 140, 159, 183, 212, 333, 336
Midlands, English 129, 134, 159, 191, 209, 220, 246, 266, 333, 371, 375, 395, 412
Midland Valley (Scotland) 356
Milford series 196
mineralogy, in soil classification 52, 56, 70
mineral soil (horizons), definition 83
mining (colliery) spoil 62, 127
moder 3–4, 8, 132, 144, 157, 287, 409
mole drainage see drainage
Molinia caerulea; see purple moor-grass
mollic epipedon 30, 68, 70–1, 107, 115, 123, 132, 178, 290, 293, 313, 333, 360
Mollisols 67–8, 70, 115, 123, 209, 293, 302
molybdenum 66
Monaghan (Co.) 169
Moor Gate series 249, 252
moorland (vegetation) 9, 22, 244, 274–5, 284, 355, 381–2, 418, 421, 423
mor 3–4, 8, 157, 210, 244
Moray Firth area 144, 246, 254, 271
Moretonhampstead series 239
Moreton series 183
Moreton-in-Marsh, Gloucs. 372
mosses (see also *Sphagnum* and *Rhacomitrium*) 117, 144, 147–8, 151, 253, 257, 294, 368, 414, 425
moulding (biological ripening) 409, 413
mountain-top detritus 121, 254
mudstones 104, 142, 196, 220, 223, 241, 274, 287, 307
mull 3–4, 7–8, 132, 138, 157, 196–7, 210, 225, 409
Mulltaggart series 258
Musselburgh area, Scotland 390
Mylorstown series 366
Myreton series 354

Naburn series 162
Nardus stricta; see mat-grass
Neolithic period 22, 158, 405
Netherlands see soil classification
Newchurch series 318
Newent series 202
New Forest, Hants. 255, 366, 368
Newmarket series 140
Newport series 54, 147, 159
Newtondale series 128
Newton Stewart, Galloway 153
nickel 195
night soil 386, 419

nitrogen, in soil 3, 60, 66, 94, 130, 146, 156, 248, 410, 419
nodules, definition **89**
Norfolk 144, 161, 163, 165, 167, 190, 255–6, 315, 324, 326, 337, 366, 411, 416
Northamptonshire 175–6, 221–2, 334–5, 397
Northaw Great Wood, Herts. 371
North Downs 137, 188
Northern Ireland 43, 65, 234
Northmoor Terrace (of Upper Thames) 188, 330
Northumberland 128, 235, 238, 280–2
North Wales 116, 274
North York Moors 266, 271, 276, 280
North Yorkshire 119, 125, 177, 218, 295, 310, 355, 357–8
Norway spruce *Picea abies* 23, 341, 354
Nottinghamshire 161–2, 183, 263–4, 297

Oak (*Quercus robur*; *Quercus petraea*) 22, 123, 197, 221, 226, 228, 230, 238, 260, 267, 288, 366, 372
Oak series 375
Ochrepts 66–7, 69, 72, 76, 156, 194, 233
    Cryochrepts 238
    Dystrochrepts 69–70, 173, 194, 198, 200, 205
    Eutrochrepts 66, 69–70, 76, 139, 141, 169, 173, 175, 179–91, 183–4, 186, 189, 192–4, 199, 202, 205, 208–9, 332
    Fragiochrepts 194, 207, 238–9, 242
ochric brown soils *sols bruns ocreux* 233
ochric epipedon 68, 293
Old Red Sandstone 37, 40, 196, 206, 212, 236, 355, 388
Ollerton series 165
Onecote series 359
opencast mining 394–6
organic detritus muds 400, 402, 411, 413
organic horizons **100**
organic matter, determination of 40, 60, **94**
organic soil materials 82, 89–90
organic soils, major group of (*see also* peat soils) 42, 51, 61
Orkney Islands 299, 386
Orthents 67, 126, 129, 139, 143, 152
    Cryorthent 154
    Udorthents 120, 128, 136–7, 141–3, 148, 154–5, 398
Orthods 67, 246, 248, 254–5
    Cryorthods 252, 254, 259
    Fragiorthods 252, 259
    Haplorthods 235, 240–1, 250–1, 269
Outer Hebrides 299
Oxford Clay 39, 336–7, 375
Oxford district 229, 375
Oxfordshire 40, 186, 188, 231, 318, 330–1
oxic horizon 5, 69

packing density **97**, 111, 218, 221, 276, 361
paleo-argillic B horizon 109, 112, 211, 372
Panbride association 159
Panholes series 183
papules **92**, 102, 109, 228–9, 371–2
pararendzinas 47, 116, 129, **142–3**
Park Gate series 363
particle-size analyses 40, **93–4**
particle-size (textural) classes 52, 70, **83–4**
Paternias 152, 155
patterned ground 13, 25–6, 121, 165, 379, 381
Peak District 275
peat 4, 8, 14, 45, 59, 61, 63, 83, 89, 100, 160, 272, 297, 299, 378–9, 400–28
    amorphous peat **89**, 425
    basin peat (bogs) 42, 62, 65–6, 406
    drummy peat 410, 416–7
    dystrophic peat 51, 423
    eutrophic peat (mires) 61, 402, 408, 419
    fen-carr peat 402, 408, 410–12, 420, 422
    fibrous peat **89**, 99–100, 409–11, 419–20, 426
    flushed peat (bogs) 61–2, 409, 419, 421, 424, 428
    grass-sedge peat 411, 413
    hill peat 42, 406
    humified peat **90**, 100, 410, 414–5, 417–9, 425–6, 428
    mesotrophic peat (mires) 61, 402, 407–8, 427
    oligotrophic peat (mires) 402, 404, 409, 418–21, 424, 426, 428
    ombrotrophic peat (mires) 402, 404, 407
    reedswamp (*Phragmites*) peat 410, 412, 422
    sedimentary peat **90**, 100, 411, 424
    semi-fibrous peat **89**, 100, 409–11, 418–9, 421, 426
    Sphagnum-Eriophorum (*Calluna*) peat 420, 422–4, 427–8
    Sphagnum peat 62, 402, 404, 418, 420–1, 426
peat (organic) soils 51, 61, 75, 77, 80, 115, 271, 290, **400–28**
    earthy peat soils 51, 406, 411, 413–5, 417–9, 427–8
    raw peat soils 51, 406, 411, 420–28
peaty topsoil 73, **107**, 110, 285, 297, 300, 303, 310, 325, 329, 336, 341, 352, 378
Pebble Gravel 371
Pebbly Clay Drift 371
pedon 28, 70
pedotubules 91
peds **4**, 85–6, 102–5, 112 *et seq*.
Peebles-shire 235, 237
pelo-characteristics **112–3**, 171, 182, 195, 211, 303, 329, 341, 360
pelosols 48, 51, 73, 112
    argillic pelosols 48, 209, 223, 225
    calcareous pelosols 48, 181, 191–3, 328, 336–7
    non-calcareous pelosols 48, 194, 208–9

Penrith, Cumbria 381
Pennines 124–5, 177, 194, 255, 355, 405–6, 420–1
Permian (Magnesian) limestone 129, 134, 139, 183, 185
Perthshire 201
Peterhead association 223
pH 2, 7, 40, 60, 111, 145, 158, 276, 361, 407–9, 411–12, 418, 427–8
Phaeozems 72, 76, 123, 159, 169, 178, 181, 194, 209
    Calcaric Phaeozems 178, 180
    Gleyic Phaeozems 296, 300
    Haplic Phaeozems 124, 162–3, 185, 191
phosphorus, in soil 6, 60, 71, 131, 248, 281, 387
phosphorus retention capacity 108, 195, 234
Phragmites; see reed
physical ripening **4**, 301, 323
Pinder series 366
pine(s) 22, 165, 245, 249
    Austrian pine *Pinus nigra nigra* 132
    Corsican pine *Pinus nigra laricio* 132, 145, 248, 281
    lodgepole pine *Pinus contorta* 341, 419
    Scots pine *Pinus sylvestris* 23, 145, 163, 167, 248–50, 253, 267, 269, 354, 418
placic horizon 68–9, 101, 108, 243, 261, 272
plaggen epipedon 68, 107, 386
plaggen soils 201, 237, 385–91
Plaggepts 67, 115, 143, 152, 156, 159
Planosols 72, 77, 290, 359, 364
    Dystric Planosols 369, 373
    Eutric Planosols 367–8
plantations, of conifers 163, 237, 250, 277, 294, 354
plasma separations **91**, 102, 208, 228
plateau gravels 260, 266, 280
Plateau (Northern) Drift 229
poaching, of topsoil by stock 98, 172, 284, 341, 395
plough-pan 149, 294
ploughwash *see* sheetwash
podzolic B horizon **108**, 109, 144, 156, 162, 243, 245
podzolic soils 49, 243 *see also* grey-brown podzolic soils; brown podzolic soils
podzolization 6, **8**, 146, 157, 161, 165, 210, 229, 233, 246, 266, 371
podzols 41, 49, 59–62, 65, 68, 71–77, 146, 157, 159, 233, 236, 240, **243–89**, 388, 394
    alpine podzols 61, 254, 258–9
    ferric (iron) podzols 61, 147, 246–9, 254
    humic cryptopodzols 49, 245, 254, 258
    humoferric (humus-iron) podzols 46, 49, 61, 246–7, **248–54**, 257, 259, 280, 284
    humus podzols 46, 49, 61, 121, 147, 246, **254–59**, 280, 352
    paleo-argillic podzols 49, 259–61
    peaty (gleyed) podzols 46, 61, 65–6, 270, 272, 279
    subalpine podzols 61, 248

Gleyic Podzols (FAO) 77, 261, 264–6, 269–70, 285–6
Humic Podzols (FAO) 76, 248, 254, 256–7
Leptic Podzols (FAO) 76, 236, 240–1, 252
Orthic Podzols (FAO) 76, 248–52, 254–5, 258–9
Piacic Podzols (FAO) 77, 272, 282–3
Podzoluvisols 72, 76, 209–10
    Gleyic Podzoluvisols 359, 363, 380, 383
Poldar series 311
polypedon 28
poplar *Populus* spp. 302, 319
potassium, in soil 5, 96, 131
potential evapotranspiration **16**
potential soil water (moisture) deficit 16–18, 98
potential water deficit (PWD) 16–18
Poundgate series 270
Powys 174, 349–50; series 128
Preseli series 124
Preston, Lancs. 155
Prickwillow series 417
Priddy series 280
prominent A horizon **106**, 110, 115, 122, 124, 126, 131, 152, 156, 159, 161, 169, 171, 194, 301, 328–9, 360
Psamments 67, 76, 115, 143, 152, 156, 159
    Quartzipsamments 149, 165
    Udipsamments 151, 164, 167–8
Pseudogley 291, 361, 364
Puccinellia maritima; *see* sea poa
purple moor-grass *Molinia caerulea* 261, 269, 271, 281, 284–5, 341, 356–7, 372, 404, 418–20, 424–5, 428
pyrite 105, 111, 127, 316, 325–6, 356, 375, 416–7
pyrophosphate extraction **94–5** *see also* aluminium, carbon, iron
pyrophosphate index **89**, 420–22, 425

quarries (quarrying) 1, 135, 385, 394
Quaternary stages **14**
Quorndon series 343

radiocarbon dating 186, 188, 203, 274, 402, 404
Ragdale series 378
rainwash 144 *see also* sheetwash, slopewash
raised beach (deposits) 14, 149, 169, 246, 361, 390
raised bog (moss) 42, 62, 400, **402–4**, 406, 409, 412, 418–9, 420–2, 426, 428
ranker-like alluvial soils 47, 116, 152, 154–5
Rankers 46–7, 51, 61, 63, 72, 74, 76, 115–16, 118, **120–9**, 132, 142, 246, 254, 352, 394, 404
    brown rankers 127–9
    humic rankers 47, 118, 122–4, 126
Rathangan series 366
Rathcannon series 223
raw alluvial soils 47, 152
raw sands 47, 115

raw skeletal soils 47, 116
raw soils 47, 61, 73
Reach series 336
Reading, Berks. 361
Reading Beds. 147, 228
Reaseheath series 266
reddish (haematitic) materials 56, 161, 208, 212, 220, 223, 307, 345, 355, 365
red fescue *Festuca rubra* 144, 151, 166, 185, 299, 322
Redlodge series 251
reed *Phragmites australis* 320, 402, 410–11
reference section, in peat soils 407–8, 412, 419, 426
Regosols 61, 65, 72, 76, 115, 120, 128, 129, 139, 143, 152, 173
    Calcaric Regosols (FAO) 136–7, 141–2, 151
    Dystric Regosols (FAO) 128, 148–9, 155
Rendolls 67, 76, 115, 129, 131, 133–4
rendzina-like alluvial soils 47, 116, 152, 155, 292, 301, 311, 316, 319
Rendzinas 47, 51, 61–2, 65, 71–2, 74, 76, 80, 115–16, 124, **129–43**, 165, 182, 328, 394
    gleyic rendzinas 47, 134, 331, 336
retained water capacity **108**
Revidge series 126
Rhacomitrium; see woolly fringe moss
Ribble series 155
ridge-and-furrow microrelief 25, 209
Rigosols 394
Ripon, N. Yorks. 355
Risley Moss, Cheshire 420
river drift (terrace deposits) 14, 182, 187, 194, 203, 211, 217, 259, 281, 336, 342, 379, 399
Rockliffe series 305
roddons (rodhams) 170, 313, 320
Romney Marsh, Kent 45, 169–70, 178, 301, 307
Romney series 180
roots, frequency and size of **89**
Ross and Cromarty 391
Rothamsted Experimental Station 43, 228
Rothbury, Northumberland 281
Rougemont series 217
rubbed fibre 89–90, 99
rubification 6, 14, 229
Rudham series 140
Rufford series 366
rushes *Juncus* spp. 265, 299, 353, 356–8, 368, 376, 425

St. Albans series 226
St. Lawrence series 332
Salicornia spp. *see* glass-wort
salinity (saline soil) **96**, 301, 307, 321–6
Salisbury, Wilts. 225
Salix herbacea; see least willow
sallow *Salix cinerea* 146, 402
Salop series 366
salt-making (mediaeval and Roman) 394

salt marshes 301, 321–3
sand couch grass *Elymus farctus* 143
'sanded' soils 388–90
sand-pararendzines 47, 80, 116, 143, 151, 300
sand-rankers 47, 80, 116, 143, 148–9
Sandringham Sands 255
sand sedge *Carex arenaria* 144, 147
sandstone 128, 158, 161, 183, 196, 201, 211–3, 220, 236, 240–1, 246–7, 249, 267, 280, 287, 349, 352, 355, 390, 396
Sandwich series 151
Sannan series 241
Saprists 68, 410, 413
    Medisaprists 118, 414–5, 418, 425–6, 428
saprolite 197
sea aster *Aster tripolium* 322, 324
seablite *Suaeda maritima* 322
sea lyme grass *Leymus arenarius* 143–4
sea poa *Puccinellia maritima* 322–3
sea purslane *Halimione portulacoides* 322–5
seaweed 150, 386
sedges *Carex* spp. 299, 368, 402, 423, 425
Selborne series 212
sepic plasmic fabric 208, 228, 375
serpentine (serpentinite) 56, 195
Severn (river) 169, 172, 203, 349
schists 225, 247, 249
Schoenus nigricans; see bog rush
screes 12, 115, 119, 121–2, 127, 132, 135, 255
shales 127, 172, 183, 195–6, 287, 345, 349, 352, 355–6, 364, 380, 390, 396
Shannon, Co. Clare 312; series 315
sheeps fescue *Festuca ovina* 118, 124, 133, 166, 237, 294
sheetwash (slopewash, ploughwash) 137, 186, 201
Sheldwich series 228
shell marl 313
Sherborne series 139, 141
Sherwood Sandstone 263
shingle (beach) 63, 143, 152–3
Shirdley Hill Sand 263
Shirrell Heath series 250
Shotford series 328
shrink-swell capacity (shrinkage potential) 113, 176, 208, 224, 307, 376
Shropshire 38, 40, 123, 197, 221, 249, 366
siltstones 104, 196, 206, 211–2, 220, 267, 274, 345, 364, 366
Sitka spruce *Picea sitchensis* 23, 237, 273, 277, 341, 376, 419
skeletans 88, 91, 210, 381
Skiddaw series 118, 123
skirtlands, skirt soils 336, 409
slaking, of topsoil 159, 211, 294, 312, 361, 409
slates 128, 195–6, 247, 342, 390
slickensides 88, 102, 112, 307, 376

slowly permeable subsurface horizon **110**, 182, 195, 208, 211, 234, 241, 329, 341, 344, 349, 352
slopewash *see* sheetwash
smeary consistence **87**, 233, 235
smectite (and interstratified mica-smectite) 96, 176, 180, 195, 223, 228, 307, 317, 334, 376
Snowdonia 255
sodium fluoride test 102, 233, 235
Soham series 183
soil associations 34, 36, 43, 45, 56, 59–60, 65
soil classification **26–33**
   Canada 290, 406, 422
   England and Wales 46–56
   FAO **71–3**, 76–7 *et seq.*
   Forestry Commission 61–4
   France 46, 182, 194, 262, 290, 361
   Germany 155, 194, 262, 361, 385, 394
   Great Britain (1927–40) 39–43
   Netherlands 30, 46, 158, 290, 297, 301, 323, 361, 385, 388–9, 406
   Republic of Ireland 65–6
   Scotland 56–61
   United States (Soil Taxonomy) 29–30, 39, **66–71**, 76–7 *et seq.*
   used in this book **73–7** *et seq.*
soil colour **82**
soil complexes 34, 59
soil, concepts of 1–3, 46, 73, 116
soil consistence **86–7**
soil creep 114, 182, 186
soil erosion *see* erosion and wind erosion
soil micromorphology **90–2** *et seq.*
soil-moisture (water) deficit **16–18**, 152, 159, 161, 360
soil-moisture (water) regime **15–18**, 30, 68–70, 80, 144–5, 220, 261, 272, 290, 333, 340, 349
soil phases 55, 73, 254, 273
soil-profile development 3–9
soil series, concepts of 29, 39–40, 46, 52–8, 60, 65–6, 70
soil strength 86
soil structure **85–6**
soil-survey methods 35–6, 39–40, 43–5, 56–65
soil temperature *see* temperature
soil texture 40, **83–4**
soil-water state **85**, 86
soil-water regime *see* soil-moisture regime
solifluxion (gelifluction) 12, 121, 211–2
Sollom series 264–5
Sols bruns 46, 194
Sols lessivés 46
solum **99**, 210, 339, 342, 359
Somerset 127, 226, 412
Somerset Levels (Moors) 306, 402, 409, 412
Sourhope association 234
Southampton series 54, 260

South Downs 130, 186
Southern Uplands (Scotland) 255
Southport, Lancs. 144
South Wales 395–6
South Yorkshire 412, 422
Spartina townshendii; *see* cord-grass
Sphagnum (bog moss) 125–6, 402, 418, 421, 423, 425, 427
spodic horizon 68–70, 108, 243, 272, 285, 288
Spodosols 67–8, 233, 235, 243, 272, 285
stagnogley-podzols 262, 269–70
stagnohumic gley soils 50, 292, 354–9, 380–4, 404
stagnogley soils 49, 291–2, 359, 364, 371
   cambic stagnogley 344, 346–8
   paleo-argillic stagnogley 371–5
   pelo-stagnogley 348–51, 375–8
   typical stagnogley 364, 367–70
Stanway series 370
Staffordshire 223–4, 296, 342, 367
Staugley 364
stiff sedge *Carex bigelowii* 258
stiff sedge-fescue grassland 255
Stirling association 306
'stonebrash' soils 139
stoneworts *Chara* spp. 313
stoniness **85**
Stourport, Hereford and Worcester 147
Stour valley, Essex 414
Stow series 209
Strathmore 201
Strathspey 249
Street series 366
stress cutans 88, 91, 317, 366
Studland, Dorset 144
Suaeda maritima; *see* seablite
subalpine brown soils 233
subsoiling 273, 349, 360
Sudbury, Suffolk 218
Suffolk 218, 326–7, 411, 416
sulphidic materials **111**, 301, 323, 325–6, 402, 407, 411
sulphur, in soil 95, 356, 416
sulphuric horizon 69, 111, 326, 356, 416
Summertown-Radley Terrace 188
Surrey 38
Sussex (East and West) 38, 45, 212, 214, 266–7, 313, 324, 361–2
Sutton series 54
Swale (river) 177; series 178
sword sedge *Cladium mariscus* 402, 410

Tavistock, Devon 342
Tedburn series 350
Teign (river) 172; series 154

temperature 15–16, 18, 30, 68–9, 70, 80, 146, 163, 352
Tentsmuir Forest, Fife 294
terrestrial raw soils 47, 73
Tertiary beds 183, 225, 228, 246, 254–5, 266–7, 281, 361, 366, 371
texture, see soil texture
textural classes see particle-size classes
Thame (river) 316
Thames (river) 178, 188, 229–30, 316, 399
Thames series 319
Thetford Forest, Norfolk 161, 165
Thick earth soils 385
thick man-made A horizon **107**, 115, 385
thin ironpan 8, 46, 68, 101, **108**, 233, 243, 245, 261, 270–6, 284, 287
till 11–14 *et seq.*
Tipperty association 60, 223
Tiree 149
Toberbride series 366
tonguing, of albic E horizon 210, 228–9
total exchangeable bases TEB 95
total extractable iron **95**
Towans series 151
trace elements 60, 66
trafficability 329, 360
Trent (river) 169, 171, 178, 399, 416
Trent series 173
Trichophorum caespitosum; *see* deer-grass
Tring, Herts. 136
tufa 105, 110, 187–8, 311, 313
tufted hair-grass *Deschampsia caespitosa* 318, 354, 376
Tuin earth soils 389
Tunbridge Wells Sands 206
Turbary Moor series 428
Tydd series 327
Tyne (river) 172
Tywi (river) 172, 304

Udalfs 67, 76, 156, 209
　Fraglossudalfs 210
　Glossudalfs 210
　Hapludalfs 214–17, 219–20, 222–3, 227
　Paleudalfs 211, 228, 231–2
Udolls 67, 76, 123, 159
　Argiudolls 209
　Hapludolls 124, 162, 169, 173, 178, 180–1, 185, 191, 194, 296, 300, 393
Udults 67, 76, 209
　Paleudults 211, 232
Uffington series 181
Uist 149
Ultisols 67–8, 70–1, 209, 211, 228–9, 371
ultrabasic rocks 195, 339
Ulster 234

Umbrepts 67, 72, 76, 115, 120, 159, 194
　Fragiumbrept 239
　Haplumbrepts 123–4, 162–3, 173, 202, 241
umbric epipedon 68, 70, 107, 115, 293, 342, 360
United States *see* soil classification in
unripened gley soils 47, 292, 301, 321, 324
Upper Chalk 132, 135, 139, 183, 225, 228
Upper Greensand 182, 188, 212, 381
Upton series 134–7, 140
Usher series 178

Vaccinium myrtillus; *see* bilberry
Vaccinium vitis-idaea; *see* cowberry
Vale of Aylesbury 333
Vale of Evesham 40
Vale of York 218, 262, 294
valley (flushed) bogs 404, 418
vegetable (market-garden) crops 159, 165, 410
vermiculite 96, 195, 197, 208, 226, 228–9, 236
vertic features 333, 349
Vertic subgroups 112–13, 176, 307, 319, 334, 376
Vertisols 67, 71–2, 112, 307
Verulamium (St. Albans) 387
Vinny series 202
vivianite 105, 316
Von Post scale **90**, 421
Vyrnwy valley, Powys 174

Wales 40, 64, 116, 196, 235, 284, 424
Wallasea series 310
Wantage series 134
Wareham Forest, Dorset 281
warping (warpland) 169–70, 178
warp soils 46
Warwickshire 192
Wash (bay) 169, 178, 301–2, 311, 323, 406
wastage, of peat 263, 379, 402, 407, 409–10, 419
waste materials 120, 127, 134, 146, 394, 399
Waterford (Co.) 236
water regime *see* soil-moisture regime
Waveney valley 326
wavy hair-grass *Deschampsia flexuosa* 148, 277, 372
weathered B horizon 109
weathering 5–8, 14, 157, 169, 181, 183, 191, 194, 197, 212, 233–5, 274, 345, 349, 369, 394
Welsh Borderland 123, 172, 196, 220
Welshpool, Powys 349
Wenallt series 357
Westmeath (Co.) 366, 420
West Yorkshire 421, 423
Wetton series 124
Wexford (Co.) 65, 159, 225, 246, 366
Wharfedale 310
Whitelinks series 300
Whitsome association 223
Wicklow Mountains 172

Wigtownshire 311
Wilcocks series 358
Willingham series 54, 316
willow *Salix* spp. 302
Wiltshire 132–5, 141, 225–6
Winchester series 225
wind erosion 146, 149, 169, 294, 297, 304, 409–10, 419
Windsor series 54
windthrow (hazard) 63, 341
Winter Hill series 422–4
Wisbech series 54, 314
Witham Fen 320, 416
Woburn Experimental Farm 203
Woking series 149
Wolstonian stage (deposits) **14**, 225, 228, 360, 366, 371–2
woolly fringe moss *Rhacomitrium lanuginosum* 258, 423

Worcester (shire) 147, 203
Worcester series 223
workability, of soil 98, 159, 182, 234, 329, 360, 410
Worlington series 164
Wye, Kent 137–8
Wye (river) 172, 212
Wyre series 177

Yarner series 289
Yarner Wood, Devon 287
Yarrow association 159
Yatton series 134
Yeollandpark series 344
Yorkshire 127, 134, 183, 190
Yorkshire Wolds 139

zonal soils 29–30, 65